Dinosaur Systematics

To the memory of
Dr. Charles M. Sternberg

The editors and publisher
would like to recognize
the enormous contribution of
the Tyrrell Museum of Palaeontology,
Alberta Culture and Multiculturalism,
for assistance in bringing this volume
to fruition. This included an enormous
amount of time and effort to produce
the camera-ready copy.

Dinosaur Systematics

Approaches and Perspectives

Edited by
Kenneth Carpenter
Denver Museum of Natural History

and
Philip J. Currie
Tyrrell Museum of Palaeontology

Published by the Press Syndicate of the University of Cambridge
The Pitt Building, Trumpington Street, Cambridge CB2 1RP
40 West 20th Street, New York, NY 10011-4211, USA
10 Stamford Road, Oakleigh, Melbourne 3166, Australia

First published 1990
First paperback edition 1992
Reprinted 1993, 1995, 1996

Library of Congress Cataloging-in-Publication Data is available

A catalogue record for this book is available from the British Library

ISBN 0-521-36672-0 hardback
ISBN 0-521-43810-1 paperback

Transferred to digital printing 2004

Contents

Contents

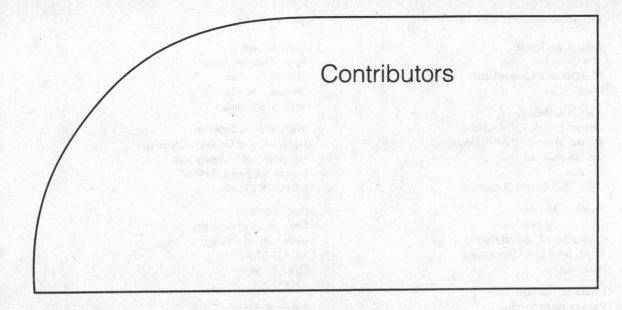

Contributors

Michael K. Brett-Surman
Department of Paleobiology
National Museum of Natural History
Smithsonian Institution
Washington, DC
20560 USA

Kenneth Carpenter
Department of Earth Sciences
Denver Museum of Natural History
City Park
Denver, Colorado
80205 USA

Ralph E. Chapman
Scientific Computing, A.D.P.
National Museum of Natural History
Smithsonian Institution
Washington, DC
20560 USA

Alan J. Charig
Department of Palaeontology
British Museum (Natural History)
Cromwell Road
London
SW7 5BD United Kingdom

Edwin H. Colbert
Museum of Northern Arizona
Route 4, Box 720
Flagstaff, Arizona
86001 USA

Walter P. Coombs, Jr.
Department of Biological and Physical Sciences
Western New England College
1215 Wilbraham Rd.
Springfield, Massachusetts
01119 USA

Philip J. Currie
Tyrrell Museum of Palaeontology
P.O. Box 7500
Drumheller, Alberta
T0J 0Y0 Canada

Peter Dodson
School of Veterinary Medicine
University of Pennsylvania
Philadelphia, Pennsylvania
19104-6046 USA

Dong Zhiming
Institute of Vertebrate Paleontology
          and Paleoanthropology
Academia Sinica
P.O. Box 643
Beijing
People's Republic of China

Mark B. Goodwin
Museum of Paleontology
University of California
Berkeley, California
94720 USA

John R. Horner
Museum of the Rockies
Montana State University
Bozeman, Montana
59717 USA

Thomas M. Lehman
Department of Geosciences
Texas Tech University
P.O. Box 4109
Lubbock, Texas
79409 USA

John S. McIntosh
278 Court St.
Middletown, Connecticut
06457 USA

Angela C. Milner
Department of Palaeontology
British Museum (Natural History)
Cromwell Road
London
SW7 5BD United Kingdom

Ralph E. Molnar
Queensland Museum
Queensland Cultural Centre
South Brisbane, Queensland
Australia

David B. Norman
Paleontology Section
Nature Conservancy Council
Northminster House
Peterborough
PE1 1UA United Kingdom

John H. Ostrom
Peabody Museum
Yale University
New Haven, Connecticut
06520 USA

Michael A. Raath
Port Elizabeth Museum
P.O. Box 13147
Humewood
South Africa

J. Keith Rigby, Jr.
Department of Geology
Notre Dame University
South Bend, Indiana
46556 USA

Loris Russell
Royal Ontario Museum
100 Queens Park
Toronto, Ontario
M5S 2C6 Canada

William A. S. Sarjeant
Department of Geological Sciences
University of Saskatchewan
Saskatoon, Saskatchewan
S7N 0W0 Canada

Paul C. Sereno
Department of Anatomy
University of Chicago
1025 E. 57th Street
Chicago, Illinois
60637 USA

Robert E. Sloan
Geology & Geophysics
University of Minnesota
Minneapolis, Minnesota
55455 USA

David B. Weishampel
Department of Cell Biology and Anatomy
Johns Hopkins University
School of Medicine
Baltimore, Maryland
12105 USA

Peter Wellnhofer
Bayerische Staatssammlung für Paläontologie
        und historische Geologie
Richard-Wagner Strasse 10
8 München 2
Federal Republic of Germany

Foreword:

Charles Mortram Sternberg and the Alberta dinosaurs

LORIS RUSSELL

It is my privilege to begin the volume by discussing the contributions to paleontology made by Charles M. Sternberg, or Charlie, as we called him. It has been said that without his discoveries and his interpretation of them, the magnificent Tyrrell Museum of Palaeontology might not exist today.

But C. M. Sternberg was not the first to find and describe the paleontological riches of the Red Deer badlands (Russell 1966). The fossil fields of the Red Deer River valley occur in two main areas (Fig. F.1). One of these is Dinosaur Provincial Park, northeast of Brooks, Alberta. It exposes the Judith River Formation. The other area is the Red Deer River valley from the vicinity of Drumheller northward. Here are displayed the rocks that used to be called the Edmonton Formation. It has now been promoted to group status and divided into a number of formations. There are other places in Alberta where fossil vertebrates occur, and these are also shown in Figure F.1, but none is as spectacular as the Red Deer River valley.

Of Sternberg's predecessors, Barnum Brown of the American Museum of Natural History is noteworthy. He was extraordinarily successful, and by 1912 had amassed a very important collection (Brown 1911). Because of his advanced techniques, he was able to collect many of his specimens as associated skeletons. The Geological Survey of Canada was a bit red-faced about this, because they had known for 21 years that there was a great fossil field here, and had not done much about it. So they looked around for somebody to work on their behalf and collect their share of these rich deposits. The man eventually selected was Charles Hazelius Sternberg of Lawrence, Kansas. He had been a professional fossil collector for over 30 years (Sternberg 1909). It was ironic that the Survey had to go to Kansas for somebody to

compete with Brown, because Brown was also from Kansas. In hiring Sternberg, the Survey not only got one experienced collector, they got four, the father and his three sons, George Friar, the eldest, Charles Mortram, the middle, and Levi, the youngest. They already had

Figure F.1. Map of Alberta showing principal fossil sites: asterisks for dinosaurs, circles for mammals. The asterisks on Red Deer River represent, respectively, the Edmonton (or Drumheller) badlands, and Dinosaur Provincial Park. Drumheller is situated where the railway crosses the Red Deer River.

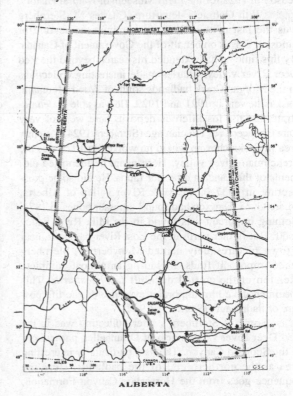

ALBERTA

In Dinosaur Systematics: Perspectives and Approaches, *Kenneth Carpenter and Philip J. Currie, eds. Copyright © Cambridge University Press, 1990.*

served as assistants to their father in the fossil fields of Kansas and Wyoming.

Although their field work was in Alberta, the Sternbergs' home base was Ottawa, in what was then called the Victoria Memorial Museum, now part of the National Museums of Canada. They began their Canadian field work in 1912, in the Drumheller area, and their first important find was an excellent skeleton of a duckbilled dinosaur in the tributary valley of Michichi Creek. Later the specimen was mounted in the Victoria Memorial Museum, the first such display in Canada. It was described by Gilmore (1924) as *Thespesius edmontoni*, but now we recognize it as a species of the genus *Edmontosaurus* (named by Lawrence Lambe 1920), under the name *Edmontosaurus edmontoni*.

In 1912 Barnum Brown moved down the river to the Steveville Ferry, in what is now Dinosaur Provincial Park. Next year, 1913, he was followed by the Sternberg family, who set up camp at Steveville Ferry. About an hour after the Sternberg party had landed, son Charles found the skeleton of a carnosaur, which became the type of *Gorgosaurus libratus* Lambe (1917). Sternberg senior and his son Levi left the Geological Survey of Canada in 1916. In 1919 Levi joined the staff of the Royal Ontario Museum, where, under the direction of Dr. W. A. Parks, he built up an impressive collection of Alberta dinosaurs. In 1918, George Sternberg also left the Geological Survey of Canada. He spent the summers of 1920 and 1921 in the Red Deer River badlands, collecting for the University of Alberta, and with a final season in 1922 for the Field Museum of Natural History.

With George's departure, Charles M. Sternberg was the last of the family to continue the search for dinosaur fossils on behalf of the Government of Canada. By this time he had extended his search beyond the Red Deer River valley. Charles made interesting collections in the Morgan Creek badlands south of Wood Mountain, Saskatchewan, in 1921 and 1929. He was able to demonstrate that the fossiliferous deposits here were of very Late Cretaceous or Lancian age (Sternberg 1924). Another area in Saskatchewan where he worked in 1928 was the Frenchman River valley. Here again he identified elements of the Lancian dinosaur fauna. He did some prospecting in 1937 in the Milk River valley of Alberta, where G. M. Dawson had found dinosaur bones in 1874. Nothing important was found in the Milk River valley itself. But farther north in the Lost River badlands, near a town picturesquely named Manyberries, Sternberg made important finds, including an almost complete skeleton of the crested duckbill *Lambeosaurus*. This specimen gave him important evidence on the life posture of hadrosaurs (Sternberg 1942).

Some of Sternberg's later collecting was in the Red Deer River valley north of Drumheller, particularly in the Scollard Canyon. This is a particularly interesting area at the northern end of the badlands. Here the sequence goes from the Horseshoe Canyon Formation,

up through the Whitemud and Battle (thin but conspicuous formations), into the Scollard Formation. The demonstration that the Scollard dinosaur fauna was of Lancian age was one of Sternberg's most important biostratigraphic discoveries (Sternberg 1949). His collections here included the skull of a new species of *Triceratops*, the last word in Lancian index fossils. So we know, thanks to Sternberg's work, that the lower part of the Scollard Formation is equivalent in age to the Lance and Hell Creek formations of the western United States.

Charles Sternberg's last important field assignment was laying out the initial displays in Dinosaur Provincial Park (Fig. F.2). I like to tell young people that this dinosaur was put in a cage to prevent it from escaping.

In discussing the systematic work of Charles Sternberg, I shall not follow a strictly chronological order, as so many of the projects on which he worked were continued at intervals, and often more than one at a time. Charles got into descriptive work by writing a paper to supplement an unfinished paper by Lawrence Lambe. Lambe had died in 1919, and Sternberg was encouraged to write a supplementary study of *Panoplosaurus mirus*, an armored dinosaur. This he did, and it was his first scientific paper (Sternberg 1921). He went on from there to describe many important finds, most of them his own. This included the two fine skeletons of *Chasmosaurus* in the National Museum of Natural Sciences. He was intrigued by the presence of peculiar cavities in the skull roof of ceratopsians, which seemed to be closing as he traced them through the geological record. He wrote an important paper on this feature (Sternberg 1927), pointing out that early ceratopsians had a wide-open orifice exposing the primary skull roof. But this became progressively restricted, and in the last of the sequence, *Triceratops*, the skull roof was completely closed. Significant intermediate stages are seen in the skulls of *Styracosaurus* and *Centrosaurus* of

Figure F.2. *In situ* display of dinosaur skeleton, Dinosaur Provincial Park, Alberta. Left to right: Roy L. Fowler, Grace E. Russell, Gordon Gyrmov. L.S.R. photo, 1964.

the Judithian Stage, and in *Anchiceratops* of the Edmontonian Stage. This investigation was typical of Sternberg. Not only did he describe meticulously the anatomy of his specimens, but he also placed them in their chronological sequence based on their stratigraphic position and succession, which in most cases he himself had established.

In the Scollard Formation, Sternberg found three skeletons of *Leptoceratops*, a small hornless ceratopsian that Brown had found and named (Brown 1914). Sternberg's specimens enabled him to point out that *Leptoceratops* was an extremely primitive ceratopsian, even more primitive than *Protoceratops* (Sternberg 1951). Also in the Scollard Formation, Sternberg found an excellent skeleton of *Thescelosaurus*, a small ornithopod dinosaur. On the basis of this specimen he published (1940) some important conclusions on the relationships of the Hypsilophodontidae and similar light-limbed, primitive ornithopods that might be called collateral ancestors of the duckbilled dinosaurs.

Figure F.3. Hadrosaur skulls: *Saurolophus osborni* Brown (after Brown, 1912); *Edmontosaurus regalis* Lambe (after Lambe, 1920). Drawn by L.S.R.

Saurolophus

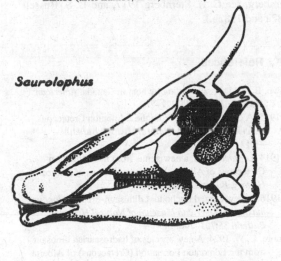

1 metre

Edmontosaurus

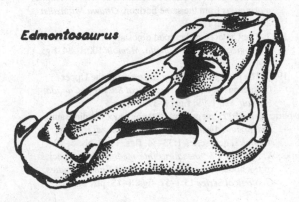

Sternberg's most spectacular find, I think, was the skull of *Pachyrhinosaurus*, from the St. Mary River Formation near Fort MacLeod in southern Alberta (Sternberg 1950). Some years later, a skull of the same strange ceratopsian was found in the Red Deer River valley north of Drumheller by local collectors, and more recently this animal has been found in Grande Prairie (Alberta) and Alaska. The facial roof is covered by a thick excrescence of hard bony material where normally there would be three horns. As Sternberg recognized, this is not secondary bone, but is analogous to the great thickening of the cranial roof in *Stegoceras* (Sternberg 1933).

Some of Sternberg's most interesting work was on the duckbilled dinosaurs (Fig. F.3). Sternberg pointed out that the peculiar spiked version that Brown (1912) named *Saurolophus*, is like an *Edmontosaurus* (Lambe 1920) with a spike on its head, but otherwise is a typical hadrosaurine.

In contrast to the flat-headed hadrosaurines, there are the extraordinary crested hadrosaurs or lambeosaurines. The cranial roof is surmounted by a more or less blade-shaped crest, which is actually made up of the nasal and premaxillary bones. The nasal passage wanders up into this crest and then down into the throat region, bifurcating along the way. Many workers have speculated on the function of this remarkable development. Sternberg was the first, in my opinion, to offer a plausible explanation (Sternberg 1935, p. 9). He said that the inverted U-shaped passage was an air trap, which prevented water from going down into the throat during under-water feeding. There have been other explanations offered since, but I still think Sternberg had the right idea (see also Sternberg 1964).

Typical of the crested duckbills is *Lambeosaurus* (Parks 1923). Originally, Lambe referred it to his genus *Stephanosaurus* (Lambe 1914). In this animal the nasal passages go up into the strange crest and then back down into the throat. Some people have restored the lambeosaurine crest as kind of cockscomb, but it is a wide structure, with space inside for the nasal passages. Figure F.4 shows the model that Sternberg made to demonstrate the course of these passages. They are not reservoirs for air, they are just winding pipes. In *Lambeosaurus magnicristatus* the air passages seem very complicated, but actually they just go up into the enormous crest and then come back down again.

One of the last major topics that Sternberg wrote about was the posture of hadrosaurs (Sternberg 1942). He took exception to the bipedal pose that we see in most restorations. On the basis of the posture in which most duckbilled dinosaur skeletons are found in the rock and on the structure of their pelvis, Sternberg concluded that the hadrosaurs were essentially quadrupedal (Sternberg 1965).

Sternberg's last paper (1970) dealt with what we now call taphonomy, that is, what happens to a carcass after the animal dies and is buried. He had long been interested in this subject, beginning, no doubt, with his participation in the discovery of the so-called dinosaur mummy in Wyoming in 1908. This was the famous specimen (Osborn 1912) that showed for the first time the contours of the somewhat desiccated but nevertheless authentic hadrosaur body. Up to this time very little

had been recorded about fossil skin impressions, and I shudder to think how much had been destroyed by unknowingly whittling down through it to get at the bones. In later years Charlie described many areas of skin impressions of duckbilled and horned dinosaurs (Sternberg 1925).

Figure F.5 shows Charles M. Sternberg in his prime, a highly successful collector and an enthusiastic student of dinosaurs. For a man who graduated from a Kansas high school, and made no pretence of an academic background, he did very well. In 1948 he was promoted to the rank of Assistant Biologist, the equivalent of curatorial status in the National Museum of Canada. In 1949 he was elected a Fellow of the Royal Society of Canada, not just for digging up dinosaurs, but because of his many contributions to the knowledge of dinosaurian anatomy and systematics. In later years Charles Sternberg received an honorary degree of Doctor of Laws from the University of Calgary, and an honorary Doctor of Science from Carleton University in Ottawa. These academic honors recognized that Charles Mortram Sternberg was a distinguished scientist.

(For additional biographical information on C. M. Sternberg, see C. H. Sternberg 1917, and L. S. Russell 1982a and 1982b.)

Figure F.4. Restoration of the head of *Corythosaurus excavatus* Gilmore by C.M. Sternberg (Sternberg, 1942).

Figure F.5. Charles M. Sternberg, National Museum of Canada Photo No. 101651.

References

Brown, B. 1911. Fossil hunting by boat in Canada. *American Museum Journal* 11:272–282.

　　1912. A crested dinosaur from the Edmonton Cretaceous. *American Museum of Natural History, Bulletin* 31:131–136.

　　1914. *Leptoceratops*, a new genus from the Edmonton Cretaceous of Alberta. *American Museum of Natural History, Bulletin* 33:567–580.

　　1916. A new crested trachodont dinosaur, *Prosaurolophus maximum. American Museum of Natural History, Bulletin* 35:701–708.

Gilmore, C. W. 1924. A new species of hadrosaurian dinosaur from the Edmonton Formation (Cretaceous) of Alberta. *Geological Survey of Canada, Bulletin* 38:13–26.

Lambe, L. M. 1914. On a new genus and species of carnivorous dinosaur from the Belly River Formation of Alberta, with description of the skull of *Stephanosaurus marginatus* from the same horizon. *Ottawa Naturalist* 28:13–20, pl. 1.

　　1917. The Cretaceous theropod dinosaur *Gorgosaurus. Geological Survey of Canada, Memoir* 100:1–84, figs. 1–49.

　　1920. The hadrosaur *Edmontosaurus* from the Upper Cretaceous of Alberta. *Geological Survey of Canada, Memoir* 120:1–79, figs. 1–39.

Osborn, H. F. 1912. Integument of the iguanodont dinosaur *Trachodon. American Museum of Natural History Memoirs*, new series 1:33–54, figs. 1–13, pls. 5–10.

Parks, W. A. 1923. *Corythosaurus intermedius*, a new species of trachodont dinosaur. *University of Toronto Studies, Geological Series* 15:1–57, figs. 1–13, pls. 1–6.

Russell, L. S. 1966. Dinosaur hunting in western Canada. *Royal Ontario Museum, Life Sciences Contribution* 70:1–37.

    1982a. Charles Mortram Sternberg 1885–1981. *Royal Society of Canada, Proceedings*, series 4, 20:132–135.

    1982b. Charles Mortram Sternberg, 1885–1981. *The Canadian Field-Naturalist* 96:483–486.

Sternberg, C.H. 1909. *The Life of a Fossil Hunter* (New York: Henry Holt and Co.).

    1917. *Hunting Dinosaurs in the Badlands of the Red Deer River, Alberta, Canada* (Lawrence, Kansas: C. H. Sternberg).

Sternberg, C. M. 1921. A supplementary study of *Panoplosaurus mirus*. *Royal Society of Canada, Transactions*, series 3, 15:93–102.

    1924. Notes on the Lance Formation of southern Saskatchewan. *Canadian Field-Naturalist* 33:66–70.

    1925. Integument of *Chasmosaurus belli*. *Canadian Field-Naturalist* 39:108–110.

    1927. Homologies of certain bones in the ceratopsian skull. *Royal Society of Canada, Transactions*, series 3, 21:135–143.

    1933. Relationships and habitat of *Troodon* and the nodosaurs. *Annals and Magazine of Natural History*, series 10, 11:231–235.

    1935. Hooded hadrosaurs of the Belly River Series of the Upper Cretaceous: a comparison, with descriptions of new species. *National Museum of Canada, Bulletin* 77:1–37.

    1940. *Thescelosaurus edmontonensis*, n. sp., and classification of the Hypsilophodontidae. *Journal of Paleontology* 14:481–494.

    1942. New restoration of a hooded duck-billed dinosaur. *Journal of Paleontology* 16:133,134.

    1949. The Edmonton fauna and description of a new *Triceratops* from the Upper Edmonton Member; phylogeny of the Ceratopsidae. *National Museum of Canada, Bulletin* 113:33–46.

    1950. *Pachyrhinosaurus canadensis*, representing a new family of the Ceratopsia, from southern Alberta. *National Museum of Canada, Bulletin* 118:109–120.

    1951. Complete skeleton of *Leptoceratops gracilis* Brown from the Upper Edmonton Member on Red Deer River, Alberta. *National Museum of Canada, Bulletin* 123:225–255.

    1964. Function of the elongated narial tubes in the hooded hadrosaurs. *Journal of Paleontology* 38:1003–1004.

    1965. New restoration of hadrosaurian dinosaur. *National Museum of Canada, Natural History Papers* 30:1–5.

    1970. Comments on dinosaurian preservation in the Cretaceous of Alberta and Wyoming. *National Museums of Canada, National Museum of Natural Sciences, Publications in Paleontology* 4:1–9.

Preface

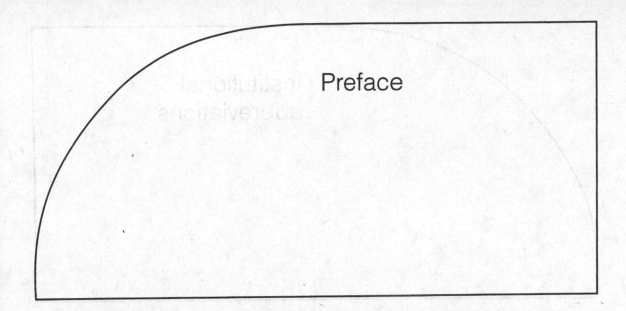

The Dinosaur Systematics Symposium was held June 3–5, 1986, at the Tyrrell Museum of Palaeontology, Drumheller, Alberta. There were over sixty-five individuals in attendance, with twenty-eight papers given. Of these, nineteen are presented in this volume. The purpose of the symposium was to examine sexual dimorphism, ontogeny, individual variation, and any other factors that may influence the taxonomic designation of a particular specimen of dinosaur. The symposium was the second in an unstructured series on aspects of dinosaurs as living creatures, the first being held the previous month (Gillette and Lockley 1989). Systematics was selected as the theme because all other studies essentially build on systematics.

The success of the symposium was made possible by the friendly attitude of all the participants. Although there were disagreements, there was no hostility. As is typical for dinosaur paleontologists, their passion for the subject led to numerous discussions, well into the evening, over beer at the local watering hole. Thus, it is not surprising that one participant called the symposium "the Woodstock of dinosaur paleontology."

Because so much current systematic work on dinosaurs is being done on Upper Cretaceous material, the symposium and this volume are dedicated to Charles Mortram Sternberg (1885–1981). Although some of the species he established are shown to be invalid by papers in this volume, he did try to recognize variation in the material he studied. In fact, he seems to have anticipated many of the conclusions reached by the symposium. For example, he correctly recognized that differences in the robustness of skeletons of the ceratopsian *Chasmosaurus* were due to sexual dimorphism, and that different sizes of *Leptoceratops gracilis* represented ontogenetic stages. One taxonomic problem he apparently paid little attention to was individual variation, a factor that can be considerable and important as is pointed out by several papers in this volume.

All of the manuscripts in this volume were peer-reviewed, which greatly improved the quality of the papers. We would like to extend thanks to the reviewers: W. Abler, M. Benton, W. Coombs, P. Dodson, J. Farlow, P. Galton, J. Horner, T. Lehman, M. Lockley, R. Molnar, D. Norman, J. Ostrom, K. Padian, G. Paul, M. Raath, K. Rigby, Jr., A. Russell, D. Russell, W. Sarjeant, P. Sereno, D. Spalding, H. Sues, R. Thulborn, and D. Weishampel.

We are grateful to Linda Reynolds and Heather Whitehead (Ampersand Editorial Associates, Calgary) for their editorial assistance and efforts in bringing this volume to completion.

Finally, thanks are due to the staff at the Tyrrell Museum of Palaeontology for their hard work in making the symposium possible. Most of the clerical work for both the symposium and this volume was done by Rebecka Kowalchuk and Linda Culshaw. Although unforeseen difficulties slowed the publication of this volume, we hope that the results were worth the wait.

Kenneth Carpenter and Philip J. Currie
Drumheller, Alberta, 1990

Institutional abbreviations

AMNH	American Museum of Natural History, New York
ANSP	Academy of Natural Sciences of Philadelphia
BHI	Black Hills Institute of Geological Research, Hill City
BM(NH)	British Museum (Natural History), London
BNHM	Beijing Natural History Museum, Beijing
BSP	Bayerische Staatssammlung für Paläontologie und historische Geologie, Munich
CM	Carnegie Museum of Natural History, Pittsburgh
CUP-FMNH	Catholic University of Peking Collection, Field Museum of Natural History
CV	Municipal Museum of Chungqing
DMNH	Denver Museum of Natural History, Denver
FMNH	Field Museum of Natural History, Chicago
GI SPS	Geological Institute, Section of Palaeontology and Stratigraphy, Academy of Science of the Mongolian People's Republic, Ulan Bator
GSI SR	Geological Survey of India, Calcutta
HMN	Humboldt Museum für Naturkunde, Berlin
IG	Institute of Geology, Beijing
IVPP	Institute of Vertebrate Palaeontology and Palaeoanthropology, Academica Sinica, Beijing
KU	University of Kansas Museum of Natural History, Lawrence
LACM	Museum of Natural History of Los Angeles County, Los Angeles
MCZ	Museum of Comparative Zoology, Harvard University, Boston
MNA	Museum of Northern Arizona, Flagstaff
MOR	Museum of the Rockies, Bozeman
NMC	National Museums of Canada, National Museum of Natural Sciences, Ottawa
OMNH	Oklahoma Museum of Natural History, Norman
PIN	Palaeontological Institute, USSR Academy of Sciences, Moscow
QG	Queen Victoria Museum, Salisbury
ROM	Royal Ontario Museum, Toronto
SDSM	South Dakota School of Mines, Rapid City
SM	Staatliches Museum für Naturkunde, Stuttgart
SM-FUND	Seemann Collection, Staatliches Museum für Naturkunde, Stuttgart
SM-1911	Fraas Collection, Staatliches Museum für Naturkunde, Stuttgart
TMM	Texas Memorial Museum, Austin
TMP	Tyrrell Museum of Palaeontology, Drumheller
UA	University of Alberta, Edmonton
UCMP	University of California Museum of Paleontology, Berkeley
UM	University of Montana, Missoula
UNM	University of New Mexico, Albuquerque
USNM	United States National Museum (now National Museum of Natural History), Washington DC
UT	Universität Tubingen, Geologische-Paläontologisches Institut, Tubingen
UTEP	University of Texas at El Paso, Centennial Museum, El Paso
YPM	Yale Peabody Museum, Yale University, New Haven
ZDM	Zigong Dinosaur Museum, Zigong
ZPAL	Institute of Paleobiology, Polish Academy of Sciences, Warsaw

Introduction: on systematics and morphological variation

KENNETH CARPENTER AND
PHILIP J. CURRIE

In recent years, dinosaurs have captured the attention of the public at an unprecedented scale. At the heart of this resurgence in interest is an increased level of research activity, much of which is innovative within the field of paleontology. Whereas earlier studies emphasized basic morphology and taxonomy, modern studies develop our understanding of what dinosaurs were like as living animals. More than ever before we understand how their bodies worked, how they behaved, how they interacted with their surroundings and with each other, and how they changed over time. Nevertheless, these studies still rely on certain basic building blocks, including knowledge of anatomy and taxonomic relationships.

One of the aspects that we understand better than before is ontogenetic, sexual, and individual variation within a species. This helps us to evaluate our understanding of dinosaurs as biological species. Studies in progress are giving us a good understanding of all forms of variability for one or more species of each of the major groups of dinosaurs, including theropods (Chapters 6, 7), hypsilophodonts (Horner and Weishampel 1988), hadrosaurs (Horner and Makela 1979; Dilkes 1988; Horner and Weishampel 1988), iguanodonts (Norman 1987), protoceratopsians (Brown and Schlaikjer 1940; Kurzanov 1972; Dodson 1975b; Maryańska and Osmólska 1975), ceratopsians (Chapters 16, 18; studies in progress on *Centrosaurus* and *Pachyrhinosaurus*), and ankylosaurs (Maryańska 1971; studies in progress on new *Pinacosaurus* material from China).

A major shift occurred late in the nineteenth and early twentieth centuries, spurred on in part by the revolution that occurred in biology (Allen 1969). At that time, experiments in genetics produced some insights into variation seen in the natural world. Gilmore was

one of the first to recognize the importance of this research when he raised the possibility that not all of the species of *Camptosaurus* he recognized years earlier were valid, "… *C. medius* … may eventually be found to represent the female of the larger *C. dispar*, and that its fully adult development may be represented in *C. browni*." (Gilmore 1925, p. 392). It is this de-emphasis of morphology (however, see Rainger 1981, for an opposing view) and the recognition that a morphological species is not necessarily the same as a biological species that concerns many dinosaur paleontologists today. The First International Dinosaur Systematics Symposium was an attempt to gather as many dinosaur paleontologists as possible to unravel some of the problems surrounding the systematics of the dinosaurs. It was hoped that by examining specific groups, patterns might emerge to assist in the recognition of true species, sexual dimorphism, individual variation, and ontogeny.

It is difficult to imagine what it was like for those early paleontologists who had little or no comparative material and too few descriptive papers to rely upon. Consider, for example, what Joseph Leidy was faced with when F. V. Hayden presented him with a handful of teeth. It was clear to Leidy, an anatomist, that at least four distinct morphs were present to which the names *Palaeoscincus costatus*, *Trachodon mirabilis*, *Troodon formosus*, and *Deinodon horridus* were applied (Leidy 1856). Leidy seems to have considered them as biological entities, hence subject to variation, because the type material of *Trachodon mirabilis* also includes two teeth of a ceratopsian (Leidy 1860, pl. 9, figs 16–20). Leidy writes, "Two additional specimens, (figs. 16–20) found with the preceding, may perhaps belong to a different animal, but it is quite possible also that they belong to a different part of the jaws of the same animal" (Leidy 1860). But as Leidy began to acquire more dinosaur specimens, he seems to have made a subtle, but important, shift in the way he viewed the fossils. In 1865, he

In Dinosaur Systematics: Perspectives and Approaches, *Kenneth Carpenter and Philip J. Currie, eds. Copyright © Cambridge University Press, 1990.*

illustrated and described *Hadrosaurus foulkii* in considerable detail. He also described and compared numerous other hadrosaur specimens to *Hadrosaurus foulkii*, but refused to give these specimens a name. As Leidy admitted, "When first examined and compared with the corresponding bone of *Hadrosaurus* the differences which were observable ... [were] not very remarkable" (Leidy 1865). Considering how similar the postcrania of hadrosaurs are, his caution in not referring these specimens to *Hadrosaurus foulkii* is surprising. He seems to have been influenced more by the differences this time than he was years earlier when he included ceratopsian teeth with *Trachodon*.

But it was Cope and Marsh who emphasized subtle morphological differences as grounds for naming new species. This was undoubtedly due, in part, to their intense rivalry. An example is Cope's naming five species of *Laelaps* based on teeth from a small area of the Judith River Formation, Montana. His descriptions (Cope 1876a,b) emphasized minor differences among the teeth. Others following in their footsteps maintained this methodology (e.g., Lambe 1902 for species of *Monoclonius* and Huene 1907–8 for *Plateosaurus*). It was inevitable that this strict adherence to morphological tradition, which resulted in a profusion of names, would be questioned. As early as 1898, Osborn wrote concerning sauropods, "It is a priori improbable that so many different genera of gigantic Saurians of similar size co-existed. It is against the principles of evolution that closely similar types of equal size should occupy the same territory at the same time. It appears moreover to the writer that the evidence which has been brought forward to demonstrate such an exceptional condition is inadequate ..." (Osborn 1898). Hatcher wrote, "it does not seem at all improbable that some of the remains which have been referred to *Pleurocoelus*, *Astrodon* or *Elosaurus* may in reality belong to the young of some of these genera of the large sauropoda [i.e., *Brontosaurus*, *Morosaurus*, and *Diplodocus*]" (Hatcher 1902). Gilmore was even more specific when he wrote concerning *Leptoceratops* from the Two Medicine Formation, "the much smaller size of the National Museum specimens is in all probability due to immaturity, as indicated by the open sutures of the skull and the noncoalescence of the vertebral processes" (Gilmore 1939).

Although it is now generally conceded that size is not an adequate criterion to establish a species of dinosaur, how much emphasis to place on morphological differences is still a problem, as may be illustrated by our own experience. In 1982, one of us (Carpenter) named *Pectinodon bakkeri* for small theropod teeth from the Lance Formation of Wyoming. Estes (1964) had earlier identified similar teeth as cf. *Saurornithoides*, but comparison with the holotype of *Saurornithoides mongoliensis* showed numerous differences making such an identification improbable. Nor could the teeth be matched to any other theropod. This presented a

quandary: the teeth were distinct from any known theropod, but was this grounds for a new name or should no name be attached to them? Because so many of the teeth were known, it was decided that a new name was justified. However, it was the discovery of a nearly complete dentary several years later that demonstrated *Pectinodon* was not a valid taxon. This dentary showed that the distinct morphology of *Pectinodon* teeth was due to their being from the posterior portion of the jaw (Currie 1987), a possibility which was not realized in 1982 from the available material.

Remarkably, individual variation does not appear to have been a factor considered by most dinosaur paleontologists. Certainly they were aware of this for Dollo wrote concerning *Iguanodon bernissartensis*, "[it] is represented by a considerable number of animals and among them they offer remarkable divergences" (translation in Rainger 1985). Rainger argues, "The need to identify a fossil organism and establish its taxonomic relation to other organisms was a principle objective of virtually all paleontological studies of the time ... To fulfill these objectives, Dollo had to concentrate on specific characters, the morphological features that distinguished each species [i.e., *Iguanodon mantelli* and *Iguanodon bernissartensis*], and as a result did not take into consideration variations within a species" (Rainger 1985). Marsh, Cope, von Huene, Osborn, and other early dinosaur paleontologists had much the same attitude.

Although morphological distinction played an important role historically in the naming of new species, this was not always the case. As Gilmore noted, "The occurrence of *Leptoceratops* in the Two Medicine formation would suggest, on geological position alone, that it probably represents a species distinct from these found in the Edmonton and St. Mary River formations" (Gilmore 1939). This idea that dinosaurs in different formations had to be different species was an outgrowth of the early stratigraphic work of William Smith. Smith had found that fossils of one formation differed from those in another, thus permitting him to make long-distance correlations. Although this work was done in marine beds, the concept was eventually carried over to non-marine formations, as seen by Marsh's reference to the *Atlantosaurus* beds or the *Triceratops* beds. Marine invertebrate paleontologists were the first to discover that a single formation may have a sequence of different species. Even mammalian paleontologists recognized that the same species could have broad geographical distribution, and thus be found in different formations (an early summary presented by Osborn 1910). But for some reason, dinosaur paleontologists have been slower to accept the presence of the same species in more than one formation.

Finally, one other factor that has influenced the naming of species had less to do with morphology or stratigraphy, than it did with politics. For example, Barnum Brown (1917) named *Monoclonius cutleri* for a specimen having skin impressions found by William Cutler. Supposedly, one of the conditions Cutler imposed on Brown before giving it to him was that if the specimen proved to be a new species, it be named after him (Russell 1966).

Advanced prosauropods and primitive sauropods have long been known to resemble one another, leading von Huene (1932) to propose the term Sauropodomorpha for the two. It remains to be seen whether this is a natural group or not. Bonaparte (1986) noted several derived characters in the vertebrae of an unnamed advanced prosauropod from Argentina and those of the sauropod *Patagosaurus*. Sereno (1984) and Galton (1986), however, suggest that the Sauropodomorpha are not monophyletic.

The first prosauropod described was *Thecodontosaurus* (Riley and Stutchbury 1836) on the basis of isolated teeth. Since then, numerous genera and species have been named, especially by von Huene. Recent papers (see references in Galton 1984) have attempted to resolve some of the taxonomic problems associated with prosauropods, but there is much work yet to be done. In this volume (Chapter 3), Weishampel and Chapman use femora of *Plateosaurus* to illustrate the use of Principle Components Analysis in taxonomy. By using only a single monospecific assemblage, they have documented the degree of variation (possibly sexual and individual) in a population of prosauropods. In contrast with Galton (1984) who proposed to synonymize most *Plateosaurus* species into *P. engelhardti*, they tentatively conclude that there may be two species. They also document the presence of two morphs within a single population.

The first sauropod, *Cetiosaurus*, was described on the basis of vertebrae (Owen 1841). Since then, numerous sauropod species have been described from every continent except Antarctica. As with all other dinosaur groups, sauropod systematics has been confused by a proliferation of names. McIntosh (Chapter 4) does much to improve our understanding of all the families, genera, and species of sauropods, but especially those of the Morrison Formation. However, there are still many taxonomic problems that will have to be worked out (work in progress on Chinese sauropods by Dong and Russell).

Abundant material exists for studying the variability of many sauropods [juveniles being studied by Britt (1988) and Carpenter; massive numbers of individuals from single quarries in China including Zigong (Dong 1987a) and Xinjiang (Dong 1987b)], but these studies are hampered by the size of the material, and a lack of workers.

One of the first dinosaurs discovered and formally described was a carnivorous form (theropod) known as *Megalosaurus*, described by Buckland in 1824. Although it was the first to be discovered, it is still poorly known. Theropods tend to be rarer than herbivorous dinosaurs, and, as many of the species are small in size, their remains were often destroyed by scavengers and other natural processes. Nevertheless, good specimens are being recovered around the world, giving every indication of a remarkable range of diversity (Currie 1989a). Whereas we once assumed that all theropods could be assigned to only two lineages (Coelurosauria, Carnosauria), it is now realized that this approach was too simplistic and that many lineages developed over their long history (Gauthier and Padian 1989). New and remarkable species of theropods continue to be found (Bonaparte 1985; Chapter 9) that do not fit the mold. Although most species of theropods are known from only a few specimens, some forms – such as *Allosaurus* (Madsen 1976), *Coelophysis* (Colbert 1989; Chapter 6), and *Syntarsus* (Rowe 1989; Chapter 7) – are represented by numerous specimens from single localities, and can be used to evaluate the range of variation possible within single species of theropods. Other species (Chapter 10) may be represented by numerous specimens from wider geographic and stratigraphic ranges, which makes it more difficult to assess their relationships and variability. Finally, many theropods are only known from parts of their skeletons, although these may be diagnostic enough to provide some sense of variability in at least part of the animal (Chapter 8).

The Ornithopoda is a somewhat problematic group that at one time seemed to include all bipedal ornithischians. Accepted this way, the group is paraphyletic. Santa Luca (1980) attempted to redefine the Ornithopoda as those ornithischians having an obturator process, a condition that he considered derived. This was accepted by Maryańska and Osmólska (1985), but not as the sole criterion by Norman (1984) or Sereno (1984, 1986). There still is no agreement as to the composition of the Ornithopoda. Norman (1984) includes fabrosaurs, hypsilophodonts, iguanodonts, hadrosaurs, and, surprisingly, ceratopsians. Sereno (1984, 1986) includes heterodontosaurs, hypsilophodonts, iguanodonts, and hadrosaurs in his concept of the Ornithopoda, whereas Maryańska and Osmólska (1985) include only the fabrosaurs, iguanodonts, and hadrosaurs.

The first ornithopod named was *Iguanodon* (Mantell 1825) on the basis of several teeth. Ornithopods are one of the most abundant dinosaurs in Cretaceous strata, so it is not surprising that there is a surfeit of names. In Chapter 11, Norman continues his studies of iguanodontid systematics, and concludes that *Vectisaurus* is a juvenile *Iguanodon*. This gives Norman a basis for a cladistic analysis of the family Iguanodontidae.

Hadrosaurs were also first named for isolated teeth from the Judith River Formation of Montana (Leidy 1856). Morphometric analysis can be used to investigate the classification and phylogeny of hadrosaurs (Chapter 12), although even standard anatomical studies can produce some interesting results (Chapter

13). The greatest potential for the analysis of variation within single species of hadrosaurs remains with bone-bed material (such as the spectacular bone-bed near Choteau, Montana, being worked by the Museum of the Rockies, which may include the remains of more than 10,000 *Maiasaura* individuals) and nesting sites. Hatchlings of *Maiasaura* (Horner and Makela 1979) and embryos of *Hypacrosaurus* (Currie and Horner 1988) show variability even within a single nest.

Pachycephalosaurs, an intriguing group of dinosaurs, were among some of the first discoveries in western North America (Lambe 1902; Baird 1979), although it was a long time before there was enough skeletal material available to know what kind of animal they in fact represented (Gilmore 1924). Unfortunately, considerable confusion was introduced at the same time by the misconception that *Stegoceras* and *Troodon* were synonymous, when in fact the latter is a carnivorous dinosaur (Currie 1987). Over the years, a tremendous number of partial skulls have been recovered from around the world (Galton 1971; Maryańska and Osmólska 1974; Sues 1980; Sues and Galton 1987; Chapter 14), which have shown that pachycephalosaurs were successful in their diversity and numbers. Because of the number of skull caps recovered, it has been possible to look at the variation in pachycephalosaur species (Chapter 2; Chapman et al. 1981).

Stegosaurs are still poorly understood in spite of a good bone-bed known for *Kentrosaurus* (Hennig 1915; Janensch 1925; Galton 1982a), as much of the material has not been completely prepared or described. Juvenile specimens of *Stegosaurus* (Galton 1982b) and related forms (Galton 1981, 1983) provide ontogenetic information on this group of dinosaurs. With all of the new stegosaur material being recovered in China (Chapter 19), there is some hope for a better understanding of variability of at least mature specimens.

Hylaeosaurus armatus has the distinction of being the first ankylosaur named (Mantell 1833) and also of being the first dinosaur named from more than a few bones or teeth. This seemingly good start for ankylosaurs was upset by the naming of *Palaeoscincus costatus* (Leidy 1856) on the basis of a single tooth. This caused innumerable taxonomic problems when skulls and partial skeletons of ankylosaurs were discovered years later. The teeth associated with this new material were frequently compared to *P. costatus*, although the taxonomic validity of using ankylosaur teeth has never been demonstrated. Coombs (Chapter 20) examines this crucial problem and concludes that there is so much positional and individual variation that teeth cannot be used for ankylosaur taxonomy. This is an interesting contrast to the situation in theropods (Chapter 8), and shows that generalities ("all dinosaur teeth are taxonomically useful" or "no dinosaur teeth are taxonomically useful") should be avoided. Ankylosaur armor is commonly found in Cretaceous sediments, and

in Chapter 21 Carpenter demonstrates that it can be taxonomically useful to distinguish two sympatric species of ankylosaur. Ankylosaurs have generally always been relatively rare, isolated finds, making it difficult to sort out variability and taxonomy. The discovery of six sibling *Pinacosaurus* specimens in China (Currie 1989b) opens up the potential for studies on individual and ontogenetic variation, building on work already done by Maryańska (1971).

The first ceratopsian teeth figured were identified as *Trachodon mirabilis* by Leidy (1856), but were subsequently assigned to the Ceratopsia (on the basis of anatomy) and to *Monoclonius* (Hatcher et al. 1907) for stratigraphic reasons. This foreshadowed the taxonomic problems that developed as more genera and species were described, and much work is still needed rectify the situation. Papers in this volume (Chapters 15, 16, 17, 18) independently show that many characters previously used to diagnose species are due to ontogeny or sexual dimorphism. Sereno (Chapter 15) concludes that an understanding of the ontogeny of a species is necessary before it can be properly diagnosed. Using this tool, he concludes that there are four species of *Psittacosaurus*, two of which have only recently been named (Sereno and Chao, 1988; Sereno et al. 1988).

The problem of distinguishing *Monoclonius* and *Centrosaurus* is discussed by Dodson (Chapter 17). Both were initially named for fragmentary material and, consequently, have been the source of much confusion when complete skulls were later found. In his analysis of the skulls, Dodson uses biometric analysis as in previous studies on lambeosaurine hadrosaurs (Dodson 1975a) and *Protoceratops* (Dodson 1975b). He concludes, in contrast with Lehman (Chapter 16), that *Brachyceratops* is a valid taxon.

Centrosaurine ceratopsians promise to give us some of the best information on individual, sexual, and ontogenetic variation for any group of dinosaurs because of the discovery of numerous monospecific bone-beds in Alberta and Montana. Work in progress on *Centrosaurus apertus* (Currie 1981; Currie and Dodson 1984; Ryan, M., pers. comm.), a new species of *Pachyrhinosaurus* (Tanke 1988; Langston, W., pers. comm.), and *Styracosaurus* (Rogers and Sampson 1989) is showing considerable differences between juveniles and adults, and a range of variation amongst mature specimens. Once these studies are further along, we should be able to better understand the systematic position of small centrosaurines like *Avaceratops* (Dodson 1986), *Brachyceratops* (Gilmore 1917), and *Monoclonius* (Dodson and Currie 1988).

Chasmosaurine ceratopsian variation (ontogenetic and sexual) is exemplified by work done on a monospecific *Chasmosaurus* bone-bed. Lehman (Chapter 16) uses his results to assess the status of other ceratopsian species, and reaches the same conclusion as Ostrom and Wellnhoffer (Chapter 18) that all specimens of *Triceratops* can be assigned to a single species. The latter study is restricted to specimens from Wyoming, and it remains to be seen whether the size gradient noted by Sloan (1976) indicates a distinct species in Canada.

The first dinosaur ichnofossils were named by Hitchcock in 1836. These were the species *giganteus, tuberosus, ingens, diversus, tetradactylus, palmatus,* and *minimus* [generic designations were not made until 1845 (Lull 1953)]. What was Hitchcock's intent in proposing these names? "When I speak of species here, I mean species in oryctology, not ornithology. And I doubt not, that in perhaps every instance, what I call a species in the former science, would be a genus in the latter; that is to say, these different tracks were made by birds that were generically different" (Hitchcock 1836). Although Hitchcock acknowledged that "these names, implying only a resemblance, leave the real nature of the tracks open to discussion ...," he advocated using the Linnaean system: "I propose the term Ichnolite ... to be the name of the Class [and] I would divide this Class into Orders ..." (Hitchcock 1841). He later modified his intent when he wrote four years later, "hitherto names have been given to the footmarks and not to the animals. But since all geologists now admit that these impressions are real tracks, this paper attempts to name the animals that made them, and to classify and describe them, so far as it can be done from the data hitherto obtained" (Hitchcock 1845). But as more and more footprints began to accumulate, it became clear that naming footprints was not necessarily naming the animal. As Lull noted, "It has been deemed wise ... to keep the ichnite genera and species apart in their nomenclature from that applied to the actual bones, for it is at once apparent ... that footprint genera and species do not necessarily correspond in limitations or numbers with those of the actual animals which made the impressions ..." (Lull 1953). Thus ichnology went the full circle, from names used to identify the footprints to names identifying the trackmakers, and back to names identifying the footprints.

The use of names in ichnology is not without problems as Lull observed, "names given in Ichnology must be kept in a separate series which, were the trackmakers actually known, might result in two names given to the same animal, regrettable but unavoidable ..." (Lull 1953). It is the problem of using Linnaean classification on footprints that Sarjeant examines in Chapter 22. He advocates treating them as sedimentary structures that should not be named using the Linnaean system despite the provisions of the International Code of Zoological Nomenclature [International Commission on Zoological Nomenclature, 1985, Art. 1a, 1d, 10d, 13b, 23g, 42b(i), 66 and 67m]. It remains to be seen, however, whether dinosaur ichnologists adopt Sarjeant's proposals. They will certainly spark debate, which is good for the field.

Anyone attempting a serious study of dinosaurs is eventually faced with a plethora of names, many based upon material so fragmentary that it is doubtful the specimen(s) would be collected today (e.g., the teeth of the holotype of *Trachodon mirabilis* Leidy 1856). Nevertheless, many of these names are intimately linked with the early growth of dinosaur paleontology, both in North America and Europe (e.g., holotypes of *Hadrosaurus foulkii*, Leidy 1858, and *Iguanodon* Mantell 1825). But lest we be too critical of these early paleontologists, it should be realized that the plethora of names is the logical outcome of the morphological tradition of vertebrate paleontology. This tradition emphasizes differences among specimens, often at the expense of normal biological variation. Hence, Gilmore's (1909) recognition of four species of *Camptosaurus* (*C. dispar, C. browni, C. nanus,* and *C. medius*) from a single quarry in the Morrison Formation is understandable.

So where do we stand today? Dodson (1987) has pointed out that about 40% of the 265 genera of dinosaurs currently recognized have been described since 1969. That is the highest percentage of named taxa since the early days (pre-World War II) of dinosaur paleontology. This high percentage is certainly due to the maturity of the "baby boom" generation and the present renaissance of dinosaur paleontology. There are more dinosaur paleontologists alive today than there have been altogether in the past (about 110 since 1969 compared to 66 between 1824 and 1939). But the contributions made by the pioneers of the past are often not adequately appreciated by the present generation. It is all to easy to scoff because of the systematic "mess" we have inherited, forgetting that if we can see farther and clearer, it is because we stand on the shoulders of giants (Marsh, Cope, Sternberg, Brown, Osborn, von Huene, and Nopcsa to name a few).

As long as dinosaur paleontologists continue to collect and analyze new material, it seems unlikely that dinosaur taxonomy will ever have a final resolution. We will certainly not approach the accuracy of biologists working with living animals unless we have access to karyological or albumin immunological data. Nevertheless, it is hoped that the results of the First International Dinosaur Systematics Symposium presented in this volume will do much to improve our understanding of dinosaur species.

References

Allen, G. 1969. T. H. Morgan and the emergence of a new American biology. *Quarterly Review of Biology* 44:168–188.

Baird, D. 1979. The dome-headed dinosaur *Tylosteus ornatus* Leidy 1872 (Reptilia: Ornithischia: Pachycephalosauridae). *Notulae Naturae* 456:1–11.

Bonaparte, J. 1985. A horned carnosaur from Patagonia. *National Geographic Research* 1:149–151.

1986. The early radiation and phylogenetic relationships of

the Jurassic sauropod dinosaurs, based on vertebral anatomy. *In* Padian, K. (ed.), *The Beginning of the Age of Dinosaurs* (Cambridge: Cambridge University Press), pp. 247–258.

Britt, B. 1988. A possible "hatchling" *Camarasaurus* from the upper Jurassic Morrison Formation (Dry Mesa Quarry, Colorado). *Journal of Vertebrate Paleontology* 8 (Supplement to no. 3):9A (Abstract).

Brown, B. 1917. A complete skeleton of the horned dinosaur *Monoclonius,* and a description of a second skeleton showing skin impressions. *American Museum of Natural History Bulletin* 17:281–306.

Brown, B. and Schlaikjer, E. M. 1940. The structure and relationships of *Protoceratops. New York Academy of Sciences, Annals* 40:133–266.

Buckland, W. 1824. Notice on the *Megalosaurus,* or great fossil lizard of Stonesfield. *Geological Society of London, Transactions* 1:390–396.

Carpenter, K. 1982. Baby dinosaurs from the Late Cretaceous Lance and Hell Creek formations and a description of a new species of theropod. *Contributions to Geology, University of Wyoming* 20:123–134.

Chapman, R. E., Galton, P. M., Sepkoski, J. J. Jr. and Wall, W. P. 1981. A morphometric study of the cranium of the pachycephalosaurid dinosaur *Stegoceras. Journal of Paleontology* 55:608–618.

Colbert, E. H. 1989. The Triassic dinosaur *Coelophysis. Museum of Northern Arizona Bulletin* 57:1–160.

Cope, E. 1876a. Description of some vertebrate remains from the Fort Union beds of Montana. *Academy of Natural Sciences of Philadelphia, Proceedings* 1876:248–261.

1876b. On some extinct reptiles and Batrachia from the Judith River and Fox Hills beds of Montana. *Academy of Natural Sciences of Philadelphia, Proceedings* 1876:340–359.

Currie, P. 1981. Hunting dinosaurs in Alberta's great bone bed. *Canadian Geographic* 101(4):34–39.

1987. Bird-like characteristics of the jaws and teeth of troodontid theropods (Dinosauria, Saurischia). *Journal of Vertebrate Paleontology* 7:72–81.

1989a. Theropod dinosaurs of the Cretaceous. *In* Padian, K. and Chure, D. J. (eds.), *The Age of Dinosaurs,* Paleontological Society, Short Course Notes in Paleontology, no. 2, pp. 113–120.

1989b. Long distance dinosaurs. *Natural History* 6/89:60–65.

Currie, P. J. and Dodson, P. 1984. Mass death of a herd of ceratopsian dinosaurs. *In* Reif, W. E. and Westphal, F. (eds.), *Third Symposium on Mesozoic Terrestrial Ecosystems,* short papers (Tübingen: Attempto Verlag), pp. 61–66.

Currie, P. J. and Horner, J. R. 1988. Lambeosaurine hadrosaur embryos (Reptilia: Ornithischia). *Journal of Vertebrate Paleontology* 8 (Supplement to no. 3):13A (Abstract).

Dilkes, D. W. 1988. Relative growth of the hind limb in the hadrosaurian dinosaur *Maiasaura peeblesorum. Journal of Vertebrate Paleontology* 8 (Supplement to no. 3):13A (Abstract).

Dodson, P. 1975a. Taxonomic implications of relative growth in lambeosaurine hadrosaurs. *Systematic Zoology* 24:37–54.

1975b. Quantitative aspects of relative growth and sexual

dimorphism in *Protoceratops. Journal of Paleontology* 50:929–940.

1986. *Avaceratops lammersi:* a new ceratopsid from the Judith River Formation of Montana. *Academy of Natural Sciences of Philadelphia, Proceedings* 138:305–317.

1987. Dinosaur Systematics Symposium. *Journal of Vertebrate Paleontology* 7:106–108.

Dodson, P. and Currie, P. J. 1988. The smallest ceratopsid skull – Judith River Formation of Alberta. *Canadian Journal of Earth Sciences* 25:926–930.

Dong, Z. M. 1987a. *Dinosaurs from China* (Beijing: Ocean Press).

1987b. Untitled section on saurischian dinosaurs. *In* Zhao, X. J. et al. (eds.), *Stratigraphy and Vertebrate Fossils of Xinjiang* (Beijing: Institute of Vertebrate Paleontology and Paleoanthropology). [in Chinese]

Estes, R. 1964. Fossil vertebrates from the Late Cretaceous Lance Formation, eastern Wyoming. *University of California Publications in Geological Sciences* 49:1–180.

Galton, P. M. 1971. A primitive dome-headed dinosaur (Ornithischia: Pachycephalosauridae) from the Lower Cretaceous of England and the function of the dome of pachycephalosaurids. *Journal of Paleontology* 45:40–47.

1981. *Crateosaurus pottenensis* Seeley, a stegosaurian dinosaur from the Lower Cretaceous of England, and a review of Cretaceous stegosaurs. *Neues Jahrbuch für Geologie und Paläontologie* 161:28–46.

1982a. The postcranial anatomy of the stegosaurian dinosaur *Kentrosaurus* from the Upper Jurassic of Tanzania, East Africa. *Geologica et Palaeontologica* 15:139–160.

1982b. Juveniles of the stegosaurian dinosaur *Stegosaurus* from the Upper Jurassic of North America. *Journal of Vertebrate Paleontology* 2:47–62.

1983. A juvenile stegosaurian dinosaur, *Omosaurus phillipsi* Seeley, from the Oxfordian (middle Jurassic) of England. *Geobios* 16:95–101.

1984. Cranial anatomy of the prosauropod dinosaur *Plateosaurus* from the Knollenmergel (Middle Keuper, Upper Triassic) of Germany. 1. Two complete skulls from Trossingen/Württemberg with comments on diet. *Geologica et Palaeontologica* 18:139–171.

1986. Prosauropod dinosaur *Plateosaurus (= Gresslyosaurus)* (Saurischia: Sauropodomorpha) from the Upper Triassic of Switzerland. *Geologica et Paleontologica* 20:167–183.

Gauthier, J. and Padian, K. (eds.) 1989. *The Age of Dinosaurs,* Paleontological Society, Short Course Notes in Paleontology, no. 2, pp. 1–210.

Gilmore, C. 1909. Osteology of the Jurassic reptile *Camptosaurus,* with a revision of the species of the genus, and description of two new species. *U.S. National Museum Proceedings* 36:197–302.

1917. *Brachyceratops,* a ceratopsian dinosaur from the Two Medicine Formation of Montana. *U.S. Geological Survey, Professional Paper* 103:1–45.

1924. On *Troodon validus,* an ornithopodous dinosaur from the Belly River Cretaceous of Alberta, Canada. *University of Alberta, Department of Geology Bulletin* 1:1–43.

1925. Osteology of ornithopodous dinosaurs from Dinosaur National Monument, Utah. *Carnegie Museum Memoir* 10:385–409.

1939. Ceratopsian dinosaurs from the Two Medicine Formation, Upper Cretaceous of Montana. *U.S. National Museum Proceedings* 87:1–18.

Hatcher, J. 1902. Discovery of remains of *Astrodon* (*Pleurocoelus*) in the *Atlantosaurus* beds of Wyoming. *Annals of the Carnegie Museum* 2:9–14.

Hatcher, J. B., Marsh, O. C. and Lull, R. S. 1907. The Ceratopsia. *U.S. Geological Survey Monograph* 49:1–300.

Hennig, E. 1915. *Kentrosaurus aethiopicus* der Stegosauride des Tendaguru. *Sitzungsberichte der Gesellschaft Naturforschender Freunde, Berlin* 1915:219–247.

Hitchcock, E. 1836. Ornithichnology. Description of the footmarks of (*Ornithichnites*), birds on New Red Sandstone in Massachusetts. *American Journal of Science* 29:307–340.

1841. *Final report on the geology of Massachusetts*, Pt. III (Amherst and Northampton), pp. 301–714.

1845. An attempt to name, classify and describe the animals that made the fossil footmarks of New England. *Proceedings, Annual Meeting, Association of American Geologists and Naturalists, New Haven, Connecticut*, 6:23–25.

Horner, J. R. and Makela, R. 1979. Nest of juveniles provides evidence of family structure among dinosaurs. *Nature* 282:297–298.

Horner, J. R. and Weishampel, D. B. 1988. A comparative embryological study of two ornithischian dinosaurs. *Nature* 332:256–257.

Huene, F. von. 1907–8. Die Dinosaurier der europäischen Triasformation mit Berücksichtigung der aussereuropäischen Vorkommnisse. *Geologische und paläontologische Abhandlungen*, Supplement 1:1–419.

1932. Die fossile Reptil-Ordnung Saurischia, ihre Entwicklung und Geschichte. *Monographie zur Geologie und Paläontologie* 4:1–361.

Janensch, W. 1925. Ein aufgestelltes Skelett des Stegosauriers *Kentrurosaurus aethiopicus* E. Hennig aus den Tendaguru-Schichten Deutsch-Ostafrikas. *Palaeontographica* Supplement 7:257–327.

Kurzanov, S. M. 1972. Sexual dimorphism in protoceratopsians. *Paleontological Journal* 1972(1):91–97.

Lambe, L. 1902. New genera and species from the Belly River series (mid-Cretaceous). *Contributions to Canadian Palaeontology* 3(2):25–81.

Leidy, J. 1856. Notices of remains of extinct reptiles and fishes discovered by Dr. F. V. Hayden in the bad lands of the Judith River, Nebraska Territory. *Academy of Natural Sciences of Philadelphia, Proceedings* 8:72–73.

1858. Remarks concerning *Hadrosaurus*. *Academy of Natural Sciences, Proceedings* 1858:215–218.

1860. Extinct Vertebrata from the Judith River and Great Lignite formations of Nebraska. *American Philosophical Society, Transactions* 11:139–154.

1865. Cretaceous reptiles of the United States. *Smithsonian Contributions to Knowledge* 14, pp. 1–135.

Lull, R. S. 1953. Triassic life of the Connecticut Valley. *Connecticut Geological and Natural History Survey, Bulletin* 81:1–336.

Madsen, J. H. 1976. *Allosaurus fragilis*: a revised osteology.

Utah Geological and Mineral Survey, Bulletin 109:1–163.

Mantell, G. 1825. Notice on the *Iguanodon*, a newly discovered fossil reptile, from the sandstone of Tilgate Forest, in Sussex. *Philosophical Transactions of the Royal Society* 115:179–186.

1833. *Memoir on the* Hylaeosaurus, *a newly discovered fossil reptile from the strata of the Tilgate Forest*. (Geology of South-east England, London).

Maryańska, T. 1971. New data on the skull of *Pinacosaurus grangeri* (Ankylosauria). *Palaeontologia Polonica* 25:45–53.

Maryańska, T. and Osmólska, H. 1974. Pachycephalosauria, a new suborder of ornithischian dinosaurs. *Palaeontologia Polonica* 30:45–102.

1975. Protoceratopsidae (Dinosauria) of Asia. *Palaeontologia Polonica* 33:133–181.

1985. On ornithischian phylogeny. *Acta Palaeontologica Polonica* 30:137–150.

Norman, D. B. 1984. A systematic reappraisal of the reptile Order Ornithischia. *In* Reif, W. E. and Westphal, F. (eds.), *Third symposium on Mesozoic Terrestrial Ecosystems*, short papers (Tübingen: Attempto Verlag), pp. 157–162.

1987. A mass-accumulation of vertebrates from the Lower Cretaceous of Nehden (Sauerland), West Germany. *Royal Society of London, Proceedings, series B* 230:215–255.

Osborn, H. 1898. Additional characters of the great herbivorous dinosaur *Camarasaurus*. *American Museum of Natural History, Bulletin* 10:219–233.

1910. The Age of Mammals in Europe, Asia and North America. (New York: Macmillan).

Owen, R. 1841. Report on the British fossil reptiles. *Report of the eleventh meeting of the British Association of Science* 11:60–204.

Rainger, R. 1981. The continuation of the morphological tradition: American paleontology, 1880–1910. *Journal of the History of Biology* 14:127–158.

1985. Paleontology and philosophy: a critique. *Journal of the History of Biology* 18:267–287.

Riley, H. and Stutchbury, S. 1836. A description of various fossil remains of three distinct saurian animals discovered in the autumn of 1841, in the Magnesian Conglomerate on Durdham Down, near Bristol. *Geological Society of London, Proceedings* 2:397–399.

Rogers, R. R. and Sampson, S. D. 1989. A drought-related mass death of ceratopsian dinosaurs (Reptilia: Ornithischia) from the Two Medicine Formation (Campanian) of Montana: behavioral implications. *Journal of Paleontology* (Supplement to no. 3), 9:36a.

Rowe, T. 1989. A new species of the theropod dinosaur *Syntarsus* from the Early Jurassic Kayenta Formation of Arizona. *Journal of Vertebrate Paleontology* 9:125–136.

Russell, L.S. 1966. Dinosaur hunting in western Canada. *Royal Ontario Museum Life Sciences Contribution* 70:1–37.

Santa Luca, A. 1980. The postcranial skeleton of *Heterodontosaurus tucki* (Reptilia, Ornithischia) from the Stormberg of South Africa. *South African Museum, Annals* 79:159–211.

Sereno, P. C. 1984. The phylogeny of the Ornithischia: a reappraisal. *In* Reif, W. E. and Westphal, F. (eds.), *Third*

Symposium on Mesozoic Terrestrial Ecosystems, short papers (Tübingen: Attempto Verlag), pp. 219–226.

——— 1986. Phylogeny of the bird-hipped dinosaurs (Order Ornithischia). *National Geographic Research* 2:234–256.

Sereno, P. C., and Chao, S. C. 1988. *Psittacosaurus xinjiangensis* (Ornithischia: Ceratopsia), a new psittacosaur from the Lower Cretaceous of northwestern China. *Journal of Vertebrate Paleontology* 8:353–365.

Sereno, P. C., Chao, S. C., Cheng, Z. W., and Rao, C. G. 1988. *Psittacosaurus meileyingensis* (Ornithischia: Ceratopsia), a new psittacosaur from the Lower Cretaceous of northeastern China. *Journal of Vertebrate Paleontology* 8:366–377.

Sloan, R. 1976. The ecology of dinosaur extinction. *In* Churcher, C. S. (ed.), *Essays on palaeontology in honor of Loris Shano Russell* (Royal Ontario Museum, Life Science Miscellaneous Publication), pp. 134–154.

Sternberg, C. M. 1927. Horned dinosaur group in the National Museum of Canada. *Canadian Field-Naturalist* 51:67–73.

——— 1951. Complete skeleton of *Leptoceratops gracilis* Brown from the Upper Edmonton member on Red Deer River, Alberta. *National Museum of Canada, Bulletin* 123:225–255.

Sues, H. D. 1980. A pachycephalosaurid dinosaur from the Upper Cretaceous of Madagascar and its paleobiogeographical implications. *Journal of Paleontology* 54:954–962.

Sues, H. D. and Galton, P. M. 1987. Anatomy and classification of the North American Pachycephalosauria (Dinosauria: Ornithischia). *Palaeontographica A* 198:1–40.

Tanke, D. H. 1989. Centrosaurine (Ornithischia: Ceratopsidae) paleopathologies and behavioral implications. *Journal of Vertebrate Paleontology* (Supplement to no. 3), 9:41a.

Methods

1 Clades and grades in dinosaur systematics

PAUL C. SERENO

Terms in themselves are trivial, but taxonomies revised for a different ordering of thought are not without interest. (Gould and Vrba 1982)

Abstract

Dinosaur systematics provides examples of two approaches to the definition of supraspecific assemblages. The *clade*, or monophyletic taxon (*sensu stricto*, Hennig 1966), can be diagnosed by shared derived character states, or synapomorphies, and constitutes a "real" entity in the history of organismic diversity. The *grade*, or paraphyletic taxon, can be diagnosed only by both the presence and absence of synapomorphies and is delineated on the basis of morphologic distance.

The cladistic method for the characterization of monophyletic groups by synapomorphy is briefly described. A comparable method for qualitative measurement of morphologic distance is not available. Phylogenetic classification, based on monophyletic groups, is compared to traditional classification, based on a combination of monophyletic and paraphyletic groups. A historical trend toward a phylogenetic classification is observed.

Finally, some common, but mistaken, criticisms of cladistic methods are discussed. Parsimony is held to be a fundamental criterion for discriminating among hypotheses, and the construction and maintenance of a phylogenetic classification is presented as both a practical and a heuristic endeavor. Explicit phylogenetic hypotheses proposed recently for the Dinosauria promise a better understanding of the interrelationships within this diverse taxon.

Introduction

Observing similarities and differences in design among living organisms is an ancient pastime. By comparison, the concept of evolution – that organismal diversity may have arisen by some mechanism of splitting with change – has had a much shorter history, although it long predates Darwin and Wallace (Lovejoy

In Dinosaur Systematics: Perspectives and Approaches, *Kenneth Carpenter and Philip J. Currie, eds. Copyright © Cambridge University Press, 1990.*

1959). At its core, evolutionary theory is designed to elucidate the causal mechanisms that generate variation in structural design, or the *pattern* in organic form.

The advent of evolutionary theory and, more particularly, the theory of natural selection as the mechanism of evolutionary change did not radically alter systematic patterns established by nonevolutionary systematists. Natural groups fashioned by nonevolutionary systematists survived evolutionary reinterpretation largely untouched. Nonevolutionary taxonomies persisted, despite the profound implications of evolutionary theory.

The resilience of systematic pattern to explanatory reinterpretation, although perhaps counterintuitive, is a necessary precondition for noncircular reasoning: evolutionary theory is subservient to systematic pattern because the former is an attempt to explain the existence of the latter. Detailed structural similarities, commonplace at any level of organismal design, signal the operation of some unifying *process*, and an evolutionary process involving descent with modification is consistent with the prevailing hierarchic pattern of morphologic features. The pattern, nevertheless, is based on the primary data of character distributions and should be established as free as possible from notions about how it arose (Eldredge and Cracraft 1980). An underlying theme in this paper and one that figures prominently in recent systematic debate is that *the discovery and systematization of morphologic pattern should remain as free as possible from evolutionary assumption.* Insistence on a priori belief in the general concept of evolution alone (e.g., de Queiroz 1985) does not diminish the independence of pattern recognition.

The distinction between natural and artificial groups can be traced back to the earliest systematic observations of plants and animals in classical times (Nelson and Platnick 1981). The goal of most efforts to classify has been the establishment of natural groups, with the assumption, implicitly or explicitly stated, that there is one true arrangement to be discovered, be it the result of innate knowledge, the plan of creation, or the product of a unique pattern of descent with modification. To most pre-Darwinian systematists, a natural group comprised an assemblage of organisms that share a set of significant features. An artificial group brought together organisms at the convenience of the systematist on the basis of superficial similarities. For example, natural classification places whales and bats with other mammals rather than with ray-finned fishes and birds, respectively. In general, natural groups were believed to represent "real" entities in nature, not simply taxonomic artifacts.

Within an evolutionary framework, a natural group became an historical entity, a branch of the tree of life, comprising an ancestor and all of its modified descendants. As articulated by Darwin, "Community of descent is the hidden bond which naturalists have been unconsciously seeking, and not some unknown plan of creation, or the enunciation of some general propositions, and the mere putting together and separating objects more or less alike" (1859: 420). Despite general preference for elucidation of natural relations, a methodological procedure for their discovery was not formulated during the century following Darwin's remarks. One of the great achievements of modern systematics has been the development of a general method for delineating natural groups (Hennig 1966).

A persistent current in systematics – one that is also clearly defined and explicated in Darwin's writings (Ghiselin 1985) – proposes that *morphologic distance* between taxa should be recognized and combined in some fashion with phylogenetic affinity. The resulting classification, a mixture of natural and unnatural taxa, has been defended as biologically more informative and meaningful than taxa based solely on phylogenetic affinity (Simpson 1961; Mayr 1969, 1974, 1981). Unfortunately, there has never been a consistent rationale or procedure for qualitative measurement of morphologic distance and its incorporation into a classification. *Shared apomorphy may constitute the most effective systematic criterion* for the establishment of a general reference system in comparative biology.

In this paper, I present a cladistic perspective which is a personal perspective; cladistics is not a monothetic methodology despite general consensus on central propositions. In the first section, I discuss briefly the validity of systematic criteria (i.e., synapomorphy, morphologic distance) for the delineation of supraspecific taxa. In the second section, I argue the case for a cladistic (phylogenetic) classification, underscoring inadequacies in current diagnoses and the historical trend toward cladistic classification. In the third and final section, I address some common misinterpretations of cladistic methodology. Examples are drawn from ornithischian systematics.

Systematic criteria

The taxonomic subdivision of living and fossil organisms will never be resolved completely by objective criteria due to the nature of morphology and its particularization by the systematist. Subjectivity will always play a substantial role in distinguishing the units of morphology used in taxonomic analysis. No methodology (phenetics included) has succeeded in specifying a procedure that would direct systematists, working independently, to subdivide a given organism into the same morphologic units – characters and character states – all of which carry equivalent information. Verbalization and communication of character information per se is neither a trivial nor an exact undertaking. Despite these noteworthy caveats, my strong impression is that in dinosaur systematics in particular, the paucity of character analysis far outweighs the prevalence of character ambiguity.

Two systematic criteria (synapomorphy and morphologic distance) for the delineation of supraspecific taxa are outlined below, and the weaknesses of these criteria are discussed.

The clade: synapomorphy as the sole criterion

Without question, inferred phylogenetic affinity has been the most influential factor in the recognition of supraspecific taxa. As a consequence, most supraspecific taxa are depicted as discrete natural entities in phylogenetic diagrams. But phylogenetic affinity cannot be observed directly among living or fossil organisms. Rather it is inferred from novel character states – *synapomorphies* (homologies *sensu* Patterson 1982) – shared by members of a group. Patterns in the distribution of synapomorphies have long received phylogenetic interpretation, namely that the synapomorphies of a given group arose in a common ancestor and were inherited by the descendants. This phylogenetic interpretation, however, is clearly secondary; it is not a precept but rather a postscript to the observation of pattern in the distribution of synapomorphies (Platnick 1979; Nelson and Platnick 1981; Patterson 1982, 1983).

Cladistics is concerned with the establishment of clades, or *monophyletic groups*. In phylogenetic terms, a monophyletic group includes a common ancestor and all of its descendants (Farris 1974). In this paper the term *monophyletic* will carry this definition, and not the broader definition employed elsewhere (e.g., Ashlock 1971; Mayr 1974). Although lively debate continues between cladists and traditional systematists over procedures for constructing a classification, it is important to realize that nearly *all systematists now agree that phylogenetic affinity can only be inferred from synapomorphies and not from symplesiomorphies* (shared primitive character states), and that this realization is due primarily to the influence of Hennig (Ashlock 1974; Mayr 1974). Symplesiomorphies, in turn, are regarded as synapomorphies at a more inclusive level in the taxonomic hierarchy.

When considering phylogenetic relations between fossil taxa, synapomorphies are identified by *outgroup comparison*, a method which assesses the distribution of character states (Hennig 1966; Watrous and Wheeler 1981; Wiley 1981; Maddison et al. 1984). Given a character that exhibits two states within a monophyletic group (ingroup), the state that also occurs in the nearest relatives (outgroups) is considered primitive (plesiomorphic), whereas the state that occurs only within the group is considered derived (apomorphic). The importance of considering the character states in *two* immediate outgroups before assigning character polarity has recently been demonstrated (Maddison et al. 1984).

For example, among dinosaurs (the ingroup), the predentary bone is found only in ornithischians and is absent in saurischians. Because the bone is also absent in archosaurs (the outgroup) closely related to dinosaurs, the *presence of a predentary* is a synapomorphy for the Ornithischia and was early recognized as such by Baur (1891). Quite obviously, a given character, such as the predentary, can only be considered apomorphic, and thus indicative of phylogenetic affinity, at an appropriate level of generality. The presence of the predentary is plesiomorphic *within* Ornithischia, because the predentary is present in any ornithischian that could serve as an outgroup. A character state that is a synapomorphy at one level in a natural hierarchy is a symplesiomorphy at a less inclusive level.

If unique synapomorphies accompanied the appearance of each new lineage during the phylogenetic history of a clade and were conserved in all descendants, the selection of an appropriate hierarchic arrangement of taxa would involve little controversy. However, analysis of the majority of actual taxa, such as the Ornithischia, demonstrates that no single arrangement of terminal taxa is completely congruent with all character data. Some character incongruence, or *homoplasy*, is present in every possible arrangement of taxa. Cladists generally invoke *parsimony* as a criterion to decide among alternative arrangements. The arrangement that proposes the least number of homoplasies (reversals, parallelisms, convergences) is preferred – not because evolution necessarily proceeds along the simplest path, but because simplicity is the general criterion among the sciences for preference of one explanation over endless alternatives. Computerized parsimony procedures are now routinely used in phylogenetic analysis of large or complex data sets (e.g., Swofford 1985).

From a cladistic perspective, a *common ancestor* for a given taxon must exhibit all synapomorphies of the taxon but must lack *autapomorphies*, or character states peculiar to itself. It also must lack the synapomorphies at higher levels within the taxon. An ancestral species, therefore, is a paraphyletic assemblage (when there is more than one individual) and can be diagnosed only by invoking the absence of synapomorphies. In theory the heirarchic branching pattern of descent may extend below the species level to monophyletic interbreeding populations (Hennig 1966; de Quieroz and Donaghue 1988). In practice, ancestral species are rarely encountered in the fossil record. Modern cladistic method employs synapomorphy as the sole systematic criterion for the delineation of monophyletic taxa. Character polarity is commonly determined by outgroup comparison, particularly when the analysis involves extinct organisms. Character incongruence is reduced to a minimum by application of parsimony procedures. When presented at length, the data and assumptions involved in cladistic analysis and classification are open to the scrutiny and criticism of the systematic community.

The grade: morphologic distance

Some systematists prefer to supplement cladistic information with inferences about other evolutionary phenomena before drawing the boundaries between supraspecific taxa. Mayr argues that cladistics, which utilizes only phylogenetic information, ignores:

other important evolutionary phenomena such as the existence of "grades" and of highly specialized side lines, all of the phenomena of mosaic evolution, and all the causal factors in evolution. How rapidly a branch diverges, how it changes in relation to the "sister group," how many different characters it acquires, which new adaptive zone it has invaded ... . (1974, p. 106)

He suggests that cladistics:

denies the other aspects of evolutionary change such as rate of evolution, adaptive radiation, the occupation of new adaptive zones, mosaic evolution, and many other macroevolutionary phenomena. (1974, p. 106)

Furthermore, in his opinion cladistics:

exclude[s] most of evolutionary theory (for example, inferences on selection pressures, shifts of adaptive zones, evolutionary rates, and rates of evolutionary divergence). (1981, p. 515)

These evolutionary phenomena appear to express in different ways the degree of perceived overall *morphologic distance* between groups. In a diagram of three taxa A, B, and C (Fig. 1.1), B and C share a more recent common ancestry, as evidenced by two synapomorphies, and would be classified together by genealogy. A and B, however, are closer phenetically, because they exhibit far fewer autapomorphies than occur in taxon C. In other words, A and B share many symplesiomorphies with respect to taxon C and, on the basis of overall morphologic similarity, would be classified together. As an example among living vertebrates, A, B, and C can be replaced by Lepidosauria, Crocodilia, and Aves, respectively. Traditionally Lepidosauria and Crocodilia are placed together in the Reptilia based on overall similarity, whereas birds are set aside as a separate taxon, Aves, even though birds share derived characters with crocodiles.

Traditional systematists recognize morphologic distance in classifications by raising the *rank* of the divergent taxon over that of its nearest relative (sister group). The nearest relative is then grouped together with several more distant relatives in a coordinate taxon of equal rank. Raising the rank of the morphologically divergent taxon transforms the coordinately ranked primitive taxon into a *paraphyletic* group – a group that includes the common ancestor but excludes some of the descendants.

Among ornithischians, A, B, C in the previous example may be replaced by *Camptosaurus, Iguanodon,* and hadrosaurs, respectively (Fig. 1.2). Traditionally *Camptosaurus* and *Iguanodon* are placed together in the Iguanodontidae presumably on the basis of overall similarity, whereas hadrosaurs are set aside in the Hadro-

Figure 1.1. Diagram of relationships between taxa A, B, and C. Numbers indicate apomorphic characters (after Mayr 1981).

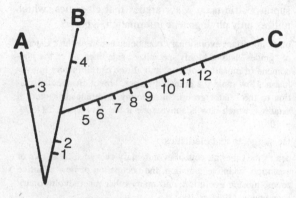

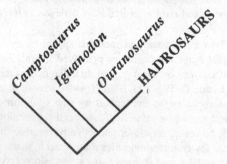

Figure 1.2. Postulated phylogenetic relationships of some advanced ornithopods (Sereno 1986) with traditional classification of the same taxa below.

FAMILY Hadrosauridae

FAMILY Iguanodontidae

Ouranosaurus

Iguanodon

Camptosaurus

sauridae even though they share derived characters with *Iguanodon*. Thus, for reasons never fully explained but possibly related to their distinctive morphology, hadrosaurs are classified in the Family Hadrosauridae at a rank above their nearest known relative, which is then grouped with other primitive genera in the coordinate taxon Iguanodontidae. The Hadrosauridae represents the monophyletic higher grade and the Iguanodontidae the paraphyletic lower grade.

Unreal taxa

As outlined above and conceded by its foremost practitioners, the designation of paraphyletic groups in traditional (evolutionary) systematics involves an element of art. The spectrum of evolutionary phenomena said to be neglected in cladistic classification appear to be reducible to a subjective qualitative estimate of morphologic distance. The adaptive zone, for example, appears to have little to do with ecologic space, which is difficult to assess at the species level and much more abstract, perhaps meaningless, for a higher taxon. Entry into a *new* adaptive zone is postulated for a taxon exhibiting a sufficient number of synapomorphies that separate it from its less derived relatives in the *old* adaptive zone (Rosen 1974). The occupants of the old adaptive zone, a paraphyletic group, can only be characterized by the absence of these synapomorphies and the presence of more general synapomorphies that unite both groups.

The paraphyletic Reptilia, for example, is characterized as "those vertebrates with amniote eggs whose species lack mammary glands and feathers" (Ashlock 1979, p. 448). Mammary glands and feathers, synapomorphies for mammals and birds respectively, are absent in Reptilia as well as all other non-mammalian and non-avian organisms. The amniote egg is present in Reptilia (in its traditional paraphyletic sense). Besides knowing that reptiles are amniotes, we know only that they lack diagnostic features of other amniote groups. In this sense, Reptilia is "artificial" and exists only as a systematic construct in the shadow of its monophyletic neighbors. The monophyletic taxa Mammalia and Aves, in contrast, exhibit many synapomorphies and exist as real historical entities outside the mind of the systematist.

It may be possible to recognize paraphyletic groups as real entities based on biologic or ecologic criteria. Van Valen has argued that "adaptively unified" paraphyletic groups are commonplace in nature and can be recognized. He suggested, furthermore, that it would be possible, although perhaps not desirable, to base a classificatory scheme exclusively on "ways of life" (1978, pp. 289, 293). The Insectivora has been used as an example of an adaptively unified paraphyletic group, from which all other placental mammals evolved, and has been characterized as follows: "almost always small for mammals, always terrestrial or at most barely semi-aquatic or gliding, almost always with a diet predominantly of individually caught invertebrates, almost always (where known) with less flexible learning than most other mammals" (Van Valen 1978, p. 289).

Are these "unified adaptations"? And what method was invoked to exclude "opossums, caenolestids, notoryctids, lizards, toads, [and] pitcher plants" (Patterson 1982, p. 58)? Because every trait is granted a margin of uncertainty – "almost always" – how is it possible to exclude the most primitive members of descendant placental lineages that are almost, but not quite, "insectivores"? And consider the many extinct paraphyletic taxa that lack modern analogs, such as iguanodonts; assessing their adaptive unity is a formidable, if not fanciful, task.

As in the case of the adaptive zone, the concept of the evolutionary grade is based on morphologic distance. The higher grade is characterized by synapomorphy and the lower grade is delimited by the absence of synapomorphy. Consider the traditional view of ornithopod evolution:

There is tacit approval of a very generalized evolutionary scheme for the ornithopod dinosaurs .... This relies mainly on the concept of grades of anatomical organization progressing from primitive to advanced ornithopods via four levels. The most primitive ornithopods, and indeed those implicated in the ancestry of the Ornithischia as a whole, are the "fabrosauroids" ... – small, bipedal ornithischians which show no obvious cranial or post-cranial specializations. Derived from the former group are the "hypsilophodontoids" ... – a morphologically conservative group of small to medium-sized ornithopods; these in turn are considered to be ancestral to several, supposedly iteratively derived, groups of larger and more anatomically specialized ornithopods called "iguanodontoids" .... Finally, the "hadrosauroids" ... representing the most sophisticated ornithopods, are characterized by their dental and cranial specializations ... and are derived from the "iguanodontoids." (Norman 1984a, p. 521; references in ellipses.)

The terms "fabrosauroid," "hypsilophodontoid," and "iguanodontoid," constitute paraphyletic assemblages, which can be characterized only by the absence of synapomorphies that diagnose other taxa and the presence of synapomorphies that apply to more inclusive taxa. The taxa within each grade, therefore, cannot be recognized by synapomorphy and do not represent real historical entities. They cannot, within reason, be construed as adaptively unified, nor can they, as supraspecific taxa, be ancestral to one another.

The identification of ancestral groups, nonetheless, is commonly viewed as the most informed formulation of relationships. Fabrosaurs represent the basal family from which all other ornithischian dinosaurs have been derived and hypsilophodonts constitute the mainline or central stock from which the iguanodonts, hadrosaurs, and ceratopsians arose. An ancestral taxon tells us (1) that the relationships within the ancestral taxon and between the ancestral taxon and the descendants are unknown and (2) that the descendant taxa may be monophyletic. The ancestral taxon, therefore, adds

no positive information about the relationships of its members.

Because grade groups exist in the shadow of every robust monophyletic group, many potential grade groups within the Ornithischia have not been proposed to date. It has been suggested that heterodontosaurs, ornithopods (*sensu stricto*), pachycephalosaurs, and ceratopsians share several unique characters among ornithischians such as asymmetrical distribution of enamel in the cheek teeth, a relatively large predentary, a fully open acetabulum, and a fingerlike lesser trochanter (Sereno 1986). The absence of these synapomorphies in other ornithischians opens the possibility for a new grade group – including *Lesothosaurus, Scutellosaurus, Scelidosaurus*, stegosaurs, and ankylosaurs – none of which show any development of these characters. This new grade of ornithischians lacks as many synapomorphies of its closest relatives as do the fabrosauroids and, by this comparison, would be equally deserving of recognition as a grade.

Ashlock has attempted to formalize Mayr's concept of grade, appealing to the relative size of the bordering gaps: "A higher taxon is a monophyletic group of species (or a single species) separated from each phylogenetically adjacent taxon of the same rank by a gap greater than any found within these groups" (1979, p. 446; *monophyletic* here subsumes both monophyletic and paraphyletic as defined in this chapter). Within these guidelines, *Lesothosaurus, Scutellosaurus*, and *Scelidosaurus*, for example, would constitute an excellent grade group, judging from the phylogeny of Sereno (1986). On one side, they (and all other ornithischians) are separated by a sizeable gap (i.e., large number of synapomorphies) from all nonornithischians. On the other side, numerous synapomorphies accumulate on route to any other ornithischian subgroup, including ankylosaurs and stegosaurs.

Besides suggesting new ornithischian grades, Ashlock's definition of a paraphyletic group undermines the traditional concept of an "iguanodont" grade, because some of the gaps within the "iguanodont" assemblage appear to exceed the gap separating "iguanodonts" from more primitive ornithopods (Sereno 1986).

Arbitrary boundaries

In the absence of any alternative method, counting the number of synapomorphies that separate taxa may be the least subjective method for estimating qualitative morphologic distance. Granting for the moment that such estimation of morphologic distance between taxa is reasonable, how much distance is required before a systematist is allowed to raise the rank of the derived taxon (thereby creating a paraphyletic sister group)? No precise answer is available. The amount of difference in overall similarity deemed sufficient to define a morphologic gap between two closely related taxa (i.e., a gap deserving discordant rank) can only be set by arbitrary convention. It has been suggested, for example, that the size of the gap necessary for discordant rank of a taxon should be inversely related to the number of included species (Mayr 1969, p. 92). Taxa with few species, therefore, would require a larger gap than taxa with many species. Unfortunately, this convention, which is designed to constrain the number of higher taxa, is both arbitrary and relative. How wide must the gap be to justify raising the rank of the hadrosaur taxon above that of its nearest relative? or how wide must the gap be to justify subordinal, rather than familial, rank for the pachycephalosaurs?

Inherent instability

Stability per se is not necessarily a virtue in biological classification. As aptly stated by Gaffney, "temporal stability of classifications often reflects ignorance of relationships and lack of work" (1979, p. 103). Ideally, the classificatory scheme should be reshaped to reflect new information and reanalysis of relationships.

At the same time, instability should not be an *inherent*, or anticipated, ingredient in classification. The classification should remain stable if the phylogenetic relationships between taxa remain unaltered. Taxa delineated by morphologic distance, however, are unstable if intermediates are presumed to have existed at any time (Wiley 1981, p. 244). The morphologic gap, potentially, can be bridged by the discovery of intermediate forms.

In advanced ornithopods, for example, the relatively recent discovery of *Ouranosaurus* has substantially decreased the number of derived characters that previously defined the Family Hadrosauridae (Fig. 1.2). The relatively greater space between the predentary and first dentary tooth, transversely expanded premaxillary snout with reflected rim, flaring anterior end of the jugal, distinct transverse narrowing of the cranium from the postorbital region posteriorly, anteriorly curving paroccipital and squamosal processes, and caudal neural spines that exceed their respective chevrons in length are a sampling of characters that *previously* were unique to hadrosaurs, but now characterize a group including *Ouranosaurus* and hadrosaurs (Sereno 1986). The discovery of a single well-placed intermediate between *Ouranosaurus* and hadrosaurs could reduce the number of derived characters for hadrosaurs to the number that currently diagnose the more inclusive group including *Ouranosaurus* and hadrosaurs. In this event, the traditional familial rank for hadrosaurs, elevated above that of its nearest relative, could not be justified on the basis of a morphologic gap (however, see Horner this volume).

Morphologic gaps: real or artificial?

Unconformities, poorly exposed horizons, and unfossiliferous strata can create a morphologic gap between closely related taxa. The morphologic gap, then, is available for interpretation as a shift into a new adaptive zone or as an unrecorded phase of rapid evolution that gave rise to a higher grade.

It is well known, for example, that terrestrial beds of the Lower Cretaceous and lowermost Upper Cretaceous are more poorly exposed, less fossiliferous, and consequently more poorly sampled than terrestrial beds in the Upper Cretaceous (Ostrom 1970). The advanced ornithopod *Ouranosaurus* was discovered in beds of probable Aptian age in Niger (Taquet 1976). The earliest known hadrosaur fossils with a secure date are the fragmentary remains of a very small individual from the Santonian of southern North America (Kaye and Russell 1973). There is no evidence to date that any of the hadrosaur remains from Asia are pre-Santonian (Maryańska and Osmólska 1981, *contra* Rozhdestvensky 1966). Therefore, no remains of hadrosaurs or immediate hadrosaur precursors are known between the Aptian (*Ouranosaurus*) and the Santonian, a span of approximately 30 million years. Given the available fossil record, it is certainly possible, if not probable, that the morphologic gap between *Ouranosaurus* and hadrosaurs is due to uneven sampling, i.e., poor representation of terrestrial vertebrate faunas in the upper Lower Cretaceous and a gap in the fossil record of advanced ornithopods in the lower Upper Cretaceous. It is possible that geologic or taphonomic factors, rather than evolutionary phenomena, are responsible for the discordant rank of the Hadrosauridae and the creation of a paraphyletic Iguanodontidae.

Coarse chronology

Chronologic information is necessary to demonstrate that a morphologic gap is due to an increase in the rate of morphologic change, rather than simply an unfossiliferous interval. To advance a hypothesis of rapid morphologic evolution between hadrosaurs and their sister group, *Ouranosaurus*, and to justify raising the rank of the hadrosaurs, it is necessary to show that the first occurrences of both taxa are separated by a relatively short time interval, assuming an excellent fossil record for both taxa. Calibrating the temporal component in the fossil record, however, can be a severe limiting factor in estimation of rates of morphologic change (Dingus and Sadler 1982; Behrensmeyer and Schindel 1983). Few radiometric ages are available for fossiliferous terrestrial deposits of Mesozoic age, and biostratigraphic correlations are generally extremely coarse and subject to major revision (e.g., Olsen and Galton 1984). These stringent requirements prohibit a reliable assessment of the rate of morphologic divergence between sister taxa for most, if not all, Mesozoic terrestrial vertebrates. Many cited examples of rapid divergence at the origin of a clade amount to no more than a qualitative impression. For terrestrial vertebrates of the Mesozoic, such as ornithischians, an accelerated rate of morphologic change should not be offered as justification for elevating the rank of a monophyletic taxon.

Balancing taxa: the tidy classification

An asymmetrical cladogram, in which a diverse apomorphic "crown" group is followed by a sequence of plesiomorphic monotypic sister taxa (Fig. 1.3), is one of many possible branching configurations (Farris 1976a). Classification of the entire clade according to its phylogenetic history is considered undesirable in traditional systematics because it would require a separate rank for each of the hierarchical levels, with single genera or species opposing much more diverse groups at each level. One solution to the problem is to divide the clade into two groups of equal rank, lumping all of the relatively primitive taxa together in one group and the more derived taxa in another.

This classificatory solution is often coincident with the allotment of taxa to adaptive zones and grades. The new adaptive zone or higher grade has been achieved by the successful speciose group with the less diverse primitive relatives remaining in the old adaptive zone or lower grade. The rank of the monophyletic speciose group is raised relative to its sister taxon, which is placed with the remaining primitive relatives in a paraphyletic *balanced* taxon. Although not explicitly stated, the high rank of the Family Hadrosauridae compared to its nearest relative may be due to the difference in diversity between hadrosaurs, with over 30 species, and the individual genera or species that constitute the nearest outgroups.

Balancing, or partially balancing, coordinate taxa by the number of included species or genera has intuitive appeal, whether motivated by convenience or by perceived biological attributes. Such systematic housecleaning results in a tidy classification with fewer names to remember. It also recognizes, by elevation of rank, successful groups that have achieved their diversity via "adaptive radiation." In practice, delineating taxa by their diversity or by their perceived evolutionary success is a subjective endeavor (Raikow 1986, 1988).

Some groups are considered speciose *without any consideration of the branching pattern within the group*. If one assumes, a priori, that the family Hadrosauridae, with over 30 species, should be compared to its nearest known relative, the single species *Ouranosaurus nigeriensis*, these sister taxa appear very unbalanced. One or

Figure 1.3. Diagrammatic asymmetrical cladogram with diverse terminal monophyletic clade.

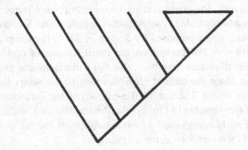

more primitive hadrosaur species, however, may constitute sister taxa to remaining hadrosaurs. Why include the most primitive hadrosaurs in the speciose group? And why exclude *Ouranosaurus*? The problem may be conceptualized by imagining a species-level cladogram of advanced ornithopods in which equally-spaced terminal branches are left unlabeled. Given only the branching pattern without any character information, it may well be impossible to see the speciose subgroup. At present, the notion that the Hadrosauridae, rather than a more inclusive or less inclusive taxon, is a particularly speciose group deserving higher rank appears to be based on a perceived morphologic gap between hadrosaurs and their nearest relatives and a priori selection of hadrosaurs as the group for comparison.

The use of morphologic distance (in its various guises) as a systematic criterion is alluring if only for its immediacy: common sense dictates that those species that look most alike be classified together. The measure of morphologic phenetic distance involves an assessment of the amount of similarity, both primitive and derived, in comparisons between any two taxa. The objective measure of quantitative morphologic distance and its exclusive use in classification is the goal of phenetic taxonomy, which unlike traditional systematics rejects any input from phylogenetic analysis (Sneath and Sokal 1973). Quantitative distance, however, has proven to be problematic. Despite a panoply of elegant algorithms, phenetic systematics has fallen short of its stated goal of providing a stable and maximally informative classification in most applications (Farris 1977, 1979, 1980, 1981).

The main objective for all schools of systematics is the establishment of a general reference system for comparative biology (Brooks 1981). The basic principles of the cladistic method for analysis of phylogenetic relationship were outlined in the previous section. By comparison, a reasonably explicit methodology for the measurement of morphologic distance and its incorporation into a classification is not available, even in the most recent versions of evolutionary systematics (e.g., Mayr 1981). Without recourse to specific operational guidelines, the traditional school has been caricatured as a monument to idiosyncrasy. The lack of a coherent procedure and the bending to authority and tradition are telling symptoms of intellectual retrenchment. The time has passed when the construction and maintenance of paraphyletic taxa can proceed unchallenged.

Classification

As discussed above, the justification of paraphyletic groups and their inclusion in traditional classification is a fundamental difference from cladistic classification. Comparison of actual classifications, however, demonstrates that differences in approach run deeper, because diagnoses of taxa that are widely recognized as monophyletic in the two classificatory schemes differ substantially. It is clear, then, that there is more than one concept of a proper diagnosis. Which characters should or should not be included?

The diagnosis: synapomorphy

The diagnoses of taxa widely acknowledged by all systematists as monophyletic differ markedly under traditional and cladistic treatment. Symplesiomorphies often outnumber synapomorphies in traditional diagnoses of monophyletic taxa. If synapomorphy provides the entire basis for monophyly, then what purpose is served by the inclusion of symplesiomorphy? In traditional supraspecific diagnoses, symplesiomorphies constitute relevant information that aids the process of taxonomic identification. In this way the diagnosis functions both as a statement about phylogeny and as a non-phylogenetic identification key (Ax 1987). Phylogenetics and identification keys, however, have different objectives and conflicting demands.

Seeley's (1887) original diagnosis of the Ornithischia, in cladistic perspective, includes three synapomorphies, two symplesiomorphies, and three ambiguous characters. Three additional synapomorphies, contributed by Baur (1891) and Marsh (1895), brought the total to five synapomorphies (palpebral, elongate premaxillary process, predentary, elongate preacetabular process, biramus pubis), which were accompanied by a short and variable list of symplesiomorphies.

Nearly 60 years later, Romer (1956, pp. 624–627) provided a lengthy diagnosis for the Ornithischia that included four additional synapomorphies, an unwieldy list of symplesiomorphies, and the outstanding autapomorphies of several ornithischian subgroups. Romer's diagnosis of the Ornithischia, now over 100 characters in length, functioned as an abbreviated description of the variation in form throughout the group. Rather than listing diagnostic characters that appeared in the common ancestor of the group, the diagnosis had become a broad-based summary, encompassing characters with distributions far beyond the Ornithischia, such as external nares separate, as well as characters found only in particular ornithischian subgroups, such as supplementary rostral bone present in one group.

The unselective range of characters listed in many traditional diagnoses converges on phenetic description, which purposefully ignores the distinction between plesiomorphic and apomorphic similarity. Clustering of characters from many levels of the taxonomic hierarchy within a single diagnosis results in a highly redundant classification, particularly if the procedure is uniformly applied. Of course, listing characters indiscriminately, without any indication of their level of generality, completely obscures their phylogenetic significance. In Romer's diagnosis of the Ornithischia, the relatively small number of valid, but unacknowledged, synapomorphies are hidden by a myriad of characters that apply to a wide range of taxa, some more general and some less general than the Ornithischia. Such conflation of character information obscures apomorphy and provides no comparative guidelines for the description of new forms.

In phylogenetic classification the diagnosis includes synapomorphies that characterize included taxa. *The character states and their level of generality in the taxonomic hierarchy are clearly specified.*

The taxon: monophyly

Because cladistic classification consists of monophyletic taxa based on synapomorphy, the phylogenetic interpretation of taxa is uncomplicated. The hierarchic arrangement of groups nested within other groups in a cladistic classification reflects the branching sequence in a corresponding cladogram.

In traditional classification, the phylogenetic significance of taxa and the characters listed in diagnoses is less matter of fact. Firstly, taxa can be either monophyletic or paraphyletic. Without specific annotation, it is impossible to ascertain immediately which groups are monophyletic. Secondly, expansive diagnoses include characters from many levels of the taxonomic hierarchy, rather than only the synapomorphies of a particular group. Thirdly, the assignment of rank on the basis of morphologic distance, rather than phylogenetic affinity, obscures the phylogenetic branching sequence expressed in the Linnean system of hierarchical rank. Under these ambiguous circumstances, taxa in a traditional classification are susceptible to misinterpretation.

The Family Fabrosauridae, for example, is based on a long list of ornithischian symplesiomorphies, such as thecodont dentition, presence of a predentary, radius and ulna of subequal length, slender fibula, long and narrow metatarsus (Galton 1978, p. 155). Although none of the listed characters is apomorphic for the included taxa, the Fabrosauridae has been frequently depicted as a monophyletic taxon in phylogenetic diagrams (Galton 1972, 1974, 1978; Thulborn 1974; Milner and Norman 1984; Norman 1984b). Other paraphyletic groups diagnosed on symplesiomorphy alone, such as the Iguanodontidae and Protoceratopsidae, have also been depicted graphically as monophyletic.

Systematic history: toward a phylogenetic classification

When robust synapomorphies have been included in the diagnosis of a higher taxon in historical classificatory schemes, the taxon and its membership have remained remarkably stable in subsequent classifications. The order Ornithischia, proposed by Seeley in 1887, is exemplary. With very few exceptions, the fossil remains of ornithischian dinosaurs have been allied in a monophyletic group ever since the earliest classifications of dinosaurs (e.g., Cope 1866). The historical continuity in the definition of the Ornithischia (= Orthopoda, Predentata) may be due to the early recognition of synapomorphies for the group and their inclusion in the diagnosis.

Seeley's Saurischia was diagnosed on symplesiomorphy alone and has witnessed a less salubrious history. It did not gain general acceptance in most early classifications. The widespread use of the Saurischia after the turn of the century is usually attributed to the prodigious publications of Huene. In the recent literature, the Saurischia has been interpreted as polyphyletic (Charig et al. 1965), paraphyletic (excluding birds, Ostrom 1976; excluding theropods and birds, Thulborn 1975) and monophyletic (including birds, Gauthier 1986), with uncertain membership for several primitive genera (Brinkman and Sues 1987).

The major structural changes in the classification of ornithischians over the years have involved a shift toward a phylogenetic system. Unnatural groups have been disbanded in favor of monophyletic subgroups. After the discovery of distinct armored forms within the ranks of the Stegosauria, a separate taxon, the Ankylosauria, was erected to accommodate this new clade. At that time, no synapomorphies were known that would encompass the entire group (Nopcsa 1915, 1929; Romer 1927). In similar fashion, pachycephalosaurs (Maryańska and Osmólska 1974) have been removed from the Ornithopoda, a paraphyletic assemblage long billed as the "central stock" of ornithischian phylogeny.

When robust synapomorphies have been discovered in a taxon which, by tradition, had been relegated to a paraphyletic "ancestral" group, the taxon has been reclassified in an appropriate monophyletic group. Psittacosaurs, for example, were long classified in the Ornithopoda until it was suggested that the bone at the anterior end of the snout is the ceratopsian rostral (Romer 1956, 1968). Now there is general consensus that psittocosaurs belong with other ceratopsians *despite the equal, or greater, phenetic similarity between psittocosaurs and primitive ornithopods, like Hypsilophodon.*

If indeed there is a historic trend in ornithischian systematics, it is toward the establishment of a cladistic classification. Monophyletic groups are generally preserved whereas paraphyletic groups are often split or reorganized into monophyletic groups. I am not aware of a single case in which an ornithischian taxon, widely acknowledged to be monophyletic, has been sacrificed to benefit an unnatural taxon.

Cladistics: the invalid critique

Whether or not one adopts cladistic classificatory practices, the cladistic method has gained recognition as a general procedure to assess phylogenetic relationship. Few practitioners, if any, would claim that cladistics has cleansed phylogenetic analysis of subjectivity. Distinguishing the descriptive subunits of morphology, or *character definition*, involves subjective decisions. Assessing linkage between characters, or *character correlation*, usually involves subjective or ambiguous decisions. Estimating the relative importance of characters, or *character weighting*, is in large part (or perhaps

entirely) subjective, whether it is applied consciously or not. These subjective limitations apply *equally* to cladistic and traditional methods of phylogenetic inference, although they are often levied as criticisms particular to cladistic analysis (e.g., Panchen 1982, p. 316).

The explicit summary of character data is the most advantageous aspect of a cladogram. This expediency backfires with imprudent character analysis. All too often in cladistic analysis, purported synapomorphies are assembled in haste, without recourse to firsthand observation. As Panchen has aptly remarked, "The first obvious fault of cladism in general is perhaps a temperamental and, one may hope, a temporary one. This is the temptation to produce new and revolutionary groupings or organisms, or to resurrect old and discredited ones, just to demonstrate that cladism is getting results" (1982, p. 320). Indeed, the general drawback in cladistic diagnoses is not the absence of synapomorphy (as in traditional diagnoses), but rather the presence of misinformation. Without question, a significant proportion of available cladistic character data on ornithischians will be discarded or revised by careful review of pertinent fossil materials and proper documentation of character distributions. Nevertheless, this situation is the result of *faulty cladistic analysis* rather than a fault inherent in cladistic method. Critical review and analysis of available ornithischian character data will eventually lead to a significant improvement in the understanding of ornithischian phylogeny.

It is often heard that the nomenclature and ranking in cladistic classification are impractical because (1) only dichotomous relationships are permitted, and (2) every node must be named. These criticisms simply do not apply to cladistic classification as it is practiced today.

Unresolved relationships among three or more taxa in a cladogram may suggest the best approximation of a real phylogenetic pattern or, alternatively, merely insufficient information. In cladistic classification all taxa must be monophyletic, whether or not the relationships of the subordinate taxa are fully resolved.

Every node in a cladogram need not be named. There is an extensive literature with pragmatic suggestions for a phylogenetic classification within the Linnean system (see Wiley 1981). I support the spirit that classifications should attempt to keep novelty and redundancy at a minimum, while at the same time maintaining enough taxonomic names to reflect in some general way current knowledge of organismic diversity. Restricting suprageneric nomenclature to a few familiar names for easy recall does not impress me as a heuristic approach to the general reference system in biology. If the additional names for monophyletic groups in cladistic classification gain no currency, they will be forgotten.

If categorical rank above the genus must be retained (and there are persuasive arguments for abandoning suprageneric rank: Griffiths 1976; Gauthier et al.

1988), then it seems prudent to avoid redundant (mono-typic) taxa and, when possible, to retain traditional monophyletic taxa in their historical rank (Farris 1976b).

The use of parsimony as a criterion in cladistic analysis to select among alternative phylogenetic arrange-ments has been a cause of consternation among some systematists (e.g., Panchen 1982, pp. 317, 320). Two assertions are entertained: (1) parsimony cannot be used because the course of evolution is not parsimonious and (2) parsimony is invalid when homoplasy is common.

The statement that evolution is not parsimonious, if taken at face value, seems equivalent to stating that homoplasy occurs in phylogeny. But the existence of homoplasy is universally acknowledged. Cladists, in fact, have endeavored to locate character incongruencies more precisely through the application of parsimony. The criticism "evolution is not parsimonious" may not do justice to the complaint, which by context might be rephrased: "the most parsimonious phylogeny may not be correct because *the course of evolution could have been more complicated*."

This criticism comes with surprise to most cla-dists, who pride themselves in admitting as little evolu-tionary assumption as possible into their systematic methodology – particularly an assumption claiming that anything about evolution is simple. As in other sciences, cladists employ parsimony as a criterion to choose one hypothesis over numerous alternatives. *Parsimony mini-mizes assumed homoplasies rather than assuming that homoplasy is minimal* (Farris 1983; Sober 1985). Parsimony is a fundamental criterion in science for dis-criminating among possible explanations. To dispose of parsimony on the impressionistic grounds that "the course of evolution could have been more complicated" is to accept as equally probable a potentially limitless number of ad hoc alternative explanations.

The second criticism of parsimony as applied to phylogenetic data is more serious. Does parsimony assume that homoplasy is infrequent? Or, to put it in other words, if the level of homoplasy is high – say over 50 per cent – is another method (e.g., compatibility, maximum likelihood, or covariance) more suitable for phylogenetic analysis? These questions must be addressed in the absence of absolute knowledge of the one true phylogeny (which we can only approximate with these methods).

Figure 1.4. Two possible cladograms for taxa A, B, and C.

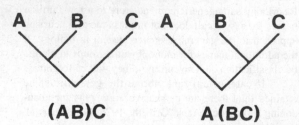

One justification of parsimony without recourse to prior assumptions about homoplasy relies on *likeli-hood*, the probability that a set of observations would have occurred given a particular hypothesis. The likeli-hood of a particular cladistic hypothesis given some character data, therefore, is a measure of how well the character data supports the hypothesis (Edwards 1972; Sober 1985). For example, a rich deposit with an abun-dance of articulated dinosaur skeletons is discovered and it represents a new clade of ornithischians with three new species A, B, and C. After establishing char-acter polarity (by comparison to the two nearest out-groups), it is observed that species A and B share ten synapomorphies, while species B and C share only two synapomorphies. Two competing hypotheses of rela-tionship (AB)C and A(BC) (Fig. 1.4) are constructed. We would like to be able to say that hypothesis (AB)C is more likely, that it could more easily account for the observed data than the alternative *irrespective of knowl-edge about the probability of homoplasy*. Or in other words, the observed character distributions are more likely to have occurred given the preferred hypothesis (AB)C.

This second justification of parsimony over other methods that require prior evolutionary assumptions is currently a controversial issue (Farris 1983; Felsenstein 1983; Sober 1985; Felsenstein and Sober 1986; Forster 1986). It has yet to be demonstrated conclusively that parsimony methods assume prior knowledge of the probability of homoplasy.

Prospectus

The Dinosauria has not yet been subject to a thor-ough cladistic analysis in which character distribution, character polarity, and missing data are discussed at length. Recently, however, several preliminary phyloge-netic analyses have been published that attempt to account for the relationships of the major subgroups (Norman 1984b; Sereno 1984, 1986; Cooper 1985; Maryańska and Osmólska 1985; Gauthier 1986). Although the suggested relationships and degree of reso-lution vary among these hypotheses, several similarities are apparent, such as the monophyly of Dinosauria, the close relationship between stegosaurs and ankylosaurs and between pachycephalosaurs and ceratopsians. More importantly, these hypotheses represent the first analy-ses of Dinosauria with explicit character data and, in this way, offer a fresh alternative to poorly substantiated phylogenetic trees.

Acknowledgments
I express my gratitude to the students and staff of the Department of Vertebrate Paleontology of the American Museum of Natural History for providing a stimulating and challenging environment during my graduate education. Comments by Jim Hopson and Bob Zanon helped to improve the final draft. Funding for this research was provided by the National Science Foundation (BSR83-05228 Doctoral Disser-

tation Improvement Grant), National Geographic Society, American Museum of Natural History, Sigma Xi, and the Department of Geological Sciences of Columbia University.

References

Ashlock, P. D. 1971. Monophyly and associated terms. *Systematic Zoology* 20:63–69.

1974. The uses of cladistics. *Annual Review of Ecology and Systematics* 5:81–99.

1979. An evolutionary systematist's view of classification. *Systematic Zoology* 28:441–450.

Ax, P. 1987. *The Phylogenetic System. The Systematization of Organisms on the Basis of their Phylogenesis* (New York: John Wiley & Sons).

Baur, G. 1891. Remarks on the reptiles generally called Dinosauria. *American Naturalist* 25:434–453.

Behrensmeyer, A. K., and Schindel, D. 1983. Resolving time in paleobiology. *Paleobiology* 9:1–8.

Brinkman, D. B., and Sues, H. D. 1987. A staurikosaurid dinosaur from the Upper Triassic Ischigualasto Formation of Argentina and the relationships of the Staurikosauridae. *Palaeontology* 30:493–503.

Brooks, D. R. 1981. Classifications as languages of empirical comparative biology. *In* Funk, V. A. and Brooks, D. R. (eds.), *Advances in Cladistics: Proceedings of the First Meeting of the Willi Hennig Society* (New York: Botanical Garden), pp. 61–70.

Charig, A. J., Attridge, J., and Crompton, A. W. 1965. On the origin of the Sauropods and the classification of the Saurischia. *Proceedings of the Linnaean Society of London* 176:197–221.

Cooper, M. R. 1985. A revision of the ornithischian dinosaur *Kangasaurus coetzeei* Haughton, with a classification of the Ornithischia. *Annals of the South African Museum* 95:281–317.

Cope, E. D. 1866. On anatomical peculiarities of some Dinosauria (on the anomalous relations existing between the tibia and fibula in certain of the Dinosauria). *Proceedings of the Academy of Natural Sciences of Philadelphia* 1866:316–317.

Darwin, C. 1859. *On the Origin of Species by Means of Natural Selection, or the Preservation of Favoured Races in the Struggle for Life* (London: John Murray).

de Queiroz, K. 1985. The ontogenetic method for determining character polarity and its relevance to phylogenetic systematics. *Systematic Zoology* 34:280–299.

de Queiroz, K., and Donaghue, M. 1988. Phylogenetic systematics and the species problem. *Cladistics* 4:317–338.

Dingus, L., and Sadler, P. M. 1982. The effects of stratigraphic completeness on estimates of evolutionary rates. *Systematic Zoology* 31:400–412.

Edwards, A. W. F. 1972. *Likelihood* (Cambridge: Cambridge University Press).

Eldredge, N. and Cracraft, J. 1980. *Phylogenetic Patterns and the Evolutionary Process* (New York: Columbia University Press).

Farris, J. S. 1974. Formal definitions of paraphyly and polyphyly. *Systematic Zoology* 23:548–554.

1976a. Expected asymmetry of phylogenetic trees. *Systematic Zoology* 25:196–198.

1976b. Phylogenetic classification of fossils with recent species. *Systematic Zoology* 25:271–282.

1977. On the phenetic approach to vertebrate classification. *In* Hecht, M. K., Goody, P. C., and Hecht, B. M. (eds.), *Major Patterns in Vertebrate Evolution* (New York: Plenum Press), pp. 823–850.

1979. On the naturalness of phylogenetic classification. *Systematic Zoology* 28:200–214.

1980. The information content of the phylogenetic system. *Systematic Zoology* 28:483–519.

1981. Distance data in phylogenetic analysis. *In* Funk, V. A., and Brooks, D. R. (eds.), *Advances in Cladistics: Proceedings of the First Meeting of the Willi Hennig Society* (New York: Botanical Garden), pp. 3–23.

1983. The logical basis of phylogenetic analysis. *In* Platnick, N. I., and Funk, V. A. (eds.), *Advances in Cladistics, Volume 2: Proceedings of the Second Meeting of the Willi Hennig Society* (New York: Columbia University Press), pp. 7–36.

Felsenstein, J. 1983. Methods for inferring phylogenies: a statistical view. *In* Felsenstein, J. (ed.), *Numerical Taxonomy* (Berlin: Springer-Verlag), pp. 315–334.

Felsenstein, J., and Sober, E. 1986. Parsimony and likelihood: an exchange. *Systematic Zoology* 35:617–626.

Forster, M. R. 1986. Statistical covariance as a measure of phylogenetic relationship. *Cladistics* 2:297–313.

Gaffney, E. S. 1979. An introduction to the logic of phylogeny reconstruction. *In* Cracraft, J., and Eldredge, N. (eds.), *Phylogenetic Analysis and Paleontology* (New York: Columbia University Press), pp. 79–111.

Galton, P. M. 1972. Classification and evolution of the ornithopod dinosaurs. *Nature* 239:464–466.

1974. The ornithischian dinosaur *Hypsilophodon* from the Wealden of the Isle of Wight. *British Museum of Natural History Bulletin (Geology)* 25:1–152.

1978. Fabrosauridae, the basal family of ornithischian dinosaurs (Reptilia: Ornithopoda). *Paläontologische Zeitschrift* 52:138–159.

Gauthier, J. A. 1986. Saurischian monophyly and the origin of birds. *In* Padian, K. (ed.), *The Origin of Birds and the Evolution of Flight. Memoirs of the California Academy of Sciences* 8:1–55.

Gauthier, J., Estes, R., and de Queiroz, K. 1988. A phylogenetic analysis of Lepidosauromorpha. *In* Estes, R. and Pregill, G. (eds.), *Phylogenetic Relationships of the Lizard Families* (Stanford: Stanford University Press), pp. 15–98.

Ghiselin, M. T. 1985. Mayr versus Darwin on paraphyletic taxa. *Systematic Zoology* 34:460–462.

Gould, S. J., and Vrba, E. S. 1982. Exaptation – a missing term in the science of form. *Paleobiology* 8:4–15.

Griffiths, G. C. D. 1976. The future of Linnaean nomenclature. *Systematic Zoology* 25:168–173.

Hennig, W. 1966. *Phylogenetic Systematics*. (Urbana: University of Illinois Press).

Kaye, J. M. and Russell, D. A. 1973. The oldest record of hadrosaurian dinosaurs in North America. *Journal of Paleontology* 47:91–93.

Lovejoy, A. O. 1959. The argument for organic evolution before "The Origin of Species," 1830–1858. *In* Glass, B., Tampkin, O., and Strauss, W. L. Jr. (eds.), *Forerunners of Darwin: 1745–1859* (Baltimore: Johns Hopkins Press), pp. 356–414.

Maddison, W. P., Donoghue, M. J., and Maddison, D. R. 1984.

Outgroup analysis and parsimony. *Systematic Zoology* 33:83–103.

Marsh, O. C. 1895. On the affinities and classification of the dinosaurian reptiles. *American Journal of Science* (series 3) 50:483–498.

Maryańska, T., and Osmólska, H. 1974. Pachycephalosauria, a new suborder of ornithischian dinosaurs. Results of the Polish–Mongolian palaeontological expeditions. Part V. *Palaeontologia Polonica* 30:45–102.

1981. Cranial anatomy of *Saurolophus angustirostris* with comments on the Asian Hadrosauridae (Dinosauria). Results of the Polish–Mongolian palaeontological expeditions. Part IX. *Palaeontologia Polonica* 42:5–24.

1985. On ornithischian phylogeny. *Acta Palaeontologica Polonica* 30:137–150.

Mayr, E. 1969. *Principles of Systematic Zoology* (New York: McGraw-Hill).

1974. Cladistic analysis or cladistic classification? *Zeitschrift für zoologischer Systematik und Evolutionsforschung* 12:94–128.

1981. Biological classification: toward a synthesis of opposing methodologies. *Science* 214:510–516.

Milner, A. R., and Norman, D. B. 1984. The biogeography of advanced ornithopod dinosaurs (Archosauria: Ornithischia) – a cladistic-vicariance model. *In* Reif, W. and Westphal, F. (eds.), *Third Symposium of Mesozoic Terrestrial Ecosystems*, short papers (Tübingen: Attempto Verlag), pp. 145–150.

Nelson, G. and Platnick, N. I. 1981. *Systematics and Biogeography: Cladistics and Vicariance* (New York: Columbia University Press).

Nopcsa, F. B. von 1915. Die Dinosaurier der Siebenburgischen Landesteile Ungarns. *Mitteilungen Jarhbuch Ungarische Geologische Reichsanstalt* 23:1–26.

1929. Dinosaurierreste aus Siebenburgen. *Geologica Hungarica Seria Palaeontologica* 1:1–76.

Norman, D. B. 1984a. On the cranial morphology and evolution of ornithopod dinosaurs. *Symposium of the Zoological Society of London* 52:521–547.

1984b. A systematic reappraisal of the reptile Order Ornithischia. *In* Reif, W., and Westphal, F. (eds), *Third Symposium of Mesozoic Terrestrial Ecosystems*, short papers (Tübingen: Attempto Verlag), pp. 157–162.

Olsen, P. E., and Galton, P. M. 1984. A review of the reptile and amphibian assemblages from the Stormberg of Southern Africa with emphasis on the footprints and age of the Stormberg. *Palaeontologia Africana* 25:87–110.

Ostrom, J. H. 1970. Stratigraphy and paleontology of the Cloverly Formation (Lower Cretaceous) of the Bighorn Basin area, Wyoming and Montana. *Peabody Museum Bulletin* 35:1–234.

1976. *Archaeopteryx* and the origin of birds. *Biological Journal of the Linnean Society* 8:91–182.

Panchen, A. L. 1982. The use of parsimony in testing phylogenetic hypotheses. *Zoological Journal of the Linnean Society* 74:305–328.

Patterson, C. 1982. Morphological characters and homology. *In* Joysey, K. A., and Friday, E. A. (eds.), *Problems in Phylogenetic Reconstruction* (Systematic Association Special Volume No. 21) (London: Academic Press), pp. 21–74.

1983. How does phylogeny differ from ontogeny? *In*

Goodwin, B. C., Holder, N., and Wylie, C. C. (eds.), *Development and Evolution* (British Society for Developmental Biology, Symposium 6) (Cambridge: Cambridge University Press), pp. 1–31.

Platnick, N. I. 1979. Philosophy and the transformation of cladistics. *Systematic Zoology* 28:537–546.

Raikow, R. J. 1986. Why are there so many kinds of passerine birds? *Systematic Zoology* 35:255–259.

1988. The analysis of evolutionary success. *Systematic Zoology* 37:76–79.

Romer, A. S. 1927. The pelvic musculature of ornithischian dinosaurs. *Acta Zoologica* 8:225–275.

1956. *Osteology of the Reptiles* (Chicago: University of Chicago Press).

1968. *Notes and Comments on Vertebrate Paleontology* (Chicago: University of Chicago Press).

Rosen, D. E. 1974. Cladism or gradism?: a reply to Ernst Mayr. *Systematic Zoology* 23:446–451.

Rozhdestvensky, A. K. 1966. New iguanodonts from central Asia. The phylogenetic and taxonomic interrelationships of late Iguanodontidae and early Hadrosauridae. *Paleontologicheskii Zhurnal* 1966:103–116. [in Russian]

Seeley, H. G. 1887. On the classification of the fossil animals commonly named Dinosauria. *Proceedings of the Royal Society of London* 43:165–171.

Sereno, P. C. 1984. The phylogeny of the Ornithischia: a reappraisal. *In* Reif, W., and Westphal, F. (eds.), *Third Symposium of Mesozoic Terrestrial Ecosystems*, short papers. (Tübingen: Attempto Verlag), pp. 219–226.

1986. Phylogeny of the bird-hipped dinosaurs (Order Ornithischia). *National Geographic Research* 2:234–256.

Simpson, G. G. 1961. *Principles of Animal Taxonomy* (New York: Columbia University Press).

Sneath, P. H. A. and Sokal, R. R. 1973. *Numerical Taxonomy* (San Francisco: W. H. Freeman and Company).

Sober, E. 1985. A likelihood justification of parsimony. *Cladistics* 1:209–233.

Swofford, D. L. 1985. *PAUP: Phylogenetic Analysis Using Parsimony, Version 2.4* (Champaign: Illinois Natural History Survey).

Taquet, P. 1976. Géologie et paléontologie du Gisement de Gadoufaoua (Aptien du Niger) (Paris: Centre Nationale de la Recherche Scientifique) *Cahiers de Paléontologie* 1976:1–191.

Thulborn, R. A. 1974. A new heterodontosaurid dinosaur (Reptilia: Ornithischia) from the Upper Triassic Red Beds of Lesotho. *Zoological Journal of the Linnean Society* 55:151–175.

1975. Dinosaur polyphyly and the classification of archosaurs and birds. *Australian Journal of Zoology* 23:249–270.

Van Valen, L. 1978. Why not to be a cladist. *Evolutionary Theory* 3:285–299.

Voss, E. G. 1952. The history of keys and phylogenetic trees in systematic biology. *Journal of the Scientific Laboratory, Denison University* 1952:1–25.

Watrous, L. E. and Wheeler, Q. D. 1981. The out-group comparison method of analysis. *Systematic Zoology* 30:1–11.

Wiley, E. O. 1981. *Phylogenetics: The Theory and Practice of Phylogenetic Analysis* (New York: John Wiley & Sons).

2 Shape analysis in the study of dinosaur morphology

RALPH E. CHAPMAN

Abstract

Morphometric and shape analysis methods can provide important information on the paleoecology, functional morphology, evolution, ontogeny, sexual dimorphism, phylogeny, taphonomy, and reconstruction of dinosaurs. The capabilities of one method, Resistant-Fit Theta-Rho-Analysis (RFTRA), a form of landmark shape analysis, are demonstrated using examples of cranial differences in the carnosaurs *Allosaurus* and *Tyrannosaurus*, ontogenetic development in the protoceratopsid *Bagaceratops*, sexual dimorphism in *Protoceratops*, cranial asymmetry in the prosauropod *Plateosaurus*, and pachycephalosaurian cranial morphology and phylogeny. The analyses show RFTRA to be a powerful method for elucidating shape differences within a variety of contexts and, in conjunction with standard phylogenetic methods, for providing information on taxonomic relationships.

Introduction

The dinosaur paleontologist interested in the quantitative analysis of his specimens' morphology is faced with a difficult paradox. The general rarity of specimens makes it all the more important to obtain the most information possible from the specimens, but this rarity prevents many of the more powerful techniques from being applied. Dinosaur researchers, or any paleontologists studying groups with limited fossil records (e.g., therapsids), must be pragmatic in their studies, framing questions to fit those techniques that can provide useful information. Within the proper context, those techniques adapted for answering these questions can provide a powerful set of tools for the analysis of dinosaur function, growth, evolution, and taphonomy.

This paper reviews in detail the types of questions accessible to dinosaur paleobiologists given the proper equipment and programs, but concentrates on one major method, Resistant-Fit Theta-Rho-Analysis.

In Dinosaur Systematics: Perspectives and Approaches, *Kenneth Carpenter and Philip J. Currie, eds. Copyright © Cambridge University Press, 1990.*

This technique is particularly adapted for use by vertebrate paleontologists, as will be shown by a series of example analyses of dinosaurs. The examples touch upon only a small subset of those approaches available, and include discussions of ontogenetic development in *Bagaceratops*, sexual dimorphism in *Protoceratops*, possible taphonomic and biological asymmetry in *Plateosaurus*, and morphology and phylogeny in pachycephalosaurs. For the latter, I would like to stress that shape analysis is not a substitute for the more established approaches to the quantitative analysis of phylogenies. Instead, shape analysis should be used to help develop characters for these analyses, and to help the paleobiologist evaluate the results of such analyses in the light of form and function. In some cases, similarities will be the result of, and an indicator for, recent ancestry. In others, similar shapes will provide clues for inferences of similar functions in separate lineages.

The first example, comparing the cranial morphology of *Allosaurus* and *Tyrannosaurus*, illustrates the capabilities that shape analysis provides for doing exploratory research in a short period of time. This particular analysis took less than an hour to run, from the decision to do it to the plotting of the results. Such an approach allows the researcher to test quickly observations made on similarities and differences between two or more specimens, and in the process allows access to many of the most important questions to be answered by paleobiologists. Shape analysis and morphometrics provide a tool box for use by the researcher in much the same way that cladistic methodologies and statistical procedures provide additional tool boxes. Together they provide the capability to study more rigorously the subjects of interest.

Recent years have seen a great expansion in the development and application of morphometric and shape analysis methods due to both an increasing interest in making evolutionary hypotheses more rigorous and the increasing availability of computer time and equipment. Implicit in most discussions of phylogenetic characters, taphonomy, functional morphology, and paleoecology are statements that depend on morphological inferences that can be supported or tested using morphometric procedures. Despite the high degree of sophistication present in the human visual system (Zusne 1970), there are limitations that prevent many trends in morphology from being detected or correctly interpreted, a consequence of the great complexity of the processes involved and the difficulty in evaluating shape changes independent of size. Morphometric procedures serve to help the morphologist confirm those trends that have been observed correctly and determine when possible errors have been made in others. Furthermore, many trends are evident only after a long process of study by a specialist. Morphometric methods allow the specialist to present the results of his experience to others without requiring them to spend an equally long period of time studying the material.

The morphometric methods available include techniques that make inferences using a set of geometrically unrelated linear measurements, and those that utilize the overall geometric configuration as context for the analysis. The first type can be derived directly from the study of allometry as developed by Huxley (1932) and many researchers from the nineteenth century to the present, and includes the modern school of multivariate morphometrics (e.g., Reyment et al. 1984). The second approach, comprising shape analysis *sensu stricto*, was influenced greatly by the works of D'Arcy Thompson (1915, 1917, 1942) and includes methods that concentrate on object outlines (e.g., Fourier Analysis, Eigenshape Analysis; see, for example, Clark 1981; Lohmann 1983; Rohlf 1986) and those that base their calculations on the coordinate positions of landmarks such as homologous points (e.g., Tensor Analysis, Resistant-Fit Theta-Rho-Analysis; see, for example, Bookstein 1978; Bookstein et al. 1985). The first major advancement on this general approach was made by Sneath (1967) who applied a least-squares algorithm to the superimposition of constellations of landmarks. This avoided the need for interpolation unavoidable in D'Arcy Thompson's grids and gave a direct indication of the differences in relative position of the landmarks themselves and, as a consequence, an indication of the morphological differences between the specimens being studied.

A second major advancement (Tobler 1977, 1978; Bookstein 1978, 1986; Bookstein et al. 1985 and references therein; Goodall and Green 1986) saw the application of tensor analysis techniques to the study of surface growth. This approach provides direct indications of the deformation that has occurred in regions of the surfaces being studied and is unsurpassed in providing information on allometric gradients and fields.

The third major development in landmark analysis, Resistant-Fit Theta-Rho-Analysis (RFTRA), was made by Siegel and Benson (1982; Benson et al. 1982; Siegel 1982). This method modified Sneath's least-squares methodology by applying a robust statistical approach that utilizes medians instead of mean values for its calculations. As with the least-squares method, the results provide the superimposition of the constellations of landmarks for pairs of specimens. The RFTRA algorithm, however, allows an analysis to detect localized changes in morphology when they occur, correcting a major limitation of the least-squares method. Bookstein (1986) also is able to analyze localized change using tensor approaches combined with a new method for the determination of "shape coordinates."

The quantitative analysis of dinosaur morphology has followed an evolution similar to that of the science of morphometrics in general, although the studies have been very few due to the lack of adequate numbers of specimens. Initial studies concentrated on providing standard size estimators, such as skull length or femur length, with later studies (e.g., Lambe 1918; Brown and Schlaikjer 1943) relying heavily on the use of ratios of parameters that were perceived to be of taxonomic importance. The first detailed analyses were made by Gray (1946) and Lull and Gray (1949), the former using classic allometric procedures *sensu* Huxley (1932), and the latter D'Arcy Thompson's (1915, 1917) grid transformation method, on various ceratopsian skulls. Other discussions of interest can be found in Colbert (1948), Olson and Miller (1951), Rozhdestvensky (1965) and Ostrom (1966).

The next detailed analyses were done in the mid-1970s following the development and proliferation of multivariate morphometric procedures, and building on previous works such as Jolicoeur and Mosimann's (1960) classic study of the painted turtle, Gould's (1966) review of allometric procedures and his (Gould 1967; Gould and Littlejohn 1973) factor analyses of pelycosaurs. As morphometric analyses of various reptile groups became standard in herpetological journals, Dodson (1975a,b,c, 1976) focused on the morphometric analysis of *Alligator* and *Sceloporus* as a means for assessing the utility of such procedures for the analysis of lambeosaurine hadrosaurs, *Protoceratops*, and dinosaurs in general. Dodson was able to demonstrate the utility of morphometric procedures for the analysis of variability and the subsequent implications of this variability on taxonomic philosophy. Furthermore, he was able to illustrate approaches to the study of sexual dimorphism and the discrimination and identification of taxa.

In his two papers on dinosaurs, Dodson (1975c, 1976) was able to bring the quantitative analysis of dinosaur morphology up to date and his studies served as a model for the research of Grine et al. (1978), Bradu and Grine (1979) and Tollman et al. (1980) on therapsids, the study by Chapman et al. (1981) on the pachycephalosaurian *Stegoceras*, and the study by

Weishampel and Chapman (this volume) on *Plateo-saurus*. Typically, these studies include detailed bivariate allometric analyses, eigenvector approaches, such as principal components analysis and principal coordinates analysis, and discrimination procedures. The result was the quantitative analysis and support of sexual dimorphism in a number of taxa, and strong support for the synonymization of taxa that had been oversplit.

Dodson's (1975c) study of lambeosaurine hadrosaurs suggested the presence of two sexually dimorphic genera, comprising three species, rather than 12, and that *Procheneosaurus* was a juvenile form. Dodson (1976) also was able to demonstrate a strong case for sexual dimorphism in *Protoceratops*, which had been suggested for many years but had yet to be shown quantitatively.

The next major study of dinosaurs using morphometric methods was that by Chapman et al. (1981) on 29 domes previously referred to the pachycephalosaurid genus *Stegoceras*. They used standard bivariate approaches, principal components analysis, and discriminant analysis and were able to suggest that most domes be referred to *S. validum*, rather than a number of other taxa. They also were able to demonstrate sexual dimorphism in that species, and included a classification function usable for referring new specimens to their proper sex. Preliminary results of this morphometric analysis also were used as one basis for the separation of one specimen into the new genus *Gravitholus* by Wall and Galton (1979).

In a similar study, Weishampel and Chapman (this volume) use similar procedures on femoral data to suggest that specimens of *Plateosaurus* from Trossingen (Baden–Württemburg, Federal Republic of Germany) are referable to only one or two taxa or sexes.

Finally, the shape analysis methods discussed herein were used by Chapman and Brett-Surman (this volume) to examine cranial and postcranial morphometric patterns in hadrosaurs.

The use of morphometric procedures has been very effective when applied to dinosaur material. It has provided the basis for reversing problems caused by the oversplitting of taxa in years past, and has provided important information into the intercorrelations of variables providing functional insights into the animals themselves. However, the studies so far have been restricted, for the most part, to those series of taxa where large amounts of material have been available (e.g., *Protoceratops*). These approaches, augmented with newer developments, such as RFTRA, will continue to provide insights for taxa with large amounts of available material such as those coming from bone-bed deposits (see Currie 1986, for example). However, it is important to develop approaches that can be applied when smaller numbers of specimens are available. RFTRA is one such approach.

Materials and methods

For the example analyses presented here a single morphometric method was applied, Resistant-Fit Theta-Rho-Analysis (RFTRA). For detailed discussions of this method see Siegel (1982), Siegel and Benson (1982), and Benson et al. (1982). RFTRA provides graphical results that include superimposed constellations of the landmarks, vectors indicating the direction of change of each landmark between specimens, and a distance coefficient representing an estimate of the morphological distance between the two specimens being studied based on the landmarks used. The plotted results are for a single pair of specimens, but distance matrices of RFTRA distance coefficients also can be generated for a suite of specimens and used in other morphometrical techniques, such as cluster analyses or ordinations. As yet, no attempt has been made to use RFTRA coefficients in numerical cladistic analyses. This is an avenue that remains to be explored both methodologically and philosophically.

In RFTRA, outlines of various related structures are digitized in addition to the landmark coordinates. The results show differences in the outlines (e.g., outline of the skull, orbit, fenestrae, liner objects, etc.) once the fit or superimposition has been made using the landmarks. It is important to note that these outlines are passive structures and are not used in any calculations making the superimposition or estimating the morphological distance coefficient. Theoretically, a second estimate of distance could be made after superimposition by comparing the relative position of important parts of the outlines, but this has yet to be developed using RFTRA and probably will be useful only on an ad hoc basis for individual studies.

It is important to note that there is no distortion imposed on the digitized morphologies. The original geometry of the landmarks and outlines is maintained as superimposition is calculated and the results plotted. The only changes that occur involve the overall size of the figures and the rotation and translation of their position in two dimensions.

To perform RFTRA on a series of specimens, these steps are followed.
1. The problems to be resolved using RFTRA are determined.
2. A series of photographs or drawings of specimens are selected for use in the digitization process.
3. The landmarks are identified and marked for easy identification during digitization. For each analysis, the same landmarks must be used on all specimens included.
4. The outlines to be digitized do not have to be consistent from specimen to specimen, but typically changes in these figures are of interest. They also tend to be consistent.
5. The RFTRA input programs are used to digitize the selected landmarks and outlines.
6. A "skeleton" or polygonal diagram is made for use in the graphical results. This is done by connecting landmarks to highlight areas of interest.

7. The RFTRA programs are run. These provide results for comparisons of pairs of specimens, or RFTRA distance matrices for larger groups of specimens for use with other programs and methods (e.g., cluster analysis).

8. RFTRA or other graphics are generated, providing either the superimposed figures, or results of the manipulations of the RFTRA distance matrices (e.g., dendrograms from cluster analyses).

9. The results are interpreted in light of the questions developed in (1) above. If alterations are suggested or additional questions develop, the process is repeated.

The formulation of the questions to be answered is the crucial phase of any quantitative analysis. Those that can be answered using RFTRA are many, but can be summed up by asking a more general question of how two or more morphologies differ. For typical research, the context will be one of functional morphology, ecology, sexual dimorphism, ontogeny, or phylogeny. However, the dinosaur researcher should not restrict the use of RFTRA to questions that will be used directly in publishable research. RFTRA should be used as well for exploratory analyses by a researcher who is curious, for example, about how two taxa (even unrelated forms) may differ in shape. It can also serve a more mechanistic purpose, such as determining the amount of taphonomic damage or distortion in a specimen, or assist the illustrator in evaluating the accuracy of drawings based on specimens. The ease and speed of applying RFTRA should make such exploratory analyses worthwhile.

Once the questions are determined, photographs and/or drawings of the specimens must be obtained and important structures (e.g., sutures, fenestrae, etc.) located. If the material or reconstructions are not sufficient for further analysis, then the study must be abandoned. However, careful work with illustrations, photographs, and specimens usually provides sufficient information to allow the analysis to proceed.

Landmark choice is another crucial phase, and one which is most easily misunderstood. The most useful and powerful landmarks are homologous points, those that can be located on all specimens and which have a theoretically similar developmental history. For dinosaurs, such homologous points typically would include triple junctions between cranial sutures or fenestrae, the limits of a tooth row, or similar structures. Bookstein et al. (1985) unnecessarily restricted landmarks to points that could be determined to be homologous in a rigorous sense because they were interested in analyses within a strictly phylogenetic context.

The analyses need not be restricted to rigorously homologous points, as points that are functionally or geometrically equivalent (e.g., inflection points, extreme points, points of articulation) can be used. The important step is to interpret the results in light of the types of landmarks used. If the points are homologous, then the results should be useful in any context. If functional or geometrical landmarks are used, then similarities or differences can be interpreted within functional contexts and for discrimination purposes. Much more cautiously, the similarities and differences may be within phylogenetic contexts, but with the realization that similar shapes might have been achieved by functional convergence or taphonomic distortion. With individual analyses, if most of the points are homologous points, then the researcher may choose to add other landmarks (e.g., tips of horns) to see how they vary in position within a phylogenetic context, to answer questions about taphonomic distortion, or to study differences due to sexual dimorphism. Such an exploratory approach is taken by Chapman and Brett-Surman (this volume) in studying hadrosaur pelvic elements. I have found analyses to be quite robust despite small variations or errors in the position of landmarks.

The development of a polygonal or "skeletal" diagram provides a geometric framework for easy interpretation of results. There is no one correct figure for any analysis. Instead, the researcher should experiment with a number of different polygonal figures to find the one that provides the easiest interpretation of the results. This choice will reflect a compromise between including enough connections to provide information, but limiting connections to avoid overwhelming the researcher with too complicated a plot. In general, it is best to try and delimit polygons that equate to functional regions (e.g., the snout), and it is usually best to avoid crossing connections. The use of the polygonal figures should become evident in the discussion accompanying the example analysis of carnosaur crania.

The next step is the digitization of the results. This is done using a standard digitizing pad interfaced with a microcomputer and using RFTRA software. The landmarks should be well-indicated on the figure or photographs. This step is mostly mechanical and takes relatively little time. The digitization of the landmarks usually is easier than the digitization of the figures. The result will be files including data on the object, the number of landmarks, the coordinates of the landmarks, and information on the number and types of figures and outlines with their coordinate data. Each file is assigned a name based on a specimen reference number.

Once the input files are completed, the analyses are run using an RFTRA program that will calculate the coefficients providing the proper superimposition of two figures and the distance coefficient for that fit. A second program will calculate all the necessary distance coefficients to produce a distance matrix for a variable number of specimens. This distance matrix is then used for cluster analyses or ordinations or other analyses of interest.

The penultimate step is to use a variety of graphic programs to provide the results in a usable form. For analyses of two specimens, three separate figures are provided in each graphical result. The first provides two superimposed polygonal figures showing where equivalent regions are varying in either position or shape, or both aspects. The second figure provides the polygonal diagram for the base specimen with vector arrows showing the changes of position of each individual landmark point. Large arrows indicate big changes, small arrows less important ones. If the differences between the two morphologies are localized, there should be a few large arrows and mostly small, randomly oriented arrows. If the differences are general, then the arrows will tend to be normally distributed in both magnitude and direction. Programs are available for the plotting of histograms and rose diagrams of vector sizes and directions. Differences in these distributions have important functional and evolutionary implications (Chapman in preparation). The third graphic provided is of the superimposed figures or outlines. These show where figures differ and whether other important regions of variation occur. The analyses using the distance matrices will provide standard dendrograms from cluster analyses or ordinations indicating the relative position of the specimens. The cluster analyses (UPGMA – Unweighted Pair-Group Method with Arithmetic Averages) used herein for examples and interpretation of the dendrograms have the same strengths and weaknesses inherent in the technique.

It is important to note that for the paired comparisons of specimens, the results are always presented with a base specimen, which remains unchanged in size and orientation, and a second specimen that is fit onto the base specimen by the analysis. Vectors then show changes from the base specimen to that of the second specimen.

The final stage is to utilize the graphical results to determine their usefulness and suitability to answer the questions initially posed in (1). If additional problems or questions arise, then the analyses are repeated varying the landmarks, polygonal figures, and specimens.

The programs now available for RFTRA include a general FORTRAN program published in Siegel (1982) and a series of BASIC programs written for the DOS microcomputer environment by Chapman using past programs by Benson, Siegel and Chapman as a basis. The programs are available at no charge from me and source codes can be provided to those wishing to adapt the approach to different computer systems. It is important to note that the distance coefficients calculated in the FORTRAN program published in Siegel (1982) are not normalized to the size of the figures as is the case for the coefficients presented herein. This will have the disturbing effect of providing, on the average, higher distance values for larger figures independent of object size. This can be corrected easily and assistance is available from me.

The two-dimensional limitation can be overcome, at least in part, by analyzing specimens from more than one view (e.g., dorsal, lateral). In most cases, the analysis of different views provides congruent results. However, in studies where there is an unusually high degree of variability, or a small or restricted set of landmarks, differences can be apparent. The results of all analyses must be considered for final interpretation, taking into account the nature of the data for each. Three-

Figure 2.1. Carnosaur crania, lateral views. Included are views of **A**, *Allosaurus*; and **B**, *Tyrannosaurus*; **C**, the 25 landmarks and **D**, polygonal figure for *Allosaurus*. The regions marked *sa*, *d*, and *m* refer to the surangular, dentary, and maxilla regions, respectively.

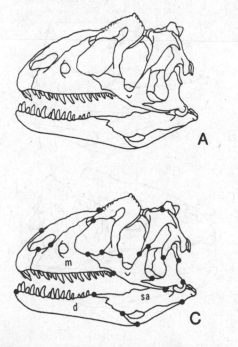
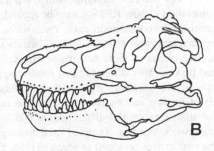
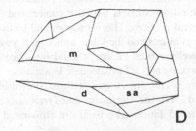

dimensional programs are being developed as three-dimensional digitizers and more advanced graphical capabilities become available.

A series of examples follow that should demonstrate the types of analyses for which RFTRA is adapted. The first example, that of carnosaur cranial differences, is given in detail and subsequent analyses are treated more briefly. Because these examples are meant to be illustrative in nature, detailed discussions of landmark choices and polygonal figure selection and development will not be given. These will be necessary for more detailed applications for phylogenetic and functional interpretation.

The usefulness and implications of applying landmark shape analysis methods within a phylogenetic context have yet to be explored fully, although clearly significant information can be provided (see, for example, Bookstein et al. 1985). The application of least-squares methods *sensu* Sneath (1967) reflect a phenetic approach. Applying RFTRA within this context also probably reflects a phenetic approach, although the ability to identify localized changes that may serve as derived characters makes this less clear. As discussions evolve regarding the relationships between homology and synapomorphy (e.g., Patterson 1982), the usefulness of landmark techniques for phylogenetic analyses should increase. For now, I suggest RFTRA be considered a phenetic approach providing dendrograms or ordinations that should help in the interpretation and development of cladograms resulting from standard phylogenetic analyses. Perhaps it may also aid in the selection of a single cladogram among a number of possible choices.

Example 1 – carnosaur crania

For the first example, which provides a detailed illustration of how RFTRA may be applied, I will use the crania of two carnosaur taxa, *Tyrannosaurus* and *Allosaurus*, to answer a simple and exploratory question: how are the crania of these two major carnosaur taxa similar and how are they different? In a research application, this may be asked within the framework of a functional analysis of feeding mechanisms, or it may be used as a basis for analyzing cranial differences within a phylogenetic framework. For this example I used drawings of crania of the two genera provided in Romer (1956, Figs. 78B and 78C). Twenty-five landmarks were identified on the lateral views of the crania, consisting mostly of triple-junction contacts between cranial bones or fenestrae, but also including contacts between sutures and the limits of the lateral profiles. The former should be excellent homologous points and the latter, although more affected by non-developmental factors, should introduce little significant variance and act as adequate landmarks.

The drawings used are shown in Figures 2.1A and 2.1B, for *Allosaurus* and *Tyrannosaurus* respectively. The 25 landmarks used are illustrated on the cranial

drawing of *Allosaurus* in Figure 2.1C and the polygonal figure used is given in Figure 2.1D. Note how the areas apparent in the polygonal figure correspond to major features of the crania. For example, regions corresponding to the surangular, dentary, and maxilla are designated by *sa*, *d*, and *m* respectively. Here, the lower jaw is analyzed attached to the cranium. It is also possible to analyze each component separately as will be done in the other examples.

The results of the RFTRA of the 25 landmarks are given in Figure 2.2, with the polygonal figures in Figure 2.2A, the vector diagram in Figure 2.2B, and the superimposed outline figures in Figure 2.2C. The distance, 2.1824, is also provided, and is surprisingly low considering that the two taxa are members of different families. However, a dorsal cranial view or analysis of other elements may indicate a much higher level of dif-

Figure 2.2. Results of Resistant-Fit Theta-Rho-Analysis of carnosaur crania using *Allosaurus* (base specimen, 2701) and *Tyrannosaurus* (specimen 2702). **A**, diagram of superimposed polygonal figures; **B**, vector diagram; and **C**, superimposed outline figures. Distance coefficient is 2.1824.

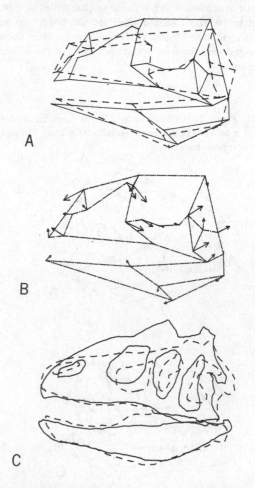

A

B

C

ference. The distance values are only directly comparable within a single study because of variations in landmark choice and number. For example, choosing a large number of tightly clustered landmarks will provide a much lower distance for the same specimens than using a smaller number of more dispersed landmarks. However, after a large number of analyses, a researcher develops a useful intuition into the values of the coefficients. The specimen numbers in the figure legends refer to files storing the data and are used for reference during the analyses.

The results presented in Figure 2.2 can be summarized as follows. For the landmarks, the changes tend to be general and not localized. However, two major trends are apparent. First, the posterior part of the jaw appears to be more robust in *Tyrannosaurus*. Second, the overall cranium appears to have a much higher degree of dorso-ventral vaulting in *Allosaurus* and, consequently, is more elongate in *Tyrannosaurus*. This is indicated by vector directions in the anterior and ventral

direction for the landmarks in the premaxilla and nasal areas, and vectors in the posterior direction for the jugal and quadratojugal landmarks. Remember, this analysis used *Allosaurus* as the base specimen (reference number 2701) and, consequently, the vectors indicate change from the *Allosaurus* condition to that of *Tyrannosaurus* (reference number 2702). The outlines confirm the landmark trends and the overall impression: that *Tyrannosaurus* has a relatively more elongate cranium with a more robust posterior portion of the lower jaw. These differences would have important implications relative to the forces that could be generated by the jaws and the feeding modes suggested for the two genera.

Example 2 – ontogenetic development in *Bagaceratops*

As an example of an application of RFTRA within an ontogenetic context, I analyzed the skulls of three specimens of the protoceratopsid *Bagaceratops rozhdestvenskyi* from the Campanian of Mongolia,

Figure 2.3. *Bagaceratops* ontogenetic series, dorsal cranial views. Included are reconstructions of **A**, small individual; **B**, intermediate individual; and **C**, large individual; and figures illustrating **D**, the 36 landmarks and **E**, polygonal figures on the intermediate specimens.

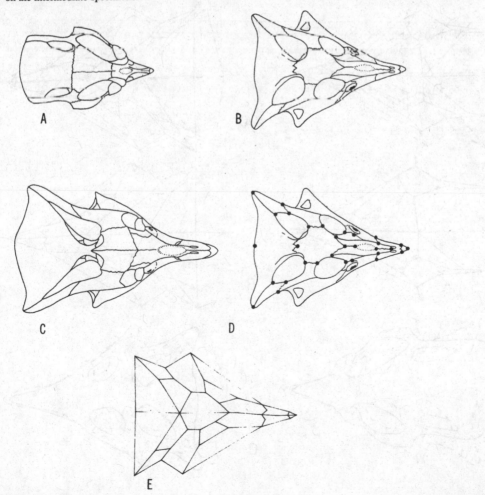

described and illustrated in Maryańska and Osmólska (1975). The three ontogenetic stages and respective specimens are illustrated in Figure 2.3A-C, redrawn from figures in that monograph. The youngest stage, approximately 45 mm in cranial length, is based on specimen ZPAL MgD-I/123, and is shown in Figure 2.3A. The middle specimen, based on the holotype skull ZPAL MgD-I/126, is illustrated in Figure 2.3B. It has a basal cranial length of approximately 124 mm. Figure 2.3C illustrates the largest stage, is based on material from 3 skulls, ZPAL MgD-I/127, 129, 133, and represents a basal cranial length of approximately 250 mm.

Maryańska and Osmólska (1975) noted five characteristic ontogenetic changes for the Protoceratopsidae based on ratio data for a series of *Bagaceratops, Protoceratops,* and *Leptoceratops.* These include (1) a relative decrease in orbit length, (2) a slight increase in snout length, (3) an initial increase in frill length followed by subsequent shortening, (4) a widening of the frill, and (5) a widening of the jugal and quadrate area. Their suggestion can be tested for the *Bagaceratops* lineage using RFTRA on the three specimens or composite figures. Only dorsal views of the crania were available, but these are sufficient to examine trends in all the characters listed above. A series of 36 landmarks were identified on the figures, illustrated in Figure 2.3D, and these produced the polygonal figure illustrated in Figure 2.3E. As with the carnosaurs, the landmarks are mostly triple junctions of cranial bones and fenestrae with a few extreme points added on the margins.

The results of the study are summarized in Figures 2.4–2.6 which compare the smallest (reference number 2501) and middle (2502) specimens, the middle and largest (2503) specimens, and the smallest and

Figure 2.4. Results of Resistant-Fit Theta-Rho-Analysis of *Bagaceratops* ontogeny for small (specimen 2501) and intermediate specimens (2502). **A,** diagram of superimposed polygonal figures; **B,** vector diagram; and **C,** superimposed outline figures. Distance coefficient is 2.2034.

Figure 2.5. Results of Resistant-Fit Theta-Rho-Analysis of *Bagaceratops* ontogeny for intermediate (2502) and large specimens (2503). **A,** diagram of superimposed polygonal figures; **B,** vector diagram; and **C,** superimposed outline figures. Distance coefficient is 1.9367.

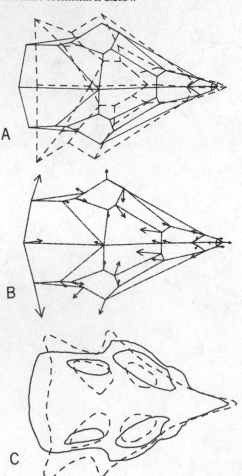

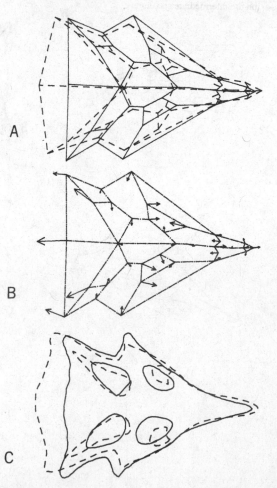

largest, respectively. The first transition, illustrated in Figure 2.4 for 2501 (base) and 2502, strongly supports all of the points listed by Maryańska and Osmólska (1975, p. 167). All figures, and especially the vectors, demonstrate a shortening and widening of the frill, a shortening of the orbit, a widening of the jugal and quadratojugal area, and a small increase in snout length. The second transition of 2502 to 2503 shown in Figure 2.5, however, complicates the picture. Here, the frill lengthens instead of continuing to shorten, and remains about the same relative width, while the jugal region shortens slightly. The orbit and snout trends continue strongly. The total change, illustrated in Figure 2.6 for specimens 2501 and 2503, shows a net change following the suggested trends for orbit length, snout length, and frill width. Frill length and jugal width show almost no net change.

The analysis provides support for three of the trends suggested by Maryańska and Osmólska (1975), but points out variation in the other two. This may be an indication of the difficulties involved with using individual ratios to suggest trends because no single measurement provides an adequate characterization of overall size. This also may be the result of the difficulties inherent in producing composite figures based on partial specimens.

This example illustrates the usefulness of bilateral symmetry for providing landmarks, and the information provided by analyzing two symmetrical sets of landmarks to obtain a more robust result. Asymmetry is common in fossil vertebrates due to taphonomic distortion and the use of both sides provides a more robust and useful analysis.

Example 3 – *Protoceratops* sexual dimorphism

The excellent series of crania available for the protoceratopsid *Protoceratops andrewsi*, from the Upper Cretaceous of Mongolia, has interested many investigators since Granger and Gregory (1923). Gregory and Mook (1925) were the first to discuss possible sexual dimorphism in the species and, since then, it has been the topic of many discussions, including Brown and Schlaikjer (1940), Kurzanov (1972), and Dodson (1976). The most rigorous of these studies is the allometric analysis by Dodson (1976) who suggested the following 12 characters as important for discrimination of the sexes: (1) frill length, (2) frill width, (3) postorbital skull width, (4) skull width across the jugals, (5) nasal height, (6) frill height, (7) orbital height, (8) length and (9) width of parietal fenestrae, (10) length and (11) width of external nares, and (12) height of the coronoid process. He presents illustrations of specimens AMNH 6438 and AMNH 6466 as examples of a mature and large "male" and "female" respectively.

If these specimens are considered to be excellent examples of the sexes, it is of interest to see if they exhibit the differences in a RFTRA analysis inferred from the bivariate and multivariate analyses performed by Dodson (1976). To do this I used illustrations in Dodson (1976, Fig. 6) and Brown and Schlaikjer (1940, Figs. 7, 18) to provide lateral and dorsal cranial views, and lateral views of the lower jaw for these two specimens. The illustrations, redrawn from these sources, are given with the landmarks and polygonal figures in Figures 2.7–2.9. For the jaws, the incomplete specimen (AMNH 6438) limited the analysis to 5 points. The anterior section is inferred for the outline figure. Fifteen and 18 landmarks were identified for the lateral and dorsal views of the skulls, respectively. They are similar to those discussed in the previous examples with the addition of the tooth row limits.

The results of the analyses using RFTRA are given in Figures 2.10–2.12, following the same format

Figure 2.6. Results of Resistant-Fit Theta-Rho-Analysis of *Bagaceratops* ontogeny for small (2501, base) and large (2503) specimens. **A**, diagram of superimposed polygonal figures; **B**, vector diagram; and **C**, superimposed outline figures. Distance coefficient is 2.3877.

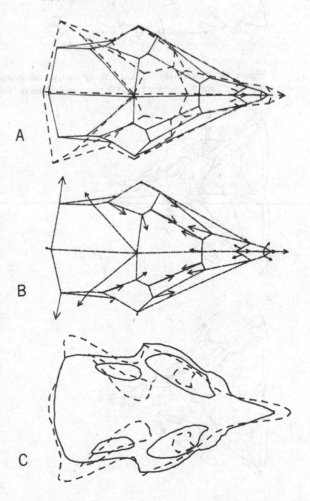

Figure 2.7. *Protoceratops* sexual dimorphism, lateral cranial views. Included are reconstructions of **A**, "female"; **B**, "male"; and **C**, landmarks and **D**, polygonal figure of "female" specimen.

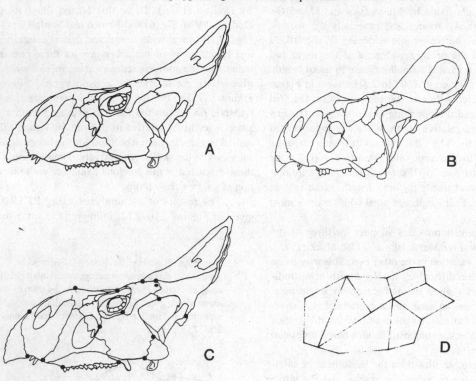

Figure 2.8. *Protoceratops* sexual dimorphism, dorsal cranial views. Included are reconstructions of **A**, "female"; **B**, "male"; and **C**, landmarks and **D**, polygonal figure of "female" specimen.

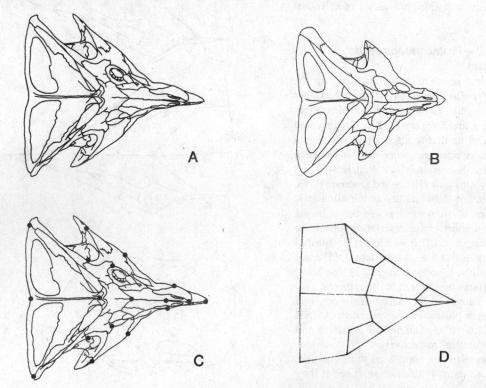

as for the previous examples. Despite the incomplete material, the "male" (AMNH 6438) does appear to exhibit a relatively higher coronoid process (Fig. 2.12), although other differences, especially those apparent in the outlines, are either insignificant or based on extrapolated material. For the 11 skull characters mentioned by Dodson (1976), the following is apparent. From Figure 2.11, frill length and width are relatively larger in the "male," supporting points (1) and (2) above, but increases in the jugal (4) and postorbital (3) widths do not appear to be significant. The parietal fenestrae (8, 9) also appear to be about the same size. From Figure 2.10, the predicted relationship is observed in the nasal height (5), orbital height (7), and length of the external nares (10). The external nares width (11) and frill height (12)

are less evident, although the latter exhibits the angular relationship suggested by Kurzanov (1972). The lack of increase in the size of the parietal fenestrae in AMNH 6438 is because the posterior region of the frill is missing and it is reconstructed from other specimens. The results of RFTRA and Dodson's (1976) study suggest that the parietal fenestrae should be much bigger than they are in the reconstruction. This illustrates the usefulness of using RFTRA and conventional morphometric methods for evaluating reconstructions based on partial specimens.

The results of RFTRA reflect a number of complicating factors. As Dodson (1976) points out, individual specimens will exhibit variance that will, at times, not support an overall population trend. However, I believe the results also illustrate the problems encountered in attempting to remove size as a major factor using conventional methods. Despite the failure of some of the characters to exhibit the predicted relationship, the absolute values for each of these characters does

Figure 2.9. *Protoceratops* sexual dimorphism, lateral view of lower jaw. Included are reconstructions of A, "female"; B, "male"; and C, landmarks and D, polygonal figure of "female" specimen. Dashed lines in B represent projected areas not based on real material and not used for any calculations.

Figure 2.10. Results of Resistant-Fit Theta-Rho-Analysis of *Protoceratops* sexual dimorphism for lateral cranial views of AMNH 6466 (specimen 2832, base) and AMNH 6438 (2831). A, diagram of superimposed polygonal figures; B, vector diagram; and C, superimposed outline figures. Distance coefficient is 1.9499.

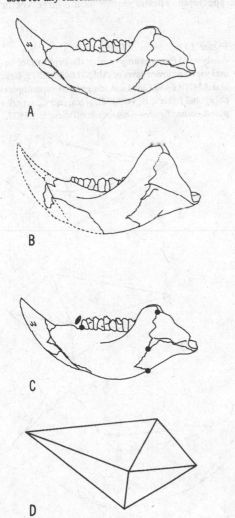

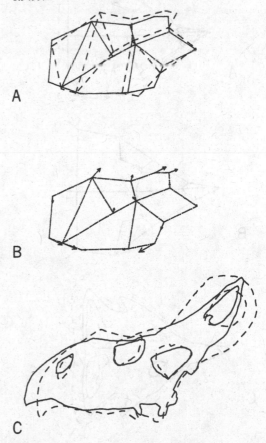

support the prediction because of the larger size of the "male" specimen. The ability of RFTRA to remove more easily confounding size factors and indicate important relative differences is well-illustrated by this example.

Example 4 – *Plateosaurus* asymmetry

One interesting capability allowed by RFTRA and other shape analysis methods is the comparison of equivalent halves of a bilaterally symmetrical organism. If growth is perfect, the two halves should be identical. However, developmental and environmental histories combine to produce a biological component of asymmetry, and taphonomic distortion frequently introduces a large amount. RFTRA is a useful method for the analysis of taphonomic distortion and graphical methods allow the researcher to correct for certain types of dis-

tortion and obtain a more accurate picture of the original specimen.

To give a brief example, two lateral views of a single skull of the prosauropod *Plateosaurus engelhardti* from the Upper Triassic of Germany, described and illustrated by Galton (1985) were used (specimen HMN XXIV). Using two separate views requires that one be reversed to allow a proper analysis. This can be done either mechanically using a light table before digitization or algorithmically using a computer after the figure is digitized. Figure 2.13 gives drawings of the two views, redrawn from Galton (1985, Fig. 1.3A,B, page 124), the 17 landmarks selected, and the polygonal figure used for the analysis.

The results are shown in Figure 2.14. The differences, yielding a distance coefficient of 1.4308, were large, as might be expected from distortion due to compaction in a form with a laterally compressed cranium. The differences, however, are not randomly distributed and reflect a moderate amount of differential crushing and an anomalous contact of the external nares, maxilla, and nasal bones, possibly due to damage of the specimen. This is shown as the large vector in

Figure 2.11. Results of Resistant-Fit Theta-Rho-Analysis of *Protoceratops* sexual dimorphism for dorsal cranial views of AMNH 6466 (2802, base) and AMNH 6438 (2801). **A**, diagram of superimposed polygonal figures; **B**, vector diagram; and **C**, superimposed outline figures. Distance coefficient is 1.5283.

Figure 2.12. Results of Resistant-Fit Theta-Rho-Analysis of *Protoceratops* sexual dimorphism for lateral views of lower jaws of AMNH 6466 (2872, base) and AMNH 6438 (2871). **A**, diagram of superimposed polygonal figures; **B**, vector diagram; and **C**, superimposed outline figures. Distance coefficient is 1.2877.

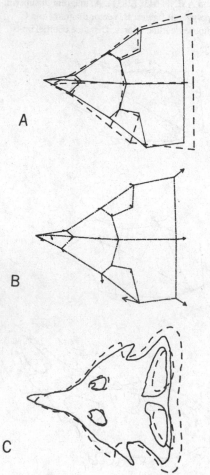

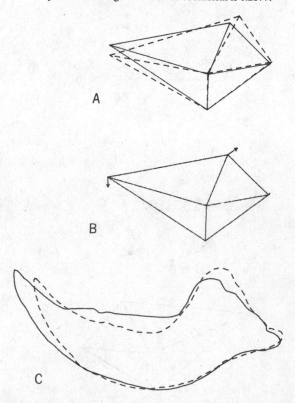

Figure 2.13. *Plateosaurus* cranial asymmetry, lateral views. Included are reconstructions of **A**, left and **B**, right side of cranium with **C**, 17 landmarks and **D**, polygonal figure used for analysis.

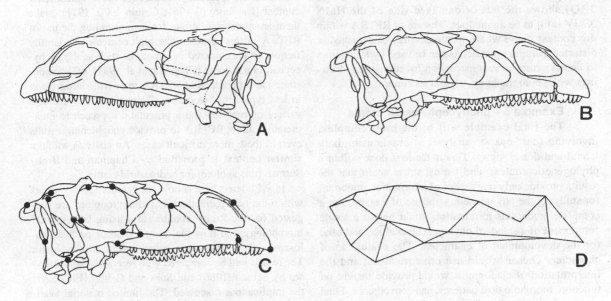

Figure 2.14. Results of Resistant-Fit Theta-Rho-Analysis of cranial asymmetry in *Plateosaurus* for HMN XXIV (2602, base, right lateral view) and HMN XXIV (2601, left lateral view). **A**, diagram of superimposed polygonal figures; **B**, vector diagram; and **C**, superimposed outline figures. Distance coefficient is 1.4075.

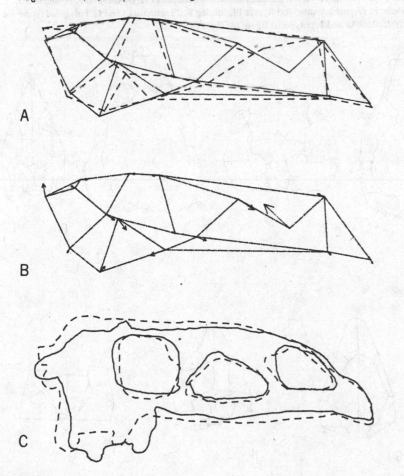

Figure 2.14B. Comparison of these specimens with other specimens (e.g., AMNH 6810, Galton 1985, Figs. 3P,Q) shows the left or damaged side of the HMN XXIV skull to be anomalous. The use of RFTRA within this context shows where specimens may be damaged or distorted. Similar approaches can be used by illustrators to detect errors in reconstruction by comparing drawings with photographs or original specimens.

Example 5 – pachycephalosaur crania

The final example will be the most complex, involving four separate analyses of crania using both lateral and dorsal views. This is the first done within a phylogenetic context, and I must stress again that the results provide only one aspect of information important for studying the phylogenetic structure of a taxon. For a complete study, this information would be but a single component of a standard phylogenetic analysis and used for the development of characters. The evaluation of characters chosen would use other methods, and the interpretation of cladograms would provide insight on function, morphological patterns, and convergence. Final conclusions should be based on the congruence of all the available information.

The distinctive cranial morphology of pachycephalosaurs, apparently adapted for use in intraspecific combat (Davitashvili 1961; Galton 1970, 1971), make them an appropriate group for demonstrating the use of RFTRA in this context. The high degree of variability frequently encountered in structures such as the pachycephalosaurid dome (Chapman et al. 1981), will demonstrate the capabilities of RFTRA to provide information in less typical or "worse-case" situations. The use of a variety of analyses and a pragmatic approach to interpretation allow RFTRA to provide enlightening results even in these more difficult cases. An analysis within a similar context is provided by Chapman and Brett-Surman (this volume) for hadrosaurid crania.

Of interest here is how cranial morphology varies within the pachycephalosaurs, what groupings are suggested for the Pachycephalosauria using only cranial morphology, and how the domed skull of pachycephalosaurs differs from other, more typical ornithischians. The results will be compared with recent cladistic analyses by Sereno (1986) and Sues and Galton (1987), and the implications discussed. The limited material available for the group dictated the complex structure of the analysis; incomplete specimens did not allow single

Figure 2.15. Pachycephalosaurids, dorsal cranial views. Included are reconstructions of **A**, *Goyocephale*; **B**, *Stegoceras*; **C**, *Homalocephale*; **D**, *Prenocephale*; **E**, *Hypsilophodon*; and figures illustrating **F**, 23 landmarks and **G**, polygonal figure for full set of landmarks; and **H**, 20 landmarks and **I**, polygonal figure for abbreviated set.

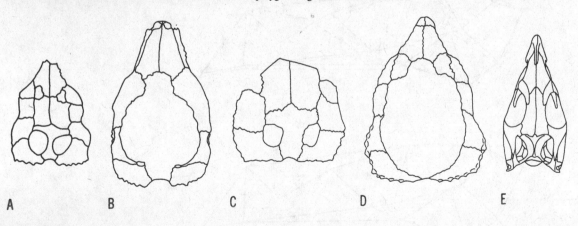

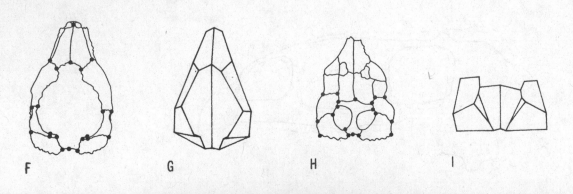

analyses for dorsal and lateral cranial views. Instead, two different analyses were run for each, utilizing different suites of landmarks and specimens as was appropriate.

The specimen illustrations used include the dorsal view of specimen GI SPS 100/1501, referred to *Goyocephale* by Perle et al. (1982), and a series of crania illustrated by Maryańska and Osmólska (1974), including dorsal and lateral views of *Stegoceras* (UA2), *Homalocephale* (GI SPS 100/51) and *Prenocephale* (ZPAL MgD-I/104) (Figs. 2.15, 2.16). Unfortunately, no specimen of *Pachycephalosaurus* could be used due to taphonomic damage and extensive fusion of the cranial bones resulting in difficulties in identifying landmarks. For comparison, an illustration of the cranium of *Hypsilophodon* (BMNH R197 and R2477) was taken from Galton (1973) and redrawn for Figs. 2.15 and 2.16. *Hypsilophodon* was chosen because it represents an early euornithopod (*sensu* Sereno 1986), and provides a relatively unspecialized representative of the group referred to as the Cerapoda by Sereno (1986), which includes the pachycephalosaurs and most other ornithischians. *Hypsilophodon* also was used by Sues (1978), who applied D'Arcy Thompson's grid transformation method to analyze differences between that taxon and *Stegoceras*. The results, as would be expected, are quite similar, although changes, such as the reduction in relative orbit size, are much more easily observed in the RFTRA results (see Fig. 2.24 and compare with Sues 1978, Fig. 5).

Two analyses were made of dorsal views of the crania. The first used the more complete skulls, including *Goyocephale* (Fig. 2.15A), *Stegoceras* (Fig. 2.15B), *Prenocephale* (Fig. 2.15D), and *Hypsilophodon* (Fig. 2.15E). The 23 landmarks were chosen as with the other examples and avoided the missing premaxilla on the skull of *Goyocephale*. The landmarks and polygonal figure used are illustrated in Figures 2.15F and G, respectively. The second dorsal analysis used only *Goyocephale* and *Homalocephale* in an attempt to document similarities in these two members of the Homalocephalidae (*sensu* Sues and Galton 1987). The 20 landmarks and polygonal figure used for this comparison are illustrated in Figures 2.15H and I, respectively.

The results of the latter analysis are illustrated in Figure 2.17 and support the close relationship between *Homalocephale* and *Goyocephale* suggested by Sues and Galton (1987). The differences between the two forms are minor despite a high degree of distortion evident as cranial asymmetry, either of taphonomic or biological origin. The major difference is in the relatively smaller supratemporal fenestrae in *Homalocephale*.

Figure 2.16. Pachycephalosaurids, lateral cranial views. Included are reconstructions of **A**, *Tylocephale*; **B**, *Stegoceras*; **C**, *Homalocephale*; **D**, *Prenocephale*; **E**, *Hypsilophodon*; and figures illustrating **F**, 18 landmarks and **G**, polygonal figure for complete set of landmarks; and **H**, 16 landmarks and **I**, polygonal figure for abbreviated set.

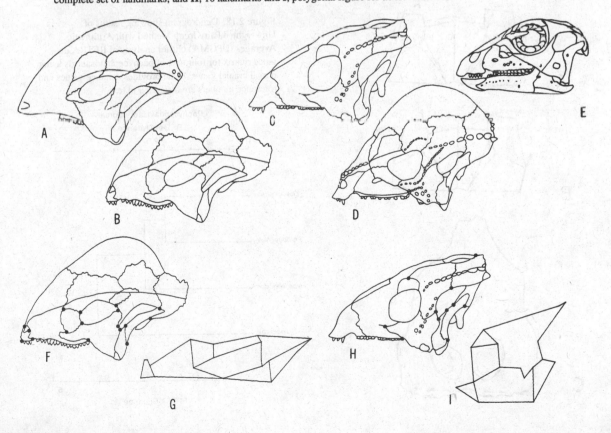

Given the use of single specimens and large amounts of asymmetry, this supports a very high degree of similarity between these two taxa.

The results of the more detailed dorsal analyses are illustrated as a dendrogram in Figure 2.18, and in standard RFTRA graphics in Figures 2.19–2.21. The dendrogram shows that *Prenocephale* (reference number 3003) and *Goyocephale* (3001) have the most similar overall cranial shapes (RFTRA distance = 2.2176), joined next by *Hypsilophodon* (3002) and, finally, by *Stegoceras* (3004). Comparison of *Hypsilophodon* with *Goyocephale* (Fig. 2.19) gives results indicative of those obtained by comparing all the pachycephalosaurians with *Hypsilophodon*: the supratemporal fenestrae show marked reduction, there is an expansion of the fronto-parietal regions, and the cranium in general is wider in the pachycephalosaurians. The latter reflects the change from a laterally compressed cranium to one more dorsoventrally compressed for combat. The other features are characteristic of the Pachycephalosauria in general (e.g., Sereno 1986; Sues and Galton 1987). As expected, the

nearest-neighbor (the form most similar) to *Hypsilophodon* is *Goyocephale* (distance = 2.6929). However, *Prenocephale* is only slightly more different (distance = 2.7963) and *Stegoceras* is quite distant (4.315). In fact, the results of most analyses show that *Stegoceras* has the most unusual cranial morphology. A comparison of *Goyocephale* with *Prenocephale* (Fig. 2.20) shows mostly an extension of the trends observed from *Hypsilophodon* to *Goyocephale*. Comparison of *Prenocephale* with *Stegoceras* (distance = 3.1521), however, shows that the major difference between these two forms is a greater parietal expansion of the dome in *Prenocephale* accompanied by closing of the supratemporal fenestrae, and an elongation of the frontal region of the dome in *Stegoceras*.

Two sets of lateral analyses also were run to accommodate relatively incomplete material. The first utilized 18 landmarks and was run only on *Hypsilophodon*, *Prenocephale*, and *Stegoceras* (Figs. 2.22–2.24). *Stegoceras* is again the specimen most distant from the others based solely on landmarks. It exhibited distances of 3.3029 and 2.7948 with *Hypsilophodon* and *Prenocephale*, respectively. Those two taxa show a RFTRA difference between them of only 1.7807. However, the landmarks selected did not include any points that reflect the expansion of the dome because of difficulties in identifying homologous points. The outline figures, however, suggest a much closer overall similarity between *Stegoceras* and *Prenocephale*. Major differences

Figure 2.17. Results of Resistant-Fit Theta-Rho-Analysis of cranial dorsal views of *Goyocephale* (3101, base) and *Homalocephale* (3102) using abbreviated set of dorsal landmarks. A, diagram of superimposed polygonal figures; B, vector diagram; and C, superimposed outline figures. Distance coefficient is 1.7151.

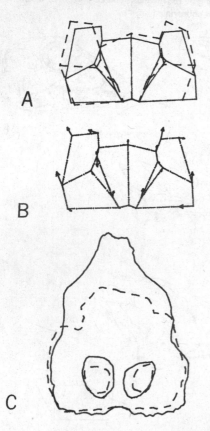

A

B

C

Figure 2.18. Dendrogram showing results of Unweighted Pair-Group Method with Arithmetic Averages (UPGMA) cluster analysis of RFTRA distance matrix for four taxa of pachycephalosaurids using dorsal cranial views. Scale provides distance values and reference numbers are discussed in text.

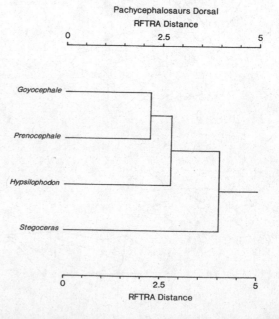

Pachycephalosaurs Dorsal
RFTRA Distance

0   2.5   5

Goyocephale

Prenocephale

Hypsilophodon

Stegoceras

0   2.5   5
RFTRA Distance

between those two taxa reflect the differences mentioned for the dorsal view, and the development of the squamosal shelf in *Stegoceras*.

    The more abbreviated analysis utilized only 16 landmarks that were concentrated toward the posterior end of the cranium. This was run using *Prenocephale* (reference number 3302), *Stegoceras* (3303), *Homalocephale* (3301), and *Tylocephale* (3304). The dendrogram (Fig. 2.25) exhibits two clusters, one with *Prenocephale* and *Homalocephale*, and the other with *Stegoceras* and *Tylocephale*. This is partially misleading because *Tylocephale* showed its greatest similarity with *Prenocephale*, but the latter taxon's even higher similarity with *Homalocephale* resulted in the pattern evident in Figure 2.25. The results of two individual analyses are illustrated for *Prenocephale* and *Homalocephale* in Figure 2.26, and for *Stegoceras* and *Tylocephale* in Figure 2.27. The differences between the latter two taxa

are along the same lines as those between *Stegoceras* and *Prenocephale*, reflecting an intermediate position for *Tylocephale*. *Prenocephale* and *Homalocephale* show significant differences in the outline figures not expressed in landmarks, directly related to the expansion of the dome.

    The results of applying RFTRA to the pachycephalosaurids demonstrate the problems and successes to be expected in such an analysis. The high degree of damage and asymmetry in the specimens, combined with a high degree of variability characteristic of the group (Chapman et al. 1981), and the small number of specimens, makes this a difficult application for any quantitative method. However, the results do provide important insights into the taxa studied. The relatively small distance coefficients observed between *Hypsilophodon* and either *Homalocephale* or *Prenocephale*

Figure 2.19. Results of Resistant-Fit Theta-Rho-Analysis of dorsal cranial views for *Hypsilophodon* (3002, base) and *Goyocephale* (3001). A, diagram of superimposed polygonal figures; B, vector diagram; and C, superimposed outline figures. Distance coefficient is 2.6929.

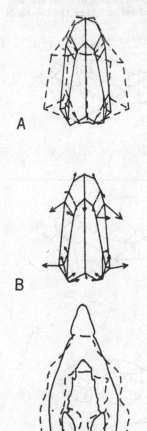

Figure 2.20. Results of Resistant-Fit Theta-Rho-Analysis of dorsal cranial views for *Goyocephale* (3001, base) and *Prenocephale* (3003). A, diagram of superimposed polygonal figures; B, vector diagram; and C, superimposed outline figures. Distance coefficient is 2.2176.

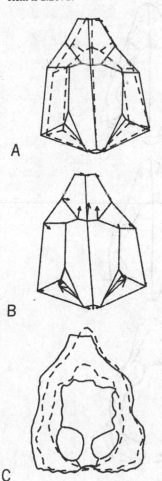

supports the position that pachycephalosaurids are spe-cialized (in the development of the dome) modifications of a basal cerapodan morphology. The high similarity apparent between *Homalocephale* and *Goyocephale* provides support for the Homalocephalidae as discussed in Perle et al. (1982) and Sues and Galton (1987). The biggest surprise is the apparent unusual morphology of *Stegoceras*, even for the pachycephalosaurids, and its possible position as the most modified member of that group. The degree of difference in *Stegoceras* is not as distinctive as that found for *Parasaurolophus* in the analysis of hadrosaurs by Chapman and Brett-Surman (this volume), but still suggests that detailed analyses, especially of *Pachycephalosaurus*, be attempted. The intermediate position of *Tylocephale* between *Stegoceras* and *Prenocephale* provides partial support for cladistic grouping of *Tylocephale* with *Stegoceras* suggested by

Sues and Galton (1987). But the material limits any pos-sibility for evaluating that problem with confidence.

The analyses also reflect the difficulties encoun-tered in studying groups like the pachycephalosaurians with little material and high variability. The fit of *Stegoceras* with *Prenocephale* was variable in different views (Figs. 2.21, 2.23), and the final interpretation had to take both results into consideration. This happens rarely among other taxa studied. The RFTRA algorithm provides best results when change is concentrated locally and peripherally, as is the case frequently with transitions of interest to paleobiologists. The central location of the change, in the frontal and parietal areas, in this example, provided a worse-case application that still provided insight given a pragmatic approach and different analyses. Similarly, the last set of analyses were more variable due to the concentration of land-marks to the posterior. Without any points anchoring the snout, the fit often can rotate around these posterior points to a small degree. This was evident in analyses

Figure 2.21. Results of Resistant-Fit Theta-Rho-Analysis of dorsal cranial views for *Prenocephale* (3003) and *Stegoceras* (3004, base). **A**, diagram of superimposed polygonal figures; **B**, vector diagram; and **C**, superimposed outline figures. Distance coeffi-cient is 3.1521.

Figure 2.22. Results of Resistant-Fit Theta-Rho-Analysis of lateral cranial views for *Hypsilophodon* (3201, base) and *Prenocephale* (3202). **A**, diagram of superimposed polygonal figures; **B**, vector diagram; and **C**, superimposed outline figures. Distance coeffi-cient is 1.7807.

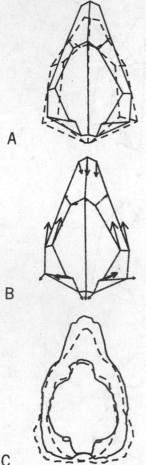

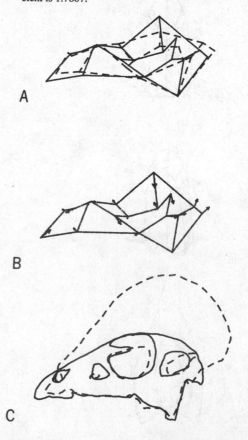

that overlapped with others using the more uniform distribution of landmarks. In general, RFTRA provided important insights despite this worse-case scenario.

Conclusions

The examples presented demonstrate the usefulness of Resistant-Fit Theta-Rho-Analysis for analyzing morphological change or variation in dinosaurs. The examples concentrated on cranial studies but the method is applicable to any elements where landmarks are identifiable (Chapman and Brett-Surman, this volume, for example). The different forms of graphical output available provide a variety of information for interpretation. The polygons and vectors supply direct landmark information and the figures demonstrate variability in additional structures. RFTRA is ideal for the analysis of ontogenetic growth or sexual dimorphism, as is demonstrated in the examples for *Bagaceratops* and *Protoceratops*, respectively. Furthermore, the analysis of *Plateosaurus* demonstrates capabilities for measuring taphonomic and biological asymmetry, and indicates that RFTRA will be

useful for helping evaluate the quality and accuracy of drawings and reconstructions.

Phylogenetic studies can be more difficult because of added problems in the selection of common landmarks between different taxa. However, RFTRA was effective in providing information on patterns of morphological variability in the application by Chapman and Brett-Surman (this volume) on hadrosaur crania and pelves. Significant insight was also gained in the much more difficult analysis of pachycephalosaur crania discussed above.

No single morphometric method can answer all questions. I suggest that RFTRA be considered by a researcher studying any relevant problem, but for a full analysis of a taxon, RFTRA should be only one of a number of applied methods. The allometric and multivariate morphometric approaches used by Dodson (1975a,b,c, 1976) and Chapman et al. (1981) certainly have provided and will continue to provide information that is valuable in itself and as support for projects using

Figure 2.23. Results of Resistant-Fit Theta-Rho-Analysis of lateral cranial views for *Prenocephale* (3202) and *Stegoceras* (3203, base). **A**, diagram of superimposed polygonal figures; **B**, vector diagram; and **C**, superimposed outline figures. Distance coefficient is 2.6405.

Figure 2.24. Results of Resistant-Fit Theta-Rho-Analysis of lateral cranial views for *Hypsilophodon* (3201, base) and *Stegoceras* (3203). **A**, diagram of superimposed polygonal figures; **B**, vector diagram; and **C**, superimposed outline figures. Distance coefficient is 3.3029.

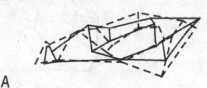

A

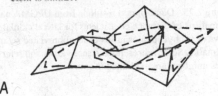

A

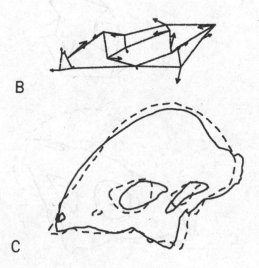

B

C

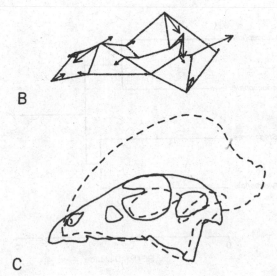

B

C

a wide range of methods. Bookstein et al. (1985) recently have discussed methods (e.g., sheared principal components analysis) for providing a more conventional analysis of shape, independent of confounding size factors.

Other developments in landmark shape analysis, specifically the tensor analysis methods discussed by Bookstein (1978) and Bookstein et al. (1985), should be applied when surficial growth and allometric fields and gradients are of interest and homologous points (*sensu* Bookstein et al. 1985) are available. Similarly, the "shape coordinates" method developed by Bookstein (1986) should be attempted in a wide range of applications.

Outline methods probably only will have limited applicability in dinosaur morphometrics because outline information typically is contained in the results of a tensor analysis or RFTRA. Furthermore, many of these techniques utilize a limited number of (or no) landmarks, and would provide less effective or potentially incorrect information in many cases. Typically, in dinosaur studies, sufficient landmarks should be available for the use of the landmark methods, which are more powerful when used correctly.

I suggest that recent developments in technology and morphometrics should make a work station with shape analysis capabilities a standard piece of laboratory equipment for dinosaur study and preparation. The

cost of such a system no longer is prohibitive, being less than that of standard word-processing workstations three or more years ago, and continuing advancements will reduce costs further while increasing their sophistication and usefulness. As imaging capabilities evolve and become more accessible, related approaches will enhance significantly the speed and efficiency of our preparation methods and may soon become a great aid in exploration itself. The combination of shape analysis procedures, such as those discussed herein, and artificial intelligence should provide a mechanism for the development of expert systems for automated assistance in specimen identification and new possibilities in the study of the functional morphology of dinosaurs.

Acknowledgments

Many individuals provided thought-provoking and interesting feedback during the course of this study. Mike

Figure 2.25. Dendrogram resulting from UPGMA cluster analysis of distance coefficients for lateral cranial views of pachycephalosaurids using abbreviated set of landmarks. Scale represents distance values and references numbers are explained in text.

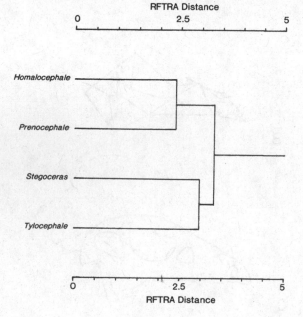

Figure 2.26. Results of Resistant-Fit Theta-Rho-Analysis of lateral cranial views using abbreviated number of landmarks for *Homalocephale* (3301, base) and *Prenocephale* (3302). **A**, diagram of superimposed polygonal figures; **B**, vector diagram; and **C**, superimposed outline figures. Distance coefficient is 2.3885.

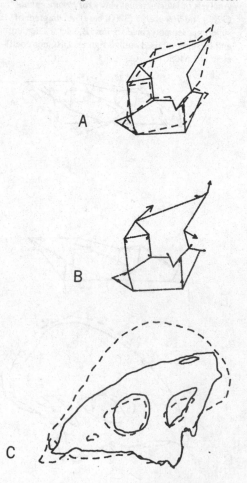

Brett-Surman and Linda Deck provided important input at all stages. Discussions with Hans-Dieter Sues over the past six months were much appreciated. Other helpful conversations occurred with Richard Thorington, Lee-Ann Hayek, Nick Hotton, and Bill DiMichele of the National Museum of Natural History, Peter Galton of Bridgeport, Mark Goodman of Berkeley, and many of the participants in the Dinosaur Systematics Conference held in Drumheller during June, 1986. I thank Jennifer Clark for providing the illustrations. I also appreciate the editing support and organizational abilities of Phil Currie and Kenneth Carpenter. My morphometric knowledge benefitted greatly from past contacts with Jack Sepkoski, Fred Bookstein, Alan Cheetham, Andy Siegel, and, especially, Richard Benson. Finally, Peter Dodson was the first to analyze dinosaurs morphometrically using modern approaches and provided me with inspiration and helpful constructive criticism during my first attempts to study dinosaur morphology (Chapman et al. 1981). I dedicate to him any usefulness this paper may have for dinosaur morphometry.

Figure 2.27. Results of Resistant-Fit Theta-Rho-Analysis of lateral cranial views using abbreviated number of landmarks for *Tylocephale* (3304) and *Stegoceras* (3303, base). **A**, diagram of superimposed polygonal figures; **B**, vector diagram; and **C**, superimposed outline figures. Distance coefficient is 2.9518.

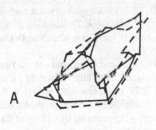

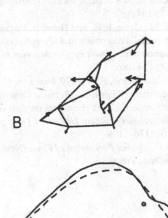

References

Benson, R. H., Chapman, R. E., and Siegel, A. F. 1982. On the measurement of morphology and its change. *Paleobiology* 8(4):328–339.

Bookstein, F. L. 1978. *The Measurement of Biological Shape and Shape Change* (New York: Springer-Verlag).

1986. Size and shape spaces for landmark data in two dimensions. *Statistical Science* 1(2):181–242.

Bookstein, F. L., Chernoff, B., Elder, R., Humphries, J., Smith, G., and Strauss, R. 1985. Morphometrics in evolutionary biology. *Academy of Natural Sciences, Philadelphia, Special Publications* 15:1–277.

Bradu, D., and Grine, F. E. 1979. Multivariate analysis of diademodontine crania from South Africa and Zambia. *South African Journal of Science* 75:441–448.

Brown, B., and Schlaikjer, E. M. 1940. The structure and relationships of *Protoceratops*. *New York Academy of Science Annals* 40:133–266.

1943. A study of the troodont dinosaurs with the description of a new genus and four new species. *American Museum Natural History Bulletin* 82:115–150.

Chapman, R. E., Galton, P. M., Sepkoski, J. J., Jr., and Wall, W. P. 1981. A morphometric study of the cranium of the pachycephalosaurid dinosaur *Stegoceras*. *Journal of Paleontology* 55(3):608–618.

Clark, M. W. 1981. Quantitative shape analysis: a review. *Mathematical Geology* 13(4):85–92.

Colbert, E. H. 1948. Evolution of the horned dinosaurs. *Evolution* 2(2):145–163.

Currie, P. J. 1986. Dinosaur fauna. *In* Naylor, B. G. (ed.), *Fieldtrip Guidebook to Dinosaur Provincial Park, Dinosaur Systematics Symposium, Tyrrell Museum of Palaeontology, Drumheller, Alberta, Canada* (Drumheller: Tyrrell Museum of Palaeontology).

Davitashvili, L. A. 1961. *The Theory of Sexual Selection*, (Moscow: Izd-vo Akademia Nauk SSSR).

Dodson, P. 1975a. Functional and ecological significance of relative growth in *Alligator*. *Journal of Zoology* 175:315–355.

1975b. Relative growth in two sympatric species of *Sceloporus*. *American Midland Naturalist* 94:121–150.

1975c. Taxonomic implications of relative growth in lambeosaurine hadrosaurs. *Systematic Zoology* 24(1):37–54.

1976. Quantitative aspects of relative growth and sexual dimorphism in *Protoceratops*. *Journal of Paleontology* 50(5):929–940.

Galton, P. M. 1970. Pachycephalosaurids – dinosaurian battering rams. *Discovery*, Yale University 6:23–32.

1971. A primitive dome-headed dinosaur (Reptilia: Pachycephalosauridae) from the Lower Cretaceous of England and the function of the dome in pachycephalosaurids. *Journal of Paleontology* 45(1):40–47.

1973. Redescription of the skull and mandible of *Parkosaurus* from the Late Cretaceous with comments on the family Hypsilophodontidae (Ornithischia). *Royal Ontario Museum, Life Sciences Contributions* 89:1–29.

1985. Cranial anatomy of the prosauropod dinosaur *Plateosaurus* from the Knollenmergel (Middle Keuper, Upper Triassic) of Germany. *Geologica et Palaeontologica* 19:119–159.

Goodall, C. R., and Green, P. B. 1986. Quantitative analysis of surface growth. *Botanical Gazette* 147(1):1–15.

Gould, S. J. 1966. Allometry and size in ontogeny and phylogeny. *Biological Reviews* 41:587–640.

1967. Evolutionary patterns in pelycosaurian reptiles: a factor analytic model. *Evolution* 21:385–401.

Gould, S. J., and Littlejohn, J. 1973. Factor analysis of caseid pelycosaurs. *Journal of Paleontology* 47:886–891.

Granger, W., and Gregory, W. K. 1923. *Protoceratops andrewsi*, a pre-ceratopsian dinosaur from Mongolia. *American Museum Novitates* 72:1–6.

Gray, S. W. 1946. Relative growth in a phylogenetic series and in an ontogenetic series of one of its members. *American Journal of Science* 244:792–807.

Gregory, W. K., and Mook, C. C. 1925. On *Protoceratops*, a primitive ceratopsian dinosaur from the Lower Cretaceous of Mongolia. *American Museum Novitates* 156:1–9.

Grine, F. E., Hahn, B. D., and Gow, C. E. 1978. Aspects of relative growth and variability in *Diademodon* (Reptilia: Therapsida). *South African Journal of Science* 74:50–58.

Huxley, J. S. 1932. *Problems of Relative Growth*. (New York: Dover Publications).

Jolicoeur, P., and Mosimann, J. E. 1960. Size and shape variation in the painted turtle, a principal components analysis. *Growth* 24:339–354.

Kurzanov, S. M. 1972. Sexual dimorphism in protoceratopsians. *Paleontological Journal* 1972:91–97.

Lambe, L. M. 1918. The Cretaceous genus *Stegoceras*, typifying a new family referred provisionally to the Stegosauria. *Royal Society of Canada Transactions*, series 3, 12(4):23–36.

Lohmann, G. P. 1983. Eigenshape analysis of microfossils: a general morphometric procedure for describing changes in shape. *Mathematical Geology* 15(6):659–672.

Lull, R. S., and Gray, S. W. 1949. Growth patterns in the Ceratopsia. *American Journal of Science* 247:492–503.

Maryańska, T., and Osmólska, H. 1974. Pachycephalosauria, a new suborder of ornithischian dinosaurs. *Palaeontologia Polonica* 30:45–102.

1975. Protoceratopsidae (Dinosauria) of Asia. *Palaeontologia Polonica* 33:133–181.

Olson, E. C., and Miller, R. L. 1951. Relative growth in paleontological studies. *Journal of Paleontology* 25:212–223.

Ostrom, J. H. 1966. Functional morphology and evolution of the ceratopsian dinosaurs. *Evolution* 20(3):290–308.

Patterson, C. 1982. Morphological characters and homology. *In* Joysey, K. A. and Friday, A. E. (eds.), *Problems of Phylogenetic Reconstruction*. Systematics Association Special Volume 21 (London: Academic Press).

Perle, A., Maryańska, T., and Osmólska, H. 1982. *Goyocephale lattimorei* gen. et sp. n., a new flat-headed pachycephalosaur (Ornithischia, Dinosauria) from the Upper Cretaceous of Mongolia. *Acta Palaeontologica Polonica* 27(1/4):115–127.

Reyment, R. A., Blackith, R. E., and Campbell, N. A. 1984. *Multivariate Morphometrics*, 2nd edn. (New York: Academic Press).

Rohlf, F. J. 1986. Relationships among Eigenshape Analysis, Fourier Analysis, and analysis of coordinates. *Mathematical Geology* 18(8):845–854.

Romer, A. S. 1956. *Osteology of the Reptiles*. (Chicago: University of Chicago Press).

Rozhdestvensky, A. K. 1965. Growth problems in some Asian dinosaurs and some problems of their taxonomy. *Paleontologicheskiy Zhurnal* 13:95–109. [in Russian]

Sereno, P. C. 1986. Phylogeny of the bird-hipped dinosaurs (Order Ornithischia). *National Geographic Research* 2(2):234–256.

Siegel, A. F. 1982. Geometric data analysis: an interactive graphics program for shape comparison. *In* Launer, R. L., and Siegel, A. F. (eds.), *Modern Data Analysis* (New York: Academic Press).

Siegel, A. F., and Benson, R. H. 1982. A robust comparison of biological shapes. *Biometrics* 38:341–350.

Sneath, P. H. A. 1967. Trend-surface analysis of transformation grids. *Journal of Zoology* 151:65–122.

Sues, H.-D. 1978. Functional morphology of the dome in pachycephalosaurid dinosaurs. *Neues Jahrbuch für Geologie und Paläontologie, Monatshefte* 1978(8):459–472.

Sues, H.-D., and Galton, P. M. 1987. Anatomy and classification of the North American Pachycephalosauria (Dinosauria: Ornithischia). *Palaeontographica* A, 98:1–40.

Thompson, D. W. 1915. Morphology and mathematics. *Royal Society of Edinburgh, Transactions* 50:857–895.

1917. *On Growth and Form* (Cambridge: Cambridge University Press).

1942. *On Growth and Form*, 2nd edn., in two volumes. (Cambridge: Cambridge University Press).

Tobler, W. R. 1977. Bidimensional regression. A computer program. Published by the author. (Santa Barbara: Geography Department, University of California).

1978. Comparison of plane forms. *Geographic Analysis* 10(2):154–162.

Tollman, S. M., Grine, F. E., and Hahn, B. D. 1980. Ontogeny and sexual dimorphism in *Aulacephalodon* (Reptilia, Anomodontia). *South African Museum Annals* 81(4):159–186.

Wall, W. P., and Galton, P. M. 1979. Notes on the pachycephalosaurid dinosaurs (Reptilia: Ornithischia) from North America, with comments on their status as ornithopods. *Canadian Journal of Earth Sciences* 16(6):1176–1186.

Zusne, L. 1970. *Visual Perception of Form* (New York: Academic Press).

II Sauropodomorpha

3 Morphometric study of *Plateosaurus* from Trossingen (Baden–Württemberg, Federal Republic of Germany)

DAVID B. WEISHAMPEL AND
RALPH E. CHAPMAN

Abstract

Eleven femoral measurements on 33 specimens of European anchisaurid taxa are analyzed to gain insight into the taxonomic structure of the genus *Plateosaurus*. Principal Components Analysis indicates that the taxon as a whole is fairly homogeneous with respect to gross femoral dimensions. Within the general pool of material, two morphs may be indicated on the second Principal Component. These morphs, also analyzed using bivariate analyses, are discernable in relative dimensions of the proximal and distal femoral articulations, and the size and shape of the fourth trochanter. The two morphs may be sexual in nature, implying that locomotory regimes may have been different between males and females.

Introduction

Plateosaurus has long been regarded as one of the best known taxa among the earliest dinosaurs. *Plateosaurus* material is known from Upper Triassic deposits in the Federal Republic of Germany, the German Democratic Republic, Switzerland, and France. It ranges in quality from isolated bones and teeth to mass accumulations often comprising an outstanding number of complete individuals. Most localities are clustered along the Neckartal in Baden–Württemberg and Pegnitztal of Bavaria, both of the Federal Republic of Germany, although other sites are found along the Franche Comte in eastern France and in the northern Harz Mountains along the border between the German Democratic Republic and the Federal Republic of Germany.

One of the first dinosaurs to be named (Meyer 1837), *Plateosaurus* has been the subject of several monographic treatments of anatomy and taxonomy (Huene 1907/08, 1926, 1932), as well as shorter treatments (Plieninger 1850; Pidancet and Chopard 1862). Most recently, Galton (1984, 1985a) has redescribed

In Dinosaur Systematics: Perspectives and Approaches, *Kenneth Carpenter and Philip J. Currie, eds. Copyright © Cambridge University Press, 1990.*

much of the existing cranial material pertaining to the genus. Because of the occurrence of several mass accumulations, taphonomic and paleoecologic studies of this Late Triassic plateosaurid also have been made (Jaekel 1913; Huene 1928; Seeman 1933; Weishampel 1984b; Sander personal communication). Lastly, it is well known that *Plateosaurus* and the other prosauropods represent the earliest radiation of large-bodied herbivores, exploiting high browsing as an important aspect of their feeding (Bakker 1978; Galton 1984, 1985b; Weishampel 1984a).

Despite these "honorable mentions", *Plateosaurus* has the dubious distinction of having more referred species and synonymized genera and species attached to it than any other dinosaurian genus. These referrals and synonymies are listed in Table 3.1. A number of taxa based on isolated teeth are no longer referred to *Plateosaurus* and/or other plateosaurids (Galton 1985a), while the majority of skeletal remains have been synonymized with *Plateosaurus engelhardti* Meyer 1837 (Galton 1984, 1985a). Galton's sweeping revision of plateosaurid prosauropods is the first attempt to make biologic sense of the quagmire of prosauropod taxa from the Late Triassic of Europe, but the synonymies themselves consist only of a few remarks found within the descriptive portions of his papers. The most problematic issue in Galton's taxonomy is that his synonymy of taxa is based on incongruent material. That is, Galton's discussions are based on a detailed restudy of the previously published cranial material of *Plateosaurus*. However, the species to which he refers this material, *P. engelhardti*, originally consisted of isolated vertebrae, tibia, and a fragmentary femur without any associated skull or even fragmentary cranial material.

To bring the postcranial elements of *P. engelhardti* (and the holotype specimens of other European plateosaurid species) under the perview of Galton's revision, we have undertaken a two-fold morphometric study. First, we attempt to assess skeletal variability within the plateosaurid material collected from Trossingen (Baden–Württemberg, Federal Republic of Germany). See Huene (1923, 1929), Seemann (1932a, 1932b, 1933, 1941), and Weishampel (1984b) for the history of collecting at Trossingen. This site has yielded the single richest assemblage of European plateosaurid

material, and it has been argued (Weishampel 1984b, in preparation) on taphonomic grounds that the assemblage represents a monospecific assemblage of an unspecified species of *Plateosaurus*. All of this material comes from the Knollenmergel (Upper Triassic; Norian in age).

Second, we incorporate data obtained from holotypic plateosaurid specimens from other European

Table 3.1. *Partial history of the taxonomy of* Plateosaurus *and allied anchisaurids from the European continent*

Original taxon	Taxonomic history
1830–1900	
Plateosaurus engelhardti Meyer, 1837	holotype for the genus *Plateosaurus*
Smilodon crenatus Plieninger, 1846	referred to *Zanclodon crenatus* (Huene 1907/08)
	removed from Prosauropoda Galton (1985a)
Gresslyosaurus ingens Rutimeyer, 1856	referred to *Plateosaurus engelhardti* (Galton 1985a)
Megalosaurus cloacinus Quenstedt 1858	referred to *Plateosaurus cloacinus* (Huene 1907/08)
	referred to *Gresslyosurus cloacinus* (Huene 1932)
	removed from Prosauropoda Galton (1985a)
Dimodosaurus poligniensis Pidancet and Chopard, 1862	referred to *Plateosaurus poligniensis* (Huene 1907/08)
Megalosaurus obtusus Henry 1876	referred to *Plateosaurus obtusus* (Huene 1907/08)
	removed from Prosauropoda (Galton 1985a)
Smilodon laevis Plieninger 1846	referred to *Zanclodon lavis* (Plieninger 1847)
	referred to *Zanclodon plieningeri* (Fraas 1896)
	referred to *Zanclodon quenstedti partim* (Koken 1900)
	referred to *Plateosaurus quenstedti partim* (Huene 1932)
1900–1930	
Zanclodon schutzii Fraas 1900	referred to *Teratosaurus schutzii* (Huene 1932)
Thecodontosaurus elizae Sauvage 1907	referred to *Plateosaurus elizae* (Huene 1907/08)
	taken out of Prosauropoda (Galton 1985a)
Gresslyosaurus robustus Huene 1907/08	referred to *Plateosaurus robustus* (Huene 1932)
	referred to *Plateosaurus engelhardti* (Galton 1985a)
Pachysaurus ajax Huene 1907/08	referred to *Pachysauriscus ajax* (Kuhn 1959)
	referred to *Gresslyosaurus ajax* (Steel 1970)
	referred to *Plateosaurus engelhardti* (Galton 1985a)
Pachysaurus magnus Huene 1905b	referred to *Pachysauriscus magnus* (Kuhn 1959)
	referred to *Gresslyosaurus magnus* (Steel 1970)
	referred to *Plateosaurus engelhardti* (Galton 1985a)
Plateosaurus erlenbergiensis Huene 1905b	referred to *Plateosaurus engelhardti* (Galton 1985a)
Plateosaurus quenstedti Huene 1905a	referred to *Plateosaurus engelhardti* (Galton 1985a)
Plateosaurus ornatus Huene 1905b	
Gresslyosaurus plieningeri Huene 1905b	referred to *Plateosaurus plieningeri* (Huene 1932)
	referred to *Plateosaurus engelhardti* (Galton 1985a)
Plateosaurus reiningeri Huene 1905a	referred to *Gresslyosaurus reiningeri* (Huene 1926)
	referred to *Pachysaurus reiningeri* (Huene 1932)
	referred to *Pachysauriscus reiningeri* (Kuhn 1959)
	referred to *Plateosaurus engelhardti* (Galton 1985a)
Gresslyosaurus torgeri Jaekel 1911	referred to *Plateosaurus plieningeri* (Huene 1932)
Plateosaurus longiceps Jaekel 1913	referred to *Plateosaurus quenstedti* (Huene 1932)
	referred to *Plateosaurus engelhardti* (Galton 1985a)
Plateosaurus trossingensis Fraas 1913	renamed *Plateosaurus fraasianus* (Huene 1932)
	referred to *Plateosaurus engelhardti* (Galton 1985a)
1930–1940	
Pachysaurus giganteus Huene 1932	referred to *Pachysauriscus giganteus* (Kuhn 1959)
	referred to *Gresslyosaurus giganteus* (Steel 1970)
Pachysaurus wetzelianus Huene 1932	referred to *Pachysauriscus wetzelianus* (Kuhn 1959)
	referred to *Gresslyosaurus wetzelianus* (Steel 1970)

localities with those from Trossingen in order to understand the morphometric properties of European plateosaurids as a whole. Most of these localities come from the Knollenmergel or equivalents, but some (those yielding the holotypes of *Sellosaurus gracilis*, *S. fraasi* and *Teratosaurus minor*) are found in the subjacent Stubensandstein. Trends in both quarry and regional analyses are discussed in terms of sexual dimorphism, taphonomic biases, and species-specific differences among specimens. In the former analyses, we are successful in documenting relative uniformity of specimens with respect to population variability. In the latter, we are less successful, but point to a variety of approaches that will enable a better understanding of the relationship among *Plateosaurus* species within the area.

Our study and that of Chapman and Brett-Surman (this volume) represent the first attempts at applying morphometric analyses to the study of dinosaur postcranial material. Such an approach stems from the burgeoning field of dinosaurian morphometrics, which is discussed in detail by Chapman (this volume); we refer readers to that paper for a general review. Papers most relevant to this study include an application of Resistant-Fit Theta-Rho-Analysis, a form of landmark shape analysis, to the study of cranial asymmetry in *Plateosaurus*, and the allometric, bivariate and multivariate studies of lambeosaurine hadrosaurids and the ceratopsian *Protoceratops* by Dodson (1975, 1976), and of cranial variation in the pachycephalosaurid *Stegoceras* by Chapman et al. (1981).

Material and methods of analysis

The morphometric analyses presented herein were applied to provide insight into the number of prosauropod taxa that are represented in the specimens collected from the Trossingen quarry (Huene 1932; Weishampel 1984b). The data pertain to femoral measurements illustrated in Figure 3.1. We have chosen data from the femur because it is one of the best represented elements in the collections of Trossingen material, there are many landmarks that are measurable with confidence (Fig. 3.1), and virtually all the plateosaurid type material found elsewhere in Europe includes at least half of a femur, if not the entire element. Individual data related to gross size (ex: FEMURL, FDIAM), muscle scar dimensions (ex: CDFEMBL, FMTBEXTL), and approximate lever arms acting around the hip joint (ex: HIPTROCH).

All data were collected using dial callipers (± 0.5 mm). Data were collected for the 33 Trossingen quarry specimens listed in Table 3.2 and accessory data were obtained for 11 holotype specimens of plateosaurid taxa collected from elsewhere in Europe (Table 3.3). Each specimen is given a reference letter in either upper or lower case that will be used in the discussions herein. Included in Table 3.2 are the reference letters, museum identification information, current taxonomic designa-

tions, our interpretation of the taxonomic position of each specimen, and an indication of the techniques we used to come to our decision.

Three general groups of analyses were performed. The first used all 44 specimens, the second used only the 33 quarry specimens, and the final used a subset of 22 quarry specimens that provided complete data for use in the Principal Components Analysis (PCA); these specimens are indicated in Table 3.2.

The data available for the prosauropod material from the quarry at Trossingen is typical of that available to most vertebrate paleontologists in that there are a significant number of missing values in the data matrix, and that there is a relatively high degree of variability present, apparently the result of taphonomic deformation and breakage. To accommodate such an incomplete and highly variable data set, we decided to use only well-established morphometric methods that have been successfully applied previously to this type of data (e.g.,

Figure 3.1. a. Caudal view of left *Plateosaurus* femur. b. Medial view of same femur. Measurements indicated by reference number. 1: FEMURL (femur length), 2: FPROXW (proximal femoral width), 3: FPROXB (proximal femoral breadth), 4: FDISTW (distal femoral width), 5: FDISTB (distal femoral breadth), 6: HIPTROC (approximate lever arm between hip joint and 4th trochanter), 7: FDIAM (femoral diameter), 8: CDFEMBL (length of M. caudifemoralis brevis attachment site), 9: CDFEMLH (height of M. caudifemoralis longus attachment site), 10: CDFEMLW (width of the M. caudifemoralis longus attachment site), 11: FMTBEXTL (approximate length of the M. femorotibioexternus attachment site). Scale bar = 10 cm.

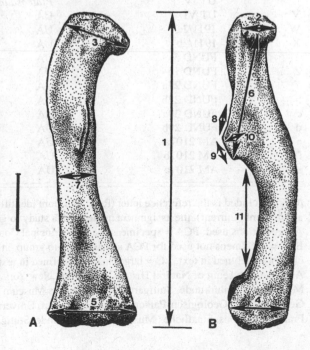

univariate and bivariate methods, Principal Components Analysis) rather than more complex methods (e.g., factor analysis, multidimensional scaling) that might further complicate interpretation. With problematic data, it is easier to interpret results if the behavior and biases of the morphometric techniques are well-known. In addition, the use of the bivariate analyses and PCA makes the results more comparable with those of Chapman et al. (1981), who studied a similar problem for *Stegoceras*.

The first step was to transform the raw measurements to logarithms (base 10) for use in the bivariate and multivariate analyses. The reasons for such a step are well established in morphometrics and have been discussed in detail by Sokal and Rohlf (1969) and Bookstein et al. (1985). Simply, logarithmic data tend to provide more useful results in bivariate analyses, reduce problems associated with deviations from homoscedasticity, and more realistically mirror biological growth processes (see Bookstein et al. 1985).

Table 3.2. *Specimens from the Trossingen Quarry used for the morphometric analysis*

Ref.	Museum No.	Current	This	Basis
A	UT I	*Plateosaurus quenstedti*	SC	PCA
B	UTIIe	*Plateosaurus erlenbergiensis*	SC	PCA
C	UT V	*Pachysaurus wetzelianus*	SC	PCA
D	SM 13200	*Plateosaurus fraasianus*	LC	PCA
E	FUND 8	UA	LC	BSP
F	FUND 10a	UA	LC	PCA
G	FUND 10b	UA	LC	PCA
H	FUND 14	UA	LC	PCA
I	FUND 29	UA	LC	PCA
J	FUND 33	UA	LC	PCA
K	FUND 61	UA	UA	—
L	FUND 65a	UA	LC	PCA
M	FUND 65b	UA	SC	PCA
N	FUND 65c	UA	LC	PCA
O	FUND 65d	UA	LC	PCA
P	AH 2106a	UA	LC	PCA
Q	AM 2109	UA	LC	BSP
R	AM 6810	UA	SC	PCA
S	AM 21560	UA	LC	BSP
T	UT III	*Plateosaurus robustus*	SC	BSP
U	UT IV	*Plateosaurus plieningeri*	LC	PCA
V	UT VI	UA	SC	PCA
W	1911/2a	UA	LC	PCA
X	1911/2b	UA	LC	PCA
Y	FUND 5	UA	UA	—
Z	FUND 15	UA	SC	BSP
a	FUND 27a	UA	LC	PCA
b	FUND 27b	UA	LC	PCA
c	FUND 29a	UA	LC	BSP
d	FUND 29b	UA	UA	—
e	AM 2107a	UA	LC	BSP
f	AM 2107b	UA	LC	PCA
g	AM 2107c	UA	SC	BSP

Note: Included is the reference letter (Ref.), museum identification number (Museum No.), current taxonomic assignment (Current), the assignment made by this study to group (This), and the basis for this assignment (Basis). Abbreviations used: PCA = specimens used for Principal Components Analysis and assigned to group on that basis, BSP = specimens not used for PCA and assigned to group on the basis of the Bivariate Scatter Plots, SC = small cluster as deigned in text, LC = large cluster as defined in text, UA = unassigned. Museum abbreviations: AM: American Museum of Natural History, New York, New York, USA; FUND: Seemann collections, Staatliches Museum für Naturkunde, Stuttgart; SM: Staatliches Museum für Naturkunde, Stuttgart, Federal Republic of Germany; UT: Geologisch-Paläontologisches Institut, Universität Tübingen, Federal Republic of Germany; 1911: Fraas collections, Staatliches Museum für Naturkunde, Stuttgart, Federal Republic of Germany.

Although morphometric studies typically start at low levels (e.g., univariate and bivariate analyses) and progress to more sophisticated techniques (e.g., PCA), such an approach can tend to provide too much information and too little guidance in determining important groupings. For our present study, we wished to note whether there were apparent groupings of specimens, presumably reflecting different taxa or sexual dimorphs. Given the 11 variables used, it would be difficult to select pairs of variables out of the possible 55 combinations and try to detect biologically meaningful clustering. Instead, after examining the general behavior of the variables, we chose to perform a Principal Components Analysis (see Davis 1973; Chapman et al. 1981) and allow the results to suggest these bivariate analyses that might provide meaningful or interesting results.

The Principal Components Analysis (PCA) was run using a subset of the available data to avoid having specimens with missing measurements that might otherwise tend to obscure the interpretation of the analyses. Of the 33 Trossingen specimens, 22 were usable in the PCA (Table 3.2), which was run using a correlation matrix based on logarithmic data. As such, the PCA was run utilizing the same methodology as that used by Chapman et al. (1981) in their study of *Stegoceras* crania. This approach has been the standard and most frequent way of analyzing data with PCA. However, Bookstein et al. (1985) argue convincingly that analyses using variance/covariance matrices also can provide important insight. Both methods were applied here and provided congruent results that suggest the patterns in the data are quite robust. We present only the results of the PCA using the correlation matrix, but these should indicate any important trends that were present also in the variance/covariance analysis. As with any PCA, the contributions of the variables to the Principal Components (PC's) are summarized in the loadings (Table

3.5) and the relations of the specimens are summarized in bivariate scatter plots of the Principal Component scores (e.g., Fig. 3.2).

It is not the intention of this paper to discuss allometric relationships among the variables. Instead, bivariate analyses were made simply to allow us to infer information about the 11 quarry specimens not usable in the PCA and to evaluate relationships suggested by the PCA. In one case, a Reduced Major Axis (RMA) was fit to the bivariate data following the methodology of Imbrie (1956) and Rayner (1985) to provide a reference line (an average trend for the quarry specimens) for use in bivariate space. Placement of specimens above or below this line in the Bivariate Scatter Plot (BSP) was used to infer membership in one of the two groupings evident in the PCA.

All analyses were run using the statistical program SYSTAT (Wilkinson 1986), using the DTA, FACTOR, STATS, and GRAPH modules. The analyses were run on an IBM PC/AT microcomputer.

Results of the analyses

As indicated earlier, the data matrix used in this study is typical of those encountered in vertebrate paleontology in that it contains a significant number of missing values that are not randomly distributed among the specimens. This is usually the result of missing elements or other taphonomic processes. The matrix for quarry material included 33 specimens across 11 variables. It was approximately 92.6% complete (336 of a possible 363 entries) with variables ranging from 97% completeness (32 of 33 specimens represented) for FPROXW, FDISTB, and FDIAM to 87.9% for FEMURL, FPROXB, and FMTBEXTL. Most specimens (22 of 33, 67%) were completely represented. Of the 11 others, a single specimen (FUND29b, "d") was represented by only four measurements. The data avail-

Table 3.3. *Other specimens used in the analyses not from the quarry at Trossingen*

Ref.	Taxon	Authors
h	*Plateosaurus engelhardti*	Meyer 1837
i	*Dimodosaurus poligniensis*	Pidancet and Chopard 1862
j	*Zanclodon quenstedti*	Koken 1900
k	*Gresslyosaurus robustus*	Huene 1907/08
l	*Plateosaurus erlenbergiensis*	Huene 1907/08
m	*Plateosaurus reiningeri*	Huene 1907/08
n	*Gresslyosaurus torgeri*	Jaekel 1911
o	*Sellosaurus gracilis*[*]	Huene 1907/08
p	*Sellosaurus fraasi*[*]	Huene 1907/08
q	*Teratosaurus minor*[*]	Huene 1907/08
z	*Pachysaurus ajax*	Huene 1907/08

Note: Included are references numbers (Ref.), original taxonomic assignments (Taxon), and author of original description (Authors). Asterix indicates taxa from the Stubensandstein of the Federal Republic of Germany.

able for the non-quarry specimens were incomplete and could be used only in some bivariate comparisons. Summary statistics for each variable are given in Table 3.4. Highest variability is apparent for FPROXW, FDISTW and CDFEMLW. Lowest values are evident for variables considered to be general size indicators (e.g., FEMURL HIPTROCH).

The results of the PCA using a correlation matrix for log-transformed variables are summarized in Table 3.5, which lists the loadings and eigenvector information, and in Figure 3.2, which provides a plot of the scores illustrating the relative positions of the specimens. The results are similar in many ways to those obtained by Chapman et al. (1981) for *Stegoceras*. The first Principal Component (PC) accounted for 58.9% of the total variance and clearly is a factor reflecting size differences among the specimens. Such a relationship to size is indicated by the relatively high positive loadings for all the variables. The highest loadings for this PC are for those variables that exhibited the lowest overall variance (e.g., FEMURL, HIPTROCH). A single measurement on either of these variables would be an excellent indicator of overall size. The lowest loadings are for FDISTB and FPROXW, the major variables controlling the second PC, which accounted for 13.05% of the variance in FDISTW and CDFEMLW (+) versus FPROXW and FDISTB (–). The other PCs have eigenvalues less than 1.0, thereby contributing less variance than a single variable, and were not considered further.

The factor scores plot shown in Figure 3.2, providing the relative positions of the specimens for the first two PCs, illustrates two interesting trends. First, there are three major size classes indicated by clusters on the first PC. Specimen "C" is the largest individual by far, and is a single specimen cluster. A second group of moderate to large individuals includes six specimens (I, X, F, W, H, U). The other specimens make up the cluster of small to moderate individuals.

The second trend is expressed as two groupings or clusters for the quarry specimens along PC2. Here, a small cluster includes six specimens (A, B, C, M, R, V) with positive scores of approximately 0.5 or larger, suggesting higher relative values for FDISTW and CDFEMLW and smaller relative values for FDISTB and FPROXW. The larger cluster of 16 specimens (D, F, G, H, I, J, L, N, O, P, R, U, W, X, a, b, f) reflects the opposite trend. These specimens all have PC2 scores that are negative or approximately zero.

Table 3.5. *Variable loadings and eigenvector information resulting from the Principal Components Analysis of the correlation matrix calculated from standardized data for 22 specimens from the Trossingen quarry with complete data*

Variable	PC 1	PC 2
FEMURL	0.950	0.071
FPROXW	0.547	– 0.591
FPROXB	0.864	0.079
FDISTW	0.669	0.461
FDISTB	0.501	– 0.836
HIPTROCH	0.945	– 0.049
FDIAM	0.817	0.042
CDFEMBL	0.730	0.137
CDFEMLH	0.794	– 0.061
CDFEMLW	0.625	0.369
FMTBEXTL	0.850	0.031
VARIANCE	6.483	1.435
% VARIANCE	58.937	13.049

Note: Abbreviations: PC = Principal Component. Variables are as in Figure 3.1.

Table 3.4. *Summary statistics for variables used in the analysis*

Var. No.	Var.	N	Mean	S.D.	Minimum	Maximum
1	FEMURL	29	2.809	0.066	2.674	2.985
2	FPROXW	32	1.880	0.077	1.702	2.137
3	FPROXB	29	2.174	0.050	2.073	2.326
4	FDISTW	30	2.087	0.081	1.961	2.320
5	FDISTB	32	2.148	0.078	1.969	2.281
6	HIPTROCH	31	2.483	0.056	2.354	2.604
7	FDIAM	32	1.848	0.068	1.761	2.021
8	CDFEMBL	30	1.741	0.068	1.617	1.891
9	CDFEMLH	31	2.016	0.074	1.855	2.162
10	CDFEMLW	31	1.644	0.081	1.492	1.811
11	FMTBEXTL	29	2.376	0.061	2.258	2.544

Note: Data are for 33 Trossingen quarry specimens only. Data are in logarithms where appropriate. Variables are illustrated in Figure 3.1. Var. No. = variable number, S.D. = standard deviation, *N* = number of observations.

The relationships of the large and small clusters and the possible positions of the 11 quarry specimens not included in the PCA is illustrated further by the BSP in Figure 3.3 for FDISTB versus CDFEMLW. The Reduced Major Axis serves to separate the two clusters seen in Figure 3.2 for PC2. The small cluster specimens have positions above the RMA and the larger cluster specimens are near and below it. This BSP also allows us to suggest to which of the two clusters eight of the other 11 quarry specimens may belong. Joining the small cluster (SC) are three new specimens (T, Z, g) and joining the larger cluster (LC) are five more specimens (E, Q, S, c, e). Three specimens (K, Y, d) lack critical measurements and remain unassigned to the clusters.

Given the nature of the data used in our analyses, we choose to interpret these results conservatively and assume that separate taxa only can be justified if sufficient characters can be provided or morphometric distinctness demonstrated. These data only provide insight for the measurements we employed; other data certainly can provide further insight for grouping or separating specimens.

Given such a rationale, the quarry specimens form a single major cluster which is subdivided into two groups by PC2. This subdivision can be interpreted as either general population variability, including sexual dimorphism, or separate taxa. We choose to interpret this as intra-specific variation because of the relatively low amount of variation represented by PC2 and the low degree of distinctness of the clusters. The data are insufficient to support sexual dimorphism for now but would suggest possible differences in the locomotory regimes of the sexes if dimorphism is supported by subsequent analyses. See Table 3.2 for specimen reference information.

It would be appropriate to investigate other *Plateosaurus* mass accumulations (e.g., Halberstadt [German Democratic Republic] and Frick [Switzerland]) to see if the morphometric patterns are reproduced in data from these sites. If such dimorphism fails to be evident in subsequent analyses, then it probably is unimportant given the low variability accounted for by PC2. If the clusters continue to be apparent in subsequent analyses, then such partitioning would support either different taxonomic assignments or sexual dimorphism, depend-

Figure 3.2. Bivariate plot of scores for Principal Component 1 versus Principal Component 2 from the Principal Components Analysis using correlation matrix calculated from a standardized data matrix for 22 specimens from Trossingen quarry. Letters are reference letters from Table 3.1. FACTOR(1) = PC1; FACTOR(2) = PC2.

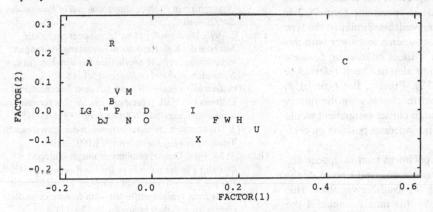

Figure 3.3. Bivariate Scatter Plot of FDISTB (x-axis) versus CDFEMLW (y-axis) for all specimens with data for both measurements. Letters are reference letters from Tables 3.1 and 3.2. Line is Reduced Major Axis for all quarry specimens. The equation for the RMA is $Y = 1.04X - 0.59$. Specimens above the line have affinities to the small cluster of quarry specimens. Those near or below have affinities to the large cluster.

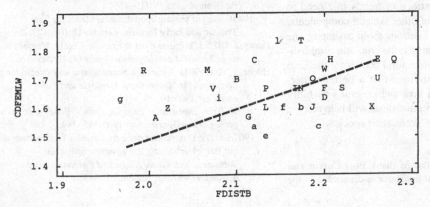

ing on the structure and distinctness of the clusters themselves. The latter would be favored if functional interpretations can be made supporting the morphological differences between the two clusters within the context of sexual dimorphism as was done for *Stegoceras* crania by Chapman et al. (1981). Such an interpretation of sexual dimorphism based on femoral data likely implies perhaps subtle differences in locomotor grade between males and females, or varying structure to support differing mean body weights. The two morphs are separated by differences in shape of both proximal and distal articular surfaces and by the size of the Mm. caudifemoralis longus and brevis, all obviously related to the kinematics of femoral excursion. In the event that these functional interpretations do not suggest intraspecific biological differences, a taxonomic separation may well be justified. Analyses of other elements and material from other mass accumulations may well enhance one or the other of these interpretations.

Because of the paucity of data available for the non-quarry material, the relationships of these specimens to those from Trossingen are not appreciably clarified by our analyses. For instance, the holotype of *P. engelhardti* ("h") is too incomplete to cluster with the Trossingen material in all of the significant bivariate plots (see PCA discussion). However, it does tend to cluster with the small cluster specimens from PC2 in those BSP's where data are available. Similarly, the type of *P. erlenbergiensis* ("l") also tends to cluster with this group on various BSP's. This is interesting because specimen "B" of that group also has been referred to that species (Huene 1932). Finally, the type of *P. reiningeri* ("m") does tend to cluster with the quarry specimens in general although cluster assignment would be difficult because of its intermediate position on critical plots.

All the other type specimens tend to appear different from the quarry clusters on one or more of the various BSP's plotted using the available variables. This interpretation is tenuous at this time because of the paucity of data available for these type specimens. In some cases, this conclusion is based on a single BSP which is problematical because even quarry specimens can appear different on the basis of single plots. These data simply provide suggestions of trends that need to be clarified using analyses of other skeletal components.

At this stage, we are cautious about assigning one or more names to the Trossingen material and establishing affinities with other plateosaurid taxa from the Late Triassic of Europe. It is obvious that a better understanding of non-Trossingen taxa and general variability of other anchisaurid mass accumulations will help resolve the taxonomic affinities of *Plateosaurus* species.

Acknowledgments

The authors would like to thank Phil Currie and Kenneth Carpenter for editorial assistance and patience in the preparation of this manuscript. Further advice and assistance was provided by Mike Brett-Surman, Linda T. Deck and Hans-Dieter Sues of the Smithsonian Institution. Thanks are extended to F. Westphal (UT), R. Wild (SM), and E. S. Gaffney (AMNH) for permission to examine material in their care. This work was begun by Weishampel during the tenure of a NATO postdoctoral fellowship at the Universität Tübingen; many thanks go to NATO for their support and to W.-E. Reif for his supervision of the project.

References

Bakker, R. T. 1978. Dinosaur feeding behaviour and the origin of flowering plants. *Nature* 274:661–663.

Bookstein, F. L., Chernoff, B., Elder, R. L., Humphries, J. M., Jr., Smith, G. R., and Strauss, R. E. 1985. Morphometrics in evolutionary biology. *Academy of Natural Sciences, Philadelphia, Special Publications* No. 15:1–277.

Chapman, R. E., Galton, P. M., Sepkoski, J. J., Jr., and Wall, W. P. 1981. A morphometric study of the cranium of the pachycephalosaurid dinosaur *Stegoceras*. *Journal of Paleontology* 50:608–618.

Davis, J. C. 1973. *Statistics and Data Analysis in Geology.* New York: Wiley.

Dodson, P. 1975. Taxonomic implications of relative growth in lambeosaurine hadrosaurs. *Systematic Zoology* 24:37–54.

1976. Quantitative aspects of relative growth and sexual dimorphism in *Protoceratops*. *Journal of Paleontology* 50:929–940.

Fraas, E. 1896. Die schwäbischen Triassaurier nach dem Material der Kgl. Naturalien-Sammlung in Stuttgart zusammengestellt. *Mittheilungen aus dem königlichen Naturalien-Kabinet zu Stuttgart* 5:1–18.

1900. *Zanclodon schuetzii* n. sp. aus dem *Trigonodus*-Dolomit von Hall. Jahreshefte des Vereins für vaterländische Naturkunde in Württemberg 1900:510–513.

1913. Die neuesten Dinosaurierfunde in der schwäbischen Trias. *Naturwissenschaften* 1913:1907–1100.

Galton, P. M. 1984. Cranial anatomy of the prosauropod dinosaur *Plateosaurus* from the Knollenmergel (Middle Keuper, Upper Triassic) of Germany. I. Two complete skulls from Trossingen/Württ. with comments on diet. *Geologica et Palaeontologica* 18:139–171.

1985a. Cranial anatomy of the prosauropod dinosaur *Plateosaurus* from the Knollenmergel (Middle Keuper, Upper Triassic) of Germany. II. All the cranial material and details of soft-part anatomy. *Geologica et Palaeontologica* 19:119–159.

1985b. Diet of prosauropod dinosaurs from the Late Triassic and Early Jurassic. *Lethaia* 18:105–123.

Henry, J. 1876. L'Infralias dans la Franche-Comte. *Mémoires de la Société d'emulation du Doubs* (4) 10:285–476.

Huene, F. von. 1905a. Ueber die Nomenklatur von *Zanclodon*. *Zeitschrift für Mineralogie, Geologie und Paläontologie* 1905:10–12.

1905b. Trias-Dinosaurier Europas. *Zeitschrift der deutschen geologischen Gesellschaft* 1905:345–349.

1907/08. Die Dinosaurier der europäischen Triasformation mit Berücksichtigung der aussereuropäischen Vorkommnisse. *Geologische und Paläontologische Abhandlungen, Supplement* 1:1–419.

1923. Exkursion nach Trossingen. *Paläontologische Zeitschrift* 5:369–373.

1926. Vollständige Osteologie eines Plateosauriden aus dem schwäbischen Keuper. *Geologische und Paläontologische Abhandlungen, N.F.* 15:139–179.

1928. Lebensbild des Sauriervorkommens im obersten Keuper von Trossingen. *Palaeobiologica* 1:103–116.

1929. Die Plateosaurier von Trossingen. *Umschau* 44:6.

1932. Die fossile Reptilordnung Saurischia, ihre Entwicklung und Geschichte. *Monographien zur Geologie und Paläontologie* 4:1–361.

Imbrie, J. 1956. Biometrical methods in the study of invertebrate fossils. *American Museum of Natural History Bulletin* 108:211–252.

Jaekel, O. 1911. Die Wirbeltiere. Eine Übersicht uber die fossilien und lebenden Formen. (Berlin: Borntraeger Verlag), pp. 1-252.

1913. Über die Wirbeltierfunde in der oberen Trias von Halberstadt. *Palaontologische Zeitschrift* 1:155–215.

Koken, E. 1900. [Review of E. Fraas: Triassaurier]. *Neues Jahrbuch für Mineralogie, Geologie und Paläontologie* 1900:303.

Kuhn, O. 1959. Ein neuer Microsaurier aus dem deutschen Rotliegenden. *Neues Jahrbuch für Geologie und Paläontologie, Monatshefte* 1959:424–426.

Meyer, H. von. 1837. Mitteilung an Prof. Bronn (*Plateosaurus engelhardti*). *Neues Jahrbuch für Mineralogie, Geologie und Paläontologie* 1837:317.

Pidancet, J., and Chopard, S. 1862. Note sur un Saurien gigantesque appartenant au Marnes irisées. *Comptes-rendus hebdomadaires des séances de l'Académie des Sciences* 54:1259–1262.

Plieninger, T. 1846. Über ein neues Sauriergenus und die Einreihung der Saurier mit flachen, Schneidenden Zähnen in eine Familie. *Jahreshefte des Vereins für vaterländische Naturkunde in Württemberg* 2:148–154.

1847. Nachträgliche Bemerkungen zu dem Vortrage (S. 148) über ein neues Sauriergenus und die Einreihung der Saurier mit flachen, Schneidenden Zähnen in eine Familie. *Jahreshefte des Vereins für vaterländische Naturkunde in Württemberg* 2:247–254.

1850. Ueber ein Saurierskelett im obersten Keupermergel (*Belodon*?). *Jahreshefte des Vereins für vaterländische Naturkunde in Württemberg* 5:171–172.

Quenstedt, F. A. 1858. *Der Jura.* (Tübingen), pp. 1-842.

Rayner, J. M. V. 1985. Linear relations in biomechanics: the statistics of scaling functions. *Journal of Zoology* 206:415–439.

Rutimeyer, L. 1856. Reptilienknochen aus dem Keuper. *Allgemeine schweizerische Gesellschaft für die gesammten Naturwissenschaften, Verhandlungen* 41:62–64.

Sauvage, H.-E. 1907. Vertébrés. *In* Thiéry, P., Sauvage, H.-E., and Crossman, M. (eds.) *Note sur l'Infralias de Provencheres-sur-Meuse* 365:6–17.

Seemann, R. 1932a. Verlauf und Ergebniss der Trossinger Sauriergrabung. *Jahreshefte des Vereins für vaterländische Naturkunde in Württemberg* 88:52–54.

1932b. Die Sauriergrabung bei Trossingen. *Heimatblätter vom oberen Neckar* 102:1453–1455.

1933. Das Saurischierlager in den Keupermergeln bei Trossingen. *Jahreshefte des Vereins für vaterländische Naturkunde in Württemberg* 89:129–160.

1941. Merkwürdige Lebensspuren in den Trossinger Keupermergeln und ihre Bedeutung für die Erklärung des Saurischierlagers. *Jahresberichte und Mitteilungen des Oberrheinischen geologischen Vereins, N.F.* 30:42–47.

Sokal, R. R., and Rohlf, F. J. 1969. *Biometry.* San Francisco: Freeman.

Steel, R. 1970. Saurischia. *Handbuch der Paläoherpetologie* 14:1–87.

Weishampel, D. B. 1984a. Interactions between Mesozoic plants and vertebrates: fructifications and seed predation. *Neues Jahrbuch für Geologie und Paläontologie, Abhandlungen* 167:224–250.

1984b. Trossingen: E. Fraas, F. von Huene, R. Seemann, and the "Schwäbische Lindwurm" *Plateosaurus. In* Reif, W.-E. and Westphal, F. (eds.), *Third Symposium on Mesozoic Terrestrial Ecosystems,* short papers. Tübingen: Attempto Verlag, pp. 249-253.

Wilkinson, L. 1986. *SYSTAT: The System for Statistics.* Evanston, Illinois: SYSTAT, Incorporated.

4 Species determination in sauropod dinosaurs with tentative suggestions for their classification

JOHN S. MCINTOSH

Abstract

This paper is divided into two parts. The first sketches sauropod classification, in which vertebral characters play a major role. Six families are recognized: Vulcanodontidae, Cetiosauridae, Brachiosauridae, Camarasauridae, Diplodocidae, and Titanosauridae. The most perplexing question centers about the relationship of the Upper Cretaceous titanosaurids with the other families. The second part seeks, not altogether successfully, general criteria for the separation of sauropod species. Thirty general characters that have proved successful in discussing sauropod taxonomy are presented, and illustrated as they apply to the six firmly established North American Jurassic genera. Species differentiation is discussed in five of the best known species for which adequate material exists.

Introduction

This report consists of two parts, the second of which, dealing with species determination in the Sauropoda, was presented at the Drumheller Conference. In response to a request that a brief discussion of overall sauropod classification be included, the major part of a talk delivered at the AAAS meeting in Missoula, Montana, in June 1985, has been added. Entries of all works referred to in the discussion of classification would swell the bibliography to a size equal to that of the rest of the report and well beyond the scope of this paper. Only those papers specifically cited are included in the references. A comprehensive bibliography is available elsewhere (Chure and McIntosh 1989).

Classification of the Sauropoda

Sauropods appeared in the Early Jurassic of most continents, excluding North America, and quickly spread throughout the world. Much new and splendid material collected recently in China (Dong, Zhou and Zhang 1983; He et al. 1984; Zhang, Yang and Peng

In Dinosaur Systematics: Perspectives and Approaches, *Kenneth Carpenter and Philip J. Currie, eds. Copyright © Cambridge University Press, 1990.*

1984), Argentina (Bonaparte 1979, 1986; Bonaparte and Powell 1980), and Morocco (Monbaron and Taquet 1981), is certain to shed much light on the early relationships of these animals and may necessitate a substantial revision of the tentative views here expressed.

Prior to 1929, students of the sauropods generally followed the pioneer work of Marsh, succeeded by that of von Huene, and divided the suborder into roughly half a dozen families. These classifications, with minor variations, agreed in general with one another. In 1929, Janensch proposed a two family scheme based on broad spatulate-tooth types as opposed to slender peg-like ones; he divided each family into several subfamilies. The proposal was quickly seized upon by Nopcsa, von Huene, and Romer, and has since become nearly universally adopted. However, there has been some disagreement as to which family a given subfamily belonged, as well as to the appropriate names of the two families themselves. Romer accepted Brachiosauridae for the broad-toothed types and Titanosauridae for the slender-toothed ones, and these names are now commonly used. One result of this scheme has been to throw together the highly specialized diplodocids and, in some ways, the more generalized Upper Cretaceous titanosaurids, as dissimilar a pair of groups as exists in the Sauropoda. A major persistent difficulty with any attempt to understand the relationships of the various sauropods with one another has been their notorious habit of losing their heads – complete skulls are known in only a fraction of the accepted genera. On the other hand, the necessity of reducing the body weight of these huge animals as much as possible led to extreme specialization of the vertebrae, and these elements are most useful in tracing the phylogeny. Nevertheless, many gaps persist in our knowledge of the majority of the genera. Indeed, in only about four genera can we claim to have anything like an overall picture of the skeletal anatomy, and in only one, *Camarasaurus*, may it be said that our knowledge is complete. When fully described, the new Chinese form *Shunosaurus* may become the second fully understood genus. It is these many gaps in our knowledge that led to Romer's (1968) classic lament, "It will be a long time, if ever, before we obtain a valid, comprehensive picture of sauropod classification and phylogeny." The many recent discoveries cited above are beginning to change this perception – hence this preliminary attempt at sauropod classification.

For many years it has been apparent to me that the two family division of the sauropods was both simplistic and unnatural, and that the suborder should be divided into at least six families: the primitive Vulcanodontidae, the generalized Cetiosauridae, the long forelimbed Brachiosauridae, the Camarasauridae and Diplodocidae with their uniquely bifurcated neural spines in the shoulder region (but with spatulate and peg teeth respectively), and finally the primitive Upper Cretaceous Titanosauridae. The following characters have proven most useful in defining the different families and genera.

1. The teeth and associated skull type: those with broad spatulate teeth are associated with more robust skulls with lateral nares, while those with peg teeth occur with the longer snouted, more slender skulls where the nares open dorsally.

2. The vertebral formula: the primitive saurischian vertebral formula is 9–10 cervicals, 15 dorsals, 3 sacrals, and a varying number of caudals. Among primitive theropods, these numbers are 9 cervicals, 14 dorsals, 1 dorso-sacral, 3 sacrals, a caudo-sacral, and a variable number of caudals. Occasionally, a dorsal will be taken into the neck to produce 10–13–1–3–1. The usual prosauropod formula is similar, except that there is one more cervical resulting in 10–15–3. Much greater variations occur in the sauropods. Riggs (1903) showed that Marsh's claim for the number of sacral vertebrae as a generic character was not valid. All sacra incorporate a modified dorsal and a modified caudal which, with the basic three, give five sacrals. Extremely old individuals and most Titanosauridae sometimes add a second modified dorsal. In the presacrals, the trend is to incorporate vertebrae into the neck from the trunk. The usual increase is by one or two, but in a few cases the number is increased by as many as six. The number of caudal vertebrae varies enormously from about thirty to over eighty.

3. The degree of excavation of the articular and lateral faces of the presacral centra, and in particular of the posterior dorsals.

4. The presence or absence of a divided neural spine in the posterior cervical-anterior dorsal region.

5. The overall height of the presacral spines and the increased height of those in the sacral region in forms with longer hind limbs, and in the pectoral region of those with abnormally long forelimbs.

6. The presence or absence on the anterior caudals of thin wing-like transverse processes (caudal ribs) resembling sacral ribs.

7. The presence of pleurocoels on the anterior caudal centra, often associated with ventral excavation.

8. The presence of gently procoelous centra in the anterior caudal region of some forms or extremely proceolous centra with large posterior balls in others.

9. The greater or lesser development of strong chevron facets on the caudal centra.

10. The presence of a whip-lash at the end of the tail with about 80 usually elongated centra, contrasted with the lack of a whip-lash in a tail of 53 or fewer caudals with short centra.

11. The presence or absence of the typical diplodocoid forked chevrons in the middle tail region.

12. The relative expansion of the distal and/or proximal end of the scapula.

13. Certain limb ratios: a) humerus:femur, b) metacarpus:radius, c) tibia:femur.

14. The relative expansion and means of articulation of the distal ends of the ischia.

15. The presence or absence of an ossified calcaneum.

16. The presence or absence of an inwardly directed process on the distal end of metatarsal I.

17. Whether the greatest metatarsal length is numbers 2 and 3 or numbers 3 and 4.

In 1965, Charig, Attridge, and Crompton broke with accepted dogma, that the sauropods had developed from bipedal prosauropod ancestors, and suggested that the ancestors were quadrupedal – perhaps members of the little known Melanorosauridae. I am inclined to accept this point of view, but will make no attempt to trace the pre-sauropod ancestry in this paper. What is clear is the emergence of primitive but true sauropods in the Lower Jurassic of India, Africa, China, and Europe. These may be assigned tentatively to the inadequately known family Vulcanodontidae. In *Vulcanodon* (Africa), the skull and presacral vertebrae are unknown. In *Barapasaurus* (India), the skull and feet are unknown, and the rest of the skeleton has received only a preliminary description. A tentative family diagnosis is: broad teeth with denticles on cutting edges, presacral vertebrae with shallow depressions on sides of centra but no true pleurocoels, presacral spines undivided, posterior dorsal centra platycoelous or amphicoelous, diapophyses directed laterally, sacrum with dorso-sacral and three sacrals, sacrum narrow, sacro-costal yoke not contributing or minimally contributing to the acetabulum, symphysis of pubes forming an "apron" shorter than in prosauropods but longer than in later sauropods, ischium elongate and slender as in prosauropods. The family may be represented in Europe and China by the ill-defined *Ohmdenosaurus* and *Zizhangosaurus* respectively.

The Cetiosauridae are represented in the Middle Jurassic by *Cetiosaurus* in Europe and Africa (Monbaron and Taquet 1981), by *Shunosaurus* (Dong, Zhou and Zhang 1983; Zhang, Yang and Peng 1984), *Datousaurus* (Dong, Zhou and Zhang 1983), and *Omeisaurus* (Dong, Zhou and Zhang 1983; He et al. 1984) in China, and by *Patagosaurus* (Bonaparte and Powell 1980) and *Amygdalodon* in Argentina. *Haplocanthosaurus* is a late survivor in the Upper Jurassic of North America. The family is characterized by a heavy skull with numerous teeth having heart-shaped crowns, cervicals with simple pleurocoels and undivided spines in the shoulder region, 13 or 14 dorsal vertebrae (with diapophyses directed upwards and outwards, and posterior dorsals platy-coelous or amphicoelous), sacral spines of only moderate height, very short caudal centra with quite prominent chevron facets, humerus:femur = .8, tibia:femur = .55 to .6, and a calcaneum present. The three Chinese forms possess forked diplodocoid middle chevrons, while the European and American forms apparently do not. It is doubtful that the "Cetiosauridae" as here defined will be found to constitute a monophyletic group when the Chinese genera, as well as *Patagosaurus* and a fine, as yet undescribed skeleton of *Cetiosaurus* from Morocco, have received adequate descriptions. Before such information is available, it would be premature to try to subdivide the family. I shall return to this discussion under the Diplodocidae below.

The Brachiosauridae represents a major advance from the cetiosaurids and requires more than the subfamily level that it is often accorded. The family is characterized by a relatively large skull with large, broad, spatulate teeth, and enormously elongated cervicals with very long posterior ribs. In *Brachiosaurus*, there are 13 cervicals and 11 or 12 dorsals. The presacrals have deep, clearly defined pleurocoels, with a complex structure of subcavities in the cervicals. The neural spines are simple, but, unlike other sauropods, are highest in the shoulder region because of the enormously long forelimb. The posterior dorsal centra are strongly opisthocoelous, there are low sacral and caudal neural spines, the caudal ribs and chevrons are simple, and the tail is short. The ratio of humerus:femur is .9 to 1.1; metacarpal II:radius = .44 to .51; tibia:femur = .58 to .59. The scapula has a greatly expanded distal end, and the shaft of the ischium is broad but not expanded distally. The family would appear to have originated in Gondwanaland.

Numerous bones of juvenile individuals from the Middle Jurassic of Madagascar have been described in a thesis by A. Ogier as "*Bothriospondylus madagascariensis*" (recently referred to a new genus *Lapparentosaurus* by Bonaparte 1986). This animal would appear to possess the characters expected of a brachiosaurid ancestor: simple vertebral characters, low posterior dorsal and sacral spines, elongate and slender forelimbs, and a greatly elongated pubic peduncle of the ilium. What would appear to be a similar animal from Patagonia named *Volkheimeria* by Bonaparte (1979) has so far received only a preliminary description. *Brachiosaurus* is known from Africa, and also from Europe and North America to where it presumably migrated in the Upper Jurassic. The family survived into the Lower Cretaceous in Europe as *Pelorosaurus* and in Argentina as *Chubutisaurus*. Remains that probably belong to the family have been reported from the Cloverly Formation of North America. A more controversial form from the

Lower Cretaceous of Maryland and Texas (and probably England as well) is *Pleurocoelus* (?*Astrodon*), known mostly from juvenile individuals. This animal certainly did not descend from *Brachiosaurus* itself. It possessed a weaker dentition, a less elongate forelimb, and was apparently a much smaller animal even as an adult. Regrettably, no neural spines are known from vertebrae in the shoulder region. The very deep pleurocoels in the presacrals and, particularly, certain developments in the caudals appear closer to *Brachiosaurus* than to *Camarasaurus*, although disarticulated skull elements suggest closer association with the latter. The teeth are more slender than in either of these forms. Pending the discovery of better material of the cranial and presacral regions, *Pleurocoelus* is provisionally included in the Brachiosauridae.

The origins of the Camarasauridae are obscure. The family has not been reported from Tendaguru. It is represented by a single genus *Camarasaurus* in North America, where it was the most common Morrison sauropod. The family is characterized by a short robust skull with broad spatulate teeth. The presacral formula is 12–12–1. The presacral centra have prominent, well-defined, very deep pleurocoels, which persist back to the sacrum, as does their opisthocoelous nature. As with the brachiosaurids, the posterior cervical ribs are long, extending beneath one or two succeeding centra; the centra themselves are of moderate length. The spines of the posterior cervicals and anterior dorsals are deeply cleft in a U-shape. The spines in the posterior dorsal, sacral, and anterior caudal regions are only of moderate length, but broadly expanded transversely. The transverse processes of the dorsal vertebrae are horizontal. The anterior caudal ribs are simple. The caudal centra are short, spool-shaped, and amphicoelous, lack pleurocoels, and have no especially well developed chevron facets. The tail is relatively short, without a whip-lash. Both ends of the scapula are expanded. The shaft of the ischium is slender and the unexpanded distal ends meet one another edge to edge. The humerus:femur ratio is .72 to .78; metacarpal III:radius = .46; tibia:femur = .58 to .63. There is a small globular calcaneum. Metatarsal I does not possess a distal process, and metatarsals II and III are coequal and the longest of the set.

There are four other genera that may be tentatively included in the Camarasauridae. The first is *Euhelopus* from the Upper Jurassic of China. This animal has often been referred to the Brachiosauridae, or to its own subfamily or family. The skull is short and resembles that of *Camarasaurus* more than that of *Brachiosaurus*, but it is a little less robust. A second factor suggesting a closer association with *Camarasaurus* is the beginning of a divided neural spine in the posterior cervical region. The forelimb is known only from a humerus found some years later at what may be the same locality where one of the cotypes of *E. zdanskyi* was collected. The ratio of its length to that of the incomplete femur is about 1 to 1, but this cannot be taken too seriously considering the questionable association. The humerus does not exhibit the slenderness of that of *Brachiosaurus* and the spines of the anterior dorsals are shorter than would be expected of an animal with an abnormally long forelimb. A puzzling factor is the high number of presacral vertebrae: 17 cervicals, 14 dorsals (not including the dorso-sacral). This feature also occurs in the recently described skeleton of *Omeisaurus* from China. The latter, however, possesses a primitive cetiosaurid skull and middle chevrons of the diplodocid type. The tail is unknown in *Euhelopus*. Should its chevrons prove to be of the diplodocid type, reassessment of that animal would be necessary, possibly to the Cetiosauridae. The only other sauropod with a comparably high number of presacrals is *Mamenchisaurus*, but, as will be discussed presently, that animal is clearly diplodocoid. A second camarasaurid is "*Apatosaurus*" *alenquerensis* (Lapparent and Zbyszewski 1951) from the Upper Jurassic of Portugal. This animal possesses 12 dorsals, even the most posterior of which are strongly opisthocoelous and with a large anterior ball. The limb and girdle bones resemble those of *Camarasaurus* with minor differences, including a somewhat greater humerus to femur ratio of .9. Unfortunately, the skull and all the presacral arches are missing, making definite identification difficult. It is possible that when more material is found, the animal will prove to be a new genus of the Camarasauridae. As discussed in the second section of this report, I refer it provisionally to *Camarasaurus ?alenquerensis*. The third animal, which appears to be closer to *Camarasaurus* than to any member of any other family, is *Opisthocoelicaudia*, known from a skeleton without head or neck from the Upper Cretaceous of Mongolia. This animal, which is provisionally referred to the Camarasauridae, has the U-shaped, deeply divided neural spines, and is barred from the Diplodocidae by its short tail with unbifurcated chevrons. The little known *Campylodoniscus* from the Upper Cretaceous of Argentina has broad teeth and might belong to this family. The forelimb originally named *Pelorosaurus becklesii* from the Lower Cretaceous of England, which is sometimes referred to *Camarasaurus*, is considered Sauropoda *incertae sedis*.

Recent discoveries from China (Dong, Zhou and Zhang 1983) and Argentina (Bonaparte 1986) are beginning to shed some light on the origins of the Diplodocidae. The main features of this family are: long snouted skull with superior nares and weak peg-like teeth confined to the front of the jaws, cervicals and anterior dorsals opisthocoelous, posterior dorsals amphicoelous, short cervical ribs, deeply divided V-shaped neural spines in and on both sides of the shoulder region, very long tail with elaborate wing-like transverse processes on anterior caudals, caudal centra gently proceolous with weak chevron facets, elongated middle and distal caudals, fore and aft expanded forked chevrons in the mid-tail region, tail ending in a whip-lash, short metacarpals, and metacarpal II or III:humerus = .32 to .37. The more advanced forms exhibit three other tendencies: 1) reduction of the number of dorsals and a corresponding increase in the number of cervicals, 2) reduction of the humerus to femur ratio correlated with 3) high spines in the sacral region.

The Diplodocidae may be divided into three subfamilies, the Diplodocinae, Mamenchisaurinae, and Dicraeosaurinae. The Diplodocinae includes the well known *Diplodocus* and *Apatosaurus* from North America, and the less well known *Barosaurus* from North America and possibly Africa. The subfamily is characterized by "cervicalization" of the presacral formula in *Diplodocus* and *Apatosaurus* at 15–10–1. In *Barosaurus* there are one or two fewer dorsals and probably a corresponding increase in the cervicals. The number of caudals is about 80. The distal end of the scapula is less expanded than in most families; the distal end of the ischium is much expanded, the humero–femoral ratio is only .66 in *Diplodocus* and *Apatosaurus*. An ossified calcaneum is absent. Metatarsals III and IV are longest. The first metatarsal possesses a prominent process on the posterior inner margin at the distal end. Some of these characters are also found in the other two subfamilies, but in some instances they are incompletely known. Until recently, the Diplodocinae was known only from the Upper Jurassic, but in 1980 Charig reported a diplodocine chevron from the Wealden of England.

The second subfamily, the Mamenchisaurinae, is more primitive than the Diplodocinae in most respects, but also more advanced in a few. The skull is unknown. The presacral formula in *Mamenchisaurus* is 19–11–1, a large increase in the total number. The cleft in the presacral spines is less pronounced than in Diplodocinae, and the sacral spines are not as high, perhaps indicating a less reduced humero–femoral ratio. Though prominent, the forking of the median chevrons does not reach the extreme development of *Diplodocus*. The English genus *Cetiosauriscus* is provisionally referred to this subfamily due to its similarity to *Mamenchisaurus* in the caudals and chevrons. But it might also be a primitive diplodocine because its humero–femoral ratio is just .66, and metatarsal I bears the typical distal process.

The subfamily Dicraeosaurinae is somewhat of an enigma. In many respects it represents the epitome of diplodocid specialization – the presacral spines are the most deeply bifid of all, the sacral spines are quite high, the wing-like transverse processes of the anterior caudals are strongly developed, and the forelimb is quite short. On the other hand, the presacral formula is a conservative 12–12–1, the dorsal diapophyses are directed upward as well as outward (reminiscent of the cetiosaurids), and most surprisingly, the dorsal centra are totally devoid of pleurocoels (likely a secondarily derived character). The Upper Cretaceous Mongolian genera *Nemegtosaurus* and *Quaesitosaurus*, known only from incomplete skulls, are provisionally referred to this sub-family, because they bear some resemblance to the incompletely known skull of *Dicraeosaurus*.

The least understood family of sauropods is the Upper Cretaceous Titanosauridae. No complete skulls are known, but several braincases from Argentina and India have been assigned to the family, principally because most of the Upper Cretaceous sauropods appear to be titanosaurids. A mandible with small slender teeth resembling those of the diplodocids was associated with the partial skeleton of a large *Antarctosaurus wichmannianus*. A single fragmentary cervical of no diagnostic value was found with this skeleton. Mostly on the basis of the jaw, the diplodocids and titanosaurids are referred to the same family, albeit separate subfamilies. The difficulty of this is that in other skeletal features, particularly those of the characteristic vertebrae, the two groups show almost no resemblance to one another whatsoever.

The Titanosauridae are characterized by presacrals with crude pleurocoels, which are often ill-defined pits of irregular shape and depth. This is quite unlike the deep pockets with sharply etched subcircular lips of the diplodocids. The presacral spines are simple throughout and, for the most part, those of the dorsal column incline backward instead of being straight upward as in the Diplodocidae. The sacrum incorporates a second dorso-sacral in all adults, and the preacetabular blades of the ilia bend strongly outward giving the bone a more nearly horizontal orientation than in other sauropods. The caudal centra are strongly procoelous with large balls on the posterior faces, and the prominent chevron facets give the ventral aspect of the caudal a sculptured appearance. The chevrons are simple. The upper end of the scapula is not expanded. The ischium is reduced in size but meets its mate in a symphysis the entire length from the distal end to the pubis. The sternal plates are huge. The ratio of metacarpal II to the radius is .50. Finally, there is, surprisingly, body armour (Bonaparte and Powell 1980).

A number of generic names have been proposed for animals of this family, but no doubt many of them will prove to be synonyms. *Titanosaurus, Alamosaurus*, and *Saltasaurus* represent valid forms, and *Antarctosaurus* may represent another, if it belongs to this family. *Laplatasaurus, Aegyptosaurus, Magyarosaurus, Microcoelus, Argyrosaurus, Hypselosaurus*, and *Macrurosaurus* may prove to be valid, but the material at this time is incomplete.

The only possible titanosaurids prior to the Upper Cretaceous are several caudal centra from the Wealden of England, and the incompletely known *Tornieria* from the Upper Jurassic of East Africa. This form is sometimes regarded as the titanosaur ancestor, but after examining all the material, I am not convinced that the procoelous caudals are titanosaurid. In addition, I am not convinced that all the material assigned to *Tornieria* belongs to a single genus. It is possible that Janensch

took all the Tendaguru sauropod material that could not be assigned to *Brachiosaurus, Barosaurus*, or *Dicraeosaurus*, and lumped it together as *Tornieria*.

No cladogram of the sauropods is given, because I am convinced that one drawn ten years from now, when all the recently discovered material has been described and evaluated, would bear little resemblance to a current effort. I will, however, summarize my current thinking on sauropod phylogeny. Sauropods such as the Vulcanodontidae may have arisen from prosauropod ancestors near the end of the Triassic or beginning of the Jurassic in Gondwanaland. Subsequently, the vulcanodontids developed into the widespread family Cetiosauridae in the Middle Jurassic. At that time, the Brachiosauridae split off in Gondwanaland. The Camarasauridae split from the Brachiosauridae early in the Late Jurassic, possibly in Europe or North America.

The picture then becomes confusing because of four factors splitting the sauropods into two classes. First, there is the division into the broad- and narrow-toothed forms, the latter comprising the Diplodocidae and apparently also the Titanosauridae. Second, there are the divided versus the undivided neural spines, the former comprising the Diplodocidae and Camarasauridae. Third, there are the normal versus the forked middle chevrons, the latter comprising the Diplodocidae and the three Chinese cetiosaurids. Fourth, there are those forms with 25–26 presacrals versus Chinese forms with 31–32 presacrals (*Euhelopus, Mamenchisaurus*, and *Omeisaurus*, each assigned to separate families).

No cladogram can accommodate all of this data without admitting parallel development of some characters. If the divided spines represent a unique character, the Diplodocidae would had to have split from the Camarasauridae after the latter split from the Brachiosauridae. This implies that the slender teeth of the former and of the Titanosauridae developed independently. On the other hand, if the slender tooth type represents a unique character, the Diplodocidae would have to have split from the Cetiosauridae before the Brachiosauridae split from the latter. The forked chevrons in the Chinese cetiosaurids suggest that the diplodocids developed from those sauropods, but published accounts assign broad teeth to them. Indeed, the presacrals of *Datousaurus* suggest a development in the diplodocid direction, and the appendicular skeleton could well have been the diplodocid precursor. However, the jaws assigned to this form have broad teeth. Bonaparte (1986) has recently suggested another possible diplodocid ancestor, *Patagosaurus*, based on the complicated laminae in the arches and spines of the presacrals. But there is support from the tail that the animal was connected with the diplodocids. There is even the remote possibility that, despite dissimilarities, the titanosaurids and diplodocids split from the cetiosaurids in the Middle Jurassic, and then from one another, perhaps via *Tornieria*, in the Upper Jurassic. Alternatively, each family split from the Cetiosauridae independently, and the weak peg dentition is a parallel development. There is the slight possibility that the peg-toothed *Antarctosaurus* skull does not belong to a titanosaurid at all but to a diplodocid, and that the titanosaurids did not have peg teeth (if the broad-toothed *Campylodoniscus* from the Upper Cretaceous of Argentina is actually a titanosaurid). However, the lack of diplodocid material from the Upper Cretaceous of South America would cast severe doubt on such a hypothesis.

Species determination

The establishment of reasonable criteria for the determination of species in sauropods is a problem that I have, for the most part, evaded in the past because of the paucity of articulated material that would render statistical methods useful. Until recently, only a single sauropod skeleton had been reported that might be said to be virtually complete, and perhaps a dozen that were even two-thirds complete. Little help could be garnered from the skulls, as they are so rarely found in the Sauropoda. Furthermore, the majority of complete skulls come from Dinosaur National Monument, and all but one of those belong to single species of *Camarasaurus* and *Diplodocus*. To these may now be added *Shunosaurus*. Because few complete or partial skulls are associated with postcranial remains, it is difficult to determine the skeletons generically, let alone specifically. The highly specialized vertebrae, particularly the presacrals, have proved the most useful for determination

of sauropod genera. When articulated series are discovered, they are often not collected, or more often destroyed or severely damaged. Still others usually end up in mounted skeletons, skillfully restored with coloured plaster the precise shade of the bones. Under multiple layers of shellac, three meters or more above floor level, their view is also obstructed by ribs. Unless performed before mounting, their detailed study is most difficult.

A large number of disarticulated limb bones and a smaller number of girdle bones have been collected, and I have studied, measured, and photographed most of these. While useful for generic identification, their lack of association with vertebrae greatly limits their usefulness for specific identification. Much more associated material must be available before a truly definitive study can sort out which characters are important in defining species, which are due to sexual dimorphism, and which are merely the result of individual variation. Until then, I shall try to state as much as I can at this time by looking at specific examples.

The only serious attempt to set down criteria for the determination of sauropod species was that of Mook (1917a). He cited five categories of characters: 1) size when associated with the same ontogenetic stage of development, 2) ontogenetic characters in association with other characters, such as size, 3) slight morphological differences exhibited in samples of sufficient size to rule out individual variation, 4) order of differentiation of parts, such as whether a dorsal is taken into the sacrum before a caudal or vice versa, and 5) loss or devel-

Figure 4.1. *Diplodocus* – caudal vertebrae, left view: **A**, caudals 17–20, USNM 10865 (after Gilmore); **B**, caudal (?) 20, type of *D. longus*, YPM 1920 (after Marsh); **C, D**, caudals 19 and 20, respectively, AMNH 223.

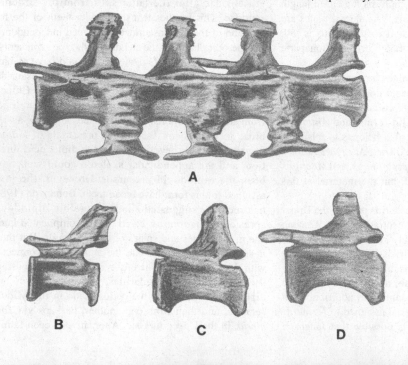

opment of characters in different parts of the skeleton. Although many of these points would appear to be almost obvious, the paper was useful in discrediting a number of characters employed by Marsh and other early writers. Among the taxonomically insignificant characters used by Marsh and others are the number of sacral vertebrae, co-ossification of the scapula and coracoid or of the distal ends of the ischia, and overall size disregarding the age of the individual. Many of Mook's criteria had already been enunciated by Riggs (1903) fourteen years earlier. The example that Mook chose to illustrate his conclusions proved to be an unfortunate one, namely a specimen representing a new species of *Apatosaurus, A. minimus.* From the shape of the distal end of the ischium and various sacral characters, particularly the height of the neural spines, it is now clear that this animal is not *Apatosaurus,* and also must be excluded from the Diplodocidae and considered Sauropoda *incertae sedis.* Lest I appear to sound overly critical of Mook's attempt, let me state that I can offer few better characters for determining sauropod species.

Before considering the possible species distinctions in five of the better represented genera, I present a table of family and generic characters that serve to distinguish the six well established Morrison genera (Table 4.1).

The first genus to be examined is *Diplodocus.* Four species have been proposed: *D. longus* (type of the genus), *D. lacustris, D. carnegii,* and *D. hayi. D. lacustris* is based on the teeth of both maxillae, the bone having decayed away. It came from the top of the Morrison, associated with the type of *Stegosaurus armatus* to which Marsh (1877) originally assigned and figured it. The teeth are well preserved, but cannot be determined at the generic level as it is possible that they might belong to *Apatosaurus.* The type specimen of *D. longus,* YPM 1920, consists of seventeen caudal vertebrae, two of which are complete. Seven others are in poor condition while for the rest only field data exists. One of the characteristic chevrons, which gave the animal its name, was also recovered. A hind limb and foot found nearby were originally thought to be associated with the tail, but are now known to belong to *Apatosaurus.*

Several partial *Diplodocus* skeletons, with presacral, sacral, and caudal vertebrae in articulation have been collected from about the same stratigraphic level of the Morrison Formation. This is stratigraphically higher than Garden Park, Colorado, whence came the type of *D. longus,* and stratigraphically lower than Morrison, Colorado, whence came *D. lacustris.* Among these skeletons are the cotypes of *D. carnegii* (CM 84 and CM 94). A careful comparison of the figured caudal (Marsh 1896) of the type of *D. longus* with the articulated tails indicates that this vertebra, the twelfth in sequence, is probably caudal 20 (Fig. 4.1). It is similar to the corresponding vertebrae of USNM 10865, AMNH 655, and CM 94, and differs in minor details from that

of AMNH 223. It is doubtful that it belongs to the closely allied genus *Barosaurus* because of the deep trough-like excavation on the ventral surface of the centrum.

Gilmore (1932) pointed out that neither of Hatcher's (1901) criteria for separating *D. carnegii* from *D. longus* were valid. One criterion was derived from Marsh's (1896) erroneous assignment to *D. longus* of an *Apatosaurus* cervical from the type locality. The other criterion involved a comparison of the tails of *D. carnegii* (CM 84 and CM 94) with that of AMNH 223 assigned by Osborn (1899) to *D. longus.* In the latter the spines of the caudal vertebrae are more nearly erect than those of the former. Gilmore dismissed this character as a matter of individual variation (Fig. 4.1).

I do not wish to discuss at this time the two skulls (USNM 2672 and 2673) found in the same quarry as the holotype tail, but which were widely separated from it, and yet arbitrarily assigned by Marsh (1884) to *D. longus.* Eventually, these skulls may prove useful in defining that species. More material will be needed from the far from exhausted Marsh Quarry 1 at Garden Park, Colorado, or at corresponding levels of the Morrison before a definitive statement can be made concerning the separation of *D. carnegii* from *D. longus.* Until such time, it would appear prudent to retain the name *D. carnegii,* based as it is on such fine material.

Diplodocus hayi (formerly CM 662) is based on a cranium and the greater part of a skeleton now mounted in the Houston Museum of Science. The specimen is apparently of a single animal because there is no duplication of elements, and all are the proper size to belong to a single individual. Holland's (1924) criteria for establishing *D. hayi* are based on contrasting features of its cranium with those of the perfect skull, *D. longus* from Dinosaur National Monument, CM 11161. A valid observation made by Holland is the placement and direction of the parasphenoid. Another feature that he did not mention is the shorter basipterygoid processes in *D. hayi.* In fact, this cranial fragment more closely resembles that of *Apatosaurus louisae* (CM 11162) than those of *Diplodocus* (CM 11161 and CM 3452). The *D. hayi* cervicals are typical of *Diplodocus.* The caudals are relatively shorter and less excavated ventrally than those of *D. carnegii.*

Although much material has been arbitrarily assigned to *Diplodocus* by Osborn (1899) and others solely on the slenderness of the bones, in only two specimens have forelimb bones been found directly associated with *Diplodocus* vertebrae – the type of *D. hayi* and USNM 10865, *D. longus* or *D. carnegii. D. hayi* is the only one with forefoot material. A third specimen, AMNH 5855, in which both fore and hind limb bones were found but no vertebrae, most likely belongs to *Diplodocus* as well (Mook 1917b). Comparisons of the ratios of the least circumference of the shaft to the length of the bone are instructive (Table 4.2). The fact that the *D. hayi* ratios lie midway between those of

Table 4.1. *Family and generic characters of six well known sauropods*

	Haplocanthosaurus	Brachiosaurus	Camarasaurus	Apatosaurus	Diplodocus	Barosaurus
1. Overall skull proportions	—	heavy	massive	light	light	—
2. Length of muzzle	—	long	short	long	long	—
3. Position of nares	—	lateral	lateral	dorsal	dorsal	—
4. Dentition	—	massive	massive	reduced	reduced	—
5. Vertebral formula: cervicals, dorsals, sacrals, caudals	?,14,5,?	13,?,5,53	12,12,5,53	15,10,5,82	15,10,5,75+	?,8-9,5,?
6. Length of middle cervical vertebrae	short	very long	moderate	short	long	very long
7. Cervical ribs	short	long, slender	long, massive	short, slender	short, slender	short
8. Neural spines in shoulder region	simple	simple	divided-U	divided-V	divided-V	divided-V
9. Pleurocoels in dorsals	moderate	large and deep	large and deep	moderate	moderate	moderate
10. Nature of posterior dorsal centra	amphi-coelous	opistho-coelous	opistho-coelous	amphi-coelous	amphi-coelous	amphi-coelous
11. Height of spines in sacral region	low	very low	moderate	high	high	high
12. Robustness of same	moderate	massive	massive	slender	slender	slender
13. Anterior caudal centra	amphi-coelous	amphi-coelous	amphi-coelous	amphi-coelous	mildly procoelous	mildly procoelous
14. ? Pleurocoels in anterior dorsals	no	no	no	no	yes	yes
15. ? Wing-like transverse processes (caudal ribs) on anterior caudals	no	no	no	yes (first 3)	yes	yes
16. Ventral excavation in antero-middle and middle caudals	no	no	no	no trough	deep concavity	gentle
17. Presence of whip-lash at end of tail	—	no	no	yes	yes	—
18. Development of chevron facets on anterior and middle caudals	great	moderate	weak	weak	weak	weak
19. Presence of typical diplodocid chevrons with anterior development	no	no	no	yes	yes-greatest	yes
20. Distal (upper) end of blade of scapula	expanded	much expanded	expanded	not expanded	mildly expanded	mildly expanded
21. Development of proximal plate of scapula	small	very great	great	great	great	great
22. Angle between blade of scapula and muscle ridge on plate	large	large	large	large	acute	somewhat acute
23. Humerus:femur	—	~1.05	~.75	~.65	~.65	~.71
24. Metacarpal 2 or 3:radius	—	~.51	~.46	~.37	~.30	—
25. Tibia:femur	—	~.58	~.61	~.63	~.69	~.73
26. Breadth of distal end of ischium	narrow	broad	narrow	expanded	expanded	expanded
27. Thickness of distal end of ischium	thin	thin	thin	very thick	thick	thick
28. Presence of ossified calcaneum	—	? yes	yes	no	no	no
29. Greatest metatarsal length, 2 and 3 versus 3 and 4	—	2 and 3	2 and 3	3 and 4	3 and 4	3 and 4
30. Presence of process on postero-distal margin of lateral face of metatarsal I	—	no	no	yes	yes	yes

Apatosaurus and *Diplodocus* would appear to be significant and, together with the other evidence previously cited, clearly establish *D. hayi* as a valid species. It may be further noted that in all casts of the *D. carnegii* skeletons distributed throughout the world, the forelimbs are molded (upsized) from the *D. hayi* skeleton.

The type specimen of *Amphicoelias altus* is clearly diplodocid, but is too incomplete to identify with either *Diplodocus* or *Barosaurus*. It was collected from the top of the Morrison and is 10 to 15% larger than the largest *Diplodocus* from older beds at Dinosaur National Monument, Como Bluff, Bone Cabin Quarry, and other sites. Additional material from the younger beds is needed before definitive statements can be made, but other giants from this time zone will be discussed in connection with *Apatosaurus* and *Camarasaurus*.

The next genus considered is *Apatosaurus* (Fig. 4.2) and its synonym *Brontosaurus* (Riggs 1903). Six species have been referred to the former: *A. ajax*, the genotype, *A. grandis*, *A. laticollis*, *A. louisae*, *A. minimus*, and *A. alenquerensis*. Two were referred to *Brontosaurus*, *B. excelsus* and *B. amplus*. *Atlantosaurus immanis* may be removed from that genus (whose type species *A. montanus* is indeterminate even to family) and placed in *Apatosaurus*. Three species, *A. grandis*, *A. minimus*, and *A. alenquerensis*, may be removed from *Apatosaurus*. Furthermore "*Atlantosaurus*" *immanis* and *Apatosaurus laticollis*, whose type specimens came from the same quarry as that of *A. ajax*, are indistinguishable

from that species. Marsh (1883) gave four characters to distinguish *B. amplus* from *B. excelsus*: 1) less massive dorsals, particularly the zygapophyses, 2) longer anterior caudals without cavities in the transverse processes, 3) thinner sternal plate, and 4) more elongate metacarpals. Examination of the single dorsal complete enough for comparison does not confirm point 1. The anterior caudals are not longer and their processes are not complete enough to determine whether cavities are present or not. The third point is considered an age char-

Table 4.2. *Ratios of minimum circumference to length*

	Humerus		Radius		
	L	R	L	R	Femur
(?)*Diplodocus*					
AMNH 5855	.43	.43	.34	.37	.36
Diplodocus carnegii or *longus.*					
USNM 10865	.45	.46		.35	.35
Diplodocus hayi					
HMS	.49	.50	.36	.40	.41
Apatosaurus excelsus					
YPM 1980	.53	.53		.43	.43

Figure 4.2. *Apatosaurus*, postmedian cervical vertebrae, side views, showing cervical ribs: **A**, *A. ajax*, YPM 1840 (after Marsh); **B**, AMNH 550 – cervical rib; **C**, *A. excelsus*, U. of Wyoming 15556 (formerly CM 563) (after Gilmore); **D**, *A. louisae*, CM 3018; **E**, AMNH 460.

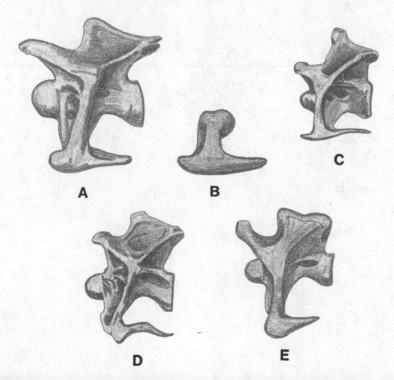

acter, and the fourth point is simply an error. The type skeleton of *B. amplus* YPM 1981 was found just 50 yards from that of *B. excelsus* YPM 1980 in the same stratum of the Morrison. There is no reason to believe that the species are not identical. This leaves *A. ajax*, *A. excelsus*, and *A. louisae* for consideration.

Apatosaurus excelsus and *A. louisae* have been given detailed descriptions by Marsh (1896), Riggs (1903) and Gilmore (1936). In his brief paper that established *A. louisae*, Holland (1915) cited five characters of the dorsal and caudal vertebrae that distinguished it from *A. excelsus*. In his exhaustive study, Gilmore (1936) found thirteen vertebral characters (differing in some respects from Holland's) and three based on the scapula. Three specimens were cited to illustrate *A. excelsus*, YPM 1980 (the type), CM 563 (now mounted at the University of Wyoming, UW 15556) and FMNH 7163 (now FMNH P 25112), and one for *A. louisae* (CM 3018, the type). While the scapulae of CM 3018 and YPM 1980 do differ in the ways cited by Gilmore, those

of *A. ajax* (YPM 1860) and of CM 563 are eroded or incomplete in key areas. The shape of the sauropod scapula shows considerable variation within a species, so until more associated material is available, it is probably better to rely on vertebral characters. It is doubtful that, when a better statistical sample is available, all of Gilmore's specialized criteria can be maintained. For example, several concern minor variations in the hyposphene, an element known to be variable within a given species. On the other hand, several of his criteria have proven significant, and attention will be directed to them. The most obvious is the absence of the anterior extension of the cervical ribs in *A. louisae* (CM 3018) in contrast to those of *A. excelsus* (YPM 1980, CM 563, and AMNH 550), which have a prominent forward extension. The latter two specimens (and probably YPM 1980 also, although the state of preservation is poor) also show the infrapostzygapophysial cavity of the mid- and posterior cervicals to be subdivided by an accessory lamina, while those of CM 3018 are not (Gilmore

Figure 4.3. *Camarasaurus*, fourth dorsal vertebra: **A**, *C. grandis*, YPM 1901 (after Marsh); **B**, *C. supremus*, AMNH 5761 (after Osborn and Mook); **C**, *C. lentus*, YPM 1910 (after Marsh).

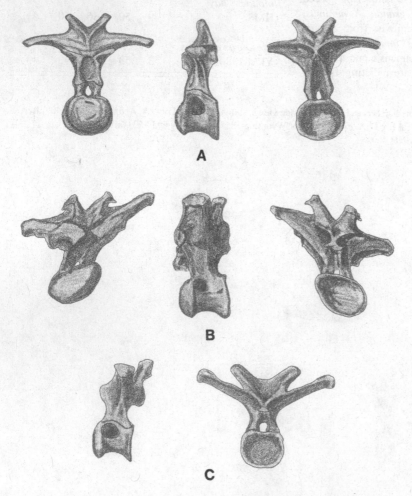

A

B

C

1936). One other apatosaur skeleton shares the absence of the cervical rib extension and the accessory lamina. It is the mounted skeleton in the American Museum, AMNH 460. Cervical 9 does show some evidence of the lamina, but it appears to be restored. None of the cervical ribs have forward extensions.

Another striking feature of the *A. louisae* skeleton, CM 3018, is the extraordinary robustness of the hind limb bones, particularly the femur. Table 4.3 lists the ratios of the least shaft circumferences to the lengths of a number of *Apatosaurus* femora. As mounted, the femur of FMNH 7163 (= P 25112) does not permit an accurate measurement of its shaft circumference, but the least breadth of the shaft shows it to fall into the slender category. One other femur may fall into the massive category, an isolated bone, University of Wyoming 3350, which appears to have a ratio .46. But its distal end is encased in a plaster base, causing some uncertainty of its true length. The evidence provided by vertebrae and femoral robustness suggests that *A. louisae* is a separate species from *A. excelsus*.

A more difficult question concerns the separation of *A. ajax* at the top of the Morrison, and *A. excelsus* from lower down. Their morphologies are similar. Both have the anterior extension of the cervical rib and the accessory lamina. Although no complete femur is known in *A. ajax*, the partial one of the type YPM 1860 and the immense type of "*Atlantosaurus immanis*" (= *A. ajax*), YPM 1840, are clearly of the more slender variety. One point of difference is that an adult *A. ajax* was 10 to

15% larger than an adult *A. excelsus*. YPM 1840 is a fully adult individual, but the type YPM 1860 is a juvenile as shown by the lack of co-ossification of either the dorso-sacral or caudo-sacral centrum to the three primary sacrals, and by the lack of co-ossification of the scapula and coracoid. Still, it is almost as large as YPM 1840! According to Mook's (1917a) criterion, this size difference would justify calling *A. ajax* and *A. excelsus* separate species. I do so provisionally, but although cognizant of the differences between mammals and reptiles, I cannot help but be reminded of the Great Dane and the Chihuahua.

Camarasaurus is the most abundant North American sauropod (Figs. 4.3–4.6). It is the only one in which the osteology is completely known, and in which a number of skulls have been found directly articulated to postcranial remains. Conclusions concerning its various species are blurred by at least four key factors:

1.   All the complete or nearly complete skulls found associated with postcranial skeletons came from one quarry, that at Dinosaur National Monument, and therefore probably belong to a single species.

Figure 4.4. *Camarasaurus*, anterior dorsal vertebrae: A, dorsal 1, *Camarasaurus grandis*, YPM 1910 (after Marsh); B, dorsal 2, *Camarasaurus lentus*, YPM 1910 (after Marsh); C, dorsal 3, *C. supremus* AMNH 5760 (after Osborn and Mook).

Table 4.3. *Ratios of minimum circumference to length in femora*

Specimen	Ratio
Apatosaurus louisae CM 3018	.473
AMNH 460	.486
Como Bluff specimens	
A. excelsus YPM 1980	.43
A. "amplus" YPM 1981	.41
AMNH 222	.42
Bone Cabin Quarry specimens	
AMNH 271	.41
(now in Greece)	
AMNH 353	.42
AMNH 370	.43
(now in Melbourne)	
AMNH 451	.41
[now BM(NH) R 3214]	
AMNH 490	.42
(now in Battle Creek, MI)	
AMNH 521	.41
(now in Amherst College)	

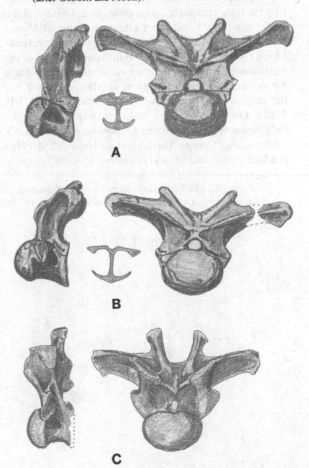

A

B

C

2.   There are half a dozen *Camarasaurus* skulls from the Cleveland-Lloyd Quarry, which exhibit small, noticeable variations (for example, the relative size of the condyle and small differences in the inclination of the paroccipital process). After careful study of this material, Jim Madsen and I have concluded that such differences are probably due to individual variation. In the future, when more material is available for comparison, the Cleveland-Lloyd specimens may be of considerable value in the study of speciation and sexual dimorphism, but their use for this purpose now is quite limited.

3.   I have been able to determine with certainty that some of the material assigned to *Camarasaurus supremus* by Osborn and Mook (1921), which would help to define the species, did not come from either Cope Quarries 1 or 2 at Garden Park, Colorado, and almost certainly does not belong to this genus. For example, the two (relatively) short, stocky humeri (AMNH 5761) almost surely belong to *Apatosaurus*.

4.   Limb bone ratios clearly indicate that the Portuguese species *"Apatosaurus" alenquerensis*, which is a camarasaurid, is a different species from any of the American camarasaurs. The problem is that the neural arches of all the presacral vertebrae are missing, as is the skull of course! The species may represent a new genus of the Camarasauridae.

What can be said about the species of *Camarasaurus* now? First, as with *Diplodocus* and *Apatosaurus*, the material from the top of the Morrison, namely Cope's type species *C. supremus*, is definitely much larger than any adult from Dinosaur National Monument, Como Bluff, Bone Cabin Quarry, Cleveland-Lloyd Quarry, or anywhere else in the middle part of the Morrison (see Table 4.4). Age variation no doubt plays a prominent part in the range of variation exhibited, but the largest femur is about 15% shorter than AMNH 5761a. On the other hand, size would appear to be the only character that separates *C. supremus* from Marsh's *"Morosaurus" lentus*. The type of the latter (YPM 1910) is a half grown animal, and the referred, nearly complete skeleton (CM 11338) from Dinosaur National Monument is only one-third the adult size (Gilmore 1932).

One character Marsh (1889) recognized to separate *"M." lentus* from *"M." grandis* (and its synonyms *M. impar* and *M. robustus*, Williston 1898) concerns the neural pedicels of the dorsal centra. In *C. grandis*, exemplified by YPM 1901 and CM 584, the pedicels are greatly elevated above the neural canal, but they are low and "normal" in *C. lentus* (YPM 1910), *C. supremus* (AMNH 5760, AMNH 5761), and in the Dinosaur National Monument specimens typified by the type of *C. "annae"* (= *lentus*) (CM 8942). This character would appear to be outside what would be expected of individ-

Table 4.4. *Femoral length for* Camarasaurus

Specimen	Length (m)
C. supremus	
AMNH 5761a	1.8
(should be 5760)	
Other species of *Camarasaurus*	
Como Bluff, Q-1	1.18, 1.175, 1.145, 1.135, 1.075, 0.955
Como Bluff, Q1a	1.430
Bone Cabin Quarry	1.075, 1.015
Dinosaur National Monument	1.150, 1.430, 1.285, 1.080, 1.010
Cleveland-Lloyd Quarry	1.390, 1.325, 1.270 (and a questionable femur, 1.540)

Figure 4.5. Left scapulae, lateral views: **A**, *Camarasaurus grandis*, YPM 1901 (after Marsh); **B**, *Camarasaurus* (?) *alenquerensis*, type; **C**, *Apatosaurus excelsus*, YPM 1980 (after Marsh).

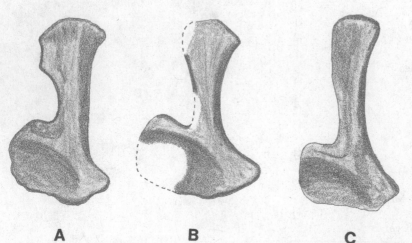

A   B   C

ual variation and establishes the validity of *C. grandis*. But separation of the Upper Morrison *C. supremus* from the Middle Morrison *C. lentus*, if valid, rests completely on size as already noted.

"*Apatosaurus*" *alenquerensis* may be provisionally referred to *Camarasaurus* because:

1. There are twelve dorsal vertebrae (*Apatosaurus* has ten) and all are opisthocoelous.
2. There is a widely expanded distal (upper) end of scapula.
3. The humerus is slender (Fig. 4.6), and has a significantly greater humero–femoral ratio.
4. The slender ischium has a less expanded distal end (Fig. 4.7).

Mook (1914) incorrectly gave the humero-femoral ratio in *Camarasaurus* as .6, probably misled by the misassignment of the humeri to *C. supremus*. In the American species, it is typically about .75, as determined from four or five individuals in which both elements have been found. This contrasts with .65, a typical ratio for *Apatosaurus* and *Diplodocus*. In *C.(?) alenquerensis* the ratio is .9, a figure that would clearly indicate a separate species. Another significant character is the much shallower pleurocoels in its dorsal centra. Except for the longer forelimb, the limb and girdle bones are in reasonable accord with those of the American species. Although this great disparity in the fore- to hind- limb ratio would normally indicate separate genera, I consider it more prudent to wait until the diagnostic presacral neural arches and perhaps even the skull are found. Sauropod systematics is already cluttered with far more invalid taxa than valid ones.

Five species of *Brachiosaurus* have been described (Figs. 4.8, 4.9), but the Saharan form *B. nougaredi* is too incomplete to allow even a generic identification. The other four include the genotype, *B. altithorax* from Colorado, *B. brancai* and *B. fraasi* from Tanzania, and *B. atalaiensis* from Portugal. Janensch (1914) originally separated the Tanzanian species on the basis of the scapula and humerus. The upper (distal) end of the scapula of *B. fraasi* is somewhat less expanded than that of *B. brancai*, and the humerus of the former is a bit more robust than that of the latter. From what is known of individual variation in the scapula of *Apatosaurus* and *Camarasaurus*, the differences here would appear to be insignificant (Table 4.5 – ratios of Tanzanian *Brachiosaurus* humeri in Berlin and London). There is seen to be considerably greater variation between many of the paratypes than between the two types. Janensch (1961) eventually synonymized the two species.

Comparison of *B. altithorax* with its more completely known African relative shows that the coracoid, femur, and sacrum of the two species are in complete accord. The American humerus appears to be a bit more robust than that of the type specimen of *B. brancai*. Its circumference to length ratio is .36 based on the preserved length of 2.040 m. However, as Riggs (1904) stated,

When found the distal end of the former humerus was exposed at the surface, broken and displaced. When the fragments had been gathered up and fitted to the portion in the matrix, the bone measured about seven feet in length .... The surface of the distal end of the humerus had flaked away in the process of weathering to a firm chalcedony core .... All traces of rugosity have likewise disappeared from the distal end, indicating that the humerus was probably some inches longer than it now stands.

I have examined the bone and completely agree with Riggs's assessment. I would estimate that when complete this humerus would be every bit as long as that of the *B. brancai* holotype specimen and with the same circumference to length ratio. Thus, the characters to separate the species must focus on the dorsal vertebrae. The neck and first four dorsals of *B. brancai* (Humboldt

Figure 4.6. Left humeri, anterior views: **A**, *Camarasaurus* (?) *alenquerensis*, type; **B**, *Camarasaurus grandis*, YPM 1901 (after Marsh); **C**, *Apatosaurus louisae*, CM 3018 (after Gilmore).

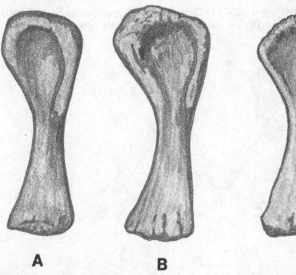

A **B** **C**

Museum S II) were preserved in serial order, while the mid- and posterior ones were somewhat displaced, and some of them were incomplete. Janensch (1950) determined that 13 cervicals and 11 dorsals were present, and the column was thus restored in the great mounted skeleton in Berlin. If eleven is the correct number of dorsals in *Brachiosaurus*, then the seven belonging to *B. altithorax* FMNH P 25107 would be numbers 5–11. The transition of dorsal 4 of Humboldt S II to dorsal 5 of FMNH P 25107 would then appear to be somewhat abrupt. The spine of number 4 is much higher and more slender in Humboldt S II. Its centrum would appear to be relatively smaller (though it is imperfect), its pleurocoel higher on the centrum, etc. These incongruities would largely disappear, however, if there were twelve

dorsals instead of eleven, a possibility that appears to me most likely. The Chicago series would then represent dorsals 6–12. Minor differences, including proportions, do occur in the posterior dorsals of the two animals and are perhaps of diagnostic value. Paul (1988) considers these differences more important than I do, and has erected the subgenus *Giraffatitan* for *Brachiosaurus brancai*.

Evaluation of *Brachiosaurus atalaiensis* is more difficult, because only a few fragmentary presacrals are available, and the only complete limb bones are a radius and a tibia. The cervical figured by Lapparent and Zbyszewski (1951) certainly cannot be a cervical of *Brachiosaurus* or any closely allied form as it is much too short. My best judgment is that it is an anterior dor-

Table 4.5. *Ratios for Tendaguru* Brachiosaurus *humeri*

Length (mm)	Proximal breadth/length	Minimum shaft breadth/length	Distal breadth/length	Circumference of shaft/length
2130[a]	.28	.12	.24	.31
2100	.26+	.11	.22+	.30
1730	.30	.12	.26	—
1700[b]	.30	.11	.26	.32
1690	.34	.14	.28	.36
1660	.33	.12	.28	.35
1610	.32	.13	.24	.33
1530	.31	.12	.27	.35
1300	.33	.13	.25	.36
1295	.27	.11	.25	.30
1165	.31	.11	.24	.34
1100	.27	.11	.23	.33
990	.29	.12	.25	.34
690	.30	.12	.27	.34

[a]Holotype of *B. brancai*.
[b]Holotype of *B. fraasi*.

Figure 4.7. Left ischia, lateral views: **A**, *Apatosaurus ajax*, YPM 1840 (after Marsh); **B**, *Apatosaurus excelsus*, YPM 1980 (after Marsh); **C**, *Camarasaurus* (?) *alenquerensis*, type; **D**, *Camarasaurus lentus*, YPM 1910 (after Marsh).

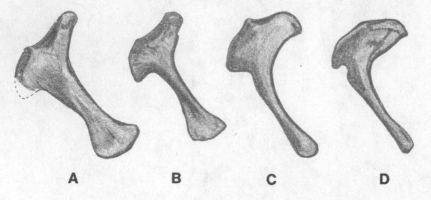

A   **B**   **C**   **D**

sal with a very high pleurocoel. Having lost all its processes, it is of little diagnostic value. The caudals, with their long but simple transverse processes and greatly reduced arches and spines, are similar to those of *B. brancai*. The limb and pelvic bones also resemble those of *B. brancai*, but measurements (Table 4.6) would suggest a somewhat shorter fore- to hind- limb ratio. A

Table 4.6. *Length (m) of limb elements in* Brachiosaurus

	Radius	Tibia	Fibula
Brachiosaurus brancai			
HM SII	1.240	—	1.190
B. brancai			
HM St 148-9	—	1.070	1.090
B. altalaiensis	1.140	1.120	—

Table 4.7. *Ratios of minimum shaft circumference to length in* Dicraeosaurus

	Femur		Fibula
	R	L	
Dicraeosaurus			
hansemanni HM m	.42	.43	.33
D. sattleri HM M	.39	.40	—
D. sattleri HM O	—	—	.30

radius to tibia ratio of 1.06 is suggested for *B. brancai* compared to 1.02 for *B. atalaiensis*. More complete material is needed before the Portuguese species can be properly assessed.

The last genus, which will be touched on very briefly, is the African genus *Dicraeosaurus* (Fig. 4.10), so completely described and figured by Janensch (1929, 1961). The two species are derived from two distinctly different levels of the Tendaguru Beds, *D. hansemanni*, the holotype species, from the middle stratum and *D. sattleri*, from the upper. In this aberrant diplodocid, the division of the presacral neural spines reached its bizarre climax with the very deep cleavage of very high spines in the shoulder region. The later *D. sattleri* shows the development of this feature to a significantly advanced degree over *D. hansemanni*. The limb and girdle ratios show that *D. sattleri* was more gracile than *D. hansemanni*. Examples of shaft circumference to length ratios of several limb bones are given in Table 4.7 and two distinct species are indicated.

Conclusions

It has not been possible to develop a definitive list of criteria for the determination of sauropod species as attempted by Mook (1917a). Mook's own criteria have not proven to be useful. To date, skulls have been of little or no use because they are so rare, and when preserved, are usually not associated with postcrania. On the other hand, in five of the better represented genera, it has been possible to distinguish species on the basis of overall limb robustness, certain limb ratios, and vertebral characters. Overall size may also be significant. It is hoped that recent discoveries in South America, and particularly in China, may eventually provide more definitive criteria.

Figure 4.8. *Brachiosaurus*, scapulae, lateral views: **A**, *Brachiosaurus brancai*, HM Sa, left; **B**, *Brachiosaurus "fraasi"*, HM Y, left and right.

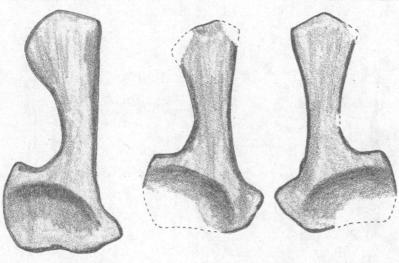

A   **B**

68

Acknowledgments

I wish to express my deepest gratitude to Kenneth Carpenter, who has redrawn all the figures. I also wish to thank Dr. Philip Currie and Dr. Dale Russell for critically reading the manuscript and offering many suggestions for improvement.

Figure 4.9. *Brachiosaurus*, dorsal vertebrae (data from Janensch and Riggs): dorsal 4, *B. brancai*: **A**, lateral; **B**, posterior; dorsal 5 (or 6), *B. altithorax*: **C**, lateral; **D**, posterior.

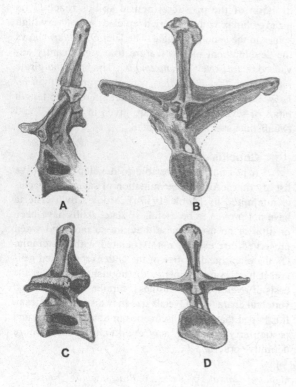

References

Bonaparte, J. F. 1979. Dinosaurs, a Jurassic assemblage from Patagonia. *Science* 205:1377–1379.

1986. The early radiation and phylogenetic relationships of the Jurassic sauropod dinosaurs, based on vertebral anatomy. *In* Padian, K. (ed.), *The Beginnings of the Age of Dinosaurs* (Cambridge: Cambridge University Press), pp. 247–258.

Bonaparte, J. F. and Powell, J. E. 1980. A continental assemblage of tetrapods from the Upper Cretaceous Beds of El Brete, northwestern Argentina. *Memoires de la Societe Geologique de France*, NS 59(139):18–28.

Charig, A. J. 1980. A diplodocid sauropod from the Lower Cretaceous of England. *In* Jacobs, J. (ed.), *Aspects of Vertebrate History* (Flagstaff: Museum of Northern Arizona Press), pp. 231–244.

Charig, A. J., Attridge, J., and Crompton, A. W. 1965. On the origins of the sauropods and the classification of the Saurischia. *Linnean Society of London Proceedings* 176:197–221.

Chure, D. J. and McIntosh, J. S. 1989. *A Bibliography of the Dinosauria (Exclusive of Aves)* (Grand Junction: Museum of Western Colorado), pp. 1–226.

Dong, Z., Zhou, S., and Zhang, Y. 1983. The dinosaurian remains from the Sichuan Basin, China. *Paleontologia Sinica, Whole Series* 162, (C), No. 23:1–145. [in Chinese with English summary]

Gilmore, C. W. 1932. On a newly mounted skeleton of *Diplodocus* in the United States National Museum. *United States National Museum Proceedings* 81:1–21.

1936. Osteology of *Apatosaurus* with special reference to specimens in the Carnegie Museum. *Carnegie Museum Memoirs* 11:175–300.

Hatcher, J. B. 1901. *Diplodocus* (Marsh): its osteology, taxonomy and probable habits, with a restoration of the skeleton. *Carnegie Museum Memoirs* 1:1–63.

He, X., Li, K., Cai, K., and Gao, Y. 1984. *Omeisaurus tianfuensis*–a new species of *Omeisaurus* from Dashanpu,

Figure 4.10. *Dicraeosaurus*, mid-dorsals, posterior views (data from Janensch): **A** and **B**, *D. hansemanni*, HM m; **C** and **D**, *D. sattleri*, HM M.

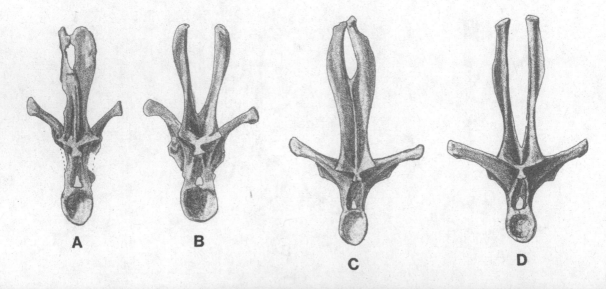

Zigong, Sichuan. *Journal of the Chengdu College of Geology*, Supplement 2:13–32. [in Chinese with English summary]

Holland, W. J. 1915. A new species of *Apatosaurus*. *Carnegie Museum Annals* 10:143–145.

——— 1924. The skull of *Diplodocus*. *Carnegie Museum Memoirs* 9:378–403.

Janensch, W. 1914. Uebersicht uber die Wirbelthierfauna der Tendaguru-Schichten, uns. *Archiv fur Biontologie* 3:81–110.

——— 1929. Die Wirbelsaule der Gattung *Dicraeosaurus*. *Palaeontographica*, Supplement VII (1) T. 2 L. 1:37–133.

——— 1950. Die Wirbelsaule von *Brachiosaurus brancai*. *Palaeontographica*, Supplement VII (1) T. 3 L. 2:27–93.

——— 1961. Die Gliedmassen and Gliedmassengulter der Sauropoden der Tendaguru-Schichten. *Palaeontographica* Supplement VII (1) T. 3 L. 4:177–235.

Lapparent, A. F. and Zbyszewski, G. 1951. Découverte d'une riche faune de reptiles dinosauriens dans le Jurassique supérieur du Portugal. *Compte Rendus de l'Académie des Sciences de Paris* 233:1125–1127.

——— 1957. Les dinosauriens du Portugal. *Services Geologiques du Portugal Memoire* (NS) 2:1–63.

Marsh, O. C. 1877. A new order of extinct Reptilia (Stegosauria), from the Jurassic of the Rocky Mountains. *American Journal of Science* 14:514–516.

——— 1880. Principal characters of American Jurassic dinosaurs. Part 3. *American Journal of Science*, series 3, 19:253–259.

——— 1883. Principal characters of American Jurassic dinosaurs. Part 5. *American Journal of Science*, series 3, 21:417–423.

——— 1884. Principal Characters of American Jurassic Dinosaurs. Part 7. On the Diplodocidae, a new family of the Sauropoda. *American Journal of Science*, series 3, 27:160–168.

——— 1889. Notice of new American dinosaurs. *American Journal of Science*, series 3, 37:331–336.

——— 1896. The dinosaurs of North America. *United States Geological Survey Annual Report* 16(1):133–244.

Monbaron, M. and Taquet, P. 1981. Decouverte du squelette d'un grand Cetiosaure (Dinosaure Sauropode) dans le bassin Jurassique moyen de Tilougguit (Haut-Atlas central, Maroc). *Comptes-rendus de l'Academie des sciences, Paris*, series D, 292(2):243–246.

Mook, C. C. 1914. Notes on *Camarasaurus* Cope. *New York Academy of Science Annals* 24:19–22.

——— 1917a. Criteria for the determination of species of the Sauropoda, with special reference to a new species of *Apatosaurus*. *American Museum of Natural History Bulletin* 37:355–358.

——— 1917b. The fore and hind limbs of *Diplodocus*. *American Museum of Natural History Bulletin* 37:815–819.

Osborn, H. F. 1899. A skeleton of *Diplodocus*. *American Museum of Natural History Memoirs* 1:191–214.

Osborn, H. F. and Granger, W. 1901. Fore and hind limbs of the Sauropoda from Bone Cabin Quarry. *American Museum of Natural History Bulletin* 24:199–208.

Paul, G. S. 1988. The brachiosaur giants of the Morrison and Tendaguru with a description of a new subgenus *Giraffatitan*, and a comparison of the world's largest dinosaurs. *Hunteria* 2(3):1–9.

Riggs, E. S. 1903. Structure and relationships of opisthocoelian dinosaurs. Part 1. *Apatosaurus* Marsh. *Field Columbian Museum Publications, Geological Series* 2(4):165–196.

——— 1904. Structure and relationships of Opisthocoelian dinosaurs. Part 2. The Brachiosauridae. *Field Columbian Museum Publications, Geological Series* 2(6):229–248.

Romer, A. S. 1968. *Notes and Comments on Vertebrate Paleontology* (Chicago: University of Chicago Press), pp. 1–304.

Zhang, Y., Yang, D., and Peng, G. 1984. New materials of *Shunosaurus* from the Middle Jurassic of Dashanpu, Zigong, Sichuan. *Journal of Chengdu College of Geology*, Supplement 2:1–12. [in Chinese with English summary]

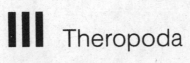

III Theropoda

5 Variation in theory and in theropods

RALPH E. MOLNAR

Abstract

The alpha taxonomy of dinosaurs encounters difficulties from the inability to apply biological species criteria to fossil material, and the prevalence of small sample sizes and of incomplete specimens. These enforce a quasi-typological praxis, in which differences in form must be used to distinguish species. The problem is recognizing taxonomically significant features that distinguish closely related species. Certain modern Australian lepidosaur and marsupial species would be indistinguishable if known only from fossil material available in small samples. Ecological arguments that one or few conspecific herbivores will occupy a given area and that the larger the body size of a terrestrial herbivore the less likely the existence of a sympatric closely related herbivore species, are not compelling. Ontogenetic variation in large theropods involves predominantly changes in proportion of skeletal structures, such as limbs, and some changes in proportions of individual bones. Individual variation involves largely changes in proportions of individual bones and also in the number of serial elements (teeth).

Introduction

The importance of variation as a theoretical concept and an integral part of the theory of evolution by natural selection is well established (e.g., Mayr 1963). The recognition of phenotypic variation in dinosaur systematics follows from this. The typological taxonomy of nineteenth- and early twentieth-century paleontology, often recognizing several contemporaneous sympatric congeneric species, no longer exists, although even then workers such as Gilmore (1925) and Parks (1935) recognized the role of variation. Taxa, such as the several species of Morrison *Camptosaurus* and of some Albertan hadrosaurs, have been reassessed in the light of our understanding of variation, not only individual but also sexual (i.e., sexual dimorphism) and ontogenetic.

In Dinosaur Systematics: Perspectives and Approaches, *Kenneth Carpenter and Philip J. Currie, eds. Copyright © Cambridge University Press, 1990.*

Such reassessment was pioneered in the USSR by Rozhdestvensky (1965) for ontogenetic variation and Kurzanov (1972) for sexual dimorphism, and in North America by Dodson (1975, 1976).

The recognition of phenotypic variation in fossil populations has two overlapping aspects, biological and taxonomic. Biological studies of variation, such as those of Kurzanov (1972) and Dodson (1976) have provided a picture of growth and sexual dimorphism in dinosaurs. Taxonomic studies, such as those of Rozhdestvensky (1965) and Dodson (1975) have applied the analysis of variation to alpha taxonomic problems. Although the biological work is of great interest, this essay will discuss only the taxonomic aspects.

Essential to the treatment of variation in fossil populations is the relationship between the biological and the paleontological species, recently reviewed by Fox (1986). Fox contends, and this view is adopted here, that the paleontological species is the biological species recognized from fossil material. It does not have independent status. However, criteria for recognizing biological species cannot be applied to fossil material because the relevant observations (e.g., reproductive isolation) cannot be made. Thus, the "paleotaxonomist" must rely on morphological distinctions to recognize species, although this is not how species are defined.

Taking variation into account in comparing two species requires large enough samples of each to allow reasonable confidence in the limits of variation: Dodson used 30 (1975) and 24 (1976) specimens. Of 171 named theropod species (excluding *nomina nuda* and those based on isolated teeth), 85% are known from five or fewer specimens. The proportion of species known from similarly few specimens is much the same for some other dinosaur groups. Such small sample sizes imply that statistical methods for detecting variation are not effective (Simpson, Roe, and Lewontin 1960).

The difficulty arising from small sample size is

complicated by the availability of only the skeleton (usually incomplete). In distinguishing species the constancy of difference is significant, rather than the degree of difference (Dodson and Dodson 1985). This implicitly recognizes that significant differences may be slight. It is suggested here that features that distinguish closely related species (distinguishing features) may not be found in the skeleton. Even when they are, they may be difficult to discern from incomplete specimens known from small samples.

With small samples one can only compare what material is available. This enforces in practice a typological-like operational species concept, especially as many theropods (at least 42% of named species) are represented by a single specimen. Thus, operationally, species are treated taxonomically in the same manner as genera, families, and higher categories. Because fewer than ten theropod species are known from more than ten specimens, this implies that most comparisons will be between taxa known from small samples. Even the t-test to compare a single specimen with a population (Simpson et al. 1960) is not applicable. Incomplete material known from small samples also makes it difficult or impossible to apply the rule that the degree of difference in form between fossil species should approximate that between living related species. Rejecting taxa known from small samples would likely allow more reliable identification of specimens, by restricting identification to nearly complete specimens, but would discard considerable information regarding paleocommunity structure and paleozoogeography.

The emphasis here is on problems of alpha taxonomy of species or genera known from samples too small for statistical analysis. Given that specimens assigned to two closely related named species (each represented by two or three individuals) share some features but differ in others, do the differing features represent variation and imply that all specimens belong to a single species? Or do they indicate membership in two distinct species? The question is how well biological species can be recognized when represented by fossil material, known from small samples. In this regard variation is a kind of alpha taxonomic "noise." Even with large sample size, it is impossible to be certain that a sample is drawn from a homogeneous population (Simpson et al. 1960). Two kinds of errors are possible. First, one species can be mistaken for two (or more) by interpreting variability as distinguishing features, and second, two (or more) species can be misidentified as one by mistaking distinguishing for variable features. Techniques for identifying features in this fashion in a rigorous and repeatable manner would be useful. Unfortunately, this appears difficult or impossible at present, but methods of increasing the objectivity of species recognition from small samples will be explored, using theropods as an example. This essay is not a comprehensive review of variation, its measurement and its statistical treatment, the placement of specimens in a phylogenetic tree or cladogram, nor a comparison of specimens not represented by corresponding material. Rather, it is concerned only peripherally with ascertaining the types of variation, whether individual, sexual or ontogenetic.

There are two possible approaches to discriminating between variable and distinguishing features. First, the possibility that arguments not dependent on morphology may circumvent consideration of variation and allow conjectures or conclusions regarding the taxonomic allocation of a specimen. Two ecological arguments used in this capacity will be examined. Second, the possibility that distinguishing features, at the specific or generic level, can be discerned without recourse to examination of large samples. Generalizations derived from Yablokov's (1966) studies of variation in living mammals will be discussed in this context and compared with similar observations of theropod dinosaurs, mostly taken from the literature. The provenance of a specimen is not examined as a potential source of information in this context, although it is often so used in practice, because the relative age and geographic locality of a specimen are not intrinsic, biological characters. This essay is not intended to present final results, but to suggest directions for further inquiry; especially to suggest that research on variation in living animals, taking into account the problems in dinosaur alpha taxonomy, is as important to understanding dinosaur taxonomy as is studying dinosaurs themselves.

Origins of variation seen in fossil material

Several sources produce the variation seen in fossil material: 1) individual, 2) sexual (i.e., dimorphism), 3) ontogenetic, 4) geographical and chronological, and 5) intraspecific population (4 and 5 are discussed by Yablokov 1966). For completeness, two other sources of apparent variation may be mentioned: postmortem deformation and pathological changes. The effects of these two latter sources usually can be determined with some confidence – but see Walker (1961) and Madsen (1976) for an instance of each that proved difficult to recognize. The study of modern tetrapods has provided an understanding of variation resulting from both sexual differences and ontogenetic development. This understanding has been applied to fossil vertebrates – samples generally being too small and restricted for any assessment of geographical and other similar variation – but other sources of variation have received less attention.

Osteologically indistinguishable species

Systematics seeks to represent the genetic relationships of organisms, and to do this individual specimens must be assigned to taxa, the most basic of which is the species. Biological individuals, unlike individual electrons, vary (Mayr 1982), so that for living species a morphological definition is not ideal. Modern species are, in principle, defined on reproduction (e.g., Dobzhansky et al. 1977) or recognition criteria (Paterson 1985). Such criteria cannot be determined from the fossil record (and often not for modern species, Fox 1986). Thus, for practical reasons morphological criteria are used. The preservation of only skeletal material in the fossil record clearly indicates that some morphological differences of the species represented may be lost.

Species may differ genetically by only small amounts (Raff and Kaufman 1983), and sibling species differ genetically but not morphologically (Mayr 1982). In both cases distinct biological species are morphologically similar. In some cases it is clear that morphological change can proceed at a much lower rate than genetic or biochemical change (Raff and Kaufman 1983, and references cited therein). Thus, one might expect tetrapod species that differ only in nonskeletal features and these species could not be distinguished in the fossil record. Indeed, tetrapod species that can be distinguished by skeletal features only with the aid of statistical tests requiring sample sizes greater than five, cannot be distinguished from a record like that available for theropods. It has been assumed that taxonomically significant differences are expressed in the skeleton. But experience with some Australian tetrapods (marsupials and lepidosaurs) suggests that this assumption is not always valid.

Several modern species of Australian tetrapods, not sibling species, are either osteologically indistinguishable, distinguishable only by features that may be overlooked as due to individual variation, or distinguishable only with large samples. These include species of agamid and scincid lizards, as well as phalangerid and macropodid marsupials (Bartholomai and Molnar 1981). They range from a length of a few centimetres and weight of a few grams (*Ctenotus*) to a length of 2 metres and a weight of about 50 kilograms (*Macropus*). Such osteologically similar forms give no indication of being restricted to any given size range or habitat.

Instances of modern osteologically indistinguishable species seem poorly known among paleontologists, so certain of these examples will be presented in detail. Species of the *Ctenotus lesueurii* group are readily distinguished by color pattern and snout profile (Czechura and Wombey 1982), but are visually indistinguishable osteologically (Czechura, pers. comm. 1988). The Late Cenozoic (Late Pliocene or Early Pleistocene) kangaroo, *Macropus mundjabus*, may be distinguished from the living *Macropus giganteus* (the grey kangaroo) by the form of upper incisors, but not of the molars, and by the fortuitous preservation of the integument revealing a distinctive pattern of hair follicles (Flannery 1980). The possibility that the pattern of the pelt was somehow affected by its unusual preservation is obviated by specimens of *M. giganteus* discovered in the same deposit with identical preservation. The postcranial osteological

differences seem trivial: a difference in the femur:tibia ratio, and in the form of pedal ungual IV (more strongly curved with a less developed dorsal crest in *M. mundjabus*, Flannery 1980). These could easily be regarded as due to individual variation, unless statistical analysis was possible.

Further examples are found among the agamid lizards, *Amphibolurus* (Czechura pers. comm. 1988), and in the macropodids, *Petrogale persephone* (Maynes 1982) and *P. inornata*. These examples all involve taxa known from few individuals, however the phalangerids *Trichosurus vulpecula* (the brushtail possum) and *Trichosurus caninus* (the mountain brushtail) are known from large numbers of specimens. They are obviously different in coat color, pinna form, and tail pelage (Strahan 1983). However, osteologically they seem indistinguishable, and unlike placental mammals they do not even differ in dental form. Cranial and postcranial elements cannot be distinguished by inspection. Ratios of cranial measurements reflecting cranial form (breadth of braincase:condylobasal length, zygomatic breadth:condylobasal length, maxillary tooth row length:condylobasal length, and horizontal diameter of foramen magnum:condylobasal length) did not reveal significant differences with Student's t-test (sample of 40 used for each species). Further statistical tests have not yet been applied. These species are broadly sympatric, and would certainly occur together in a fossil assemblage, but they do not interbreed or produce viable hybrids either in the wild or in captivity.

The foregoing examples indicate that taxonomically significant morphological differences are not necessarily found readily in the skeleton, and that seemingly minor osteological differences are not necessarily due to individual or other taxonomically nonsignificant variation. Even though more extensive statistical tests might reveal differences between the species considered here, it would be impossible to distinguish between them if they were known only as fossils from small samples. These examples suggest that two (or more) fossil species may easily be taken for one. Thus the number of recognized (valid) fossil species likely underestimate the number of biological species from which the remains are derived.

Species not distinguishable on osteological material could possibly be recognized with other methods of determining variation in a fossil population. The existence of proteinaceous residues in fossil bone (Wyckoff 1972 and Rowley et al. 1986, among others) suggests that biochemical variation could be estimated where sufficient residues remained that had not been subject to chemical alteration. This in turn would allow an independent check on variation between fossil specimens that yield appropriate chemical residues. If osteologically indistinguishable specimens had two or more distinct varieties of protein, this would suggest membership in different species.

Two arguments from ecology

One attempt to distinguish taxonomically non-significant variation from taxonomically significant difference has employed two related nonmorphological arguments. These argue that in certain cases, mistaking one for two (or more) species is more likely than the reverse. In other words, that two (or more) species have been named when only a single species is represented. It has been suggested (e.g., Galton and Powell 1980, Ostrom and Wellnhofer 1986, and this volume) that the occurrence of morphologically different fossil specimens in a small geographical area increases the likelihood that the differences are due to variation and not taxonomic distinction. (In neither case cited did this compose the mainstay of the argument, but was presented in only a supplementary capacity.) This argument seems to assume that several species of a tetrapod genus or family would not occupy the same geographic region. It also assumes that the death assemblage under consideration represents a single habitat, otherwise closely related taxa could have been allopatric.

For certain regions today this argument is valid. The North American Great Plains supports only two large native herbivores, the bison and the pronghorn. On the other hand, southeast Queensland, Australia, is (or was earlier this century) occupied by five species of Macropodidae, the largest living marsupial herbivores (Cooke 1983): *Macropus giganteus, M. parryi, M. rufogriseus, Thylogale thetis,* and *Wallabia bicolor*, as well as *Aepyprymnus rufescens* from the closely related Potoroidae. These five species from a single family, three congeneric, would be found in a single death assemblage because this area has only a single drainage, the Brisbane River.

Similar occurrences of closely related forms are found in East Africa. Maberly (1960) presents detailed locality data for bovids in the national parks of Kenya. In two of these, Nairobi and Masai Mara National Park, the fluvial drainage is such that creatures living there may reasonably be expected to make up a single death assemblage (cf. maps given in Williams and Arlott 1981). Nairobi National Park includes 17 species of bovids (of which five are rare), whereas Masai Mara National Park includes 15. In both parks there are two instances of two congeneric species being found: *Kobus dieassa, K. ellipsiprymnus, Redunca fulvorufula,* and *R. redunca* in Nairobi and *R. fulvorufula, R. redunca, Cephalophus herveyi,* and *C. monticola* in Masai Mara. *Cephalophus* and *Redunca* are small forms, but *Kobus* is not. It must be admitted that there is not universal agreement regarding the validity of all of these species (as exists for the Australian species). In this paper, those accepted by East African workers have been used. Clearly, the number of congeneric or confamilial large mammalian herbivores inhabiting a given small area depends upon which small area (and hence which herbivores) is selected. The conclusions drawn from any analogy between the distribution of modern large mammalian herbivores and extinct dinosaurian herbivores depend heavily upon the choice of the analogs.

The second, closely related, nonmorphological argument is weaker. The larger the body size (of land dwelling forms), the less likelihood there is that two closely related forms will be sympatric. The ability of two species to coexist is presumably determined by the limiting similarity of the taxa, although to my knowledge this has not been explicitly stated. Hence, the morphological differences seen in putatively congeneric sympatric dinosaurs will more likely be due to variation, rather than be of taxonomic significance. This argument, based on the work of Hutchinson (1959, 1965), has been set out in Dodson (1975).

Several objections may be raised against this argument from size. First, Hutchinson looked at tetrapods known to be sympatric because they are alive today. Fossil tetrapods are known from death assemblages and whether or not they were sympatric is inferred not observed. Indeed, one could equally conclude from these premises that two closely related, reputedly sympatric dinosaurs must have been derived from two different habitats.

Second, the assumption of uniformity in applying this argument to the Mesozoic overlooks at least two possible differences of the modern world from that of the Mesozoic. The present diversity of large tetrapods has certainly been reduced by human activity. It may also have been affected, possibly reduced, by the unusual climatic conditions of the Late Cenozoic. Finally, one may speculate from the curve of species numbers vs. area given by MacArthur and Wilson (1967), that land areas the size of Laurasia and Gondwanaland may have supported a greater diversity of land dwelling tetrapods per unit area than the present continents.

Third, the idea that diversity is governed by limiting similarity (often assumed to be directly proportional to taxonomic proximity) is itself now under attack (Strong et al. 1984; Ricklefs 1987). Until the validity of the basic concepts (such as limiting similarity) that govern the structure of animal communities has been established, it is premature to use inferences from them as starting points for paleontological interpretations.

An apparent counterexample to the argument of size derives from dinosaurian paleontology itself. Several species of large sauropod did live together, that is, are found in the same death assemblages in the Morrison Formation (Dodson, Behrensmeyer, and Bakker 1980) and Tendaguru beds (Russell, Beland, and McIntosh 1980). Such examples cast doubt on the validity of the argument from size.

Both ecological arguments approach the problem of variation obliquely. Both interpret morphological variation by applying nonmorphological concepts, which themselves are based in part on the substantially morphological concept of limiting similarity. Although it would seem preferable to avoid this excursion from morphology to ecology and back again, where there is little morphological information such excursions have value. This is not to say that such an argument is always misleading. The well known Eltonian pyramid suggests that top carnivores will have low taxonomic diversity, which is corroborated by the record at least for the Hell Creek and Judith River Formations in North America (Russell 1970; Molnar 1978).

Both the considerations given in this section and those of the previous section suggest that the mistaking of two (or more) species for one may be more common than generally believed.

Generalizations regarding morphological variation

Given small samples and their attendant difficulties, synonymies based on known trends in variation, such as that by Galton and Powell (1980) of *Camptosaurus*, based on known ontogenetic trends, are more secure than those based merely on general similarity. Evidence for regarding even minor osteological differences as indicating variation rather than interspecific difference (and vice versa) lends confidence to a synonymy. Cooper (1981) suggested that *Anchisaurus polyzelus* was the juvenile form of *Ammosaurus major* because of the smaller size and narrower feet of the former genus. Among some mammals, and seemingly *Allosaurus* (Steel 1970) at least, the feet of juvenile individuals are relatively larger (i.e., broader) than those of older individuals, not relatively more slender. The change in form required by Cooper's synonymy is opposite to this trend, hence this synonymy cannot be supported on these grounds. This example shows the utility of a knowledge of general trends or principles of morphological variation.

The secure niche afforded variation by evolutionary theory seemingly has distracted attention from the study of morphological variation, although this was pursued by Simpson (1944, 1953). The paleontological treatment of variation usually ends with the recognition of features that consistently differ between specimens as being of taxonomic value and those that do not as due to variation. A more extensive treatment of variation in modern mammals is given by Yablokov (1966). Much of his work deals with metrical variation, which cannot directly be transferred to work with small samples. Nonetheless it is of interest to summarize his work to outline the kind of results found. The following generalizations, unless otherwise cited, are from Yablokov.

The degree of metrical variability can be ranked into three classes:

1. (of 10% or less) – linear measurements of the skull and postcranial skeleton, linear measurements of the teeth, brain weight, and linear measurements of the body (among others);
2. (of 10–15%) – number of elements in the postcranial skeleton (among others); and
3. (of 15% or greater) – body weight, weights of components, and linear measurements of integumentary structures (among others). Yablokov has provided figures on the normal range of variation (by percentage) for terrestrial mammals. Characteristic variabilities are for body length 4–6%, tail length 5–10%, and body weight 12–15%. The same general ranking holds for average values of coefficients of variation as well as for percentages: linear variability in the skull is 4–5%, in teeth 5–10%, and in the postcranial skeleton 3–5%.

Some linear measurements are often more variable than others. The variability of cranial measurements usually ranges from 0.2–12% and the variability in the width of the skull usually considerably exceeds that of other cranial measurements. The variability of body length for ungulates and carnivores ranges from 1.7–7.9%, whereas that for tail length (in primates) ranges from 6.6–30%. Unfortunately the data do not allow comparison of these ranges for one and the same group of mammals. None of these results seems immediately applicable to theropod alpha taxonomic problems. The results show that some elements, and presumably some features, vary more than others. If study of theropods known from large samples could identify the more (or less) variable elements, this would aid in seeking distinguishing features. Knowledge of such characteristic variabilities could provide a guide to differences likely to be beyond the range of variation.

Linear variability may depend upon scaling, for Yablokov writes, "... it appears that during a comparison of functionally similar traits or the characters of organs of one system, there is a negative relationship between the size of the trait and the magnitude of its variability; the higher the absolute value of the trait the lower its variability." This is an interesting result, but see Yablokov for a full discussion and some caveats. If verified for dinosaurs, it would suggest that larger species might be distinguished by more subtle features than their smaller relatives. Dodson (1976) has reported some association of the greatest coefficients of variability with the most strongly allometric variables in *Protoceratops andrewsi*. He also discusses the significance of this. So far no data for theropods has been analyzed for such effects.

Different statistical populations may show differences in variability. Some mammals, such as white rats and cattle (*Bos taurus*), exhibit sexual dimorphism in their variability: females showing less than males. Slightly greater variability is shown in aquatic forms (pinnipeds and cetaceans) than in land dwelling mammals. Some taxa may have a markedly greater range of variation than others, *Henricosbornia lophiodonta* (Simpson 1937) being a species with high variability in certain features. Whereas of intrinsic interest, such results are of little help for taxa known from small samples except to counsel caution.

Two quite general principles may also be mentioned, although further work is desirable regarding the first. Structures presumed to be nonfunctional seem to exhibit enhanced variability (Simpson 1953). Work in the USSR on fossil mammals supports "the opinion of Simpson … that the variability of fossil forms is very close in magnitude to that of present-day forms; this can be fully confirmed" (Yabokov 1966:187). This conclusion reinforces the relevance of studies of modern variation for paleontology.

It would be useful to know how applicable to dinosaurs are these generalizations derived for mammals, for such generalizations would provide helpful guidelines to the taxonomic treatment of variation. This could perhaps be determined by analysis of theropod species known from large samples, such as *Allosaurus fragilis* (Madsen 1976), *Coelophysis bauri* (Colbert, this volume), or *Syntarsus rhodesiensis* (Raath, this volume). One suggested feature of low variability in theropods is the placement of the nutrient foramina in limb elements (Madsen 1976). If the reduced variability for postcranial characters reported by Yablokov were to hold for theropods, it would suggest that more detailed studies of dinosaurian postcranial skeletons are desirable, although so far the more complex cranial skeletons have attracted more attention (however, see Carpenter, this volume, on *Tyrannosaurus*).

Trends in variation among large theropods

In a purely anatomical sense, there are several components of variation. Elements may vary in form or in size. Variation in form may be subdivided into the following categories:

1a. Structures, such as limbs, composed of several bones may vary in the number of elements present, or
1b. in their proportions.
2. Serial elements, such as teeth, may vary in the number present.
3a. Individual bones may vary in the proportions of the component subunits (e.g., a transverse process may be long or short),
3b. in placement of their component subunits (e.g., a transverse process might be anteriorly, centrally, or posteriorly placed), or
3c. in orientation of their component subunits (e.g., a transverse process might extend perpendicular to the axis of the centrum or be anteriorly, posteriorly, dorsally, or ventrally inclined).
4. Elements may vary in weight or volume.
5. Individual elements may become fused.

The results of Yablokov reported in the previous section suggest that item 4 is likely to be more variable than items 1–3.

Among the large theropods some generalizations may be proposed. The following features seem to vary from individual to individual:

1. The relief of sculpture or rugosity of the cranial elements, especially the postorbital rugosity (in *Tyrannosaurus rex* this varies from almost completely absent to well developed, and shows what may be individual varia-

tion superimposed on what may be sexual dimorphism).
2. The form of some cranial joint surfaces, e.g., the laterosphenoid joint surface on the postorbital (*Tyrannosaurus rex*).
3. The height of the ascending process of the jugal relative to its overall length (*T. rex*). The height of the ascending ramus may vary by almost 50%. This is one example of a principle drawn from Yablokov that can be applied to dinosaurs, namely that some cranial measurements may show significantly higher variability than others.
4. The proportions of the jaw. In Los Angeles County Museum of Natural History 23844 (*T. rex*) the dentary is at least 4% longer, relative to the surangular length, than in American Museum of Natural History 5027 (Tabrum, pers. comm. 1969).
5. Splenial foramen "variably" closed (*Allosaurus fragilis*, Madsen 1976).
6. Maxillary and dentary tooth number: *Allosaurus fragilis,* has 14–16 maxillary and 14–17 dentary teeth (this is individual variation, and not related to age, Madsen 1976) and *T. rex* has 13–14 dentary teeth (Osborn 1912).
7. The degree of fusion of pelvic elements in *A. fragilis*: this may be sex related (Madsen 1976).
8. The development of bony processes or other structures related to muscle attachments, such as the biceps tubercle, can vary widely in different individuals independently of age (*A. fragilis*, Madsen 1976).

A list of ontogenetic changes may also be drawn from the literature ("relative length" in the changes reported by Russell is relative to femoral length):
1. Closure (and ultimately fusion of at least some) of the cranial sutures (Russell 1970). In *T. rex*, the individual (Museum of the Rockies 008) with the most developed surface sculpture and postorbital rugosity also shows fusion of elements (e.g., angular with surangular) asymmetrically, a condition not reported elsewhere among large theropods. This is probably age related.
2. Development of supraoccipital alae of the parietals (tyrannosaurids, Russell 1970).
3. Possibly a decrease in the serration count of teeth, based on evidence collected by Gallup (reported in Molnar 1978), but further work is needed.
4. Increase in relative length of presacral column (*Albertosaurus libratus*, Russell 1970).
5. Disproportionate growth of cervical neural spines (*A. fragilis*, Madsen 1976).
6. Decrease in relative tail length (*A. libratus*, Russell 1970).
7. Increase in relative size of limb girdles (*A. libratus*, Russell 1970).
8. Decrease in relative length of hindlimb, especially distally (*A. libratus*, Russell 1970). Decrease in relative size of hind limb was reported by Steel (1970) from observations by White on juvenile *Allosaurus*.
9. Decrease in relative height of the astragalar ascending process (Welles and Long 1974).

Individual variation seems limited to numbers of serial elements (individual variation, item 6) and changes in proportions of bones (items 1–3, 5, 8) and structures (item 4). Changes in proportions of individual bones seem most common. Ontogenetic variation almost entirely involves changes in proportions of structures (ontogenetic variation, items 4, 6–8) or individual bones (items 2, 5, 9). Fusion is related to ontogenetic and possibly sexual variation.

Some ontogenetic trends appear to be taxon specific. Lawson (1976) has shown that the rate of growth of the maxillary fenestra exhibits different relationships to that of the maxilla in *Allosaurus fragilis*, *Albertosaurus libratus*, and *Tyrannosaurus rex*. Molnar (1978) has similarly shown that the relative depth of the snout decreases with age in *Tarbosaurus bataar* and *Tyrannosaurus rex* but not *Albertosaurus libratus*, whereas the relative depth of the dentary also shows different trends in different taxa. This work is all based on small sample sizes, and is thus subject to considerable uncertainty.

Summary

The concept of variation, especially individual variation, is of great importance in the theory of evolution by natural selection. It has been assimilated into the practice of dinosaurian alpha taxonomy, with the result of unifying several contemporaneous apparently sympatric congeneric (and sometimes confamilial but not congeneric) species into single species.

Difficulties of paleontological alpha taxonomy include the non-utility of criteria used for modern taxonomy, the prevalence of small samples, and the prevalence of incomplete specimens. The practical problem is to distinguish taxonomically significant morphological differences from those due to variation. Two kinds of errors may result from failing to distinguish between these. First, mistaking one species for more, and second, mistaking two (or more) species for one.

Small sample sizes of taxa represented by incomplete specimens imply decreased ability to recognize distinguishing features. Certain modern Australian lepidosaurs and marsupials suggest that distinguishing features for species may be difficult or impossible to detect in taxa represented only by small samples, even with complete skeletal material. This would lead to mistaking two (or more) species for one.

Two related ecological arguments are often presented for regarding minor differences between similar putative species of fossil tetrapods as due to variation, i.e., for synonymizing the species. These are that congeneric (or confamilial) species of large tetrapods do not occur sympatrically, and that the larger the tetrapod the less likely that congeneric or confamilial species will occur sympatrically. Overlooking the difficulties in being certain that any fossil tetrapod species were sympatric, examples from eastern Australia and East Africa suggest that the first argument is not universally valid. In the comparison of dinosaurs with modern mammalian analogs, the results depend upon which analogs are chosen. In regard to the second argument, from size, differences between the Mesozoic and modern worlds suggest such a contention based on living mammals may not apply to the Mesozoic. Indeed the sauropod faunas of the Morrison Formation and Tendaguru beds suggest that it does not. Furthermore, this argument is based on an ecological concept now under challenge and reassessment.

The Australian tetrapods and the critique of the ecological arguments imply that the taxonomic diversity of any group known only from fossils is probably underestimated. So far there is no secure basis for assessing the degree of this underestimation, or even determining if it is significant.

General principles of morphological variation that could be applied in dinosaurian alpha taxonomy would be helpful. Some principles have been suggested by the work of Yablokov and of Simpson. In terms of metrical variability, not all structures are equally variable, and variabilities may be ranked. Thus some structures may be more useful than others in recognizing distinguishing features. Yablokov suggests that degree of variability is inversely proportional to absolute size. Not all groups are equally variable – at least some modern mammals show sexual dimorphism in variability of some characters. Aquatic mammals also seem more variable than terrestrial. All of these indicate areas for further work on variability in those dinosaurs known from sufficiently large samples.

For large theropods there is little data on variation but some generalizations may be proposed. Ontogenetic variation involves differences in proportions of skeletal structures, such as limbs, and, to a lesser extent, of individual bones. Individual variation also involves differences in proportion, but more often of individual bones than of skeletal structures. Differences in number of serial elements (teeth) occur as individual variations. Fusions of cranial elements likely increase with age, while fusion of pelvic elements may indicate sex (Madsen 1976). Placement of structures, specifically nutrient foramina of long bones, has been suggested to be invariant (Madsen 1976).

The understanding of osteological variation is necessary to the alpha taxonomy of dinosaurs as to that of all extinct organisms. Any variant organism cannot necessarily be regarded as belonging to a different taxon. Equally, specimens that differ slightly do not necessarily pertain to a single biological species. Minor phenotypic variations in some cases indicate different biological species, in other cases they do not. However, no two specimens should be judged conspecific (or in some cases, congeneric) on the basis of posited variation substantially different in kind or degree from variation observed in modern animals. This being the case, further work on what kinds of structures may provide distinguishing features, and what kinds may not, would aid in reducing the subjectivity of alpha taxonomy for species represented by small samples of incomplete material.

References

Bartholomai, A., and Molnar, R. E. 1981. *Muttaburrasaurus*, a new iguanodontid (Ornithischia: Ornithopoda) dinosaur from the Lower Cretaceous of Queensland. *Queensland Museum Memoirs* 20:319–349.

Cooke, B. 1983. Koala, wallabies, bandicoots, possums, *In* Davies, W. (ed.), *Wildlife of the Brisbane Area* (Brisbane: Jacaranda Press), pp. 32–42.

Cooper, M. R. 1981. The prosauropod dinosaur *Massospondylus carinatus* Owen from Zimbabwe: its biology, mode of life and phylogenetic significance. *National Museums and Monuments (Zimbabwe) Occasional Papers* B 6:689–840.

Czechura, G. V., and Wombey, J. 1982. Three new striped skinks, (*Ctenotus*, Lacertilia, Scincidae) from Queensland. *Queensland Museum Memoirs* 20:639–645.

Dobzhansky, T. F., Ayala, F. J., Stebbins, G. L., and Valentine, J. 1977. *Evolution* (San Francisco: W. H. Freeman).

Dodson, E. O., and Dodson, P. 1985. *Evolution, Process and Product*, 3rd ed. (Boston: Prindle, Weber and Schmidt).

Dodson, P. 1975. Taxonomic implications of relative growth in lambeosaurine hadrosaurs. *Systematic Zoology* 24:37–54.

1976. Quantitative aspects of relative growth and sexual dimorphism in *Protoceratops*. *Journal of Paleontology* 50:929–940.

Dodson, P. A., Behrensmeyer, A. K., and Bakker, R. T. 1980. Taphonomy of the Morrison Formation (Kimmeridgian–Ortlandian) and Cloverly Formation (Aptian–Albian) of the western United States. *Memoirs de la Societe geologique de France* 59:87–93.

Flannery, T. F. 1980. *Macropus mundjabus*, a new kangaroo (Marsupialia: Macropodidae) of uncertain age from Victoria, Australia. *Australian Mammalogy* 3:35–51.

Fox, R. C. 1986. Paleoscene #1. Species in paleontology. *Geoscience Canada* 13:73–84.

Galton, P. M., and Powell, H. P. 1980. The ornithischian dinosaur *Camptosaurus prestwichii* from the Upper Jurassic of England. *Palaeontology* 23:411–443.

Gilmore, C. W. 1925. Osteology of ornithopodous dinosaurs from the Dinosaur National Monument, Utah. *Carnegie Museum Memoirs* 10:385–409.

Hutchinson, G. E. 1959. Homage to Santa Rosalia, or why are there so many kinds of animals? *American Naturalist* 93:145–159.

1965. *The Ecological Theatre and the Evolutionary Play* (New Haven: Yale University Press).

Kurzanov, S. M. 1972. Sexual dimorphism in protoceratopsians. *Paleontologichesky Zhurnal* 1972:92–97. [In Russian]

Lawson, D. A. 1976. *Tyrannosaurus* and *Torosaurus*, Maestrichtian dinosaurs from Trans-Pecos Texas. *Journal of Paleontology* 50:158–164.

Maberly, C. T. A. 1960. *Animals of East Africa* (London: Hodder and Stoughton).

MacArthur, R. H., and Wilson, E. O. 1967. *The Theory of Island Biogeography* (Princeton: Princeton University Press).

Madsen, J. H., Jr. 1976. *Allosaurus fragilis*: a revised osteology. *Utah Geological and Mineral Survey, Bulletin* 109:1–63.

Maynes, G. M. 1982. A new species of rock wallaby, *Petrogale persephone* (Marsupialia: Macropodidae), from Prosperpine, central Queensland. *Australian Mammalogy* 5:47–58.

Mayr, E. 1963. *Animal Species and Evolution*. Cambridge, Massachusetts: Harvard University Press.

1982. *The Growth of Biological Thought*. Cambridge, Massachusetts: Harvard University Press.

Molnar, R. E. 1978. A new theropod dinosaur from the Upper Cretaceous of central Montana. *Journal of Paleontology* 52:73–82.

Osborn, H. F. 1912. Crania of *Tyrannosaurus* and *Allosaurus*. *American Museum of Natural History Memoirs* 1:1–30.

Ostrom, J. H., and Wellnhofer, P. 1986. The Munich specimen of *Triceratops* with a revision of the genus. *Zitteliana* 14:111–158.

Parks, W. A. 1935. New species of trachodont dinosaurs from the Cretaceous formations of Alberta with notes on other species. *University Toronto Studies (Geological Series)* 37:1–5.

Paterson, H. E. H. 1985. The recognition concept of species. *Transvaal Museum Memoirs* 4:21–29.

Raff, R. A., and Kaufman, T. C. 1983. *Embryos, Genes, and Evolution*. New York: Macmillan.

Ricklefs, R. E. 1987. Community diversity: relative roles of local and regional processes. *Science* 235:167–171.

Rowley, M. J., Rich, P. V., Rich, T. H., and Mackay, I. R. 1986. Immunoreactive collagen in avian and mammalian fossils. *Naturwissenschaften* 73:620–622.

Rozhdestvensky, A. K. 1965. Growth in some Asian dinosaurs and some problems with their taxonomy. *Paleontologichesky Zhurnal* 1965:95–109. [in Russian]

Russell, D. A. 1970. Tyrannosaurs from the Late Cretaceous of western Canada. *National Museum of Natural History, Publications in Palaeontology* 1:1–34.

Russell, D. A., Beland, P., and McIntosh, J.S. 1980. Paleoecology of the dinosaurs of Tendaguru (Tanzania). *Memoirs de la Societe geologique de France* 59:169–175.

Simpson, G. G., 1937. Super-specific variation in nature from the viewpoint of paleontology. *American Naturalist* 71:236–267.

1944. *Tempo and Mode in Evolution*. New York: Columbia University Press.

1953. *The Major Features of Evolution*. New York: Columbia University Press.

Simpson, G. G., Roe, A., and Lewontin, R. C. 1960. *Quantitative Zoology*. New York: Harcourt, Brace and World.

Steel, R. 1970. Saurischia, *In* Kuhn, O. (ed.), *Handbuch der Palaeoherpetologie*, 14:1–87 (Stuttgart: Gustav Fischer Verlag).

Strahan, R. (ed.) 1983. *The Australian Museum Complete Book of Australian Mammals*. Sydney: Angus and Robertson.

Strong, D. R., Simberloff, D., Abele, L. G., and Thistle, A. B. 1984. *Ecological Communities: Conceptual Issues and the Evidence* (Princeton: Princeton University Press).

Walker, A. D. 1961. Triassic reptiles from the Elgin area: *Stagonolepis*, *Dasygnathus* and their allies. *Philosophical Transactions of the Royal Society of London* 244:103–204.

Welles, S. P., and Long, R. A. 1974. The tarsus of theropod dinosaurs. *Annals of the South African Museum* 64:191–218.

Williams, J. G., and Arlott, N. 1981. *A Field Guide to the National Parks of East Africa* (London: William Collins Sons).

Wyckoff, R. W. G., 1972. *The Biochemistry of Animal Fossils* (Bristol: Scientechnica).

Yablokov, A. V. 1966. *Variability in Mammals*. Moscow: Institute of Animal Morphology, Academy of Sciences, U.S.S.R. [in Russian]

6 Variation in *Coelophysis bauri*

EDWIN H. COLBERT

Abstract

The large series of skeletons and partial skeletons of *Coelophysis bauri* collected from the Upper Triassic Chinle Formation at Ghost Ranch, New Mexico, affords an opportunity to study the ontogeny, individual variation, and possible sexual dimorphism in this species. Although truly small specimens have not been found in the deposit, it is possible to follow an incomplete growth series in this dinosaurian species involving a threefold increase in linear dimensions. By working back from this partial sequence, it is estimated that there was probably between a ten to fifteenfold increase in size from hatchling to adult, during which there were proportional changes in the size of the orbit, length of the neck, and length of the hind limb and some changes in skull proportions. Substantive differences are to be seen in individual variations, which are especially marked between two adult specimens of approximately the same size. One of these specimens has a larger skull, a longer neck, and a smaller forelimb than the other, and is further differentiated by fusion of the sacral vertebrae, not seen in its counterpart. These differences may be an expression of sexual dimorphism. A variation of fusion within the skeleton is especially marked in the ankle of *Coelophysis*, particularly as seen in the size and solidity of the ascending process of the astragalocalcaneum, the fusion of this bone to the tibia, and the fusion of the second and third distal tarsals to their respective metatarsals. Fusion or the lack of fusion in the ankle seems to be independent of size, and seemingly is a mark of individual variation.

Introduction

Ghost Ranch, New Mexico, is paleontologically famous as the site of a dinosaur quarry, developed in Upper Triassic Chinle sediments, that has yielded an amazing number of skulls, skeletons, and partial skeletons of the early theropod, *Coelophysis bauri* (Colbert 1947, 1960, 1989). This dinosaur, described by Cope on the basis of fragmentary remains from northern New Mexico, exem-

plifies the features to be seen in an early theropod, as now known from the abundant materials collected at Ghost Ranch. These fossils, all collected from a quarry measuring approximately 6 by 20 metres, and representing individuals at different stages of ontogenetic development, offer an unusual opportunity for determining variations within a dinosaurian species (Fig. 6.1). The

Figure 6.1. The Ghost Ranch *Coelophysis* quarry in 1985. Photograph by Lawrence P. Byers.

In Dinosaur Systematics: Perspectives and Approaches, *Kenneth Carpenter and Philip J. Currie, eds. Copyright © Cambridge University Press, 1990.*

purpose of the present contribution is to examine variation in *Coelophysis bauri*.

At this place it may be pertinent to make a few remarks about the taxonomic status of *Coelophysis bauri*. In a recent paper Padian expresses considerable doubt as to whether *Coelophysis bauri* has any real validity, in part because the genus "was named on scrappy material that was never adequately diagnosed or compared with other taxa" (Padian 1986, p. 58). Parenthetically it should be pointed out that the genus is not based on fossil materials, it is based upon the type species which in turn is based upon a specimen or specimens; in the case of *Coelophysis*, the species *C. bauri*, arbitrarily designated by Hay as the type (Hay 1930, p. 186).

As for the type materials of *Coelophysis bauri*, they are "scrappy," but that is the case for many types. It must be borne in mind that a type is a name carrier, and as such it may or may not be diagnostic for the species it represents. This matter was succinctly set forth by Simpson more than four decades ago:

It is a natural but mistaken assumption that types are somehow typical, that is, characteristic of the groups in which they are

placed. It is, of course, desirable that they should be typical because then they are less likely to be shifted about from group to group, carrying their names with them and upsetting nomenclature, but there is no requirement that a type be typical, and it frequently happens that it is quite aberrant. Types are almost never really average specimens within a species, or fully central species in a genus. Types were formerly, and still are by many students, supposed to be not only name-bearers but also the bases on which group concepts are erected and the standards of comparison for those concepts. They cannot possibly serve either function in modern taxonomy and the requirements of these functions are flatly incompatible with the requirement of name-bearing which types can and do serve. (Simpson 1945, pp. 29–30).

As a name-bearer, Cope's original materials of *Coelophysis bauri* are adequate. Among the numerous bones from Ghost Ranch are many specimens that can be matched in morphological details and in measurements with the specimens that Cope described as *Coelurus bauri*, *Coelurus longicollis*, and *Tanystropheus willistoni* (Cope 1887, 1889; Colbert 1964). Therefore, the Ghost Ranch fossils can be accepted on all reasonable grounds of morphological similarities (as well as

Figure 6.2. A chart of the *Coelophysis* quarry at Ghost Ranch, New Mexico, showing the disposition of blocks removed by the American Museum of Natural History in 1947 and 1948, and by the Carnegie Museum of Pittsburgh, the New Mexico Museum of Natural History, the Museum of Northern Arizona and Yale University in 1981 and 1982.

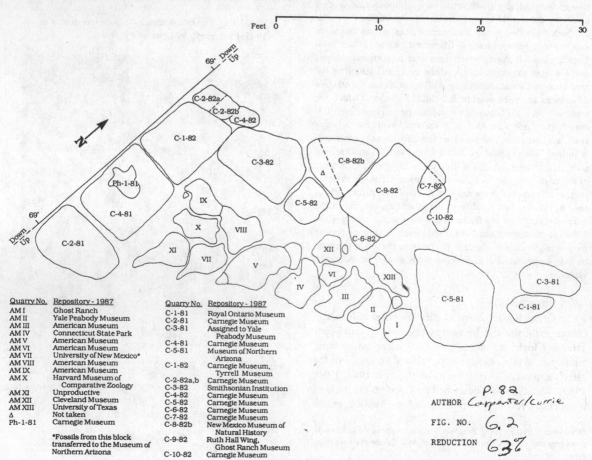

Quarry No.	Repository - 1987
AM I	Ghost Ranch
AM II	Yale Peabody Museum
AM III	American Museum
AM IV	Connecticut State Park
AM V	American Museum
AM VI	American Museum
AM VII	University of New Mexico*
AM VIII	American Museum
AM IX	American Museum
AM X	Harvard Museum of Comparative Zoology
AM XI	Unproductive
AM XII	Cleveland Museum
AM XIII	University of Texas
Δ	Not taken
Ph-1-81	Carnegie Museum

*Fossils from this block transferred to the Museum of Northern Arizona

Quarry No.	Repository - 1987
C-1-81	Royal Ontario Museum
C-2-81	Carnegie Museum
C-3-81	Assigned to Yale Peabody Museum
C-4-81	Carnegie Museum
C-5-81	Museum of Northern Arizona
C-1-82	Carnegie Museum, Tyrrell Museum
C-2-82a,b	Carnegie Museum
C-3-82	Smithsonian Institution
C-4-82	Carnegie Museum
C-5-82	Carnegie Museum
C-6-82	Carnegie Museum
C-7-82	Carnegie Museum
C-8-82b	New Mexico Museum of Natural History
C-9-82	Ruth Hall Wing, Ghost Ranch Museum
C-10-82	Carnegie Museum

on their stratigraphic and geographic occurrences) with the type materials as representative of *Coelophysis bauri*, and afford ample information about this species.

The Ghost Ranch sample

The Ghost Ranch *Coelophysis* quarry was discovered and first worked in 1947 by myself and my assistants, George Whitaker, Carl Sorensen, and Thomas G. Ierardi, at which time a series of blocks was removed and shipped to the American Museum of Natural History in New York. During the following field season the quarry was further excavated by Whitaker and Sorensen for the American Museum. As a result of these two seasons of work, 13 blocks were excavated, and of these blocks, five, number III, V, VI, VIII, and IX, were retained in New York for study and exhibition. Block I was given to Ghost Ranch, II went to Yale University, IV to the Connecticut State Dinosaur Trackway Park, VII to the University of New Mexico, XI to the Museum of Northern Arizona, X to the Museum of Comparative Zoology at Harvard University, XII to the Cleveland Museum, and XIII was assigned to the University of Texas. Most of these blocks have been prepared to various degrees of completion.

The quarry then lay fallow for many years. In 1981, according to an agreement reached between Ghost Ranch, the Carnegie Museum of Pittsburgh, the New Mexico Museum of Natural History, the Museum of Northern Arizona, and the Peabody Museum of Yale University, the quarry was reopened and worked cooperatively by those institutions. Work continued during the following field season. As a result of this second phase of quarrying operations, some 16 additional blocks of varying sizes, several being of truly huge dimensions, were removed from the quarry. Most of these blocks were shipped to Pittsburgh, but one, field number C–8–82, went direct to the Museum in Albuquerque, and another, C–5–81, to the Museum of Northern Arizona. Both of these were very large blocks. Two smaller blocks, C–1–81 and C–3–81, collected in 1981, were taken by the Royal Ontario Museum at Toronto, and by Yale University. Some material also went to the Idaho Museum of Natural History. Subsequently, about one-half of the large block, C–1–82, was sent by the Carnegie Museum to the Tyrrell Museum in Drumheller, Alberta. This block has been prepared and is now on exhibition. Still another large block, C–3–82, was sent to the Smithsonian Institution, where presently it is being prepared. Finally, a very large block, C–9–82, the last to be taken from the quarry, was removed late in 1985 and placed within the Ruth Hall Museum of Paleontology at Ghost Ranch, where now it is undergoing preparation (Fig. 6.2).

The materials of *Coelophysis bauri*, now distributed among several museums, should, taken together, offer a large array of fossils on which to base studies of variation within this species. At the moment, however, it has not been feasible to attempt research on the fossils in all of the institutions where such materials are housed. Therefore, the present contribution is based primarily upon the fossils at the American Museum of Natural History, supplemented by observations on prepared fossils at the Museum of Comparative Zoology at Harvard University, the Museum of Northern Arizona, the Tyrrell Museum, the Yale Peabody Museum, and the Carnegie Museum.

The Ghost Ranch sample is that of a large accumulation of bones representing a single species, in this respect closely resembling the Chitake River deposit of *Syntarsus rhodesiensis* (Raath this volume). It would seem that these fossils are the remains of animals buried in the bed of a pond, or perhaps in a backwater eddy of a stream. The fossils consist of some complete articulated skeletons, numerous partial skeletons, and of course many isolated elements, such as skulls and limb bones. The preponderance of articulated materials indicate that the carcasses were not transported any great distance after death, and were buried rather quickly. Moreover, the bones do not show any indications of rolling, trampling, or predation.

It seems likely that the quarry contains the record of some sort of mass death; perhaps at a river crossing or perhaps from some local catastrophe. The presence of individuals of various ontogenetic ages in the deposit would seem to indicate that in the quarry we see the remains of a herd of dinosaurs, the members of which died almost simultaneously, or within a short span of time.

The possibility that *Coelophysis bauri* was living and travelling in a herd is not to be wondered at. The analyses of dinosaurian footprints and trackways at various localities within recent years afford what seems to be strong evidence that these reptiles, of various genera and species, frequently were congregated in herds (Ostrom 1972, 1986). Such occurrences provide nice opportunities to study the population structure and behaviour in dinosaurs of different species, and of different geologic ages (Fig. 6.3).

It should be added that the Ghost Ranch Dinosaur Quarry has yielded more than the remains of *Coelophysis bauri*. Phytosaur skulls and partial skeletons have been found, as well as skeletons of other thecodont reptiles. In addition, labyrinthodont teeth, and teeth and other remains of freshwater fishes occur in the quarry. The overwhelming preponderance of fossils, however, consists of *Coelophysis* bones, and it is to these that we will here turn our attention.

Variation in size

Because there are skulls and skeletons of animals in different stages of ontogenetic development, there are of course size variations to be seen among the Ghost Ranch fossils. This illustrates the size differences that can be expected within a single dinosaurian species of this size. Frequently in the past, species of dinosaurs have been distinguished on size differences – the result of having small samples upon which to base conclusions. Perhaps ironically, a prime example of this is Cope's designation of three species of *Coelophysis*, based to a considerable extent upon size differences (Cope 1887, 1889), a specific proliferation that was accepted by von Huene (1915, p. 502): "*Coelophysis longicollis* is the largest: *Coelophysis willistoni*: the smallest species." All of Cope's fossils were collected in northern New Mexico, probably near or at the present location of Ghost Ranch.

Among the fossils from the Ghost Ranch quarry as now prepared and included in this study, the smallest skull (AMNH 7242) is approximately 80 mm in length, while the largest (AMNH 7223) is 265 mm in length (Fig. 6.4). This is something more than a threefold difference. As for postcranial bones, the smallest femur (AMNH 7234) is 118 mm in length, while the largest (AMNH 7223) is 209 mm. This is something less than a twofold difference. These obviously represent individuals at different growth stages, and can in no way be considered of specific import, as was believed by Cope when he distinguished his "*Coelophysis willistoni*" from the larger specimens in his limited sample.

Using the skulls as a measure, it seems evident that, although there is a considerable difference in size between the smallest and the largest specimens, this difference is by no means indicative of the total size range that might have been involved in an ontogenetic sequence of this dinosaur. For example, the size increase among modern crocodilians from hatchling to large adult may be on the order of a thirtyfold difference. This is based upon a comparison between a 20 centimeter-long hatchling and a six meter-long adult (Cott 1961; Dodson 1975).

Perhaps there was not so great a growth in size from hatchling to adult in *Coelophysis*. Thus, comparisons might be made with the modern Komodo dragon or ora, *Varanus komodoensis*, in which the growth from hatchling to large adult involves a fifteenfold increase in size (Auffenberg 1981).

Therefore, using the skull as a basis for measurements, and extrapolating back from an adult skull length of 250–300 mm., one might expect skulls of hatchling *Coelophysis* to be 10 mm if there was a thirtyfold increase with growth, or 20–40 mm if there was only a fifteenfold increase. The point being that the smallest known *Coelophysis* skull, with a length of 80 millimetres, is well along the ontogenetic growth curve for this dinosaur. And the particular point being made is that size differences in specimens from the Ghost Ranch quarry are to be regarded as expressions of growth and nothing more.

Figure 6.4. Skull lengths in *Coelophysis bauri* (Cope).

Figure 6.3. *Coelophysis bauri* (Cope), as restored by Margaret Colbert. New Mexico Museum of Natural History.

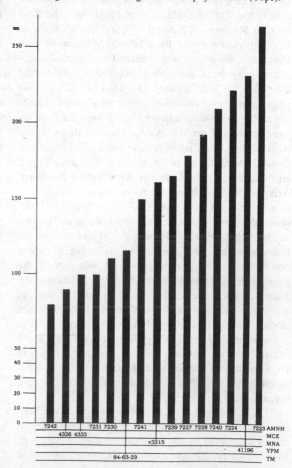

Proportional changes

Proportional changes in relation to growth are general among the vertebrates. What is the evidence in the *Coelophysis* sample from Ghost Ranch?

One might expect the skull of young *Coelophysis* individuals to be proportionately large as compared with the skull in fully adult individuals, as is so common in vertebrate ontogeny. Unfortunately, as we have seen, there are no truly small *Coelophysis* specimens known – specimens that might be considered in the hatchling range. A comparison may be made between three moderately small specimens (AMNH 7230, 7231, and MNA V 3318) in which there is a skull and vertebral series, and several large specimens (AMNH 7227, 7228, 7224, and 7223) (Fig. 6.5). In this case, the length of the skull is measured against the total length of the presacral vertebrae. This is presented in Table 6.1, where we see a

Table 6.1. Coelophysis bauri: *ratios of skull length to presacral length*

	Skull[a] (mm)	Presacral (mm)	Ratio (%)
MNA V3318	85[b]	493	19
AMNH 7231	100	477	21
AMNH 7230	110.5	440	25
AMNH 7227	179	744	24
AMNH 7228	193	760	25
AMNH 7224	222	860	26
AMNH 7223	265	910	29

[a]Premaxilla-occipital condyle.
[b]Estimated measurement.

Table 6.2. Coelophysis bauri: *ratios of orbital diameter to premaxillary quadrate length*

	Length, Premaxilla-Quadrate (mm)	Anterior-Posterior Orbital Diameter (mm)	Ratio (%)
AMNH 7242	68[a]	20	29
MCZ 4326	88	30	34
AMNH 7241	140	36	26
MNA V3315	143	32	22
AMNH 7227	159	36	23
AMNH 7224	191	37	20
AMNH 7240	198	37	19
YPM 41196	211	34	16
MCZ 4327	239	40	17
CM C-4-81	250	40	16

[a]Estimated measurement.

rather unexpected indication that the smaller specimens had proportionately smaller skulls than the larger specimens. Might this not represent a special adaption for increasing the relative size of skull in the adult to provide relatively larger jaws and consequently a more efficient bite than would otherwise be the case? If this was the case, one can envisage the hatchlings of *Coelophysis* as having relatively large skulls, with the size of the head becoming proportionately smaller during growth to the juvenile stage. Then gradually becoming proportionately and secondarily larger during progress from juvenile to adult. This assures adult individuals large jaws and, consequently, an increased power for biting, which of course would have been advantageous in an active and aggressive predator. Such positive skull allometry can be seen in recent crocodilians (Dodson 1975).

It is to be expected that the orbit in young individuals would be proportionately larger than is the case in older individuals, and here the data for *Coelophysis* show a nice relative reduction in orbit size in the sequence from smaller to larger specimens. The only marked departure from this regression series is in the second smallest specimen, probably an indication of

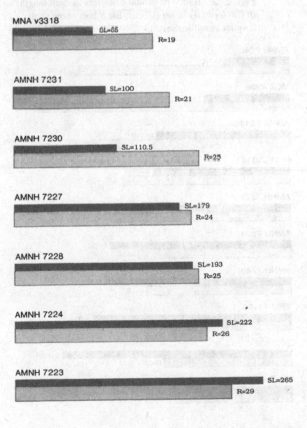

Figure 6.5. Ratios of skull lengths to presacral lengths in *Coelophysis bauri* (Cope). Black bars show skull lengths in millimeters.

SL= Skull length
R = Ratio - skull length/presacral length

individual variation (Fig. 6.6). The figures are shown in Table 6.2.

It should be expected that in a long-necked species like *Coelophysis bauri* there might be proportional differences in the length of the neck during progression from young to adult. In other words, assuming positive allometry for the neck, young animals should be expected to have proportionately shorter necks. Let us look at a sample of seven *Coelophysis* individuals in which comparisons can be made between the cervical and dorsal series (Table 6.3). In this example, the length of the dor-

Table 6.3. Coelophysis bauri: *ratios of cervical to dorsal lengths*

	Cervical (mm)	Dorsal (mm)	Ratio (%)
AMNH 7230	215	225	96
MNA V3318	184	255	72
AMNH 7231	215	262	79
AMNH 7228	430	330	130
AMNH 7227	384	360	107
AMNH 7223	485	425	114
AMNH 7224	405	455	89

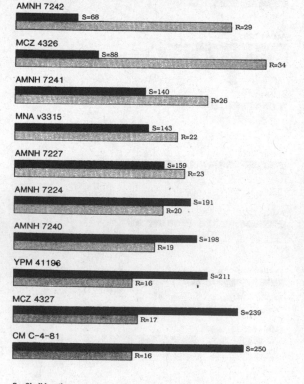

Figure 6.6. Ratios of orbital diameters to skull lengths in *Coelophysis bauri* (Cope). Black bars show skull lengths in millimeters.

S = Skull length
R = Ratio - orbital diameter / skull length

sal vertebrae is used as the standard of comparison, having a value of 100. Strangely enough, in this sequence, arranged from the shortest to the longest dorsal series, a smooth curve delineating the proportional changes in length of neck during growth is hardly apparent (Fig. 6.7). In general, the three smallest individuals do have proportionately shorter necks than do the next four larger individuals, but there are certainly large deviations from an expected series. Perhaps if there were a much larger sample available the inconsistencies seen here would be greatly reduced. All that can be said is that individual variations in this feature are considerable. Again, the differences between the two largest specimens are striking, and will be discussed below.

We may next consider variations in the proportions of the limbs of *Coelophysis bauri* (Fig. 6.8). Using the length of the presacral column as the basis of comparison, having a value of 100, and arranging a sequence

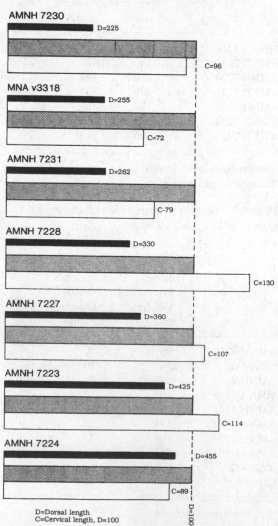

Figure 6.7. Ratios of cervical lengths to dorsal lengths (dorsal lengths = 100) in *Coelophysis bauri* (Cope). Black bars show dorsal lengths in millimeters.

D=Dorsal length
C=Cervical length, D=100

from the shortest to the longest presacral column, the proportional lengths of the forelimbs in five individuals are shown in Table 6.4. As is evident, the ratios of fore-limb sizes are fairly uniform except for that of AMNH 7224, in which the forelimb is comparatively large. Here again, as in the case of cervical length, this speci-men departs significantly from its expected position. In the one character, the neck of AMNH 7224 is signifi-cantly shorter than the neck in other adult individuals; in the present instance the forelimb is significantly larger than in other individuals.

The available sample showing hind limb length as compared with presacral length involves four speci-mens (Table 6.5). The striking fact here is that the juve-nile specimen has proportionately longer hind limbs as contrasted with the adults, an expression of negative allometry similar to that seen in crocodilians (Kalin 1955; Dodson 1975). Perhaps this reflects an ability to run rapidly to escape from larger predators.

However, when the forelimb is compared with the hind limb for three individuals (Table 6.6), we see that AMNH 7224 has a larger forelimb compared with what might be considered more "normal" individuals (this difference did not show up in the comparison of the forelimb against the presacral series, owing to the short cervical section in this specimen). Nothing else of consequence appears in the comparisons.

In summary, certain proportional changes are seen in the *Coelophysis* materials from Ghost Ranch, these being an apparent proportional increase in size of the skull from juvenile to adult, a definite proportional reduction in the orbital diameter during growth, a pro-portional lengthening of the neck, and a reduction in the proportional length of the hind limbs. The change in rela-tive orbital diameter and the length of the neck are of the kind that might be expected during ontogenetic

Table 6.4. Coelophysis bauri: *ratios of forelimb lengths to presacral lengths*

	Forelimb (mm)	Presacral (mm)	Ratio (%)
AMNH 7230	145	440	33
AMNH 7231	152	447	32
AMNH 7227	204	744	27
AMNH 7224	341	860	40
AMNH 7223	280	910	31

Table 6.5. Coelophysis bauri: *ratios of hindlimb lengths to presacral lengths*

	Hindlimb (mm)	Presacral (mm)	Ratio (%)
MNA V3318	412	439	94
AMNH 7228	559	760	74
AMNH 7224	666	860	77
AMNH 7223	680	910	75

Table 6.6. Coelophysis bauri: *ratios of forelimb to hindlimb lengths*

	Forelimb (mm)	Hindlimb (mm)	Ratio (%)
AMNH 7229	180	449	40
AMNH 7223	280	680	41
AMNH 7224	341	666	51

Figure 6.8. Ratios of fore and hind limb lengths to presacral lengths (presacral lengths = 100) in *Coelophysis bauri* (Cope). Black bars show pre-sacral lengths in millimeters.

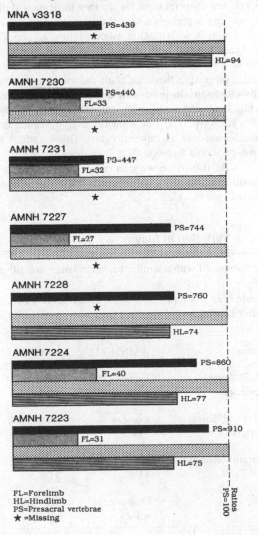

FL=Forelimb
HL=Hindlimb
PS=Presacral vertebrae
★ =Missing

growth. But the changes in skull and hind limb proportions from juvenile to adult may have been special adaptations, the first for the development of a larger and more efficient bite in the adult, and the second for the ability of rapid running during the juvenile stage when these dinosaurs may have been subject to heavy predation (not only from other large reptiles of Late Triassic age, but also from adults of their own species). In addition, there are differences between individuals that can be attributed to random individual variation.

Within the series of specimens utilized for the comparisons that have been made, two individuals (AMNH 7223, 7224) show some striking differences, and these will be considered next.

Differences between two adult skeletons

Allusions have been made several times in the foregoing discussions to differences between two of the skeletons from the Ghost Ranch quarry, namely AMNH 7223 and AMNH 7224 (see Colbert 1983, p. 69). What might be the significance of these differences (Table 6.7)? Are they random, or do they indicate a definite dichotomy – perhaps a sexual difference. And if the latter, which is which? One might venture a supposition that the skeleton with the larger skull, the longer neck, the smaller forelimbs, and the fused sacral spines is a male, supposing the large skull and the long neck were advantageous in predation and intraspecific combat, while the fused sacral spines might have been an adjunct of a strong pelvis. But what is the significance of the small forelimbs in this individual? Does a small forelimb go with a large skull, as in tyrannosaurids?

At this time we can only speculate as to the meaning of the differences seen in these two skeletons, buried side by side.

Variation in fusion

Fusion of the sacral spines in AMNH 7223 may be compared with a similar fusion in *Syntarsus*, although

in the African genus there are other rather extensive fusions within the pelvis (Raath 1969). Fusion of bones is seen also in the ankle and pes of this dinosaur – a region in the leg where considerable rigidity of structure (to prevent strains and dislocations) would have been advantageous. Thus, the astragalus and calcaneum are fused to form a single element, the astragalocalcaneum, an adaptation appearing in this dinosaur of Late Triassic age and seen in later Mesozoic theropods. Furthermore, the astragalocalcaneum of *Coelophysis bauri* may or may not be fused to the distal end of the tibia, a character that seems to be quite variable in this dinosaur. Such fusion does not necessarily seem to be a consequence of size and age. For example, in MNA V 3318, fusion between the astragalocalcaneum and the tibia is so complete that no line of junction between them can be discerned, whereas in other specimens, such as MNA V 3319 and AMNH 7224, the astragalocalcaneum is quite free. Yet, MNA V 3318 is only half as large as AMNH 7224. An additional area of variation in the ankle of *Coelophysis bauri* involves the ascending process of the astragalus. It may be present, or it may be altogether missing in some specimens. Perhaps the process, which seems to be a separate element, occasionally was lost during fossilization.

Fusion within the ankle extends beyond the astragalocalcaneum. In some specimens, for example MNA V 3328, the second and third distal tarsals are fused to the proximal ends of their respective metatarsals and to each other as well, to form proximally a solid, single surface for articulation with the transverse, cylindrical astragalocalcaneum comparable to fusion in the pes of *Syntarsus* (Raath 1969). The proximal articulating surface of the fused second and third tarsals is in alignment with the proximal articulating surface of the fourth tarsal, which is a completely separate bone. The fourth tarsal

Table 6.7. Coelophysis bauri: *a comparison of two adult specimens*

Ratio	AMNH 7223	AMNH 7224
Skull/dorsal vertebrae	Long skull 63%	Short skull 49%
Cervical/dorsal vertebrae	Long neck 114%	Short neck 89%
Forelimb/hindlimb	Small forelimb 41%	Large forelimb 51%
	Fused sacral spines	Free sacral spines

Table 6.8. Coelophysis bauri: *relation of astragalocalcaneum to tibia-fibula*

	Size	Astragalocalcaneum
AMNH 7223	Large	Free
AMNH 7224	Large	Free
AMNH 7243	Large	Free
MNA V3320	Large	Free
YPM 41412	Large	Free
TMP 84-63-33	Large	Free
TMP 84-63-35	Large	Free
TMP 84-63-37	Medium	Free
MNA V3321	Medium	Fused
MNA V3318	Medium	Fused
MCZ 4331	Medium	Free
TMP 84-63-34	Medium	Fused
TMP 84-63-21	Small	Fused
TMP 84-63-47	Small	Fused

articulates proximally with the lateral part of the astragalocalcaneum and distally with the proximal end of the fourth metatarsal. The second and third metatarsals, although fused to each other proximally and to the fused second and third tarsals, are nonetheless quite separate for almost their entire lengths. In other specimens, such as AMNH 7223, no such fusion is seen. Once again, as in the case of fusion of the astragalocalcaneum to the tibia, fusion in the distal part of the tarsus and the proximal ends of the two metatarsals seemingly is not a function of age or size. MNA V 3318, with strong fusion, is much smaller than AMNH 7223, in which there is no fusion. Although the specimens cited above as examples of fusion or nonfusion in the ankle provide especially good examples of variation in this feature, other specimens as well show this phenomenon (Table 6.8).

Variation in number of teeth

Another aspect of variation in *Coelophysis bauri* is the number of teeth present. There are in this dinosaur 26 to 30 upper teeth and 25 to 27 dentary teeth. The number of premaxillary teeth is constant, at four. Thus, variation in the number of maxillary teeth ranges between 22 and 26, and seems to be largely independent of size. Counts of maxillary teeth, arranged according to the size of the specimens studied, are listed in Table 6.9.

Conclusion

An attempt has been made to show that within a local population of the Upper Triassic theropod dinosaur *Coelophysis bauri*, there are various parts of the skeleton in which individual variation is pronounced. These involve proportional changes, particularly in the size of the skull, in the size of the orbit, in the length of the neck, in the length of the forelimb, in the length of the hind limb, in the fusion of bones (especially in the sacrum and in the ankle), and in the number of teeth. There are also variations in size, but these are primarily dependent upon ontogenetic growth rather than upon individual variability as such. Such differences in size are mentioned here because there are numerous examples in the

literature where species among dinosaurs are recognized upon size differences. It is here maintained that when a reasonably good sample is available, proportional changes, size differences, and quantitative differences (such as the number of teeth) are to be expected within a single species.

The case of two specimens from Ghost Ranch, one showing a long skull, a long neck, a small forelimb, and fused sacral spines, the other showing a shorter skull, a shorter neck, a large forelimb, and free sacral spines, is cited as a puzzling example of contrasting individual variations, for which no logical explanations other than sexual dimorphism can at the moment be deduced.

References

Auffenberg, W. 1981. *The Behavioral Ecology of the Komodo Monitor* (Gainsville: University of Florida Press).

Colbert, E. H. 1947. The little dinosaurs of Ghost Ranch. *Natural History* 61:392–399, 427–428.

　1960. Triassic rocks and fossils. *Guidebook, New Mexico Geological Society, 11th Field Conference*:55–62.

　1964. The Triassic dinosaur genera *Podokesaurus* and *Coelophysis*. *American Museum Novitates* 2168:1–12.

　1974. The Triassic paleontology of Ghost Ranch. *New Mexico Geological Society 25th Field Conference* 9:175–178.

　1983. *Dinosaurs. An Illustrated History* (Maplewood, N.J.: Hammond, Inc.).

　1989. The Triassic dinosaur *Coelophysis*. *Museum of Northern Arizona Bulletin* 57·i–xv, 1–160.

Cope, E. D. 1887. A contribution to the history of Vertebrata of the Trias of North America. *American Philosophical Society Proceedings* 24:209–228.

　1889. On a new genus of Triassic *Dinosauria*. *American Naturalist* 23:626.

Cott, H. B. 1961. Scientific results of an inquiry into the ecology and economic status of the Nile crocodile (*Crocodilus niloticus*) in Uganda and Northern Rhodesia. *Transactions of the Zoological Society of London* 29:215–337.

Dodson, P. 1975. Functional and ecological significance of relative growth in *Alligator*. *Journal of Zoology, London* 175:315–355.

Hay, O. P. 1930. *Second Bibliography and Catalogue of the Fossil Vertebrates of North America* (Washington: Carnegie Institution), Publication 390, 2:1–1074.

Huene, F. von. 1915. On reptiles of the New Mexico Trias in the Cope Collection. *American Museum of Natural History Bulletin* 34:485–507.

Kalin, J. 1955. Crocodilia. *In* Piveteau, J. (ed.), *Traite de Paléontologie* (Paris: Massen), pp. 695–784.

Ostrom, J. H. 1972. Were some dinosaurs gregarious? *Palaeogeography, Palaeoclimatology, Palaeoecology* 11:287–301.

　1986. Social and unsocial behavior in dinosaurs. *In Evolution of Animal Behavior* (Oxford: Oxford University Press), pp. 41-61.

Table 6.9. Coelophysis bauri: *number of teeth*

	Length of maxilla (mm)	Number of teeth
AMNH 7242	42	18
MVZ 4326	51.5	22
AMNH 7241	96	24
MNA V3315	97	24 (L)
		26 (R)
AMNH 7239	102	25
AMNH 7240	116	24
AMNH 7224	129	22

Padian, K. 1986. On the type material of *Coelophysis* Cope (Saurischia: Theropoda) and a new specimen from the Petrified Forest of Arizona (Late Triassic: Chinle Formation). *In* Padian, K. (ed.), *The Beginning of the Age of Dinosaurs* (Cambridge: Cambridge University Press), pp. 45–60.

Raath, M. A. 1969. A new coelurosaurian dinosaur from the Forest sandstone of Rhodesia. *Arnoldia* 4(28):1–25.

Simpson, G. G. 1945. The principles of classification and a classification of mammals. *American Museum of Natural History Bulletin* 85:1–350.

7 Morphological variation in small theropods and its meaning in systematics: evidence from *Syntarsus rhodesiensis*

MICHAEL A. RAATH

Abstract

Because they are small and their bones fragile, small theropods are usually found as poorly preserved fossils. It is, therefore, often difficult to identify morphological features that clearly and unequivocally characterize individual small theropod taxa, to distinguish them from each other. This legacy of fragility has unquestionably handicapped studies of small theropod systematics in the past. Several authors have recently published cladistic analyses of archosaur lineages, including the theropods, but the picture remains somewhat murky as their cladograms are not easily reconciled with each other. Perhaps this is also partly attributable to study material that is incomplete and poorly preserved.

The Early Jurassic (terminal Karoo) *Syntarsus rhodesiensis* is a small theropod taxon based on material that is both quantitatively adequate and qualitatively excellent as far as preservation is concerned. Partial remains of more than 30 individuals have been recovered from one of three known localities in the fine-grained aeolian Forest Sandstone Formation of Zimbabwe, southern Africa. The fossiliferous bed in this unusually rich locality suggests that a single event caused the catastrophic mass death of a socially gregarious group.

Syntarsus shows clear and consistent morphological variation that is bimodally distributed in the sample examined. The variation affects particularly the trochanters and muscle scars of the femur, but it has also been observed in other elements of the skeleton. I conclude that the morphological variation shown by the known *Syntarsus* sample reflects clear sexual dimorphism, and not taxonomic diversity. The dimorphism manifests as gracile and robust morphs, with robusticity acquired only after attainment of a body size taken to mark the onset of sexual maturity. Based on the population structure reflected in the Chitake River *Syntarsus* sample, I suggest that the robust morph might represent females.

Introduction

Much has been written on the systematic relationships of the theropods (see, for example, von Huene 1932; Romer 1966; Ostrom 1981; Paul 1984; Welles 1984; Gauthier and Padian 1985; Hecht 1985; Gauthier 1986). This contribution will focus on a theropod belonging to the group formerly referred to generally as "coelurosaurs," and more recently, more vaguely, as "small theropods." Ostrom (1981, Table 2) gives a useful summary of the systematic history of the coelurosaurs, some of the main points of which are: the taxon Coelurosauria was established by von Huene (1914) to accommodate small, lightly built, mainly bipedal carnivorous dinosaurs of coelurid affinities. By the time of his major monograph two decades later (von Huene 1932), his infra-order Coelurosauria had grown to include the following taxa (several now known to be nontheropod): Ammosauridae (*Ammosaurus* is a prosauropod), Hallopodidae (*Hallopus* is now considered a crocodylomorph), Podokesauridae, Procompsognathidae (these two families including all "Triassic coelurosaurs"), Compsognathidae, Coeluridae, Ornithomimidae, and Ceratosauridae. Three decades later Romer (1966) included the Procompsognathidae (incorporating Podokesauridae), ?Segisauridae, Coeluridae (with Coelurosauridae and Compsognathidae as synonyms), Ornithomimidae, and ?Caenagnathidae. Steel's (1970) list is comparable in most respects with Romer's, except that Steel regarded Podokesauridae as valid and he excluded Caenagnathidae. Ostrom (1981) also retained Podokesauridae, absorbing Procompsognathidae into it, and he added Compsognathidae, Dromaeosauridae, and Oviraptoridae to Steel's list of coelurosaurian families. The allocation of genera among these families is somewhat variable. Evidently, the content of the well-defined Cretaceous family of ornithomimid-struthiomimid theropods provokes the least disagreement.

To the extent that there is any consensus in recent cladistic analyses, it seems that most authors agree that Coelurosauria, as understood by earlier authors like von Huene and Romer, does not constitute a monophyletic

In Dinosaur Systematics: Perspectives and Approaches, *Kenneth Carpenter and Philip J. Currie, eds. Copyright © Cambridge University Press, 1990.*

clade. Of several recent detailed cladistic analyses, those by Gauthier (1984, 1986, and see also Gauthier and Padian 1985), Hecht (1985), and Paul (1984) are good examples (Fig. 7.1).

Gauthier's (1984, 1986) analysis concludes that theropods comprise three monophyletic clades: Ceratosauria, Carnosauria, and Coelurosauria. He has redefined Coelurosauria to include "all theropods that are closer to birds than to Carnosauria": Ornithomimidae, Deinonychosauria, *Archaeopteryx*, and birds, plus a number of less well known forms *incertae sedis* – such as *Coelurus, Compsognathus, Microvenator*, and *Ornitholestes*. Several other taxa previously classified as coelurosaurs are referred elsewhere – such as *Coelophysis* and *Syntarsus*, which are transferred to the taxon Ceratosauria together with *Ceratosaurus* and *Dilophosaurus*. Paul likewise concludes that there is no evidence that "coelurosaurs" form a clade.

An important contribution of these cladistic studies has been the compilation of detailed character lists for a wide variety of dinosaurian taxa, with useful evaluations of their significance for phylogenetic analysis. These compilations are now available for others to refine or refute. As a result, the body of knowledge is growing apace, but it seems to me that studies of theropod phylogenetic history still have some way to go before clarity is finally reached.

Other symposium participants have dealt in detail with the systematics of particular taxa of small theropods. My contribution does not attempt to provide an overall synthesis of small theropod systematics, because at this stage I consider it a vain and forlorn task. Instead, I present here information on one species of early small theropod, *Syntarsus rhodesiensis*, because it offers what I consider unparalleled opportunities for thorough morphological assessment. Perhaps through study of taxa like *Syntarsus rhodesiensis*, where the material available for study is abundant and well preserved, one might hope to establish firmly the fundamental theropod synapomorphies. That should help to fine-tune recognition of systematically useful characters, eliminate plesiomorphies from consideration, and so refine hypotheses of relationship.

The "quality factor" in theropod study material

Because small theropod bones are small and thin-walled, they are usually fragile and highly vulnerable

Figure 7.1. Cladograms of relationships within the Theropoda: **A**, after Paul (1984); **B**, after Gauthier and Padian (1985). Abbreviations: Al, allosaurs; AR, archaeopterygids; Bi, birds; CA, carnosaurs; Ce, Ceratosaurs; Co, compsognathids; CP, *Coelophysis*; De, deinonychosaurs; Di, dilophosaurs; Dr, dromaeosaurs; Eu, *Eustreptospondylus*; HD, herbivorous dinosaurs; He, herrerasaurs; Me, megalosaurs; OI, Ornithischia; On, ornithomimids; Or, *Ornitholestes*; Ov, oviraptorids; Pc, *Procompsognathus*; Pr, *Proceratosaurus*; SA, Sauropodomorpha; So, *Saurornithoides*; St, ornithomimids; Ty, tyrannosaurs; Ya, *Yangchuanosaurus*.

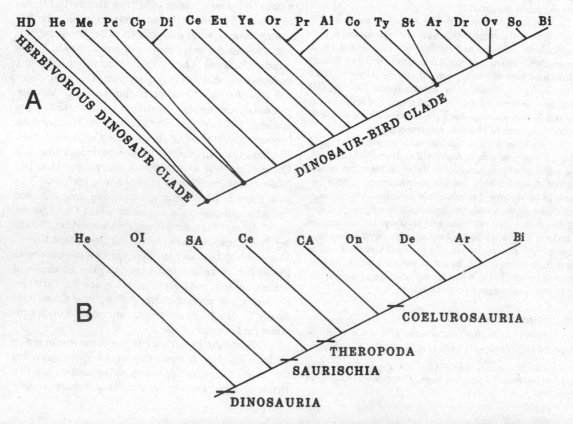

under the rigors of their taphonomic history, which means that collected material on which descriptions and taxonomic analyses must be based is often fragmentary and poorly preserved. Consequently, apomorphic character suites are often difficult to define, let alone to evaluate. The type specimens of more than half of the 40-odd small theropod species listed by Steel (1970) are based on a handful of assorted fragmentary bones each, often unassociated, and of dubious taxonomic value. This is certainly true of quite a few forms that qualify as coelurosaurs in the classical sense.

In contrast, the holotype of the small theropod *Syntarsus rhodesiensis* Raath 1969 (Fig. 7.2A) is extremely well preserved. It was collected in 1963 from exposures of the terminal Karoo (earliest Jurassic) Forest Sandstone Formation at Nyamandhlovu, near Bulawayo, Zimbabwe (Fig. 7.3, locality a). Very poorly preserved fragmentary remains of a second specimen were recovered from within a few meters of the type at

the same locality at the same time. Several features of the holotype skeleton were interpreted as indicating advanced age (Raath 1969), notably the amount of muscle-scarring and rugosity on some of the bones, and the degree of fusion shown by several elements. Because at the time of its original description, knowledge of *Syntarsus* was based essentially on only the holotype, with a negligible addition of information from the scrappy second specimen, no assessment could be made of either the ubiquity or the biological significance of the unusual features noted in the holotype.

In 1968 a well preserved isolated femur was found on the Maura River in the central Zambezi Valley of Zimbabwe, some 450 km northeast of the *Syntarsus* type locality (Fig. 7.3, locality c). This specimen – almost identical in general shape and dimensions to the holotype femur of *Syntarsus* – was considerably more gracile in that few of its muscle scars were as clearly defined as those of the holotype. Another deposit of

Figure 7.2. *Syntarsus rhodesiensis*: **A**, schematic representation of holotype to show completeness of material recovered; **B**, skeletal reconstruction. All of the postcranial skeleton behind the neck is based on the holotype, QG1; cranial and cervical osteology is based on several specimens from the Chitake River locality, central Zambezi River Valley, Zimbabwe.

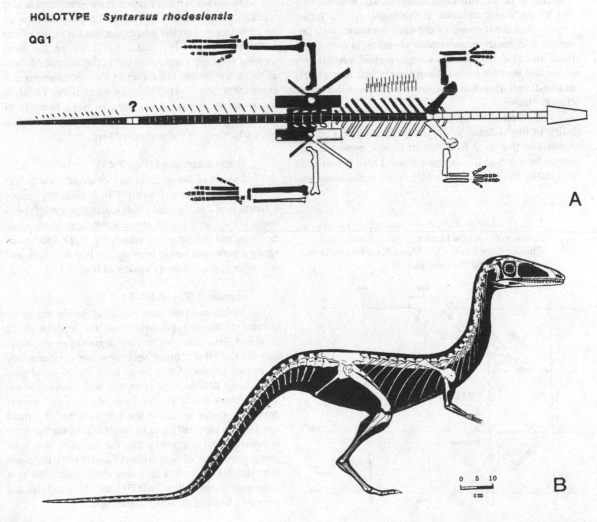

HOLOTYPE *Syntarsus rhodesiensis*

QG1

A

0   5   10
cm

B

Syntarsus bones, remarkable for its richness, concentration, and state of bone preservation, was found on the Chitake River (also in the central Zambezi Valley) some 80 km west of the Maura River locality in 1972 (Fig. 7.3, locality b). To date, the remains of more than 30 individuals have been recovered from the Chitake River bone-bed (Raath 1977), and the prepared material shows a consistent pattern of morphological variation within the sample that mirrors the differences noted between the holotype femur and the one from the Maura River. Study of the almost complete, well-articulated holotype and the abundant and exquisitely preserved material from the Chitake River locality has permitted detailed and accurate skeletal reconstruction of *Syntarsus* (Fig. 7.2B).

Because all Zimbabwean specimens of *Syntarsus*, other than the holotype and the fragmentary topotypical specimen, have come from deposits far removed from the type locality, possibly even from a separate sedimentary basin (Fig. 7.3), the question of whether the more northerly samples might represent a separate but related species of the same genus had to be addressed. To try to resolve this question, variation in comparable bones from the three localities was analyzed. The element chosen for particular attention in this study was the femur, largely because it is one of the more common, well preserved and easily recognizable elements in the collection, and also because its morphological variants are clear and unambiguous. Comparable but less clearly marked variation has also been noted in other postcranial elements.

Colbert (this volume) has commented on variability in the skeleton of the coeval and closely related American theropod from Ghost Ranch, New Mexico, which he refers to *Coelophysis bauri*. There is particular variability in the extent to which bones in the tarsus and

pelvis may be fused. Colbert concludes that most of this variation, including the degree of tarsal fusion, can be accounted for as individual variation or consequences of growth, but some aspects might relate to adult sexual dimorphism.

Morphological variation in the postcranial skeleton of *Syntarsus*

As in Colbert's *Coelophysis*, the extent to which bones might be fused in *Syntarsus* is also variable, from no fusion or very little in some specimens, to extensive fusion in others. Furthermore, extensive fusion is also associated with other striking variations affecting different parts of the skeleton of *Syntarsus*. These include robustness of the humerus, extent of development of the olecranon process on the ulna, breadth of the metacarpus, flaring of the ilium, and configuration and size of muscle trochanters on the femur. Ontogenetic age is undoubtedly important in the expression of these features, but it seems that more than age alone is involved.

Variations in pectoral girdle and forelimb

Syntarsus forelimb bones show considerable size variation. The largest elements in the collection are up to 18% larger than the corresponding elements of the holotype (Raath 1977, Table 8). However, the most striking forelimb variation is not in the absolute dimensions of the bones, but rather in their proportions and robustness (Fig. 7.4). This may parallel the variation shown by the femur, as the holotype has a humerus of robust proportions and it is associated with a femur that is clearly robust in the sense used here.

Scapulocoracoid (Fig. 7.4A)

Fusion of the scapula and coracoid, which was noted in the holotype (Raath 1969), is evidently simply a function of age. Several small (clearly juvenile) specimens show that the two bones were easily dissociated postmortem, while other relatively large specimens retain a persistent suture between the two elements, and still others are completely coalesced (Fig. 7.4A).

Humerus (Fig. 7.4B–E)

Variations in the humerus affect mainly the development of the deltopectoral crest, the breadth of the head and distal condyles, and the degree of muscle-scarring (Fig. 7.4B–E). Some specimens are comparatively short and angular with a broad head, broad distal condyles, large deltopectoral crest, and heavy muscle scars, while others are relatively long, slender, and almost devoid of visible muscle scars. The humerus of a small specimen is especially gracile while that of the holotype is robust, again suggesting that the variation is at least partly a function of age. Robust humeri have the distal condyles for the radius and ulna sharply defined by a pronounced rim (Fig. 7.4D,E), and the condyles are smoothly finished in well-formed bone.

Figure 7.3. Map of *Syntarsus rhodesiensis* localities in Zimbabwe: a = type locality (Nyamandhlovu); b = Chitake River locality; c = Maura River locality (b and c both in central Zambezi Valley).

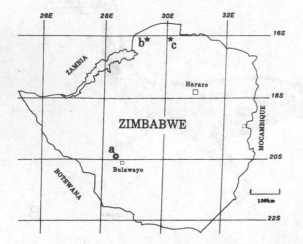

Ulna (Fig. 7.4F,G)

The greatest variation in ulnar morphology is in the degree of development of the olecranon process. Most recovered specimens have it relatively well developed, deep and robust (Fig. 7.4G), while others have almost no discernible olecranon (Fig. 7.4F). However, there is no obvious pattern to the distribution of this variation in the recovered sample, largely because relatively few associations of articulated bones have been preserved, and therefore it is not clear whether it can be

Figure 7.4. *Syntarsus rhodesiensis*, variation in forelimb elements: **A**, fused scapulocoracoids of adults (holotype QG 1 in center, medial view; others in lateral view); **B**, **C**, gracile humerus (QG 545): **B**, palmar view; **C**, anconal view; **D**, **E**, robust humerus (QG 514): **D**, palmar view; **E**, anconal view; **F**, ulna with poorly developed olecranon (QG 568); **G**, ulna with well developed olecranon (QG 514b, goes with humerus QG 514 – see **D**, **E**); **H**, right carpus and manus of slender form (QG 577), anconal view; **I**, right metacarpus of broad form (QG 573), palmar view; **J**, left carpus (broad) and manus of holotype (QG1), anconal view. (Scale bars = 2 cm.)

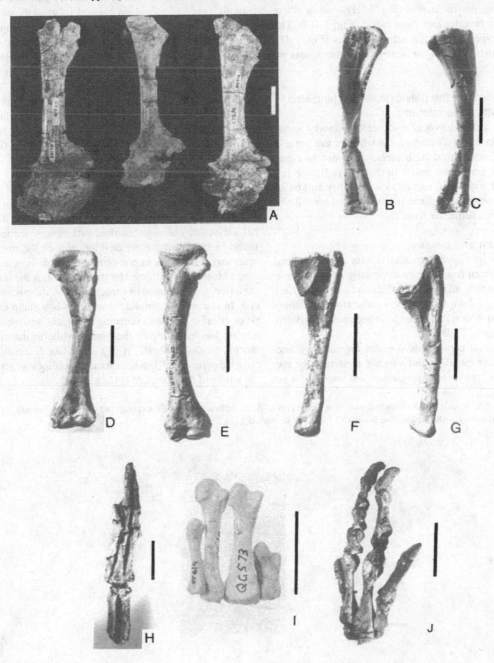

linked to other manifestations of dimorphism. Whether or not the olecranon might have ossified separately cannot be answered, because there are no unidentified pieces of bone in the collection that could conceivably be candidates.

Carpus and manus (Fig. 7.4H–J)

Patterns of variation in the carpus and manus are also difficult to assess because, apart from the holotype, few specimens are sufficiently well associated to permit detailed comparisons. In one small specimen (QG 577), the metacarpus is distinctly slender and gracile – producing a long, narrow manus (Fig. 7.4H) – while others are relatively broader and more robust (Fig. 7.4I,J). The robust holotype (QG 1) has a broad manus (Fig. 7.4J), suggesting that a broad metacarpus and robustness go together.

Variations in the pelvic girdle and hind limb (including the sacrum)

There is unequivocal evidence of bimodal variation in the femur of *Syntarsus*, but whether the same is also true of other hind limb elements is not so clear. Some of the variation noted in the pelvis (involving broader and narrower variants) is possibly linked to femoral dimorphism, but too few articulated hind limbs are available to settle the issue.

Sacrum and pelvis

A few specimens are available for study, ranging in condition from fragmentary to virtually perfect. Some represent mature adults with the bones firmly fused together (Fig. 7.5A), while others come from immature individuals in which the bones have become dissociated postmortem (Fig. 7.5B).

The sacral centra fuse together on maturity, and there is a corresponding but variable tendency for the neural spines, transverse processes, and sacral ribs to coalesce into continuous bony sheets extending the length of the sacrum. Fused sacral neural spines are invariably topped by a swollen longitudinal rim along the length of the continuous blade.

Lateral flaring of the ilium behind the acetabulum is variable. Although the sample available for study is small, it seems that smaller specimens (juveniles) have narrow dolichoiliac pelves with relatively unflared ilia (Fig. 7.5B), whereas some larger specimens have broadly flared brachyiliac pelves as represented by the holotype (Fig. 7.5A). Whether these variants represent a pelvic expression of dimorphism, or are merely part of a continuous range of variation related to growth, or are attributable instead to distortion resulting from compression, cannot be established with certainty. The hood-like supra-acetabular buttress is consistently well developed in both juveniles and adults (Fig. 7.6A–C). In this respect, and in respect to the flaring of the ilium, *Syntarsus* seems to depart significantly from *Coelophysis* (E. H. Colbert, pers. comm. 1976) (Fig. 7.6D–F), but it agrees well with the condition in *Elaphrosaurus* from the Late Jurassic of Tanzania (Janensch 1925) (Fig. 7.6G–I).

Femur

The relatively large sample of femora from the Chitake River locality shows clear bimodal morphological variation. The most marked differences are to be found in the much greater development of the muscle scars and trochanters in one femoral morph than in the other. Most clearly affected by this variation is the lesser (anterior, or iliofemoralis) trochanter at the proximal end. In one morph, termed "gracile," it is a sharp erect ridge or crest-like process facing obliquely anteromedially on the front face of the femur, while in the other morph, termed "robust," it is a shelf-like feature (the "trochanteric shelf"), which extends obliquely across the whole of the front face (Fig. 7.7).

Figure 7.5. *Syntarsus rhodesiensis*, variation in the pelvis: **A**, fused brachyiliac pelvis (holotype, QG1); **B**, smaller, unfused pelvis with dolichoiliac ilium (QG 691). (Scale bars = 2 cm.)

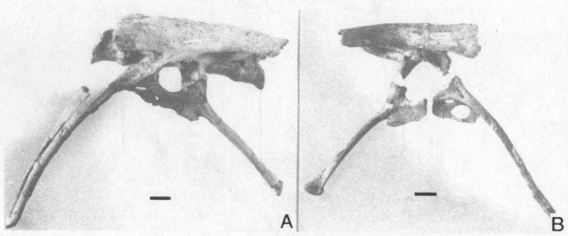

A | B

Coupled with the development of the trochanteric shelf in robust femora is the presence of a bulbous, rugose greater trochanter on the posterolateral surface of the femoral neck, and of a curved ridge just medial to the greater trochanter on the posterior surface of the proximal end (identified as an "obturator ridge": Raath 1977). Neither of these features is seen in gracile femora. The greater trochanter in the latter is flat and smooth, and there is no sign of an obturator ridge. This combination of variants is consistent, femora having always either the gracile configuration or the robust, regardless of size (Figs. 7.7, 7.8). The femur of the holotype represents the robust morph (Fig. 7.7D). The two morphs are differentiated as set out in Table 7.1. That the variation is not simply due to allometry is shown by the fact that the left femur of the holotype (QG1), which is 208 mm long, is robust, while the femur G76 is gracile yet very similar in length at 201 mm (a length difference of only 3.5%). Although few femora in the sample are complete enough for the overall length to be measured, there is a reasonable sample (more than 18) of well preserved femoral heads from which various dimensions can be taken for the purpose of comparison (Table 7.2).

Bimodal femoral variation is clearly shown if the breadth of the iliofemoralis (lesser) trochanter (one of the strongly expressed variables) is plotted against maximum transverse breadth of the femoral head, the latter taken to reflect size increase due to normal ontogenetic growth (Fig. 7.8). Note that the portion of the graph that includes gracile forms also includes all small individuals, whereas robust individuals only appear once femoral head breadth is 28 mm or more, suggesting that femoral robustness is attained only on reaching a particular size (and, therefore, age). The graph also suggests that robust specimens might grow to a larger absolute size than gracile specimens: QG 726 (the largest of the robust femora recovered) has a head 15% broader than that of QG 739, which is the largest of the recovered gracile femora.

Another feature of the femur that seems to show two distinct morphologies is the medial edge of the intercondylar groove, over which the large crural extensor tendon ran; its medial border is a well-defined sharp rim in gracile femora while in robust specimens it is more rounded, rugose, and less starkly defined.

Tarsus

The Chitake River material shows that the extent of fusion in the tarsal region, which was held to characterize the holotype (Raath 1969), actually varies con-

Figure 7.6. Comparison of the pelvis of theropods (not to same scale): **A–C**, *Syntarsus*; **D–F**, *Coelophysis*; **G–I**, *Elaphrosaurus*; sab = supra-acetabular buttress; **A, D, G**, left lateral view; **B, E, H**, dorsal view; **C, F, I**, hind view. Adapted from Raath (1977) – *Coelophysis* after Colbert (pers. comm.), *Elaphrosaurus* after Janensch (1925).

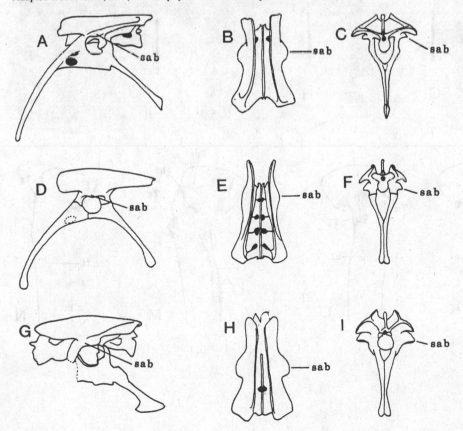

Figure 7.7. *Syntarsus rhodesiensis*, morphological variation in the femur: **A**, **B**, small gracile individual (QG 691): **A**, anterior view; **B**, posterior view; **C**, **D**: large robust individual (holotype QG1): **C**, anterior view; **D**, posterior view; **E**, **F**, medial view of proximal end showing caudifemoral muscle insertion scars and profile of lesser trochanter: **E**, gracile (QG 76); **F**, robust (QG 753); **G–J**, proximal end showing trochanter topography: **G**, **H**, gracile morph (QG 717) - **G**, anterior, **H**, posterior; **I**, **J**, robust morph (QG 753) - **I**, anterior, **J**, posterior; **K–N**, interpretation of trochanter morphology at proximal end: **K**, **L**, anterior views - **K**, robust, **L**, gracile; **M**, **N**, posterior views - **M**, robust, **N**, gracile. Abbreviations: Tg, greater trochanter; Tl, lesser trochanter; T4, fourth trochanter; Tps, "pseudotrochanter"; obt, obturator ridge; ft, insertion scar of femorotibialis muscle; cfm, insertion pit for caudifemoral muscle. (Scale bars = 2 cm.)

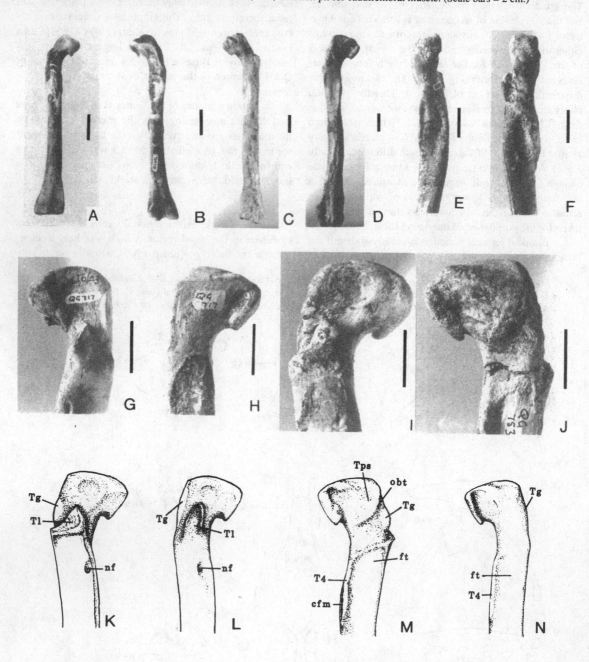

Table 7.1. *Features of femur variants in* Syntarsus rhodesiensis

Robust	Gracile
1. Greater trochanter bulbous and rugose	1. Greater trochanter flat, smooth-surfaced
2. Lesser trochanter broad, shelf-like[a]	2. Lesser trochanter narrow, crest-like
3. "Obturator ridge" present in association with "pseudotrochanter"	3. "Obturator ridge" absent; "pseudotrochanter" not clear
4. Posterior femorotibialis origin outline scarred	4. Posterior femorotibialis origin not scarred
5. Insertion pit for caudifemoralis brevis crescentic, rugose, short, sharply rimmed	5. Insertion pit for caudifemoralis brevis elliptical, smooth, elongate, not sharply rimmed
6. Medial edge of distal patellar groove rugose, not sharp-edged	6. Medial edge of distal patellar groove smooth, sharp-edged

[a]This trochanter was wrongly identified as the "greater trochanter" by Raath (1969). See Figures 7.6E-H for trochanter identification in the *Syntarsus* femur.

Figure 7.8. Bivariate plot of iliofemoralis (= "lesser") trochanter width against maximum femur head breadth, showing bimodal distribution of lesser trochanter dimensions.

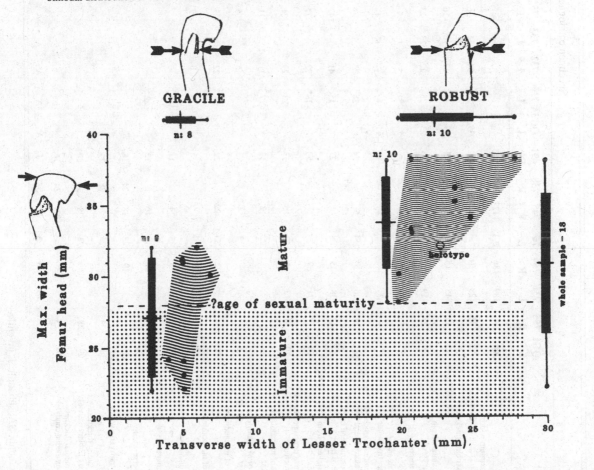

Table 7.2. Syntarsus rhodesiensis: femur dimensions (mm)

Specimen catalogue number and side

Measurement	QGI 755 L	755 R	731 L	754 R	76 L	745 R	738 R	691 R	QG3A 726 L	732 L	733 L	753 L	725 L	760 R	729 L	727 L	740 R	739 R	717 L	742 R	744 R	715 L	713 L	716 R
Overall length	208	185	189	186	201	172+	192	142																
Femur head: maximum width medial surface to greater trochanter	32	35	28	30			30	22	38	34	34	36	33	38	36	33	30	32	31	31	24	24	24	
Femur head: maximum thickness anteroposterior	15	17	15	15	12		12	11	15	17	16	16	15	16	15	15	14	15	13	13	10	12	10	14
Femur head: minimum depth at Teres ligament "hook"	20	23	19	20	20	19	19	13	23	22	23	23	20	21	20	19	18	18	20	19	14	15	12	20
Presence (+) or absence (−) of "obturator ridge"	+	+	+	+	−	−	−	−	+	+	+	+	+	+	+									+
Presence (+) or absence (−) of bulbous greater trochanter	+	+	?	−	−		−	−	+	+	+	+	+	+	+									+
Breadth of iliofemoralis trochanter (= lesser troch.) along lateral border	23	24	20	20	4	5		4	28	25		24	21	21		7	7	6	5	5	5	5	4	
Trochanter 4: proximal end from proximal surface of head	44	43	38	35	38			27	45	41		42	36	40			30		28					
Trochanter 4: distal end from proximal surface of head	74	74	71	69	70			50	78+	72		77	68	75	65								60	28
Maximum thickness (anteroposterior) of shaft at mid trochanter 4	19	23	17	19	20	16		13	23	22		21	21	21	18								15	
Greater diameter of m. caudiferm. longus tendon scar	11				22	12		12	17	11		11		13									19	
Lesser diameter of M. caudiferm. longus tendon scar	3				3			3	4	5		5		4									3	
Greater diameter of m. caudiferm brev. tendon scar	21			22	30	22	19	20	24	22		23		21	21								20	
Lesser diameter of m. caudiferm brev. tendon scar	7			6	9	9		5	10	8		9		10	8								7	
Midlength shaft diameter (transverse)	21	16	13	13	15	4	11	11	18			18	18			14			28				12	
Midlength shaft diameter (anteroposterior)	18		17	17	18	16	12	12	22				20	16					9	13			13	
Maximum transverse width of rotular surface	25			19	22	18	18	18																
Maximum distal transverse width across condyles	28	34	25		25	21	23	23						21			20							
Maximum distal lateral thickness (ant. surface to apex fibular condyle)	29	28		24	24	28	28	20																
Maximum distal medial thickness (ant. surface to apex medial condyle)	26	25				27	24	16																
Age class[a]	A	A	SA	SA	A	J	SA	J	A	A	A	A	A	A	A	A	A	A	A	A	A	J	J	SA

[a] Age class: A, adult; SA, subadult; J, juvenile.

siderably (Fig 7.9), much like the variation in the Ghost Ranch sample (Colbert this volume). In general, smaller individuals show neither the fusion of astragalus and calcaneum to each other, nor of astragalocalcaneum to the tibia as seen in larger specimens. Although the astragalus and calcaneum are always tightly fitted together and functionally united, they are not always co-ossified. Heavily muscle-scarred large individuals tend to show extensive bone-to-bone fusion. One result of this is that they develop a functionally simplified tarsal joint in which a fused tibiotarsus articulates with a fused tarsometatarsus. In these cases, only distal tarsal IV remains free in the tarsus. Again, however, there are insufficient articulated, associated specimens to establish whether this observed tendency is consistently associated with either of the two femoral morphs.

Discussion

Does the morphological variation reflect taxonomic diversity?

The *Syntarsus* bone-bed on the Chitake River is concentrated, both vertically and laterally, within a shallow and restricted layer that shows no internal indication of lamination other than irregular layers of differing color. From their geometry and distribution in the rest of the deposit, the color patches seem to be diagenetic effects of leaching and mineral migration rather than primary sedimentary features. The fossils range from well-articulated partial skeletons to completely disarticulated and scattered individual bones. Yet even the most delicate and fragile elements are perfectly preserved – such as cervical ribs, some of which are a millimetre or less in diameter but more than 12 cm long. They show

Figure 7.9. *Syntarsus rhodesiensis*, tarsal variation: **A**, unfused small left tarsus (QG 768), dorsal view; **B**, as for (**A**), proximal end view (lateral to left); **C**, stereo – isolated right astragalus and calcaneum (QG 786), proximal end view; **D, E**, extensively fused tarsus (holotype QG1): **D**, anterior view; **E**, posterior view; **F**, isolated distal tarsals III and IV: upper row, tarsal IV; lower row, tarsal III. (Scale bars = 2 cm.)

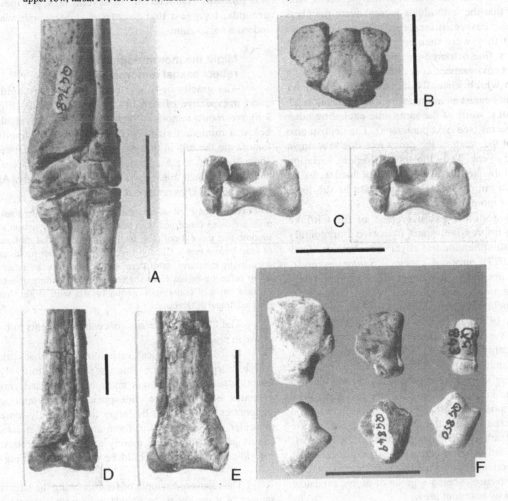

no signs of abrasion that could be attributed to fluvial transport. Large and small bones are preserved together and intimately intermixed. There is no indication of current-induced winnowing of lighter bones leaving a lag concentrate of larger elements. Some long bones were preserved oriented at steep vertical angles, cutting right across any incipient or latent bedding, but the majority lay more or less horizontally. No measurements have been taken of bone orientation, mainly because most specimens were recovered from fallen blocks of sediment no longer in their original geological context, and also because it was impractical in the field to excavate down through the 20 metre cliff face to a suitable position from which to take such measurements. Cursory examination of unprepared bone-bearing sediment blocks does not show any obvious fabric, but only a rather vague hint of preferred orientation among the long bones in the deposit; this impression of imbrication requires careful measurement to confirm or refute it.

The sedimentological evidence of the Chitake deposit and the close integration of the preserved sample suggest that the individuals present were members of a single cohesive, interacting group. This alone implies that they were members of a single species. Without very fine resource-partitioning in a resource-scarce desert environment, it is difficult to conceive of a situation in which virtually indistinguishable sister species could coexist as intimately as the Chitake occurrence indicates, while at the same time excluding other unrelated species (see next paragraph). The deposit also suggests that the death of the group was due to a single catastrophic event. Thus the morphological variation preserved in the sample seems best explained as variation within a single species, rather than as the local occurrence of more than one species.

The taxonomic exclusiveness of the Chitake bone-bed is impressive: apart from two fragmentary jaws of small sphenodontid rhynchocephalians (Gow and Raath 1977), only one species – *Syntarsus rhodesiensis* – is present in this layer. Yet in immediately adjacent sediments, several other vertebrate taxa are preserved, including quite common prosauropod dinosaurs (Raath 1977). In fact, at both the other localities from which *Syntarsus rhodesiensis* has been recovered (the type locality and the Maura River), prosauropods predominate and *Syntarsus* is rare. Remains of only two individuals are known from each place. In the Chitake bone-bed, the situation is dramatically reversed. Why are prosauropods and other components of the local fauna not mixed in with the Chitake *Syntarsus*? I have suggested (Raath 1977) that a gregarious flock of *Syntarsus* excluded other species from that place at that time, either actively by driving them off, or passively by their mere presence – being a group of active predators to be avoided by potential prey.

Ostrom (1969) has suggested that the predatory theropod *Deinonychus* hunted in packs because the remains of as many as five individuals were preserved together with the fragmentary remains of a single medium-sized ornithopod, thought to be the prey animal. Similar group hunting behavior has been suggested for other theropod genera (reviewed by Farlow 1976). Ostrom (1972) was able to demonstrate gregarious behavior in some species of bipedal dinosaurs on the basis of his analyses of fossil trackways. But he cautioned that mere accumulations of the bones of a single species at one locality, such as the famous concentration at Ghost Ranch in New Mexico, did not necessarily provide evidence of group activity by the animals concerned; that is, they need not necessarily have been behaving gregariously. In the case of the similar Chitake *Syntarsus* deposit there seems to be prima facie evidence to suggest that they were indeed behaving gregariously. There is a growing body of evidence that many dinosaurs – and especially small theropods (Hopson 1977) – were sufficiently encephalized to allow complex social behavior and that group social organization was probably characteristic of several taxa. On these grounds, I suggest that *Syntarsus rhodesiensis* was indeed a social animal.

Might the morphological variation reflect sexual dimorphism?

The gracile-or-robust bone morphology holds good irrespective of size in mature individuals (Fig. 7.8). No robust femora are known amongst individuals below a minimum size (?juveniles), whereas gracile femora are the rule in this group. This suggests that the specializations are associated with secondary sexual characters, which are manifested only at maturity. As Ager (1963: 71) has remarked:

After eating, the most widespread habits among modern animals are those concerned with sex, and there is no reason to suppose that this did not raise its allegedly ugly head millions of years before Freud. Clearly if we are to regard our fossils as once-living creatures, considerations of sex must arise, and many palaeontologists have suggested sexual dimorphism to explain pairs of contemporaneous fossils with slight, but nongradational differences.

The Chitake *Syntarsus* concentration seems to be a case in point.

If the morphological variation is indeed sexual, which morph represents which sex? It is intuitively tempting to regard the larger and more muscular forms as males, because of the widespread tendency among vertebrates for males to be larger and generally more muscular than females. But there are many exceptions to this generalization and merely to declare the robust individuals to be males would be facile and meaningless.

The recovered sample of femora, being the largest sample of a readily recognizable bone with distinctive features, was analyzed to obtain an estimate of the minimum number of individuals represented and an estimate

Figure 7.10. Size-frequency distribution of a sample of femora from the Chitake River *Syntarsus* locality. Note that all small specimens are gracile, and that robust specimens predominate amongst larger individuals. See text and Table 7.3 for further explanation.

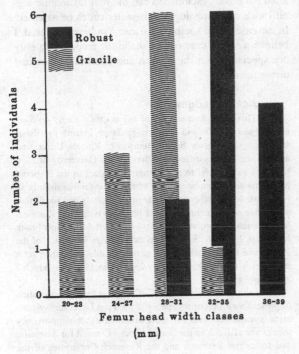

Femur head width classes (mm)

of the ratio of robust to gracile individuals in arbitrary size classes assumed to correspond roughly with age (Fig. 7.10). This preliminary analysis has been carried out only on the prepared fraction of the recovered sample, which by no means accounts for the entire sample preserved in the Chitake River deposit; many collected specimens still await preparation. Once more material is prepared, a reevaluation of the conclusions arrived at here might become necessary, but I must emphasize that in the preparation program so far there has been no conscious preferential selection of any one femoral morph for preparation over another. On the contrary, blocks of matrix have been entirely cleared of all bones as and when presented to the preparators, the aim being to recover them all for the collection sample.

The size–frequency distribution of the two main femoral variants in the sample recovered to date supports the contention that robustness develops at or near maturity. No juveniles are robust. The split takes place amongst subadults, with gracile forms predominating (Figs. 7.7, 7.10). By the time large size (= full adulthood) is reached, the ratio swings in favor of robust forms (Fig. 7.10).

To what extent the sample in the collection is statistically representative of the original population is unclear; what possible taphonomic biases might have operated are unknown. It is possible that environmental conditions prevailing at the time of death might have

Table 7.3. Syntarsus rhodesiensis: *distribution of femur morphs among "age classes" and calculation of minimum numbers (MN) – Zimbabwe sample*

Locality	Femur morph.	Side	Adult N	Adult MN	Subadult N	Subadult MN	Juvenile N	Juvenile MN	Minimun numbers per locality
Southcote	Robust	L	2	—	0	—	0	—	
		R	1	2	0	0	0	0	
Farm (Nyamandhlovu)	Gracile	L	—	0	—	0	—	0	2
		R	0	0	0	0	0	0	
Maura River (Zambezi Valley)	Robust	L	0	—	0	—	0	—	
		R	0	0	0	0	0	0	2
	Gracile	L	1	—	0	—	1	—	
		R	0	1	0	0	1	1	
Chitake River (Zambezi Valley)	Robust	L	9	—	2	—	0	—	
		R	3	9	2	2	0	0	26
	Gracile	L	1	—	4	—	3	—	
		R	7	7	1	4	4	4	
MN totals per age class				19		6		5	
Grand total MN (recovered and prepared as of Nov. 1977)						30			

selectively culled one section of the population so that it would be over-represented in a death assemblage, such as, for example, females reluctant to desert distressed young (Raath 1977). In that case, females might be expected to predominate amongst the adults in the fossil sample, which would suggest that the robust forms should be females.

As indicated above, the Chitake deposit suggests that a catastrophic event produced the bone-bed, and the lithology of the sediments indicates a generally arid environment at the time. These deposits are part of the great Gondwana desert of terminal Karoo times in the latest Triassic-earliest Jurassic (see Olsen and Galton 1984, for a discussion of the age of coeval deposits in southern Africa).

The scenario of females remaining in the vicinity of offspring in distress in a hostile environment is not implausible, but it is of course highly speculative.

If, though, the sample is anything approaching a snap-shot of the local *Syntarsus* population, and if that population was in other respects normal, a skewed sex ratio would usually imply enhanced breeding potential and the numerically predominant sex should therefore be female. This, too, would suggest that the robust forms are females.

It is by no means biologically unknown for females to be larger and generally more powerfully built than males of the same species – indeed, it is the general rule amongst raptorial birds (see, for example, Brown and Amadon 1968). However, it is also a general rule that in these birds male and female pair for life, and so in this case a sex ratio close to parity would be expected. A sex ratio favoring females is common in gregarious animals where mature males are polygamous and keep harems (many antelope and bird species). Although at this stage we do not know with certainty that *Syntarsus* was gregarious with harem-keeping males, the evidence of the Chitake bone-bed is prima facie evidence of some sort of group activity that also seems to have included some kind of parental care because of the presence of young of varying sizes. All this implies gregarious sociality.

Discussion of the underlying reasons why muscularity should be greater in one sex than the other is beyond the scope of this study. Perhaps robustness was related to a repertoire of vigorous and demanding physical activities confined to one sex – for example, intraspecific fighting, territorial defence, courtship display, nest preparation, or hunting to provide food for dependent young. None of these activities would disqualify females as candidates for the robust sex.

The excellent hypodigm of *Syntarsus* material allows studies of this kind, and it avoids the pitfalls of insecure and doubtful taxonomy (and therefore systematics) that inevitably flow from studies of inadequate type and comparative material. Without adequate well preserved material, intraspecific morphological variation appears disjunct and discontinuous. In the past, this

situation has sometimes persuaded taxonomists that they are dealing with more than one species, whereas a good series might clearly have shown this not to be the case. In my opinion, only once the limits of intraspecific variation have been established can the real taxonomic significance of morphological character suites be assessed. In the case of the southern African *Syntarsus* material, I believe a strong case can be made for recognizing only one species, which shows clear and unmistakable sexual dimorphism.

Acknowledgments

This study formed part of my doctoral thesis on *Syntarsus rhodesiensis*, and for their guidance I thank my thesis supervisors, Professor Brian Allanson of Rhodes University, and the late Professor Geoffrey Bond of the University of Zimbabwe. I am grateful to my former colleagues in the National Museums and Monuments of Zimbabwe for their assistance in fieldwork during the course of the original study and for further assistance subsequently when I have revisited the Zambezi Valley sites, particularly Des Jackson, Ted Mills, and Frank Matipano. Similarly, Euen Morrison and Tim Broderick of the Zimbabwe Geological Survey gave valuable assistance on a return visit to the Chitake River locality in 1985, for which I am deeply grateful.

I thank P. J. Currie and K. Carpenter for their invitation to present a version of this paper at the First Dinosaur Systematics Symposium and for their other valuable support; my thanks are also due to the South African Council for Scientific and Industrial Research and the Research Committee of the University of the Witwatersrand.

Edwin Colbert, Philip Currie, Thomas Lehman, John Ostrom, and Kevin Padian critically reviewed this manuscript. Their helpful comments are sincerely appreciated.

References

Ager, D. V. 1963. *Principles of Palaeoecology* (London: McGraw-Hill).

Brown, L., and Amadon, D. 1968. *Eagles, Hawks and Falcons of the World* (Feltham, England: Hamlyn).

Farlow, J. O. 1976. Speculations about the diet and foraging behavior of large carnivorous dinosaurs. *American Midland Naturalist*, 95(1):186–191.

Gauthier, J. 1986. Saurischian monophyly and the origin of birds. *In* Padian, K. (ed.), *The Origin of Birds and the Evolution of Flight. Memoirs of the California Academy of Sciences* 8:1–55.

Gauthier, J., and Padian, K. 1985. Phylogenetic, functional, and aerodynamic analyses of the origin of birds and their flight. *In* Hecht, M. K., Ostrom, J. H., Viohl, G., and Wellnhofer, P. (eds.), *The Beginning of Birds, Proceedings of the International* Archaeopteryx *Conference* (Eichstätt: Freunde des Jura-Museums) pp. 185–197.

Gauthier, J. A. 1984. *A Cladistic Analysis of the Higher Systematic Categories of the Diapsida.* Unpublished Ph.D. thesis (Berkeley, California: University of California).

Gow, C. E., and Raath, M. A. 1977. Fossil vertebrate studies in Rhodesia: sphenodontid remains from the Upper Triassic of Rhodesia. *Palaeontologia Africana* 20:121–122.

Hecht, M. K. 1985. The biological significance of *Archae-opteryx*. *In* Hecht, M. K., Ostrom, J. H., Viohl, G., and P. Wellnhofer (eds.), *The Beginning of Birds, Proceedings of the International* Archaeopteryx *Conference* (Eichstätt: Freunde des Jura-Museums), pp. 149–160.

Hopson, J. A. 1977. Relative brain size and behaviour in archosaurian reptiles. *Annual Review of Ecology and Systematics* 8:429–448.

Janensch, W. 1925. Die Coelurosaurier und Theropoden der Tendaguru-Schichten, Deutsch-Ostafrikas. *Palae-ontographica*, Supplement 7, Series 1, Pt. 1(1):1–100.

Olsen, P. E. and Galton, P. M. 1984. A review of the reptile and amphibian assemblages from the Stormberg of Southern Africa with special emphasis on the footprints and age of the Stormberg. *Palaeontologia Africana*, 25:87–110.

Ostrom, J. H. 1969. Osteology of *Deinonychus antirrhopus*, an unusual theropod from the Lower Cretaceous of Montana. *Peabody Museum of Natural History Bulletin* 30:1–165.

1972. Were some dinosaurs gregarious? *Palaeogeography, Palaeoclimatology, Palaeoecology* 11:287–301.

1981. *Procompsognathus*–theropod or thecodont? *Palaeontographica* A 175(4–6):179–195.

Paul, G. S. 1984. The archosaurs: a phylogenetic study. *In* Reif, W.-E., and Westphal, F. (eds.), *Third Symposium on Mesozoic Terrestrial Ecosystems*, short papers (Tübingen: Attempto Verlag), pp. 175–180.

Raath, M. A. 1969. A new coelurosaurian dinosaur from the Forest Sandstone of Rhodesia. *Arnoldia Rhodesia* 4(28):1–25.

1977. *The Anatomy of the Triassic Theropod* Syntarsus rhodesiensis *(Saurischia: Podokesauridae) and a Consideration of its Biology*. Unpublished Ph.D. thesis (Grahamstown, S. Africa: Rhodes University).

Romer, A. S. 1966. *Vertebrate Paleontology*, 3rd edition (Chicago: University of Chicago Press).

Steel, R. 1970. Saurischia. *In* Kuhn, O. (ed.), *Handbuch der Paläoherpetologie* (Stuttgart: G. Fischer Verlag), 14:1–87.

von Huene, F. 1914. Das natürliche System der Saurischia. *Centralblatt für Mineralogie, Geologie und Paläontologie, Jahrgang*, Abteilung B (5):154–158.

1932. Die fossile Reptil-Ordnung Saurischia, ihre Entwicklung und Geschichte. *Monographien zur Geologie und Paläontologie*, Series 1, 4(1):1–361.

Welles, S. P. 1984. *Dilophosaurus wetherilli* (Dinosauria, Theropoda), osteology and comparisons. *Palaeontographica* A 184(4–6):85–180.

8 Theropod teeth from the Judith River Formation of southern Alberta, Canada

PHILIP J. CURRIE,
J. KEITH RIGBY, JR., AND
ROBERT E. SLOAN

Abstract

Few attempts have been made in the past to identify dinosaur teeth at the species level, and consequently, many have assumed that they cannot be identified. However, a diverse assemblage of theropods from the Judith River Formation of southern Alberta have teeth that are diagnostic at the family, subfamily, generic, and even species levels. Within each taxon, up to four types of teeth can be recognized corresponding to the premaxillary, maxillary, anterior dentary, and posterior dentary regions. Overall tooth shape, cross sections, the position of anterior and posterior carinae, and the morphology of the denticles can be used to identify theropod taxa, regardless of absolute size or maturity. The teeth of *Dromaeosaurus*, *Saurornitholestes*, *Troodon*, tyrannosaurids, and a new genus and species of theropod are described. The identification of theropod teeth has the potential of refining stratigraphic determinations, extending temporal and geographic ranges, indicating relationships, and allowing paleoecological statements to be made on the relative diversity or abundance of certain taxa.

Introduction

Vertebrate paleontologists realized early in the development of the science that certain types of dinosaur teeth were distinctive enough to be diagnosed at the species level (Leidy 1856, 1860, 1868; Cope 1876a,b; Marsh 1892). As better specimens were recovered, many of these tooth genera proved to be nomen dubium. Perhaps the most famous example is that of *Trachodon mirabilis*, a species established on the basis of isolated teeth from the Late Cretaceous of Montana (Leidy 1856). The name was subsequently applied to a flat-headed duckbilled dinosaur, and even became a family (Trachodontidae–Lydekker 1888) and a popular name (trachodonts). With the discovery of better preserved and more diverse specimens from western North America however, it was realized that several species of duckbilled dinosaurs had teeth that were indistinguishable from each other. Furthermore, the type specimens of *Trachodon* turned out to be from crested forms (Lambeosaurinae), whereas the name *Trachodon* had been extended to include skeletons of flat-headed forms (Hadrosaurinae). Lambe (1918) showed the name to be a nomen dubium, and recommended that use of the name be discontinued. Another tooth genus described by Leidy, *Palaeoscincus*, is now known to be an ankylosaurian, but Coombs (this volume) has shown that ankylosaur teeth are only diagnostic at the family level.

Troodon formosus is a name that was established on the basis of a tooth originally thought to have belonged to a lacertilian (Leidy 1856). In 1901, Nopcsa recognized it as a carnivorous dinosaur. However Gilmore (1924) and subsequent workers felt that *Troodon* was a plant-eating pachycephalosaurid. Sternberg (1945, 1951) and Russell (1948) correctly identified it as a small carnivorous dinosaur, and described the first jaws of this animal. Barsbold (1974) was unaware of these references, and put *Troodon* teeth in a different family than the jaws of the same animal. The discovery of a new specimen was necessary before the teeth of *Troodon* were reunited with their jaws and skeletons (Currie 1987a). This checkered history of *Troodon* shows the danger of establishing tooth genera, although in this case the teeth were always distinctive and were always identified as *Troodon*. It is ironic that *Trachodon* and *Troodon* were originally described in the same paper (Leidy 1856) because one shows the futility of using dinosaur teeth to establish new species, while the other shows that tooth genera can be valid.

Theropod teeth are common in the Judith River Formation of southern Alberta, as well as in rocks from many other times and places. This is not surprising because most tooth-bearing theropods had 50 or more functional teeth at any one time during their life. All the teeth were replaced several times during the lifetime of the individual. Theropod teeth with roots are recovered in dentigerous bones, but are seldom found in isolation. Worn teeth with resorbed roots are far more common. These shed teeth are often associated with the carcasses of other animals (Buffetaut and Suteethorn 1989), showing that teeth in the process of being replaced tended to be lost while theropods were feeding. In some cases, the evidence suggests that the carnivores were eating carrion (Currie and Dodson 1984), but most of the time it cannot be determined whether or not the prey was killed by an active predator or simply scavenged.

In what is now Dinosaur Provincial Park, there was a strong bias against the preservation of small skeletons (Currie 1987b), and theropods are amongst the rarest of the small species. None of the dromaeosaurids, troodontids, elmisaurids, or other species of small theropods is represented by a complete skeleton. However, many types of small theropod teeth have been found in jaw fragments and in association with partial skeletons. On the other hand, there are numerous complete tyrannosaurid skeletons (Russell 1970) with complete dentitions. Because isolated teeth are much more common than skeletons (even for tyrannosaurids), the following study was undertaken to see if they could be identified.

Usually, the size of theropod denticles are recorded as number of denticles per 5 mm. However, most of the teeth dealt with in this study are too small, and in some cases do not have denticles over a 5 mm length. Those that do usually exhibit strong curvature over a 5 mm length, making it difficult to make an accurate count. Therefore we have opted to measure denticles in this study by either counting the maximum number of denticles per millimeter, or by measuring the proximodistal (in the sense of the tooth rather than the denticle) base length of the largest denticle. Total tooth length often cannot be measured accurately because of wear or breakage at the tip, or because of differences in tooth curvature. The fore-aft basal length (FABL) of the tooth has a relatively constant relationship to the total length (Farlow et al. in preparation), and can be measured easily in most teeth. FABL is therefore the standard against which all other measurements are compared in this paper.

Description

The six families of theropods currently known in the Judith River Formation are the Dromaeosauridae, Troodontidae, Elmisauridae, Caenagnathidae, Ornithomimidae, and Tyrannosauridae. Elmisauridae may be synonymous with Caenagnathidae (Currie and Russell 1988), and therefore may include only toothless species. No North American ornithomimids are known that have teeth, even though teeth are known in the central Asian *Harpimimus* (Barsbold and Perle 1984). Unfortunately, the teeth of *Harpimimus* are only simple pegs that lack

serrations (personal observation 1989), and have no significance for determining the relationships of ornithomimids. Of the toothed forms, Dromaeosauridae can be divided into the subfamilies Dromaeosaurinae and Velociraptorinae (Barsbold 1983), and the tyrannosaurids into Aublysodontinae and Tyrannosaurinae (some authors, including Paul 1988b, consider the aublysodonts and tyrannosaurs as distinct families).

Dromaeosaurinae

Dromaeosaurus albertensis is the least specialized of the well-known small theropods of the Judith River Formation. The snout is narrow, and the mandibles are relatively straight bones that meet rostrally in an acute angle. The type specimen (AMNH 5356) includes teeth from all regions of the jaws although most of these are crushed and broken. There are four premaxillary teeth, nine maxillaries, and eleven mandibulars (Colbert and Russell 1969) for a maximum total of 48 tooth positions in the mouth. The third and fifth teeth of the left maxilla, and the third tooth of the right maxilla, are more than 15 mm long when measured in a straight line from the tip to the base of the crown on the back of the tooth, and have a FABL of 7.5 mm. The longest mandibular teeth are 14 mm long, and therefore are not significantly shorter than those of the upper jaw, but the maximum FABL is only 6.7 mm. On the average, the FABL of a dentary tooth is about 9% smaller than one from the upper jaw. As in most theropods, all teeth are set in distinct sockets, and there are interdental plates lingual to the teeth (Currie 1987a). The presence of these interdental plates was not detected for a long time because they co-ossified with adjacent plates and the dentary itself. This condition is comparable with that of *Saurornitholestes* and possibly baryonychids (Charig and Milner this volume), but is distinct from the situation seen in troodontids, tyrannosaurids, a specimen that Gilmore (1924) referred to *Chirostenotes*, and most other theropods.

Dromaeosaurid maxillary and dentary teeth (Fig. 8.1) are laterally compressed so that the FABL of a tooth is as much as double the width. There are denticles on both anterior and posterior carinae, and the denticles are smallest at the proximal and distal ends of the carina. Most teeth are damaged in the holotype, but the third maxillary tooth of the type specimen has 34 denticles on the anterior carina and 45 on the posterior. The denticles are almost as high as they are long (Fig. 8.1A), and curve only slightly distally towards the tip of the tooth (Figs. 8.1B,D,P). Each denticle is relatively broad (labial-lingual) and chisel-like in form. Blood grooves extend onto the surface of the tooth from between the bases of adjacent denticles on some teeth, but are usually found only near the base of the tooth and tend to be shallow and poorly defined (Figs. 8.1A, 8.7D,G). The blood grooves are oriented perpendicular to the longitudinal axis of the tooth. Abler (in preparation) describes

Figure 8.1. *Dromaeosaurus albertensis*. **A**, SEM photograph of posterior denticles of maxillary tooth (TMP 83.36.8) in labial view. **B–P**, teeth of the holotype (AMNH 5356). **B**, anterior view of 2nd left premaxillary tooth, with mesial (above) and anterior views of enlarged denticles. **C**, labial view of 2nd left premaxillary tooth with enlargement of posterior denticles. **D**, lingual view of 2nd left premaxillary tooth with enlargement of posterior denticles. **E**, cross-section of base of 2nd left premaxillary tooth. **F**, anterodistal view of tip of 4th maxillary tooth showing characteristic twist of anterior carina. **G**, lingual view of 4th right maxillary tooth with labial (above) and lingual (below) enlargements of posterior denticles. **H**, lingual view of anterior denticles of 3rd right maxillary tooth. **I**, labial view of 2nd left dentary tooth with enlargement of posterior denticles. **J**, lingual view of 2nd left dentary tooth with enlargement (in distal aspect) of anterior denticles. **K**, enlargement of denticles along distal region of posterior carina (labial view) of 3rd left dentary tooth. **L**, anterior view of 3rd left dentary tooth. **M**, lingual view of 3rd left dentary tooth. **N**, cross-section of base of 3rd left dentary tooth. **O**, enlargement of lingual view of 3rd right dentary tooth (erupting). **P**, lingual view of 5th dentary tooth with lingual views of posterior (left) and anterior (right) denticles of newly erupted 6th dentary tooth. Centimeter scale bar is for drawings of complete teeth, millimeter scale bar is for "enlargements", and fractional scale bar is for SEM.

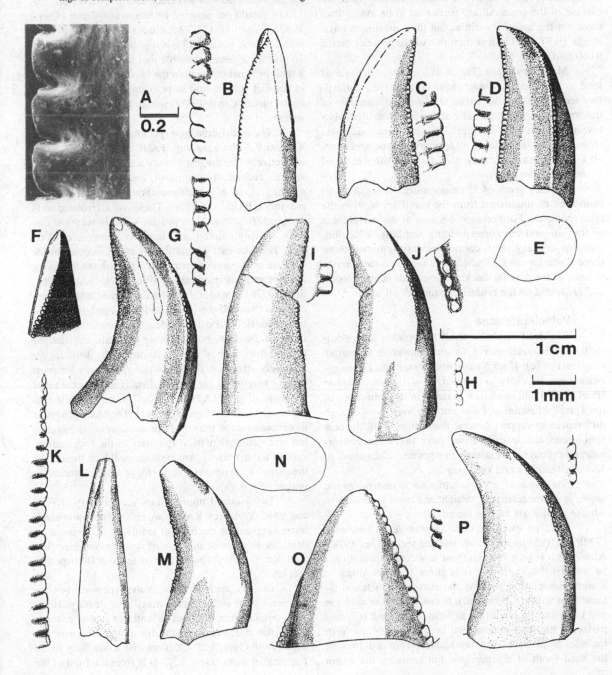

an enamel ridge on the midline of each denticle, and finds that the enamel ridges of adjacent denticles meet to form a V-shaped slot. Such ridges and slots are found in the specimen being described, as well as in most other Judithian theropods.

Both the anterior and posterior carinae of any premaxillary tooth of *Dromaeosaurus* are on the lingual side of the tooth. However, because of the narrowness of the snout, the posterior carina is posterolateral to the anterior carina, and the tooth is not D-shaped in section (Figs. 8.1E, 8.6D) as premaxillary teeth are in tyrannosaurids (Figs. 8.6U,X). Denticles on the posterior carinae of the premaxillary teeth tend to be longer than those on the anterior carinae, but their maximum basal length (0.37 mm) is less than that of the anterior carina (0.40 mm).

Maxillary teeth (Fig. 8.1G) of *Dromaeosaurus* tend to be taller and more recurved than the premaxillary teeth. The anterior carina is close to the midline of the tooth near the tip, but not far from the tip twists towards the lingual surface. This characteristic twist (Figs. 8.1F,L) is found on all teeth in the type specimen, and is the easiest way to identify the teeth of *Dromaeosaurus albertensis*.

Dentary teeth of *Dromaeosaurus* (Figs. 8.1I–P) cannot be distinguished from the maxillary teeth in the type specimen. Furthermore, because of the straightness of the jaw and the corresponding similarity of action, anterior dentary teeth cannot be distinguished from more posterior ones. Denticles are found on the anterior carinae at least as far back as the eighth dentary tooth, and are found on the posterior carinae of all teeth.

Velociraptorinae

The most common small theropod is the velociraptorine dromaeosaurid *Saurornitholestes langstoni* (Currie 1987c). Both *Saurornitholestes* and *Deinonychus* were recently synonymized with *Velociraptor* (Paul 1988a), although new material in the collections of the Tyrrell Museum of Palaeontology may show enough differences to support generic distinction. Until the new specimens are described, we have taken the conservative approach of maintaining generic distinction of *Saurornitholestes* and *Velociraptor*.

The snout of a velociraptorine is narrow, as the lower jaws are relatively straight and meet anteriorly in a loose symphysis at an acute angle.

The type specimen of *Saurornitholestes langstoni* (TMP 74.10.5) includes two isolated teeth (Sues 1978). Although it is possible that these associated teeth do not belong to the skeleton, it is more parsimonious to assume association because the skeleton was found by itself (not within a bone-bed), because both the skeleton and teeth can be identified as Velociraptorinae by comparison with other genera, and because one of the teeth includes at least part of a root and therefore could not be the shed tooth of a scavenger. Furthermore, the recent

discovery of a partial skeleton (TMP 88.121.39) of this animal confirms the association of teeth and skeleton.

An isolated premaxilla (TMP 86.36.117) has been tentatively identified as *Saurornitholestes*, and includes the bases of four premaxillary teeth as in *Deinonychus* (Ostrom 1969). No maxilla is known for *Saurornitholestes*, but given the number of mandibular teeth and the general similarity of this genus to the other velociraptorines (*Deinonychus* and *Velociraptor*), it is unlikely that there were more than ten maxillary teeth. The identification as *Dromaeosaurus* of two dentaries by Sues (1977a) was incorrect, and UA 12091 and UA 12339 should be assigned to *Saurornitholestes* (Currie 1987a). TMP 88.121.39 includes a complete dentary with 15 teeth. UA 12339 is a slightly larger dentary with 16 alveoli. It seems probable that *Saurornitholestes* had a total of about 60 teeth in the head. The teeth were set in distinct sockets and were bound lingually by interdental plates (Currie 1987a) similar to those of *Dromaeosaurus*.

The teeth of the type specimen (TMP 74.10.5) are 8.9 and 9.2 mm long with FABLs of 3.9 and 4.5 mm respectively. The longest dentary teeth in TMP 88.121.39 are the second, fourth, ninth, and eleventh from the symphysis, all of which are about 9 mm long with a maximum FABL of 5.1 mm. These and all isolated teeth assigned to *Saurornitholestes* are strongly recurved distally, sharply pointed, and laterally compressed (Fig. 8.2). They are easily identified as being velociraptorine because of the great disparity in size of denticles on the anterior and posterior carinae (Fig. 8.3N; Ostrom 1969; Sues 1977b). Anterior denticles are minute, and are usually less than half the length and base width of the posterior denticles. Posterior denticles are smaller than those of *Troodon*, but are more elongate and sharply pointed than those of *Dromaeosaurus*. The denticles are relatively straight and narrow (labial-lingual) for most of their length, but are hooked distally towards the tip of the tooth (Figs. 8.2A,H,L, 8.3N). The interdenticle slits are relatively deep (Figs. 8.7E,H,I). The blood groove is more pronounced than in *Dromaeosaurus*, although it has the same orientation (parallel to the longitudinal axes of the denticles). Amongst the Judithian theropods, the shape of the posterior denticles of velociraptorines is unique (Fig. 8.7).

Two isolated premaxillary teeth (TMP 70.37.1, and NMC 2664, Figs. 8.6E–I) are referred to *Saurornitholestes langstoni* on the basis of similarity of the posterior denticles to those of the type of *Saurornitholestes*. The teeth are flattened lingually, but are not D-shaped in section.

The two teeth found with the type specimen are not associated with either a maxilla or dentary. However, because the denticulate anterior carina extends more than halfway from the tip of the crown in the larger tooth (Figs. 8.2J–L), it may be a maxillary tooth. The smaller tooth (Figs. 8.2C–I) is from the front of the

Figure 8.2. *Saurornitholestes langstoni*. **A**, SEM of posterior denticles in lingual aspect of TMP 80.16.996. **B**, lingual view of TMP 82.24.16. Holotype (TMP 74.10.5) tooth #1 in **C**, anterior; **D**, labial; **E**, posterior; and **F**, lingual views; with **I**, cross-section of base; and enlargements of **G**, anterior and **H**, posterior denticles. Holotype (TMP 74.10.5) tooth #2 in **J**, anterior; and **K**, lingual views; with **L**, enlargement of tip in lingual view. TMP 82.19.366 in **N**, labial; and **O**, lingual views; with **P**, cross-section of base; and **M**, enlargement of labial view of distal denticles. **Q**, enlargement of posterior denticles of TMP 79.8.643 in labial view. **R**, enlargement of posterior denticles of NMC 12410 in labial view. **S**, enlargement of TMP 82.19.180 in lingual view. TMP 82.16.43 in **V**, anterior; and **U**, labial views; with **W**, cross-section of base; and **T**, enlargement of posterior denticles. Scale bars as in Figure 8.1.

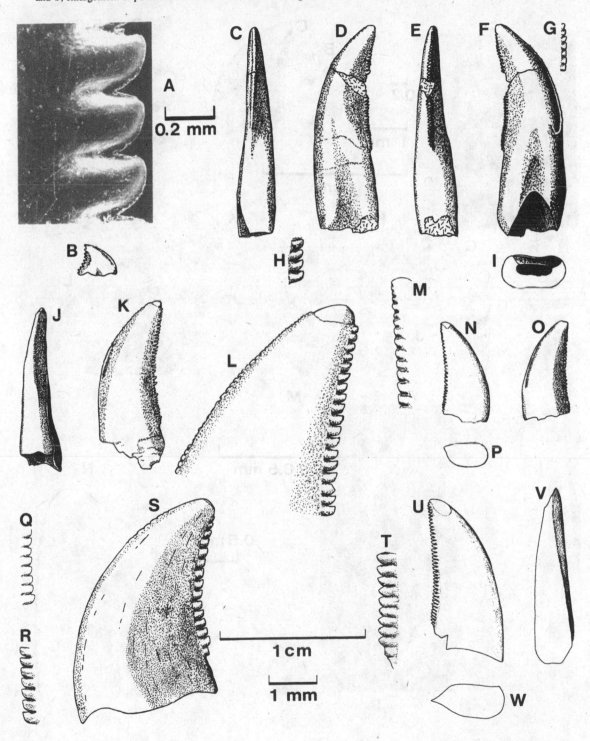

Figure 8.3. *Troodon formosus*. **A**, SEM of posterior denticles in labial view of maxillary tooth (TMP 83.45.7). TMP 85.6.186 (maxillary tooth) in **C**, lingual view; with **B**, an enlargement of posterior denticles. TMP 85.6.3 (premaxillary tooth) in **E**, lingual view; with **D**, enlargement of anteromedial denticles showing worn tips. **F**, enlargement of proximal denticles on posterior carina of TMP 83.36.215. TMP 83.36.214 (maxillary tooth) in **H**, labial; and **K**, lingual views; with **J**, cross-section of the base; and enlargements of **G**, anterior, and **I**, posterior denticles in lingual view. **L**, enlargement of distal end of mandibular tooth of TMP 83.12.11 in lingual view. **M**, SEM of posterior denticles of *Troodon formosus* (TMP 83.45.8) in posterior view. **N**, SEM of labial view of tip of tooth of *Saurornitholestes langstoni* (TMP 78.9.96). Scale bars as in Figure 8.1.

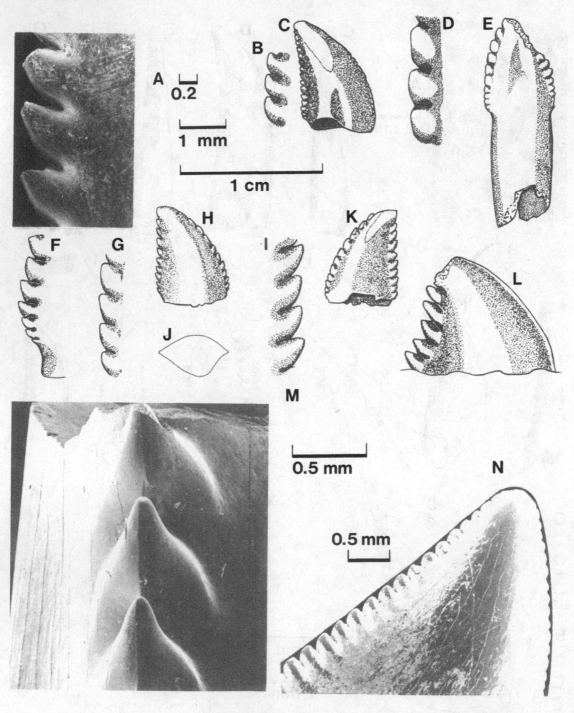

left dentary. As pointed out by Sues (1978), there are about five denticles per millimeter on the posterior carina, and seven on the anterior. Individual denticles on the posterior carina are not only broader at the base than the anterior ones, but are considerably longer and sharper as well.

Teeth in the third and fifth alveoli of the dentary of TMP 88.121.39 lack denticles on the anterior carina. However, the distal ends of the anterior carinae probably did have denticles (as in more posterior teeth of the same dentary, which have six denticles per millimeter) that have simply been worn off. Nevertheless, there are many isolated teeth of *Saurornitholestes* from Dinosaur Provincial Park that do lack anterior denticles (Figs. 8.2N,O,U,V), indicating that the presence or absence of denticles on the anterior carinae of dentary teeth is variable. There are four denticles per millimeter on the posterior carinae of TMP 88.121.39.

There are many small isolated teeth identified as *Saurornitholestes* (Fig. 8.2B), in which the posterior denticles are relatively large compared to the FABLs of the teeth, but are somewhat smaller than the denticles of larger teeth. In TMP 81.20.259, the FABL is 2.6 mm and there are seven denticles per millimeter on the posterior carina. This trend is consistent with other theropods (Farlow et al., in preparation) where young individuals tend to have fewer, smaller denticles on their teeth than more mature animals.

Troodontidae

Troodon formosus is currently the only species of troodontid recognized in the Judith River Formation (Currie 1987a), and includes specimens previously referred to *Stenonychosaurus inequalis*, *Polyodontosaurus grandis*, and *Pectinodon bakkeri*. There are three teeth preserved in the premaxilla of MOR 430, but this bone is incomplete and a fourth tooth may have been present as in *Saurornithoides* (Barsbold 1974). A maxillary fragment of *Troodon formosus*, NMC 12392, has parts of nine alveoli preserved. However, comparison with troodontid material from Asia suggests that there would have been a total of 15–20 maxillary teeth. The dentary teeth of *Troodon* are smaller and more numerous (35) than those in the upper jaw. *Troodon*, therefore, had more than 100 functional tooth positions, which is a higher tooth count than any other theropod from the Judith River Formation. Premaxillary teeth are set in sockets, whereas most of the dentary teeth are arranged in a dental groove, and are held in position by interdental bone around the roots. There are no interdental plates lingual to the teeth. Instead, the teeth are constricted between crown and root, and held in place by a partial ring of interdental bone around the constriction (Currie 1987a). The constriction would have been below the gum line on the crown of the tooth, and the enameled surface of the crown extends to the level of the constriction labially, anteriorly and posteriorly, but on some

teeth, it ends at the gum line on the lingual surface.

The front of the troodontid snout is relatively wide, and the distal ends of the lower jaws curve towards the midline as in toothless caenagnathids and ornithomimids, but in contrast with the jaws of dromaeosaurids and tyrannosaurids. The snout of a tyrannosaurid is fairly broad however, and in this sense is more similar to the snout of a troodontid than to that of a dromaeosaurid. The development of "incisiform" (in this sense, with both carina on the posterior surface) teeth is clearly related to the orientations of the jaws anteriorly, and it is not surprising that troodontids and tyrannosaurids have incisiform teeth in the premaxilla, and that troodontids have incisiform teeth at the front of the dentary.

The carinae of a premaxillary tooth (Figs. 8.3, 8.6) are both on the posterior side of the tooth, as in tyrannosaurids. In contrast with the latter family however, the premaxillary tooth of a troodontid is triangular in section rather than D-shaped (Currie 1987a). There are large denticles on both carinae of a troodontid premaxillary tooth (Figs 8.3D, E). In the type specimen (ANSP 9259), there are seven "posterior" denticles and ten "anterior" denticles (Currie 1987a). All denticles are strongly hooked with the pointed tips (Fig. 8.3M) turned toward the distal end of the tooth. The denticles are larger than those found in any other Judithian theropod, with a basal diameter measuring up to 0.7 mm in TMP 82.20.259, less than 1.5 denticles per millimeter (compare this with BHI 1281, a 9 cm long tooth of *Tyrannosaurus* that has denticles with the same basal diameter). As reported previously (Currie 1987a), the denticles on the posterior carina of premaxillary teeth tend to be longer than those of the anterior carina, but have a smaller basal diameter. The grooves between successive denticles form distinct, rounded pits, at the centres of which are found the interdenticle slits (Figs. 8.7B,F).

Maxillary teeth of *Troodon* (Figs. 8.3G–K) have long, recurved crowns, are laterally compressed, and have carinae on the rostral and posterior surfaces. TMP 65.23.32 is a centimeter long with a FABL of 5.7 mm, and TMP 88.96.2 has a FABL of 6.7 mm. In general shape, the denticles and their associated blood grooves and pits (Figs. 8.3A, 8.7B,F) are the same as the denticles of premaxillary teeth. However, maxillary denticles are smaller than premaxillary ones, and the posterior ones reach a maximum length of 0.5 mm with basal diameter of 0.5 mm (TMP 65.23.32). There can be more than thirty anterior denticles, which are smaller than the 20 or so on the posterior carina. In TMP 65.23.32, the basal diameter of the largest anterior denticle is 0.35 mm, but its height is only 0.1 mm. The "serrated" carina follow the midline of the tooth, and extend from the gum line to the tip of the tooth.

The dentary teeth of *Troodon* are considerably smaller than the maxillary ones (Currie 1987a). The

Figure 8.4. *Richardoestesia gilmorei.* Holotype (NMC 343): **A,B,C,** left; and **D,E,F,** right dentary fragments in **A,D,** lateral; **B,E,** dorsal; and **C,F,** lingual views; with lingual views of **G,** 13th right; **H,** 17th right; and **J,** sixth right dentary teeth. **I,** SEMs of labial view of denticles at mid-length of the tooth and lingual view of distal denticles. **J,** lingual view of sixth right dentary tooth. TMP 83.45.2 in **K,** labial view with enlargements of anterior and posterior denticles; **L,** enlargement of proximal denticles of posterior carina; and **M,** cross-section of the base of the tooth. TMP 80.8.298 in **N,** labial view; with **O,** enlargement of posterior denticles and cross-section of the base. TMP 84.89.274 in **P,** labial view with enlargement of posterior denticles and cross-section of the base. TMP 83.129.11 in **Q,** lingual view with enlargements of anterior and posterior denticles; and **R,** cross-section of the base of the tooth. TMP 80.16.1230 in **S,** lingual and labial views; with **T,** enlargement of proximal posterior denticles in labial view; and **U,** cross-section of base. **G, H, J,** scale bars represent 1 mm. **L–U,** scale bars as in Figure 8.1.

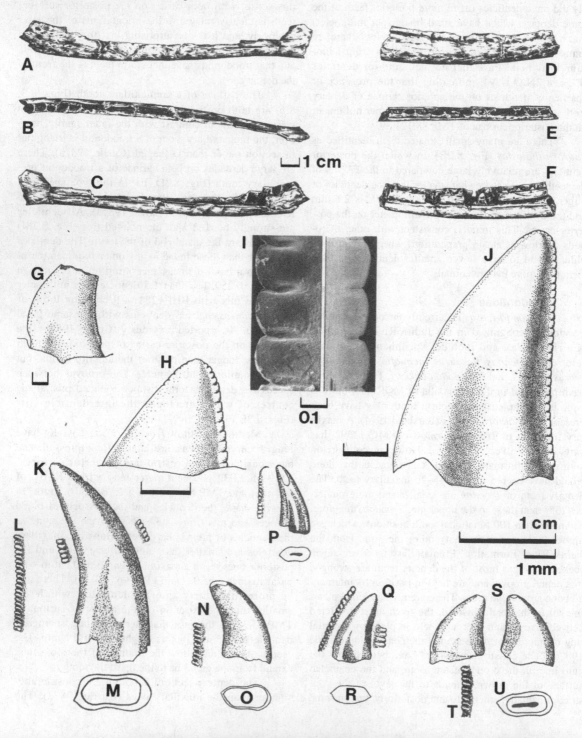

maximum length of any mandibular tooth is 6 mm with a FABL of less than 5 mm. The anterior dentary teeth, because of the curvature of the jaw, are similar to the premaxillary teeth in that both carinae are clearly on the lingual side of the tooth. At least the first half dozen or so dentary teeth have denticles on both anterior and posterior carinae, although the posterior denticles are larger than the anterior ones. The posterior denticles are larger and less numerous on these anterior teeth than they are on more posterior dentary teeth. In contrast with the premaxillary teeth, the "anterior" denticles are not wider at the base than the posterior denticles, the former having a maximum diameter of 0.5 mm compared with 0.6 mm for posterior denticles. Posterior mandibular teeth lack denticles on the anterior carina.

The teeth of juvenile troodontids have been described (Currie 1987a). The basal widths of the denticles are only slightly smaller on average than those of mature teeth, so it is not surprising that there are fewer denticles on the smaller teeth. For example, TMP 79.8.635 is a posterior dentary tooth with a length of 4 mm, and bears only eight denticles on the posterior carina, the largest of which has a basal diameter of 0.5 mm. In contrast, the 20th mandibular tooth of TMP 83.12.11 is 5 mm long, and has at least 11 posterior denticles, the largest of which has a basal diameter of 0.45 mm.

Troodontid teeth found in the Milk River Formation (ROM collections) and the Horseshoe Canyon Formation (TMP, NMC collections) are essentially identical to those of the Judith River Formation. The troodontid *Saurornithoides* teeth from Mongolia are similar in size and shape (Barsbold 1974), but the posterior maxillary teeth lack anterior denticles, making them difficult to distinguish from posterior dentary teeth except by relative size. The distinctive shape of the troodontid denticle is found in *Saurornithoides* (teeth found with IVPP 2206883, Djadokhta Formation of Bay'an Manduhu, Inner Mongolia). Teeth from the Lower Cretaceous Cedar Mountain Formation identified as troodontid (Nelson and Crooks 1987) are more likely from a velociraptorine because the denticles are too small (there are 12 posterior denticles per millimeter in their figures) and elongate. Troodontid teeth from the Maastrichtian Frenchman, Hell Creek, Lance, and Scollard Formations, as well as from the Prince Creek Formation of Alaska, are somewhat different from the Judithian teeth of southern Alberta (personal observations) and may eventually prove to represent a distinct species.

Theropoda *incertae sedis*

In 1924, Gilmore named *Chirostenotes pergracilis*, based on a pair of articulated mani. In the same paper, he described a pair of jaws found several miles away, and arbitrarily referred them to *Chirostenotes* because of their long, slender nature. The left dentary of

NMC 343 is almost complete, lacking only small parts of the rostral and caudal ends (Figs. 8.4A–C), and is 193 mm long. Gilmore's suggestion that each jaw would have had at least 18 teeth is correct, and the number would not have exceeded 20. NMC 343 represents a theropod that had more teeth in its jaws than any other known Judithian carnivore except *Troodon*. Given that there were 18–20 dentary teeth, a conservative estimate based on relative tooth counts in other theropods suggests there would have been at least 3 premaxillary and 11 maxillary teeth on each side of the skull for a total minimum of 66 teeth in the head. The maximum number would have been less than 90.

The teeth of NMC 343 are set in distinct sockets (Figs. 8.4B,E) as in all Judithian theropods except *Troodon*. The jaws are straight, and it is evident that they met in an acute angle rostrally, rather than curving to meet each other as in *Troodon*. The shallow Meckelian groove is similar to that of *Saurornitholestes*. There are interdental plates (Fig. 8.4F), in contrast with *Troodon*, but, unlike *Dromaeosaurus* and *Saurornitholestes*, the adjacent centers of interdental bone do not cover the entire base of the tooth lingually. In fact, adjacent interdental plates do not seem to touch each other, with the possible exception of the third and fourth plates on the left dentary. Amongst Judithian theropods, the interdental plates of NMC 343 are closest in appearance and morphology to those of tyrannosaurids. This is clearly a primitive characteristic however, and cannot be used by itself to indicate relationship. Nevertheless, the primitive nature of the interdental plates do show that this animal is neither a dromaeosaurid nor a troodontid.

Most of the teeth of NMC 343 fell out of the jaws before burial and fossilization. The roots of four teeth were found in the sockets, however. Three of these four broken teeth were in the process of being replaced, and developing teeth are preserved within the shells of the older teeth. At least an additional seven unerupted teeth were found within sockets where the functional teeth had either fallen out or been broken off. As pointed out by Gilmore (1924), as many as three teeth in different stages of development were found in each alveolus. The germ teeth develop lingual to the older teeth in the anterior half of the tooth socket. Each developing tooth is twisted somewhat so that the anterior carina is more medial in position than the posterior carina. As a tooth became larger, it would migrate laterally and posteriorly, and would rotate until the anterior and posterior carina were aligned. Tooth development was confined to the tooth sockets and there appears to have been little reworking of the interdental bone, in contrast with *Troodon* where there appears to have been more reworking of the interdental bone.

Two types of laterally compressed teeth were found in NMC 343. The anterior dentary teeth are elongate, relatively straight teeth (Fig. 8.4I). They are distinctive because the posterior margin of the tooth, when viewed from the side, is convex at the distal end rather than concave as it is in all other Judithian theropods. At least the first eight dentary teeth seem to fit this pattern. The average antero-posterior length – which approximates the FABL – of five anterior alveoli is 4.6 mm. The crown appears to have been at least 50% longer (proximodistally) than the FABL. The sixth right dentary germ tooth has a complete crown, although the enamel appears to have been incompletely formed on the proximal half of the tooth.

More posterior teeth of NMC 343 are distinctive in that they are more recurved (Fig. 8.4G). The average lateromedial width of a dentary alveolus is not significantly variable along the jaw. However, the base of a mid to posterior tooth is longer, with an average alveolar length (roughly equivalent to the FABL) of 5.5 mm.

There are a few denticles at the distal end of the anterior carina (Fig. 8.4I) of the sixth, eighth, and seventeenth dentary teeth, and the denticles wrap around the tip of the tooth onto the posterior carina. Each denticle is small and relatively simple in form (Figs. 8.4J, 8.7C), and hooks slightly toward the distal end of the tooth. There are interdenticle slits as in other theropods. The posterior denticles of NMC 343 are easily distinguished from those of *Dromaeosaurus*, *Saurornitholestes*, *Troodon*, and the tyrannosaurids because of their small size. There are six denticles per millimeter on the sixth dentary tooth, and five denticles per millimeter on the thirteenth. However, these denticles are short and measure less than 0.15 mm from base to tip. The denticles taper slightly, but are not hooked distally as much as they are in *Saurornitholestes* and *Troodon*. Because these teeth had not erupted by the time of death, the lack of pronounced distal hooking cannot be attributed to tooth wear.

No postcranial skeletal remains were found with NMC 343, but the same or similar teeth (Figs. 8.4N,Q,S) as those of this specimen are relatively common in the Judith River Formation. Other teeth can be referred to the same genus because of similarity of denticular size and shape, although the teeth do not necessarily have the same shape (Figs. 8.4K,P). Nevertheless, it is assumed that these teeth do represent the same species of theropod, but that they may represent teeth from the upper jaws. The labial view of the teeth in NMC 343 cannot be seen because they are covered by bone. In some isolated teeth attributed to this animal, the denticles look like "stacked bananas" in labial aspect (Figs. 8.4L,N,T) because of the blood grooves between the denticles. This unusual feature can be used to identify isolated teeth with small denticles that are different in overall tooth shape and anterior denticulation from the known mandibular teeth.

Some of the isolated teeth that can now be attributed to the same animal as NMC 343 have previously been identified as possible sebecosuchian teeth (Sahni 1972). Langston (1956) gives a list of characteristics that can be used to distinguish sebecosuchians from "carnosaurs." However, most small theropods can be distinguished from carnosaurs using some of the same characters (lateral compression, "lenticular" cross sections). Other characters cited by Langston (relative length of the root, fluting, length of denticulate portion of carina) either have not been seen in the Judithian teeth identified as sebecosuchians, or are not present. NMC 343 is unquestionably a theropod, rather than a sebecosuchian crocodile, and it is highly probable that so are all teeth from the Judith River Formation previously identified as sebecosuchian.

There is a considerable variety of teeth from Dinosaur Provincial Park that have the same small denticles seen in NMC 343. By comparison with this specimen, many of these can be identified as anterior or posterior dentary teeth. But others do not seem to be mandibular teeth. TMP 83.45.2 (Fig. 8.4K) is an elongate, fang-like tooth with denticles on both anterior and posterior carinae. We have no hesitation in referring this specimen to the same species as NMC 343 because of the minute size and shape of the denticles. The posterior denticles are hooked slightly distally, and, as in *Saurornitholestes*, are markedly larger than the anterior denticles.

When Gilmore (1924) referred NMC 343 tentatively to *Chirostenotes pergracilis*, it was hoped that additional skeletal material would be recovered to confirm or refute this referral. Although additional specimens of *Chirostenotes* have been found (Currie and Russell 1988), small theropods remain some of the rarest animals recovered and there is still no way of confirming Gilmore's identification after more than six decades. Currie and Russell (1988) presented evidence to suggest that *Chirostenotes* (exclusive of the referred dentaries) may even be synonymous with the toothless *Caenagnathus*. Russell (1984) suggested that the teeth are similar to isolated *Paronychodon* teeth, but the presence of small denticles on the posterior carinae and the absence of longitudinal ridges makes this association unlikely. However, there are many isolated teeth recovered from Dinosaur Provincial Park that are the same as those of NMC 343. Because NMC 343 is so distinctive, because teeth referrable to the same species of animal are so common, and because continued referral of NMC 343 to *Chirostenotes* is both confusing and probably misleading, we hereby propose to name a new genus and species of small theropod. It is conceivable that more complete specimens may eventually show that this animal is in fact synonymous with either *Chirostenotes pergracilis* or *Elmisaurus elegans* (Currie 1989).

Dinosauria Owen 1842
Theropoda Marsh 1881
Maniraptora Gauthier 1986
Family unknown
Richardoestesia n. gen.
Etymology. In honour of Richard Estes, whose

1964 paper on Lance Formation microvertebrate fossils demonstrated the use of theropod teeth in faunal studies.

Diagnosis. Small carnivorous dinosaur. Elongate jaw with little lateromedial curvature. 18–19 teeth per dentary. Anterior mandibular teeth relatively straight with convex posterior outline in lateral view for at least the distal half of tooth. More posterior teeth are relatively short and recurved. Denticles shorter than in other known Judithian theropods with a length of 0.15 mm. There are up to five denticles per millimeter on the posterior carina of mandibular teeth.

Genoholotype. NMC 343.

Richardoestesia gilmorei n. sp.

Etymology. In honour of G. W. Gilmore who first described this specimen in 1924.

Holotype. NMC 343, the remains of a pair of dentaries with unerupted and germ teeth.

Horizon and locality. Judith River (Oldman) Formation, Dinosaur Provincial Park, Alberta (Section 30, Twp. 20, Rge. 11, W4M).

Diagnosis. In contrast with Maastrictian teeth of *Richardoestesia*, some curvature is always present in the proximal portion of tooth.

Isolated teeth of *Richardoestesia gilmorei* are quite varied in shape and size. Although teeth of this species are common, no premaxillary teeth have been identified with certainty. TMP 81.16.194 (Figs. 8.6J–M) may be a premaxillary tooth of this species because its denticles are minute in comparison with those of premaxillary teeth in other theropods.

Teeth of *Richardoestesia* have also been identified in the Lower Campanian Milk River Formation (Russell 1935), the Scollard Formation of Alberta, the Frenchman Formation of Saskatchewan, the Hell Creek Formation of Montana, and the Lance Formation of Wyoming (Estes 1964; Carpenter 1982). Both longer straight teeth and recurved shorter teeth are known. As in *Richardoestesia gilmorei*, the serrations are often limited to the posterior carina, and individual denticles are minute. Many teeth from these formations are different, however, in that there is virtually no curvature evident in lateral view, and the teeth resemble an elongate, isosceles triangles. We suspect that the Maastrichtian teeth represent a difference species of *Richardoestesia*, which will be described later. However, the presence of identical teeth in the Lower Campanian Milk River Formation is perplexing in light of the virtual absence of this form in the Late Campanian Judith River and Horseshoe Canyon Formations.

Indeterminate small theropod teeth

Teeth of *Paronychodon lacustris* (Cope 1876a) are found throughout Upper Cretaceous beds (Russell 1935; Sahni 1972; Armstrong-Ziegler 1980; Lehman 1981; Carpenter 1982; Breithaupt 1985; Standhardt 1986) but remain enigmatic because of lack of association with skeletal remains. Junior synonyms include *Zapsalis abradens* (Cope 1876b) and *Dipriodon caperatus* (Marsh 1892). *Paronychodon* teeth are flat on one side and usually bear three or more longitudinal ridges. The other side of the tooth is convex, and can either be smooth or have longitudinal ridges as well. *Paronychodon* teeth are highly variable in shape and size. Most lack serrations but others can have denticles on either the posterior carina, or on both anterior and posterior carinae. The denticles provide the clue to identifying the true nature of these teeth. "*Paronychodon*" teeth (that is, flattened and ridged on one side and convex on the other) from Dinosaur Provincial Park usually bear serrations that identify them as *Troodon* (Figs. 8.5B,C,D; Currie 1987a, Fig. 5S), *Saurornitholestes* (Fig. 8.5A), and possibly *Dromaeosaurus* (TMP 82.19.7). One might suspect that this type of tooth is from the symphysis of the mandibles as proposed by Marsh (1892), who thought the flat surfaces of the teeth on the symphysis appressed against each other. However, many of these teeth can be identified as maxillaries or from the posterior region of the dentary. It appears more likely that these teeth represent growth abnormalities. Theropod teeth develop along the medial wall of the tooth socket and remain flattened against this wall until the root of the older and more lateral functional tooth is resorbed and the new tooth is ready to erupt. By this time, a new germ tooth has often started to develop medial to the erupting tooth. The flattened, ridged lingual surface of a "*Paronychodon*" tooth was possibly caused by prolonged contact with the medial wall of the socket. The pitted surface (Fig. 8.5D) of many of these teeth supports the notion of abnormal growth. Because most "*Paronychodon*" teeth can be referred to known genera, the name *Paronychodon lacustris* should be restricted to non-serrate forms. These tend to be more common in Maastrichtian beds, and conceivably may represent a distinct taxon of theropod.

A number of teeth in the Judith River Formation can be identified as theropods on the basis of size and shape, but lack serrations entirely (Figs. 8.5E–L). In almost all cases, the tooth surface has a chalky grey appearance, and the surface is sometimes pitted. These may be shed teeth, swallowed during feeding (Argast et al. 1987). The enamel surface of the teeth, including denticles, would have been removed by digestive acids before the teeth were expelled from the body. In some of these teeth, the bases of the denticles can still be seen as a line of circles along the anterior and posterior margins of the laterally compressed teeth.

Some small theropod teeth cannot be identified with certainty. Many of these have denticles that are morphologically similar to those of *Dromaeosaurus*. However, the anterior carina does not twist from the lingual surface to the midline of the tooth as it approaches the tip. It is possible that this type of tooth may represent a distinct species of dromaeosaurid, or a gracile, small form of tyrannosaurid.

Figure 8.5. Miscellaneous theropod teeth from the Judith River of southern Alberta. **A**, "*Paronychodon*" (*Saurornitholestes*) tooth (TMP 79.15.3) in lingual view. **B, C**, "*Paronychodon*" (*Troodon*) tooth (TMP 85.30.1) in lingual and labial views. **D**, SEM of lingual surface of TMP 79.8.635, "*Paronychodon*" (*Troodon*), scale = 0.5 mm. **E, F, G, H, I**: "digested" teeth. **E**, TMP 82.19.180. TMP 80.13.34 in **F**, labial; and **G**, lingual views; with **H**, cross-section of the base. **I**, TMP 82.20.255. **J–Q**: tyrannosaurid teeth. TMP 81.19.263 in **J**, lingual view; with **L**, enlargement of posterior proximal denticles in labial view; and **K**, cross-section of the base. TMP 82.20.47 in **M**, lingual view; with **Q**, enlargements of posterior proximal denticles, and **O**, anterior distal denticles; and **N**, cross-section of base. **P**, enlargement in lingual aspect of posterior denticles of a 6 cm long tooth (NMC 1592). Scale bars as in Figure 8.1.

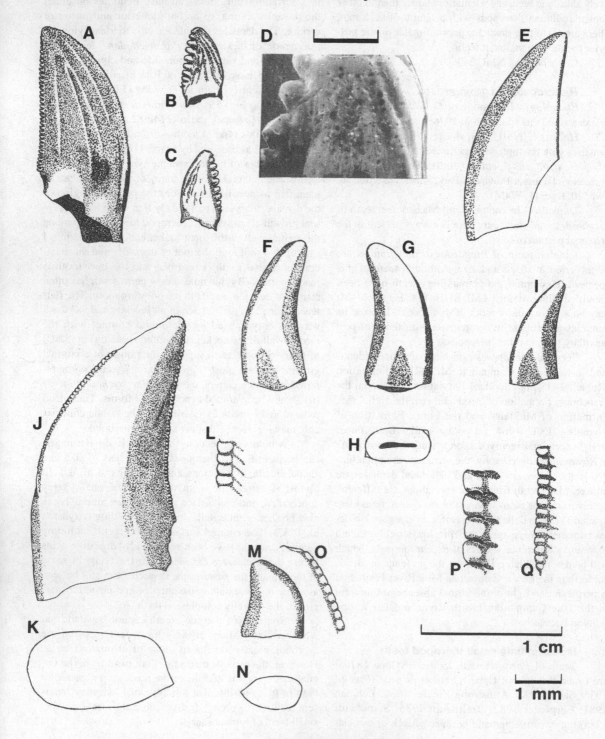

Tyrannosauridae

In going through collections of small theropod teeth, it is not surprising that a large number of juvenile tyrannosaurid teeth are found. These teeth tend to be stouter than those of dromaeosaurids or troodontids, but are still laterally compressed (Fig. 8.5N). They are simply scaled down versions of large tyrannosaurid teeth. At present, no juvenile tyrannosaurid skulls have been collected from the Judith River Formation. The smallest one known, TMP 86.144.1, is a half-grown individual with dentary teeth up to 36 mm long. In this specimen, serrations are relatively large on both anterior and posterior carinae, with three denticles per millimeter. A small tyrannosaurid maxilla (TMP 85.11.3, 25 cm long) has teeth with 2.5 denticles per millimeter. There are numerous isolated maxillary and dentary teeth that were obviously shed from much younger individuals. TMP 81.19.263 (Fig. 8.5M) is only 15.5 mm long, and bears three denticles per millimeter on the anterior carina and 3.5 denticles per millimeter on the posterior. TMP 79.10.59 is smaller (9.8 mm long with a FABL of 7.2 mm), and has 3.5 denticles per millimeter. In larger individuals, the denticles are smaller relative to tooth length, but are absolutely larger in basal diameter and denticle height (TMP 80.16.864 is a typical tyrannosaur tooth with a length of 80 mm and up to 2 denticles per millimeter along the anterior and posterior carina). As in more mature Judithian tyrannosaurid teeth, the denticles of juvenile tyrannosaurs are relatively stout and chisel shaped. The denticles of tyrannosaurids are wider labially-lingually than they are long proximodistally. This represents a compromise between the strength needed by the denticles of teeth that were biting into bone, and the need for serrations for cutting meat (Abler in prep.). Tyrannosaurid denticles (Fig. 8.7A) do not curve distally towards the tip of the tooth, but do possess sharp ridges of enamel along the midline. Long, distinctive blood grooves are found between the bases of the denticles, oriented towards the base of the tooth (Figs. 8.5L,P,Q, 8.7A). These grooves are especially evident on the lingual surface of the tooth between denticles on the posterior carina in the proximal half of the tooth. Other aspects of tyrannosaurid tooth form and function are being studied by Farlow et al. (in prep.), and by Abler (in prep.).

Aublysodon mirandus (Figs. 8.6V–X) is a peculiar type of small premaxillary tooth originally described by Leidy (1868) from the Judith River Formation of Montana. In his original description, both serrated and non-serrated teeth were identified as *Aublysodon*, although Marsh (1892) restricted the name to the non-serrated form, and described two more species. The same teeth have been found in virtually every Upper Cretaceous Formation in North America.

Considering that all *Aublysodon* teeth are less than 20 mm long, one might suspect that *Aublysodon* represents an early ontogenetic stage of known tyran-nosaurids, two of which have been described from the Judith River Formation (Russell 1970). However, there are serrated premaxillary teeth of tyrannosaurids (Figs. 8.6R–U) that are almost the same size as those of *Aublysodon*. Furthermore, the median ridge between the carina on the posterior surface of the premaxillary tooth of *Aublysodon* is more strongly developed distally than it is in either the juvenile or mature serrated tyrannosaurid teeth. This still leaves the possibility that *Aublysodon* could represent a juvenile morph of one of the large tyrannosaurid species, with juveniles of the other species having serrated premaxillary teeth. The remote possibility also exists that serrated and unserrated premaxillary teeth represent different sexes.

It is, however, more likely that *Aublysodon* represents a distinct taxa in the Judith River Formation. Identical teeth recently were recovered from the Iren Dabasu Formation at Erenhot, People's Republic of China (IVPP 170788104). The Asian "*Aublysodon*" teeth belong to *Alectrosaurus* (Perle pers. comm. 1989), a theropod related to tyrannosaurids. The Jordan theropod (Molnar 1978) was recently referred to as *Aublysodon molnaris* (Paul 1988b), and has been redescribed formally by Molnar and Carpenter (1990) as *Aublysodon*, because of its non-serrate premaxillary teeth that are D-shaped in cross-section. The maxillary teeth are serrated, laterally-compressed, blade-like structures (Molnar and Carpenter 1990), and may well turn out to be the same as the unidentified Judithian teeth referred to above as a possible dromaeosaurid or gracile tyrannosaurid.

There are differences in the teeth and denticles amongst tyrannosaurid species (Farlow pers. comm.). However, tyrannosaurid taxonomy is currently under review by a number of authors, and it would be pointless to attempt to key out these differences until the teeth can be associated with valid taxa.

Discussion

Theropod teeth from the Judith River Formation are diagnostic, and can usually be identified by size, shape, and pattern of denticulation. The denticles are diagnostic enough (Fig. 8.7) that only a portion of a tooth is usually necessary for identification. Most of the problems encountered with identification of theropod teeth relate to the absence of sufficient associated skeletal material.

Theropod teeth show little ontogenetic variation. Juvenile teeth are simply scaled down versions of teeth of more mature individuals, although juvenile teeth have fewer (but relatively larger) denticles. There does not appear to have been any significant increase in the number of teeth (Madsen 1976; Colbert 1989) as theropods matured, whereas increase in tooth rows is a common phenomena amongst ornithischians (Chapter 15).

In the Judith River Formation, a large proportion of the theropod teeth less than 2 cm long are from juvenile tyrannosaurids and *Aublysodon*. Of the remaining

Figure 8.6. Comparison of premaxillary teeth from the Judith River Formation of Dinosaur Provincial Park. First left premaxillary tooth of *Dromaeosaurus* (TMP 81.16.461) in **A**, lingual; and **B**, posterior views; with **C**, enlargement of anterior and posterior denticles in lingual view; and **D**, cross-section of base of tooth. Left premaxillary tooth (cf. *Saurornitholestes*, TMP 70.37.1) in **E**, lingual; and **F**, posterior views; with **G**, enlargement of anterior and posterior denticles in lingual view; and **I**, cross-section of base of tooth. **H**, enlargements of anterior and posterior denticles of NMC 2664 (cf. *Saurornitholestes*), a 2nd or 3rd premaxillary tooth. TMP 81.16.194 (cf. *Richardoestesia*) in **J**, lingual; and **K**, posterior views; with **L**, enlargement of posterior denticles; and **M**, cross-section of the base. NMC 1267 (*Troodon*) in **N**, lingual; and **O**, posterior views; with **P**, enlargements of denticles in lingual view; and **Q**, cross-section of base of tooth. NMC 41104 (juvenile tyrannosaurid) in **R**, lateral; and **S**, posterolingual views; with **T**, enlargement of denticles; and **U**, cross-section of base. TMP 82.19.367 (*Aublysodon*) in **V**, ?lateral; and **W**, posterolingual views; with **X**, cross-section of the base.

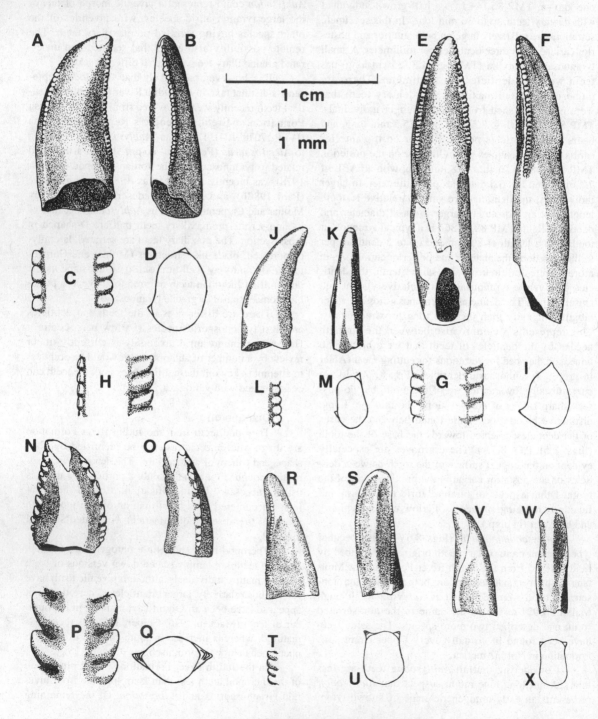

Figure 8.7. Comparison of denticles of Late Cretaceous theropods from southern Alberta. Drawings (A–E) done to same scale to show relative sizes of denticles, and all oriented so that tip of tooth is to the right. **A**, *Tyrannosaurus rex*, TMP 81.6.1 (Willow Creek Formation, Maastrichtian), lingual view of denticle mid-length along the posterior carina of a functional dentary tooth (FABL = 28.5 mm). **B**, *Troodon formosus*, TMP 82.20.320, reversed image of lingual view of eighth denticle from the distal end of the posterior carina, and (below) lingual view of eleventh denticle from proximal end of the anterior carina of an anterior maxillary tooth (FABL = 5.5 mm). **C**, *Ricardoestesia gilmorei*, NMC 343, reversed image of ninth denticle from distal end of posterior carina of germ tooth in 17th tooth position of dentary (FABL is between 5 and 5.5 mm). **D**, *Dromaeosaurus albertensis*, TMP 81.26.48, labial view of denticle from middle of posterior carina of an anterior dentary tooth (FABL = 7.0 mm). **E**, *Saurornitholestes langstoni*, TMP 78.9.96, lingual view of eighteenth denticle from distal end of posterior carina of an isolated, shed tooth. **F**, *Troodon formosus*, TMP 83.45.8, lingual view of fourth denticle from the proximal end of the posterior carina of a premaxillary tooth (FABL = 6.5 mm). **G**, *Dromaeosaurus albertensis*, TMP 81.26.48, same as **D**. **H**, *Saurornitholestes langstoni*, TMP 74.10.5 (holotype), labial view of tenth and eleventh denticles from distal end of posterior carina (FABL = 4.5 mm). **I**, *Richardoestesia gilmorei*, NMC 343, labial view of denticles on posterior carina of sixth dentary tooth (FABL = 3.7 mm). All scales = 0.2 mm. FABL = fore-aft basal length.

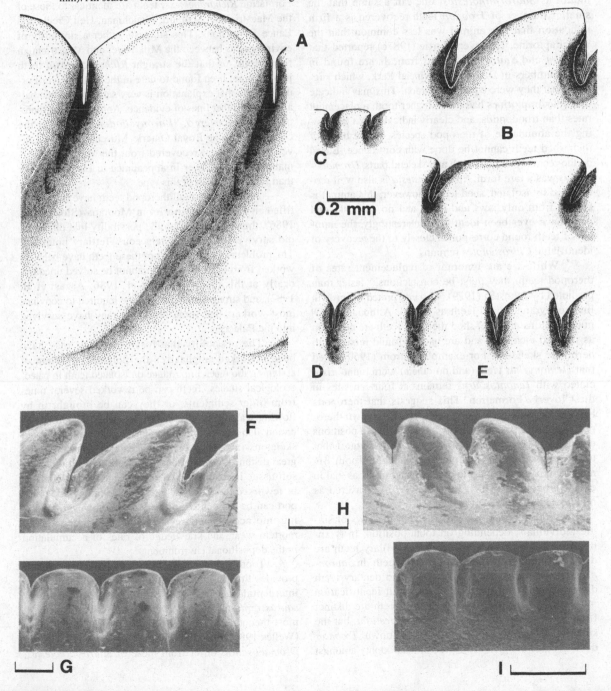

teeth studied (sample size = 424), *Saurornitholestes* teeth are the most common (47.63%), followed by *Troodon* (19.53%), *Richardoestesia* (18.05%), *Dromaeosaurus* (9.47%), "*Paronychodon*" (3.25%), and unknown/ unidentified forms (2.07%).

We do not know if troodontid and dromaeosaurid species replaced their teeth at the same rate. However, if we assume they did, then the number of teeth for each species can be divided by the approximate number of teeth in the head to give an idea of how common the small theropods were in relation to each other. Because *Troodon* had nearly twice the number of teeth in the mouth as *Saurornitholestes*, one can assume that the smaller number of *Troodon* teeth recovered is a firm indication that this animal was less common than the velociraptorine. However, Currie (1987c) reported that *Troodon* and *Saurornitholestes* frontals are found in equal numbers in Dinosaur Provincial Park, which suggests that they were equally common. This may indicate that velociraptorines had much higher tooth replacement rates than troodontids, and clearly indicates that estimating the abundance of theropod species on the basis of their shed teeth cannot be done with confidence. Based on the recovery of both teeth and skeletal parts, *Dromaeosaurus* was a rare form. *Richardoestesia* is also well represented by isolated, shed teeth. However, this animal is known from only jaws and teeth, and no other skeletal parts have ever been identified. Interestingly, the number of teeth found corresponds closely to the recovery of identifiable *Chirostenotes* remains.

While we are ignorant of replacement rates of theropod teeth, they must be conspicuously faster than predicted by Johnston (1979), who interpreted the growth lines of tyrannosaurid teeth as annual. Although theropod skeletons are rare, shed theropod teeth are common as isolated elements, and are usually found mixed with herbivore skeletons. For example, Ostrom (1969) noted that *Deinonychus* teeth (and no others) were found associated with *Tenontosaurus* remains at fourteen sites in the Cloverly Formation. This suggests that theropods lost one or more teeth with each meal. Yet known theropod skulls and jaws invariably have most tooth positions occupied. Taken together, these two facts suggest that tooth replacement was rapid and constant. Tooth life may have been of the same order of magnitude as that in extant crocodiles, which Edmund (1962) measured as nine to sixteen months.

Tooth morphology varies within the jaws of single individuals, depending on tooth position. In tyrannosaurids and *Dromaeosaurus*, premaxillary teeth are distinguishable from maxillary/dentary teeth. In *Saurornitholestes*, premaxillary, maxillary, and dentary teeth differ enough from each other to permit identification into these groupings. Anterior dentary teeth are distinct from posterior dentary teeth in *Richardoestesia*, but the situation in the upper jaw bones is unknown. *Troodon* demonstrates the greatest degree of heterodonty amongst

the Judithian theropods, with four distinct tooth morphs within a single individual.

All of the tooth types characteristic for *Saurornitholestes*, *Dromaeosaurus*, *Troodon*, *Richardoestesia*, and the Tyrannosauridae are found in the Lower Campanian Milk River Formation of southern Alberta, and up into the Upper Campanian beds of the Horseshoe Canyon Formation near Drumheller. *Richardoestia* teeth found in the Milk River Formation include some that are perfectly straight, in contrast with those from the Judith River Formation of Alberta and Montana, which are always slightly curved at the base. The Milk River Formation *Richardoestesia* teeth are identical to those of the Maastrichtian Scollard, Frenchman, Hell Creek, and Lance Formations. This suggests either a similarity of environments between the Milk River and Maastrichtian formations, or that the straight *Richardoestesia* teeth just have not been found to date in the Judith River Formation. Neither explanation is very satisfying when balanced with other lines of evidence. Amongst other theropod teeth recovered, "*Paronychodon*" teeth (in the collections of the Royal Ontario Museum and the University of Alberta) recovered from the Milk River Formation are also closer in appearance to Lancian, rather than Judithian, teeth of this type.

A variety of small theropod teeth have been identified in Paleocene deposits in Montana (Sloan et al. 1986), thereby opening up the possibility that dinosaurs did survive beyond the Cretaceous–Tertiary boundary. The problem of whether or not these teeth have been reworked from older sediments cannot be solved unequivocally at this time (Bryant et al. 1986; Argast et al. 1987), and stronger evidence will be required to convince most workers that some dinosaurs may have survived into the Paleocene.

The possibility that fossil materials may have been reworked emphasizes the potential problem of assigning too much importance on isolated teeth in paleoecological studies. Teeth can be reworked several times from older sediments, or they can be brought in by flowing water from upstream ecosystems. The latter situation also applies to both isolated bones and articulated skeletons (a floating or rolling carcass can be carried great distances downstream before decomposition of the soft tissue is complete), whereas isolated bones can also be reworked from older sediments. Reworking and transport can be discounted in many cases, however, by taking into account relative abundance, degree of postmortem wear, and knowledge of rates of accumulation in the depositional environments.

Tooth and jaw morphology in Judithian theropods provides little insight into their relationships. The fused interdental plates of *Saurornitholestes* and *Dromaeosaurus* represent a derived condition from that seen in most thecodonts (Ewer 1965) and primitive theropods (Welles 1984; Raath this volume). However, the teeth of *Dromaeosaurus* differ from those of *Saurornitholestes*,

Velociraptor, and *Deinonychus* in carina position and denticulation. These facts support inclusion of *Dromaeosaurus* and *Saurornitholestes* in separate subfamilies of the Dromaeosauridae. Troodontids are characterized by the increased disparity in size between teeth of the upper and lower jaws (this character is more extreme in *Baryonyx*, Charig and Milner this volume), by increased differentiation of teeth, by loss of the interdental plates, and by the enlarged size of the denticles. Because the front of the snout is wider than it is in most theropods, the carinae of anterior teeth have shifted to the posterior surfaces as they have in the premaxillae of tyrannosaurids. This is clearly parallel evolution related to the width of the front of the jaws, and has been accomplished in two different ways. Troodontid anterior teeth are triangular in section, whereas the premaxillary teeth of *Aublysodon* and other tyrannosaurids are D-shaped.

There are few clues as to the relationships of *Richardoestesia*. The fine denticulation and high number of teeth are reminiscent of more primitive theropods like *Coelophysis* (TMP 84.63). The form of the interdental plates is the same as in tyrannosaurids, but this is a primitive character.

In assessing relationships based on denticulation patterns, it should be remembered that the denticles perform a function related to the killing and eating of prey. It is not surprising then that denticle morphology falling within the range of variation expressed in Theropoda can also be found in such diverse forms as sharks (Frazetta 1988), lizards, thecodonts, and saber-tooth tigers (Martin 1980). The denticles in *Richardoestesia* are generalized and tell us little about diet preference. Tyrannosaurid denticles are broad and strong and are a compromise between the requirements of cutting through flesh and bone. The long, slender hooked posterior denticles of *Saurornitholestes* were well adapted to slicing flesh off of bones. Grazing tooth marks on TMP 88.121.39 show how *Saurornitholestes* teeth were used parallel to the surface of the bone rather than perpendicular to it. The shorter, broader denticles of *Dromaeosaurus*, in conjunction with the more massive skull, suggests that this animal probably was biting through the smaller bones of its prey. The enlarged, sharply pointed but broad-based denticles of troodontids probably gave the animals the option of slicing efficiently through soft material or into bone. The denticles are as large as those of tyrannosaurids, but are more sharply pointed and are hooked distally. This suggests that they were capable of cutting meat more efficiently than the large carnivores.

Tooth shape and wear facets merit more detailed study in theropods because they also provide dietary clues. The anterior dentary teeth of *Richardoestesia* are straight, whereas the more posterior ones curve back towards the throat. It is easy to imagine that these teeth would have been useful for piercing and holding insects and other soft-bodied prey. The cheek teeth of dromaeosaurids and troodontids tend to show their greatest wear at the distal end of the anterior carina, whereas premaxillary teeth show more wear on the lingual surface. In spite of their relatively small size, the premaxillary teeth of tyrannosaurids are almost invariably worn on the lingual surface. The carinae are frequently worn completely away on these teeth, making it difficult in some cases to determine whether or not there were denticles. "Wear" facets on tyrannosaurid cheek teeth are more difficult to interpret because there is little consistency in position. It appears more likely that flakes of enamel were spalling off as the animal bit into bone, and that the edges of these damaged surfaces would subsequently wear smooth. Some tyrannosaurid teeth were reduced to nubbins by breakage and wear before being shed and replaced (Carpenter 1979; Farlow and Brinkman 1987).

In conclusion, the ability to identify theropod teeth is another tool that can be used to determine the relative age of Mesozoic beds, and/or to provide information on paleoenvironments. Both temporal and geographic ranges of some theropods can be extended by recognition of their teeth. Relative abundance of theropod taxa is indicated by the numbers of teeth recovered, but this cannot be a precise indication because of our ignorance on replacement rates and, in some cases, the numbers of teeth in the skull. Some taxa are still known only by their teeth. Morphology of teeth and jaws can indicate relationships of certain taxa.

Acknowledgments

The authors would like to thank Dr. D. A. Russell, Rick Day and Kieran M. Shepherd (National Museum of Natural Sciences, Ottawa), Dr. Chris McGowan, Peter May, and Andrew Leitch (Royal Ontario Museum), A. Perle (Mongolian Peoples Republic), Peter Larson (Black Hills Institute of Geological Research), and Dr. R. C. Fox (University of Alberta) for assistance in studying collections in their institutions. Jane Danis and Jackie Wilke provided tremendous assistance in the sorting and cataloguing of theropod teeth in the collections of the Tyrrell Museum of Palaeontology. Scanning electron microscope work was done by Andrew Neuman and Linda Strong-Watson (Tyrrell Museum) using facilities at the University of Calgary with the assistance of David Harvey (Geology and Geophysics), Doris Johnston (Geology and Geophysics), and Genevieve La Moyne (Archaeology). Dr. Ken Poznikoff (Riverside Dental Clinic, Drumheller) kindly provided equipment and expertise that produced dental x-rays of the jaws of *Richardoestesia* and *Saurornitholestes*. Figures 8.4A–F were done by Linda Krause, and the rest were done by the first author. Melanie Nielsen (Delia) and Mike Todor (Drumheller) assisted with darkroom work. The manuscript was greatly improved following reviews by William Abler (Chicago), Ken Carpenter (Denver Museum of Natural History), and Jim Farlow (Indiana University–Purdue University at Fort Wayne). Production of this paper would have been highly improbable without the expert assistance of Rebecca Kowalchuk and other dedicated staff of the Tyrrell Museum of Palaeontology.

References

Abler, W. L. In preparation. On the serrated teeth of tyranno-saurid dinosaurs, with comparisons to serrated biting structures of other animals.

Argast, S., Farlow, J. O., Gabet, R. M. and Brinkman, D. L. 1987. Transport-induced abrasion of fossil reptilian teeth: implications for the existence of Tertiary dino-saurs in the Hell Creek Formation, Montana. *Geology* 15:927–930.

Armstrong-Ziegler, J. G. 1980. Amphibia and Reptilia from the Campanian of New Mexico. *Fieldiana, Geology (New Series)* 4:1–39.

Barsbold, R. 1974. Saurornithoididae, a new family of small theropod dinosaurs from Central Asia and North Amer-ica. *Palaeontologia Polonica* 30:5–22.

    1983. Carnivorous dinosaurs from the Cretaceous of Mon-golia. *Soviet-Mongolian Paleontological Expedition, Transactions* 19:5–117. [in Russian]

Barsbold, R., and Perle, A. 1984. The first record of a primitive ornithomimosaur from the Cretaceous of Mongolia. *Paleontological Journal* 18:118–120.

Breithaupt, B. H. 1985. Nonmammalian vertebrate faunas from the Late Cretaceous of Wyoming. *Wyoming Geological Association Guidebook, Annual Field Conference,* 36:159–175.

Bryant, L., Clemens, W. A., and Hutchison, J. H. 1986. [Letter.] *Science* 234: 1172.

Buffetaut, E., and Suteethorn, V. 1989. A sauropod skeleton associated with theropod teeth in the Upper Jurassic of Thailand: remarks on the taphonomic and palaeoeco-logical significance of such associations. *Palaeogeo-graphy, Palaeoclimatology, Palaeoecology* 73:77–83.

Carpenter, K. 1979. Vertebrate fauna of the Laramie Formation (Maestrichtian), Weld County, Colorado. *University of Wyoming, Contributions to Geology* 17:37–48.

    1982. Baby dinosaurs from the Late Cretaceous Lance and Hell Creek Formations and a description of a new species of theropod. *University of Wyoming, Contributions to Geology* 20:123–134.

Colbert, E. H. 1989. The Triassic dinosaur *Coelophysis. Museum of Northern Arizona, Bulletin* 57:1–160.

Colbert, E. H., and Russell, D. A. 1969. The small Cretaceous dinosaur *Dromaeosaurus. American Museum of Natural History Novitates* 2380:1–49.

Cope, E. D. 1876a. Descriptions of some vertebrate remains from the Fort Union beds of Montana. *Academy of Natural Sciences of Philadelphia, Proceedings* 1876:248–261.

    1876b. On some extinct reptiles and batrachia from the Judith River and Fox Hills beds of Montana. *Academy of Natural Sciences of Philadelphia, Proceedings* 1876:340–359.

Currie, P. J. 1987a. Bird-like characteristics of the jaws and teeth of troodontid theropods (Dinosauria: Saurischia). *Journal of Vertebrate Paleontology* 7: 72–81.

    1987b. New approaches to studying dinosaurs in Dinosaur Provincial Park. *In* Czerkas, S. J. and Olson, E. C. (eds.), *Dinosaurs Past and Present,* volume II (Los Angeles: Los Angeles County Museum), pp. 100–117.

    1987c. Theropods from the Judith River Formation of Dino-saur Provincial Park, Alberta, Canada. *Tyrrell Museum of Palaeontology, Occasional Paper* 3:52–60.

    1989. The first records of *Elmisaurus* (Saurischia, Thero-poda) from North America. *Canadian Journal of Earth Sciences* 26:1319–1324.

    In press. Saurischian dinosaurs of the Late Cretaceous of Asia and North America. *In* Mateer, N. J. and Chen, P. J. (eds.), *Aspects of Nonmarine Cretaceous Geology, Proceedings of the Conference on Nonmarine Creta-ceous Correlations, Urumqi, China, 1987* (Beijing: Ocean Press).

Currie, P. J., and Dodson, P. 1984. Mass death of a herd of cer-atopsian dinosaurs. *In* Reif, W. E., and Westphal, F. (eds.), *Third Symposium on Mesozoic Terrestrial Ecosystems,* short papers (Tübingen: Attempto Verlag), pp. 61–66.

Currie, P. J., and Russell, D. A. 1988. Osteology and relation-ships of *Chirostenotes pergracilis* (Saurischia, Thero-poda) from the Judith River (Oldman) Formation of Alberta, Canada. *Canadian Journal of Earth Sciences* 25:972–986.

Edmund, A. G. 1962. Sequence and rate of tooth replacement in the Crocodilia. *Royal Ontario Museum, Life Sciences Contribution* no. 56:1–42.

Estes, R. 1964. Fossil vertebrates from the Late Cretaceous Lance Formation, eastern Wyoming. *University of California, Publications in Geological Sciences* 49:1–180.

Ewer, R. F. 1965. The anatomy of the thecodont reptile *Eupark-eria capensis* Broom. *Royal Society of London, Philo-sophical Transactions, Series B* 248:379–435.

Farlow, J. O., and Brinkman, D. L. 1987. Serration coarseness and patterns of wear of theropod dinosaur teeth. *Geo-logical Association of America, Abstracts with Programs* 19:151.

Farlow, J. O., Brinkman, D. L., and Currie, P. J. In preparation. Size, shape and serration density of theropod dinosaur lateral teeth.

Frazetta, T. H. 1988. The mechanics of cutting and the form of shark teeth (Chondrichthyes, Elasmobranchii). *Zoomor-phology* 108:93–107.

Gilmore, G. W. 1924. A new coelurosaurid dinosaur from the Belly River Cretaceous of Alberta. *Geological Survey of Canada, Bulletin* 38:1–12.

Johnston, P. E. 1979. Growth rings in dinosaur teeth. *Nature* 278:635–639.

Lambe, L. M. 1918. On the genus *Trachodon* of Leidy. *The Ottawa Naturalist* 31:135–139.

Langston, W. Jr. 1956. The Sebecosuchia: cosmopolitan crocodil-ians? *American Journal of Science* 254:605–614.

Lehman, T. M. 1981. The Alamo Wash Local Fauna: a new look at the old Ojo Alamo Fauna. *In* Lucas, S., Rigby, K. Jr., and Kues, B. (eds.), *Advances in San Juan Basin Paleontology* (Albuquerque: University of New Mexico Press), pp. 189–220.

Leidy, J. 1856. Notices of remains of extinct reptiles and fishes, discovered by Dr. F. V. Hayden in the Bad Lands of the Judith River, Nebraska Territory. *Academy of Natural Sciences of Philadelphia, Proceedings* 8:72–73.

    1860. Extinct Vertebrata from the Judith River and Great Lignite Formations of Nebraska. *American Philosophi-cal Society, Transactions* (Series 2) 2:139–154.

    1868. Remarks on a jaw fragment of *Megalosaurus. Acad-emy of Natural Sciences of Philadelphia, Proceedings* 1868:197–200.

Lydekker, R. 1888. *Catalogue of Fossil Reptilia and Amphibia in the British Museum,* part 1 (London: British Museum).

Madsen, J. 1976. *Allosaurus fragilis*: a revised osteology. *Utah*

Geological and Mineralogical Survey, Bulletin 109:1–163.

Marsh, O. C. 1892. Discovery of Cretaceous Mammalia. *American Journal of Science* 38:81–92.

Martin, L. D. 1980. Functional morphology and the evolution of cats. *Nebraska Academy of Sciences, Transactions* 8:141–154.

Molnar, R. E. 1978. A new theropod dinosaur from the Upper Cretaceous of central Montana. *Journal of Paleontology* 52:73–82.

Molnar, R. E., and Carpenter, K. 1990. The Jordan theropod (Maastrichtian, Montana, U.S.A.) referred to the genus *Aublysodon. Geobios*, 22:445–454.

Nelson, M. E., and Crooks, D. M. 1987. Stratigraphy and paleontology of the Cedar Mountain Formation (Lower Cretaceous), East Emery County, Utah. *In* Avarett, W. R. (ed.), *Paleontology and Geology of the Dinosaur Triangle, Guidebook for 1987 Fieldtrip* (Grand Junction: Museum of Western Colorado), pp. 55–63.

Nopcsa, F. 1901. Synopsis and Abstammung der Dinosaurier. *Foldtani kozlony* (suppl.) 31:247–288.

Ostrom, J. H. 1969. Osteology of *Deinonychus antirrhopus,* an unusual theropod from the Lower Cretaceous of Montana. *Peabody Museum of Natural History, Bulletin* 30:1–165.

Paul, G. 1988a. The small predatory dinosaurs of the mid-Mesozoic: the horned theropods of the Morrison and Great Oolite – *Ornitholestes* and *Proceratosaurus* – and the sickle-claw theropods of the Cloverly, Djadokhta and Judith River – *Dienonychus, Velociraptor* and *Saurornitholestes. Hunteria* 2(4):1–9.

1988b. *Predatory Dinosaurs of the World* (New York: Simon and Schuster).

Russell, D. A. 1970. Tyrannosaurs from the Late Cretaceous of western Canada. *National Museum of Canada, Publications in Paleontology* 1:1–34.

1984. A check list of the families and genera of North American dinosaurs. *Syllogeus* 53, 35 pp.

Russell, L. S. 1935. Fauna of the Upper Milk River beds, southern Alberta. *Royal Society of Canada, Transactions* (series 3, section 4) 29:115–127.

1948. The dentary of *Troodon,* a genus of theropod dinosaur. *Journal of Paleontology* 22:625–629.

Sahni, A. 1972. The vertebrate fauna of the Judith River Formation, Montana. *American Museum of Natural History, Bulletin* 147:323–412.

Sloan, R. E., Rigby, J. K. Jr., Van Valen, L. M., and Gabriel, D. 1986. Gradual dinosaur extinction and simultaneous ungulate radiation in the Hell Creek Formation. *Science* 232:629–633.

Standhardt, B. R. 1986. Vertebrate paleontology of the Cretaceous/Tertiary transition of Big Bend National Park, Texas. Ph.D. thesis, Louisiana State University.

Sternberg, C. M. 1945. Pachycephalosauridae proposed for dome-headed dinosaurs; *Stegoceras lambei,* n. sp., described. *Journal of Paleontology* 19:534–538.

1951. The lizard *Chamops* from the Wapiti Formation of northern Alberta: *Polyodontosaurus grandis* not a lizard. *National Museum of Canada, Bulletin* 123:256–258.

Sues, H.D. 1977a. Dentaries of small theropods from the Judith River Formation (Campanian) of Alberta, Canada. *Canadian Journal of Earth Sciences* 14:587–592.

1977b. The skull of *Velociraptor mongoliensis,* a small Cretaceous dinosaur from Mongolia. *Paläontologische Zeitschrift* 51:173–184.

1978. A new small theropod dinosaur from the Judith River Formation (Campanian) of Alberta, Canada. *Zoological Journal of the Linnean Society* 62:381–400.

Welles, S. P. 1984. *Dilophosaurus wetherilli* (Dinosaur, Theropoda), osteology and comparisons. *Palaeontographica* A 185:85–180.

9 The systematic position of *Baryonyx walkeri,* in the light of Gauthier's reclassification of the Theropoda

ALAN J. CHARIG AND
ANGELA C. MILNER

Abstract

Continuing preparation of the unique specimen of *Baryonyx walkeri* has already provided much additional information on this recently described dinosaur, including the startling observation that the lower teeth are twice as numerous per unit length of jaw as are the upper teeth. The characters of the animal demonstrate conclusively that it is a theropod dinosaur, that it is not a spinosaurid, that it cannot be fitted satisfactorily into Gauthier's recent classification of the theropods, and that it is sufficiently distinctive from all other theropods to justify the present authors' earlier proposal of a new family Baryonychidae. Some fragmentary material described from the Late Cretaceous of Niger and Morocco and referred to the Spinosauridae is more likely to be baryonychid.

Introduction

Our lecture at the Drumheller Symposium on a new theropod dinosaur from the Wealden of Surrey was followed by our publication (Charig and Milner 1986) of a preliminary account of the animal, in which we named it *Baryonyx walkeri.* A few of the more important pieces were illustrated. Since then, the continuing preparation of the skeleton has revealed the existence of further elements (in particular both quadrates, the articulating region of the right mandible, an isolated dorsal neural spine, and the left radius and ulna). It has also extended our knowledge of some of the elements described previously. Conversely, we are no longer sure about the presence of any material belonging to the pubes.

Meanwhile, Gauthier has published (1986) a carefully detailed character distribution analysis (cladistic analysis) of the theropods derived from his even longer Ph.D. dissertation of 1984. *Baryonyx,* of course, is not mentioned in either of those works, both of which were written well before the publication of our preliminary description.

In Dinosaur Systematics: Perspectives and Approaches, *Kenneth Carpenter and Philip J. Currie, eds. Copyright © Cambridge University Press, 1990.*

Our present paper is therefore intended as an update and expansion of the preliminary description. It begins with brief indications of such newly revealed anatomical points as may be of interest. (We intend to give a complete osteology of *B. walkeri* as soon as the specimen is fully prepared, but that might be many months – or even years – from now.) We end with a statement of our current opinions on the animal's systematic relationships.

Bearing in mind that the present volume is concerned with all aspects of dinosaur systematics, we take this opportunity of using *Baryonyx* to test Gauthier's character distribution analysis of the Theropoda. *Baryonyx* is especially suitable for such a test because its novel features are a challenge to the arrangements for dinosaur systematics hitherto proposed. In doing this, we make also some general criticisms – both favorable and unfavorable – of Gauthier's study.

Anatomy

We have now opened all 54 of the blocks comprising the specimen and have a fairly good idea of which elements of the skeleton are preserved. It is certain, however, that there are a few more elements yet to be uncovered; furthermore, many of the elements already described will undoubtedly yield more information when their preparation is complete. We shall therefore undertake no description at this stage, instead presenting only an extended diagnosis.

Baryonyx is typified by a unique combination of characters, although we emphasize that not every individual character is necessarily unique in itself. The characters include:

1. The prenarial region of the snout extends into an extremely narrow rostrum with a spatulate, horizontal expansion at its end ("terminal rosette", Fig. 9.2). The snout is slightly downturned in lateral view (Fig. 9.1) so that the jaws have somewhat sigmoidal margins.

Figure 9.1. *Baryonyx walkeri* holotype, B.M.(N.H.) Palaeo. Dept. no. R.9951: Skull elements, arranged as far as possible in their presumed natural relative orientation; from the left side. The quadates and some of the more fragmentary lower jaw elements are not included. (Scale bar = 5 cm.)

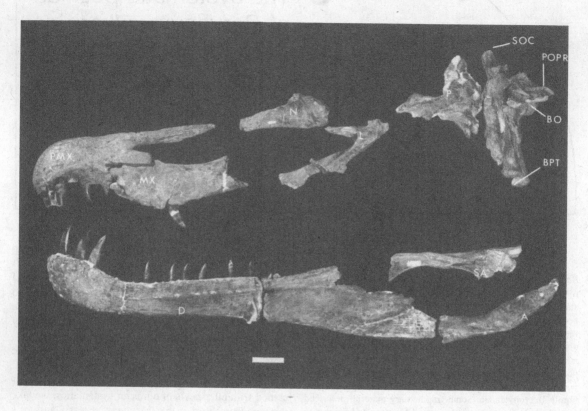

Figure 9.2. *Baryonyx walkeri* holotype, B.M.(N.H.) Palaeo. Dept. no. R.9951: Conjoined premaxillae together with left maxilla (all incomplete) in articulation; in occlusal (ventral) view.

Figure 9.3. *Baryonyx walkeri* holotype, B.M.(N.H.) Palaeo. Dept. no. R.9951: Left dentary; in occlusal (dorsal) view, to show the full complement of tooth sockets.

2.   The long, low external naris (Fig. 9.1) is situated far back on the side of the snout, its anterior end is an acute angle, ascending gently towards the rear and widening as it does so.

3.   The articulation of the premaxilla and maxilla is unfused above a subrostral notch (Fig. 9.1).

4.   There is a small median knob at the posterior end of the conjoined nasals, cruciform in dorsal view, with the anterior limb of the cross drawn forwards into a low thin median crest (Fig. 9.1).

5.   The braincase is anteroposteriorly short.

6.   The deep occiput (Fig. 9.4) has the paroccipital processes directed horizontally outwards. The basipterygoid processes descend far below the basioccipital and diverge laterally only slightly. (According to Molnar, pers. comm., this configuration seems to be unique among theropods.)

7.   In lateral view, the anterior end of the dentary (Fig. 9.1) is upturned.

8.   There is no evidence of a mandibular symphysis.

9.   All the elements in the central region of the lower jaw (Fig. 9.1) have a laminar (almost paper-thin) fragile nature.

10.   There is apparently a large mandibular fenestra (Fig. 9.1).

11.   The tooth count is remarkable with seven premaxillary, 8 or more maxillary, and 32 dentary teeth (in contrast with the usual theropod tooth count of 5, 16, and 16). The largest are the third tooth in the premaxilla and the third and fourth in the dentary. The teeth in the lower jaw are generally smaller than those in the upper and are more numerous per unit length of jaw. A portion of the dentary containing sixteen small alveoli (see Fig. 9.3) corresponds to a length of the maxilla containing only seven or eight much larger alveoli (see Fig. 9.2), i.e., the discrepancy in their numbers is approximately 2:1. Such a gross disparity in the tooth count between the two jaws is a unique phenomenon and one that we failed to comment on in our earlier paper on *Baryonyx*.

Figure 9.4. *Baryonyx walkeri* holotype, B.M.(N.H.) Palaeo. Dept. no, R.9951: Occiput; from behind. (Scale bar = 5 cm.)

It has been suggested to us that the two jaws belong to different animals. We cannot conceive, however, that that could be the case; we have no doubts whatever that the two jaws are from the same individual. This is confirmed by their commensurate size, corresponding sinuous profiles in lateral view, anterior terminal expansions in occlusal view, identical type of tooth insertion, identical type of all teeth found (including those found in isolation), identical nature of preservation, relative topographical positions in the field, and the complete absence of any other large animal remains in the vicinity.

12.   The crowns of the teeth have only slight labio-lingual flattening, are lightly fluted on the lingual side, and have fine serrations (about 7 per millimetre) on the anterior and posterior keels. The roots are exceptionally long and slender.

13.   There are no obvious interdental plates (unlike *Megalosaurus*, where they are conspicuous, high, and separate from each other).

14.   The axis is small and has a well developed hyposphene, a very small neural spine, and no spine table.

15.   The long, strongly opisthocoelous cervical vertebrae

Figure 9.5. *Baryonyx walkeri* holotype, B.M.(N.H.) Palaeo. Dept. no. R.9951: Left humerus in anterolateral view. (Scale bar = 5 cm.)

have low neural spines and lack spine tables. There are flat zygapophyses and well developed epipophyses. The ends of the centra are not offset, so when they are correctly apposed the neck does not show the typical theropod upward curve.

16. There is a lack of fusion of most of the neurocentral sutures of the cervical vertebrae. (This, however, might merely reflect a subadult life-stage).

17. At least some of the postcervical neural spines are elongate to at least a moderate degree.

18. The cervical ribs are short and crocodiloid, and possibly overlap a little.

19. The coracoid tapers posteriorly in lateral profile.

20. The humerus (Fig. 9.5) is relatively well developed. Both ends are broadly expanded but flattened, and are offset 35° against each other. The shaft is massive and almost straight. The distal end lacks both supinator process and ectepicondylar groove. (See Table 9.1 for dimensions.)

21. The radius is a little less than half as long as the humerus. The ulna is somewhat longer, with a powerful olecranon. Both epipodials are stoutly built (Table 9.1).

22. The ilium has a prominent supracetabular crest, a slender, vertically expanded anterior process, and a long, straight posterior process. There is a prominent brevis shelf and a deep, ventrally facing groove between that shelf and the medial blade.

23. The acetabulum is anteroposteriorly long.

24. The ischium has a well developed obturator process, proximally placed.

25. The femur lacks a groove on the fibular condyle.

26. The fibula has a medial groove on the proximal part of the shaft.

27. At least one pair of huge, slender talon-shaped ungual phalanges (Fig. 9.6) existed, probably on the forefeet.

These are not laterally compressed and sickle-shaped like the 2nd pedal ungual of *Deinonychus*. The ungual in question was the first element of *Baryonyx* to be discovered, and much attention has been paid to it. It is about 38% longer (in absolute terms) than the ungual borne on the 1st manual digit of *Allosaurus fragilis* (Madsen 1976, pls. 43, 44) and has a more offensive appearance. (It was the source of the nickname "Claws" given to the creature by the Press in 1983.) Proportionally, however, it is of much the same order of length as the *Allosaurus* ungual used by Madsen.

Figure 9.6. *Baryonyx walkeri* holotype, B.M.(N.H.) Palaeo. Dept. no. R.9951: **A**, "normal" digit, presumably of hand; **B**, especially large ungual phalanx. (Scale bar = 5 cm.)

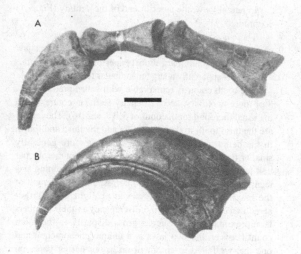

Table 9.1. *Some relevant dimensions and ratios for* Baryonyx, *compared to* Allosaurus

		Allosaurus[a]	
	Baryonyx	"M"	"G"
Dimensions (cm)			
Length of humerus	46.3	35.6	31.0
Length of radius	22.3	22.3	22.2
Length of ulna	27.8	27.8	26.3
Length of the largest ungual (manual 1), measured diagonally across the curve:	24.3	17.6	12.0
Length of the smallest ungual preserved (manual 1), measured diagonally across the curve:	13.6[b]	7.6	5.9
Ratios (%)			
Length of radius/length of humerus	48	63	72
Length of ulna/length of humerus	60	78	85
Length of largest ungual/length of smallest ungual available	179	232	203
Length of largest ungual/length of humerus	52	49	39

[a]From Madsen 1976 (M, measured from his illustrations) and Gilmore 1920 (USNM 4734, G; cited directly). The two sets of measurements differ significantly.

[b]This measurement is a cautious estimate.

(See Table 9.1 for dimensions and ratios.) It is possible that the smallest ungual preserved was not the smallest that the animal possessed. If that was the case and there was a smaller ungual, then the ratio given in Table 9.1 for *Baryonyx*, namely 179%, would be proportionately increased.

Systematic relationships
Identification of *Baryonyx* as a dinosaur and a theropod

The huge claw-bone that eventually led us to name its owner *Baryonyx* ("heavy claw") suggested to us that the animal was a theropod. All subsequent evidence has confirmed that opinion. Nevertheless some have suggested that the creature might have been a crocodilian (e.g., at the Drumheller Symposium, verbally).

We now state categorically that such a placement is out of the question. Of the following crocodylomorph/crocodilian autapomorphies listed by Gauthier, none appears to have been present in *Baryonyx*:

1. Ventromedial elongation of the coracoid (Gauthier 1984, p. 11; 1986, p. 43).
2. Reduction in size of the pubis and its near-exclusion from the acetabulum (Gauthier 1984, p. 112). [This must refer to the proximal plate-like part of the pubis, forming the rim of the acetabulum, and not – as we originally assumed – to the length of its shaft; indeed, in most early crocodilians (e.g., *Protosuchus*) the shaft is exceptionally long.]
3. Shortening of the ischium so that it is only half as long as the pubis (Gauthier 1984, p. 112). [This applies only to early crocodilians (again see *Protosuchus*) and confirms that, since the pubis was twice as long as the ischium in those forms, it cannot be the shaft of the former that is reduced.]
4. Development of a crurotarsal ankle-joint, the calcaneum possessing both an enlarged tuber and a socket (the latter to receive the peg on the astragalus) (Gauthier 1984, p. 108; 1986, p. 42). [In any case, Gauthier notes (1986, p. 42) that this character-state is found also in rauisuchians, aetosaurs, and – to some extent – in parasuchians (= phytosaurs).]

Furthermore, nearly all crocodilians (excepting only a few of the earliest) possess abundant dermal ossifications (scutes); whereas no dermal ossifications of any sort have been found in *Baryonyx*, although such evidence is negative and ipso facto cannot be verified.

On the contrary, the anatomy of *Baryonyx* affords abundant evidence of its dinosaurian relationships. Certain features have rarely been observed in crocodilians:

1. Teeth are recurved, serrated, laterally compressed blades, generally similar to those of most other theropods. Such teeth differ from typical crocodilian teeth and also from the isosceles-triangular, laterally compressed teeth of "xiphodont" (Greek "sword-toothed") crocodilians (Hsisosuchidae, Sebecosuchidae, and pristichampsine Crocodylidae).
2. Stance and gait are of the "fully improved" (Charig 1965, 1966, 1972) type. Such limb posture is evinced by the widely fenestrated acetabulum, the well developed, inturned head to the femur, and a calcaneum that in no way

resembles the highly complex "crocodiloid" calcaneum and that must have formed part of a mesotarsal ankle-joint.
3. The huge ungual would suggest that the animal walked bipedally at least part of the time, a type of locomotion that is virtually unknown in crocodilians.
4. An obturator process is present on the ischium.

Given that *Baryonyx* is a dinosaur, there is no possibility that it has any ornithischian affinities. Even if we exclude negative evidence, there is positive evidence that it lacks certain structures that are possessed by all known ornithischians (predentary teeth that evidently possessed a herbivorous function, opisthopubic pelvis, ossified intervertebral ligaments). Conversely, the cervical vertebrae of *Baryonyx* have strong epipophyses, structures that are entirely unknown in ornithischians.

Because the dentition alone is sufficient to preclude the possibility of *Baryonyx* being a sauropodomorph, we are left with the inescapable conclusion that the animal is a theropod. There is no evidence to the contrary.

Comparison of *Baryonyx* with spinosaurids and dilophosaurids

Baryonyx has been compared hitherto (Charig and Milner 1986; Paul 1988; Buffetaut 1989) with only two other genera – the gigantic *Spinosaurus* (Stromer 1915, 1934), from the basal Cenomanian of Egypt, and, less closely, with *Dilophosaurus* (Welles 1954, 1984) from the Lower Jurassic of Arizona. It has been compared also with the families based on those two genera, the Spinosauridae and the Dilophosauridae.

Apart from the Egyptian material described by Stromer, other fossil reptile remains from northern and western Africa have been referred subjectively to *Spinosaurus* or the Spinosauridae. Before we evaluate the comparison with *Baryonyx*, we must consider the correctness of those references.

Taquet (1984) described two fragmentary snouts from the Aptian of Gadoufaoua, Niger, as the mandibular symphyses of a spinosaurid. However, we noted in our earlier paper on *Baryonyx* (Charig and Milner 1986) that Taquet's snouts were almost identical to the conjoined premaxillae of our Wealden animal. Taquet himself, having seen the *Baryonyx* material, unhesitatingly accepted our conclusion (pers. comm.). (The Aptian of Gadoufaoua yielded also undescribed isolated theropod teeth, with serrations and fluting.)

Taquet also mentioned (1984) the occurrence of spinosaurid material in the Albian of Niger, Algeria, and perhaps Morocco, but he did not describe it. Bouaziz et al. (1988) described teeth from the Albian of Tunisia and referred them to *Spinosaurus* sp.

Buffetaut (1989) discusses the relationships of some new material from the Cretaceous (probably basal Cenomanian) of southern Morocco. It consists only of two jaw fragments and an isolated tooth (all registered separately and presumably not associated). One of the jaw pieces is interpreted as part of the anterior half of a right dentary of *Spinosaurus*, but "is too fragmentary for this interpretation to be considered as certain." The other is part of a large right maxilla with large, poorly preserved replacement teeth and one functional tooth. It is important to note that the only elements preserved both in this material and in Stromer's *Spinosaurus* are the teeth and their alveoli. Buffetaut has little doubt that his new specimens should be referred to *Spinosaurus* cf. *aegyptiacus* because of the circular, well separated alveoli, the absence of interdental plates, and the weakly developed lateral compression of the tooth crowns with their distinct keels (unserrated in the replacement teeth). In our opinion, however, such characters can hardly be considered on their own as diagnostic of a particular genus or even a particular family. In consequence, Buffetaut's reference should be regarded only as tentative.

Let us now evaluate the comparison of *Baryonyx* with *Spinosaurus*. We ourselves wrote (Charig and Milner 1986) that "comparison (of *Baryonyx*) with *Spinosaurus* does not suggest spinosaurid affinities." On the other hand, two subsequent authors (Paul 1988; Buffetaut 1989) have claimed a close phylogenetic relationship between the two genera. Paul (1988, p. 271) places *Baryonyx* provisionally with *Spinosaurus* in the Spinosauridae. He argues his case only on the evidence of:

1. "the slender, semiconical, crocodile-like teeth with microscopically fine serrations." [In fact the teeth of *Baryonyx* are neither semi-conical (at least not in our understanding of that term, which necessitates a straight central axis) nor crocodile-like, and those of *Spinosaurus* have no serrations at all.]
2. "the crocodile-like lower jaw with expanded tips."
3. "the seeming lack of the typical theropod S-curve in the neck." (The evidence for this in *Spinosaurus* is less than satisfactory, being based on a single, not very well preserved cervical vertebra.)

(Incidentally, characters (1) and (2) are said to be "very alike in the two species"; yet no species are mentioned, only genera.)

Paul states that his "best guess ... is that spinosaurs (spinosaurids in the correct nomenclature) are specialized, late-surviving dilophosaurs (dilophosaurids)" and admits that such a tentative opinion requires confirmation by further discoveries. In this connection, he alleges that the loose, deeply kinked premaxillary/maxillary articulation near the tip of the snout, the posteriorly positioned external naris, and the "slender teeth" of *Baryonyx* are "surprisingly similar to those of ... *Dilophosaurus*," and appear to be extreme developments of the former's snout. Further evidence of the relationship to *Dilophosaurus*, so he claims, is afforded by certain characters said to be found in both *Baryonyx* and *Spinosaurus*. These include "the long, primitive main bodies of the neck and trunk vertebrae," although the form of the trunk vertebrae is neither mentioned nor illustrated in our earlier publication on *Baryonyx*. He

also refers to the "the shallow rib cages," which is especially surprising, since (a) our earlier paper neither describes nor illustrates the ribs of *Baryonyx*, (b) Paul has not seen the specimen, (c) we ourselves, as yet, have no idea whatever of the depth of the rib cage in our animal, and (d) Welles's detailed description (1984) of *Dilophosaurus* states (p. 126) that the pectoral and dorsal ribs "seem to have been of normal size and length," while his reconstruction of the skeleton (p. 92) shows no ribs at all.

We therefore tend to treat Paul's conclusions, provisional though they are, with some degree of caution.

Buffetaut (1989) likewise believes that *Baryonyx* and *Spinosaurus* are closely related to each other. That belief depends in part upon his own reference of the new Moroccan maxilla to *Spinosaurus* – a reference that we ourselves (as already stated) can regard only as dubious. The Moroccan maxilla shows a far stronger resemblance to that of *Baryonyx*. Thus, if the Moroccan specimen is indeed baryonychid and not spinosaurid, then, in comparing it with *Baryonyx* itself, Buffetaut is merely comparing one baryonychid with another, and he cannot use their similarity as evidence of a close relationship of both of them to *Spinosaurus*.

However, even if we deprive it of the evidence of the maxilla, Buffetaut's case for the close relationship of *Baryonyx* to *Spinosaurus* seems not unreasonable – superficially at least. He concludes (1989, p. 85) that "Whether the differences are sufficient to warrant separation at the family level is not certain, however, and better knowledge of the anatomy of both *Baryonyx* and *Spinosaurus* may eventually lead to the inclusion of *Baryonyx* in the family Spinosauridae. For the time being, it may be sufficient to regard the Spinosauridae and the Baryonychidae as two closely related families sharing some peculiar adaptations not usually encountered among theropods."

We should therefore compare *Baryonyx* and *Spinosaurus* in detail, thus evaluating Buffetaut's findings. (In any case, all other possible relationships seem even less promising.) It is important to note, however, that *our* comparison of *Spinosaurus* – unlike the comparisons made by Paul and Buffetaut – is based only on the Egyptian material described by Stromer (1915, 1934). As stated above, we cannot accept that any of the other material described by Taquet (1984), Bouaziz et al. (1988), and Buffetaut (1989) is properly referred to *Spinosaurus*, or even to the Spinosauridae. Our view of the affinities of that material is presented in detail below.

The Egyptian material of *Spinosaurus* (which, of course, includes the holotype) consists only of incomplete remains. Comparison with *Baryonyx* reveals that potential synapomorphies are comparatively few and unconvincing, whereas the differences between the two genera are many and far more impressive.

Spinosaurus is characterized in particular by extremely long neural spines on the vertebrae, presumably indicating a humped back. Only one neural spine of *Baryonyx* has been found, not attached to a vertebra, but that too is clearly elongated (suggesting that *Baryonx* likewise may have had a hump). Buffetaut did not know of this new find – indeed, he had been mislead by a reconstruction of the skeleton of *Baryonyx* (Milner, Croucher and Charig 1987) into thinking that the neural spines in that genus were not elongated. On the other hand, that one isolated neural spine – complete at its upper end – is not nearly as long as those on the dorsal vertebrae of *Spinosaurus*. Likewise the neural spines on the cervical vertebrae of *Baryonyx* are so short as to be, in some cases, virtually nonexistent, whereas the one cervical neural spine of *Spinosaurus* figured by Stromer (1915), though broken off above, is still as high as the length of the centrum. In any case, the neural spines of the vertebrae are elongated in many tetrapods that are obviously not closely related at all (e.g., *Dimetrodon, Edaphosaurus, Ctenosauriscus,* "*Altispinax,*" *Acrocanthosaurus, Ouranosaurus,* some modern chameleons, and several mammals). This character, therefore, can have but little taxonomic significance. Indeed, in certain living forms it is a manifestation of sexual dimorphism, and the same may well be true of fossil tetrapods.

The two genera show some similarity in the general shape of the symphyseal end of the dentary (both in the lateral view, which shows the end bent upwards, and in the occlusal view). On the other hand, the dentary of *Baryonyx* is relatively much longer and shallower than that of *Spinosaurus* and contained twice as many teeth.

The lower jaws of both genera have alveoli that are more or less circular in occlusal view and seem to lack interdental plates (other theropods, such as megalosaurids and tyrannosaurids, possess more rectangular alveoli with interdental plates between them). The same is true of the upper jaw of *Baryonyx*, but unfortunately the original Egyptian material of *Spinosaurus* does not include any significant part of either premaxilla or maxilla.

On the other hand, the alveoli in the dentary of *Spinosaurus* (Stromer 1915, Tafel 1, Fig. 12) vary greatly in size and, in many cases, are well separated from the adjacent alveoli. Except for those in the expanded anterior tip, their walls produce perceptible bulges in the lateral margin of the dentary itself. In contrast, the alveoli of *Baryonyx* behind the "terminal rosette" are all of roughly equal size (the last few, however, diminish rapidly towards the back of the jaw), are all contiguous with their neighbors, and are not accompanied by lateral bulges.

In both genera the laterally expanded symphyseal end of the dentary bears five teeth, the largest being the 3rd and 4th. On the other hand, the respective numbers and distribution of the teeth in *Baryonyx* and *Spinosaurus* present an even more cogent argument for the systematic separation of the two genera into different families. The lower teeth of *Baryonyx* (on the evidence of the alveoli in the dentary) are more than twice as numerous as those of *Spinosaurus* (32 contrasted with 15, so that, as both have 5 teeth in the expanded tip, the straight posterior part of the dentary bears 27 in *Baryonyx* and 10 in *Spinosaurus*). Nevertheless the density of distribution of the upper teeth in *Baryonyx* confirms that their total number must have been more or less normal. *Baryonyx* is thereby the possessor of that unique feature, a lower dentition that has twice as many teeth crowded into the same length of jaw as does the upper dentition.

Both genera possess teeth with relatively long and slender roots. It is difficult to determine how unusual this condition is among theropods, for in most of the other theropod material that we were able to examine, only the crowns were preserved. In both genera too, the crowns are less flattened labio-lingually than in most other theropods.

On the other hand, the teeth of *Baryonyx* are recurved to much the same extent as those of most other theropods, whereas those of *Spinosaurus* are scarcely recurved at all or only very slightly so. And the teeth of *Baryonyx* are finely serrated along the anterior and posterior keels and fluted on the lingual surfaces of the crowns, whereas those of *Spinosaurus* have neither serrations nor fluting (Stromer 1915). It might be argued, of course, that both genera have reduced the size of the serrations, *Baryonyx* to a very small amplitude and *Spinosaurus* – apparently – to the point of complete disappearance.

We summarize our conclusions as similar to Buffetaut's, though substantive rather than provisional. All in all, the characters shared by *Baryonyx* and *Spinosaurus* suggest a phylogenetic relationship between them which, though close, is not sufficiently close to justify their being placed together in the same family. The differences between them in our opinion carry far more weight.

Comparison of *Baryonyx* with "*Altispinax*"

Baryonyx can be compared with a specimen from the English Wealden, possessing remarkably high neural spines on the dorsal vertebrae [BM(NH) Palaeontology Department R.1828]. It was originally described by Owen (1857, pp. 4–7) who, surprisingly, had already figured it without description in an earlier work (1855, pl. 19), where he referred it to *Megalosaurus bucklandi*. Later authors believed that it should be placed in the inadequately based genus "*Altispinax*" (see Steel 1970, pp. 33–35, for details of the complicated history). Not only has "*Altispinax*" been considered by some as member of the Spinosauridae, but the material in question was found in the same formation as *Baryonyx*, albeit at a probably higher level. However, there seems to be no particular similarity between the single known elongated neural spine of *Baryonyx* and the three preserved neural spines of "*Altispinax*." The latter are much longer, much heavier, with enormous lateral thickenings towards their upper ends, whereas the *Baryonyx* spine is slender and flat.

Classifications of the Theropoda

Our attempts to show a particular relationship of *Baryonyx* to *Spinosaurus* and/or *"Altispinax"* having proved fruitless, the only other thing we can try to do is to fit it more generally into a classification of the Theropoda. The problem, however, is exacerbated by the fact that no satisfactory classification of the Theropoda has ever been put forward. Older text-books (e.g., Romer 1956, 1966) retain the classic division of the group into the small, lightly built Coelurosauria and the large, heavily built Carnosauria. But it has long been realised by all present-day workers that this distinction is artificial, so that Carroll's text-book (1987) simply divides the Theropoda into families.

Walker (1964) suggested that the Carnosauria should be divided into the Megalosauroidea and the Tyrannosauroidea (the latter infraorder including the Ornithosuchidae) and he suggested also that those two categories had originated separately from the Pseudosuchia. Colbert and Russell (1969) introduced a third category, Deinonychosauria, into their classification, but that seems to offer but little improvement to the standard arrangement. Barsbold (1983) divided the Theropoda into as many as seven separate infraorders, the additional ones being Oviraptorosauria, Ornithomimosauria, Deinocheirosauria, and Segnosauria. He did not even try to suggest as to how they might be related to each other except in so far as his phylogenetic tree shows the Coelurosauria of the Late Triassic sending out broken lines in all directions – first to the Ornithischia, then to the birds, and then (simultaneously) to all six of the other theropod infraorders.

Incidentally, the subjective nature of dinosaur classification is emphasized by the fact that, while all other authorities consider the "odd man out" of the three main dinosaur groups to be either the Ornithischia (e.g., Gauthier in 1984 and 1986) or the Theropoda (e.g., Bakker and Galton in 1974), Barsbold (1983) seems to believe that the Sauropodomorpha were the first group to split off from the common stock. Thus all three possible arrangements have been put forward fairly recently. Theropod classification is rendered even more difficult by the uncertainty of the generic names upon which it is based. For example, *Megalosaurus* Buckland, 1824, is the first generic name ever given to a theropod (indeed, to any dinosaur). It has been used extensively for theropods found in many widely separated parts of the world and throughout the whole of the Jurassic-Cretaceous succession. It has been made the basis of a family name and of an even higher taxon (infraorder Megalosauroidea Walker, 1964). Yet it is based on grossly inadequate, indeterminate material. The problem is presently being studied by H. P. Powell of Oxford and S. P. Welles of Berkeley.

Gauthier's classification

The most recent classification of the Theropoda is a novel arrangement proposed by Gauthier (1984, 1986) based on a detailed character distribution analysis in a wholly "cladistic" approach where no paraphyletic taxa are permitted. Unlike many cladistic classifications, however, it avoids the ranking of taxa. (Gauthier's two works also introduce an entirely new concept, the "meta-taxon," discussion of which is unnecessary here.) The amount of work that has gone into his analysis is truly remarkable by any standards, with the literature lists citing 440 and 281 titles respectively. Furthermore, it must be said that the 1986 paper is, as far as we know, the first "cladistic analysis" within our field of study that attempts to justify synapomorphies by noting contrasting conditions in the outgroups, a practice that can only be admired and should certainly always be emulated. [It should be noted that we have taken cognizance of Panchen and Smithson's recent comments (1987, p. 343) concerning the use of the words "synapomorphy" and "autapomorphy". A *synapomorphy* is unequivocally a shared derived character-state that links two or more sister-taxa together into a larger taxon. Most authors, however, have also used the word to mean the derived character-state that typifies the single larger taxon and contrasts with a different condition in the outgroups. It is, of course, the same character-state, but when the component taxa are not mentioned it should properly be called an *autapomorphy*.] On the other hand, Gauthier's work has its flaws, some inherent in all cladistic classifications and others peculiar to his particular approach.

It has been pointed out (Charig 1981, 1982) that the establishment of any "evolutionary" classification (i.e., one in the tradition of Simpson and Mayr) and any "phylogenetic" classification (in the tradition of Hennig) requires two essential and almost wholly unrelated components. (We place those two conventional terms in quotation marks because it seems to us that their proper meanings are virtually interchangeable.). The first is an attempt to ascertain the phylogeny, which is an objective reality. The second is the translation of that phylogeny into an hierarchical classification, which is a subjective procedure in the case of the Simpsonian classification, but is more or less automatic in the case of the Hennigian. Although the phylogeny itself is an objective reality, a record of events that actually happened, its reconstruction in both cases is a subjective procedure. It requires the taxonomist to judge which shared characters constitute good useful synapomorphies and which are homoplastic, i.e., caused by such phenomena as parallelism, convergence, and reversal. Furthermore, the hierarchical classifications produced by cladists – each by a simple, direct verbalization of the cladogram or phylogenetic tree – suffer from the self-imposed prohibition of paraphyletic taxa (taxa that exclude some of their own descendants) in so far as such prohibition automatically prevents the naming and use of all ancestral taxa, all "stem taxa."

This is most easily illustrated by a simple example. Almost everyone agrees that dinosaurs gave rise to birds. The cladist, in consequence, is obliged by the tenets of his creed to consider that birds themselves are dinosaurs. So what does one call the *other* dinosaurs, the dinosaurs that are *not* birds but which include the direct ancestors of birds? One often needs to refer to that particular group of animals, yet the cladist is not allowed to give them a name. Some cladists have suggested a compromise to circumvent that difficulty when such reference is unavoidable, that is to use only "informal" names, e.g., "reptiles" but not Reptilia, and/or to use them only in quotation marks. To us, however, this seems to be begging the question. In effect one is erecting two parallel classifications: an "informal" for practical purposes for actual use, and a "formal" to represent nothing more than a verbalization of the cladogram – to be placed, as it were, in a sort of taxonomic museum and to be brought out only for ceremonial purposes.

Other important defects of the cladistic classification are inherent in the fact that nearly every cladist regards his attempt at phylogeny reconstruction as the definitive effort. Yet several different cladists working on the same group will invariably produce several different phylogenies and therefore several different classifications, of which only one (if any!) can be correct. These various new classifications usually require the introduction of a number of entirely new and unfamiliar names, or, even worse, the use of a number of old familiar names in senses entirely different from those to which we are accustomed. Such practices can lead to nothing but confusion. Moreover, each subsequent change in the perceived phylogeny will necessitate a change in the classification. Thus there is a complete lack of two essential requirements of any practical classification: uniformity and stability.

Gauthier's classification certainly suffers from all these problems. He is sufficiently confident in the correctness of his own work (he *does* have a few reservations here and there) to erect a fundamentally new classification upon it, with many new names for suprafamilial taxa. Particularly disturbing is his use of familiar names in totally unfamiliar senses – we instance Archosauria, Pseudosuchia, Ornithosuchia. (His typical "cladistic" practice of including all the birds within such groups as the Dinosauria, the Saurischia, the Theropoda, and the Coelurosauria – unfortunate through it seems to us – is obligatory for a cladist.) Conversely, he coins the new name "Avialae" for what we call Aves. His Aves excludes *Archaeopteryx* and certain other fossil forms, for his stated practice in the case of groups with living representatives is to restrict the name to the "crown group," the clade that includes only the descendants of the most recent common ancestor of the modern forms. Further, he pays no attention whatever to the meaning of names. Thus he places the true crocodiles (both living and fos-

sil) within a taxon that he calls "Pseudosuchia" – which means *false* crocodiles! Finally, some of the new classical names that he proposes (e.g., Maniraptora) and his indications of their etymology contain many errors, although it would be over-pedantic and tedious to detail them here. We must also object to his invention and repeated use of a new word "manal" – meaning, presumably, "pertaining to the hand" – in place of the familiar, correctly derived, and perfectly adequate "manual."

Apart from these problems common to all cladistic classifications, Gauthier's work has its own particular difficulties – mostly connected with his choice of autapomorphies and synapomorphies.

1. Autapomorphies found also in sister-taxon
Gauthier lists autapomorphies for a taxon, and then adds the comment, to some of the alleged autapomorphies, that the same character-state occurs also in the sister-taxon. In other cases, where Gauthier has not added such comment, the same is nevertheless true. This must surely invalidate the use of the character-state concerned as an autapomorphy for the first taxon. Thus, of five character-states listed (1986, p. 42) as autapomorphies for "Pseudosuchia" (*sensu* Gauthier), one of them (no. 2) is qualified as being present "also in Ornithosuchidae" (part of Gauthier's Ornithosuchia, sister-group of his Pseudosuchia) and two others (nos. 4 and 5) are qualified as being present "also in ornithosuchians aside from *Euparkeria*" [which is the only genus that Gauthier (p. 43) places with reservations in the Ornithosuchia!]. Conversely, of six character-states listed (p. 43) as autapomorphies for Ornithosuchia, two (nos. 3 and 6) are noted as being present in some "pseudosuchians." And, although Gauthier does not say so, the same is true of no. 2, "Centra steeply inclined in at least the first four postatlantal cervicals."

2. Synapomorphies found in only a few members of a group
He sometimes gives as a synapomorphy a character-state that is found in only a few members of the taxon concerned (usually early and/or primitive members), presumably on the assumption that that character-state was present in the ancestors of those members that do not possess it, and was subsequently lost by them in the course of evolution. It must be admitted that many systematists, Simpsonian as well as Hennigian, follow similar procedures. Nevertheless such an assumption is based on the prior assumption of a phylogenetic relationship that forms part of the phylogeny that Gauthier is trying to prove, in which case it is logically impermissible to use it as additional evidence for that same phylogeny. In short, the systematist who acts in this manner is arguing in a circle, begging the question. An example of this is the "crocodile-reversed" ankle-joint cited as autapomorphy no. 5 for the Ornithosuchia, the only one left after our comments under 1 and 2. In fact, that char-

acter-state is found only in the Ornithosuchidae and probably *Euparkeria,* being undoubtedly absent in the rest of Gauthier's "Ornithosuchia" – including all the dinosaurs and pterosaurs.

3. Autapomorphies: size- and function-related
Gauthier notes – in our opinion correctly – that the presence of the same apomorphous character-states in two or more specified taxa may sometimes be an indication, not of phylogenetic relationship, but of the common functional requirements of large size and, in some cases, of macropredacious habits too. For example, he states (1984, pp. 221–222; 1986, p. 10) that "Several carnosaur apomorphies listed below are also present in other medium to large theropods such as *Dilophosaurus* and *Ceratosaurus,*" specifying several of those apomorphies (of which we list six below) as size-related and writes that "these attributes are seen in all large saurischians." Indeed, some carnosaur autapomorphies "such as the relatively large skull and short neck, key-hole shaped orbit, enlarged teeth, and some modifications of the bones surrounding the lower temporal fenestra, appear to be related both to size and to macropredaceous habits, because they are also seen in large rauisuchians, ornithosuchids, and erythrosuchians. Further data on allometric and scaling phenomena may eventually collapse some of these carnosaur synapomorphies into a few size-related patterns of growth" (1984, p. 222). However, despite these cautionary comments of his own, Gauthier still includes the cited character-states in his formal list of synapomorphies.

4. Negative character-states used as synapomorphies
He uses negative character-states as synapomorphies to link subordinate taxa together into a larger taxon – a highly undesirable practice, for whereas similar character-states in two or more groups can be compared with each other in order to ascertain whether or not they are truly homologous, the loss of a structure in two or more groups cannot be so compared and may well have taken place independently. (On the other hand, it is perfectly legitimate to use a negative character-state, i.e., an absence, in the sister-group or an outgroup as a contrast to a positive apomorphy.) The same is generally true of reductions, although it may sometimes be possible to homologize reductions to a particular pattern, e.g., a phalangeal formula. Examples of such losses and reductions are Gauthier's autapomorphy no. 5 for "Pseudosuchia" and nos. 1, 4, and 6 for Ornithosuchia.

5. Ill-defined character-states used as apomorphies
He often employees character-states that are vague and not clearly defined. Indeed, in many cases they are incapable of clear definition. "Relatively thick-walled long bones" – relative to what? "Large neural spines and transverse processes" – does "large" mean long, or broad, or stout, or heavy, and in any case how large is "large"? On the contrary, if a character-state is to be regarded as apomorphous for a given taxon, it needs to be defined precisely and to be capable of being clearly contrasted with an opposing state of the same character, one that typifies the sister-group (in the case of an autapomorphy) or an outgroup (in the case of a synapomorphy). It should be noted that such precision and susceptibility to contrast are not required in "diagnosis" (which is really nothing more than a condensed description).

6. Failure to explain rejection of discordant characters
He shows a somewhat casual attitude towards discordant characters, categorizing them as parallelisms, convergences, or reversals when they do not fit in with his own ideas of the phylogeny but not always explaining his reasons for so doing. (In all fairness, it should be stated that he is more virtuous than most other workers in this respect.)

Despite our criticisms of Gauthier's approach, his classification represented (at the time) the most recent and most thorough analysis of the Theropoda. Indeed, no other that merits serious consideration has appeared since. We therefore decided to try to fit *Baryonyx* into his scheme. His primary division of the Theropoda is into Ceratosauria and his new taxon Tetanurae. To which of those two taxa does *Baryonyx* belong? Gauthier makes no formal list of the autapomorphies of the Ceratosauria, but instead he notes (1986, p. 9) that "The initial basis for recognition of the monophyly of this taxon stemmed from Welles's (1984) observation that one specimen referred to *Dilophosaurus* (UCMP 77270) possessed a uniquely modified trochanteric shelf (= modified anterior trochanter: see photograph of *Sarcosaurus woodi* in Charig 1976)." But we, in the Welles paper referred to, could find no mention of UCMP 77270. However, we were later informed (Molnar pers. comm.) that it is the specimen referred to on pp. 88–89 without mention of any number or name, the author considering it to represent a different new genus. Welles, moreover, based his description of *Dilophosaurus wetherilli* (not *wetherelli*, as cited by Gauthier) on a hypodigm (p. 95) consisting only of the holotype (UCMP 37302) and a referred specimen (UCMP 37303); he specifically stated (p. 136) that the holotype possessed no trochanteric shelf, while the femur of the referred specimen seems not to have been preserved. As for Charig 1976, that article contains neither any photographs whatever nor any mention of *Sarcosaurus woodi*; we think that Gauthier probably intended to refer to Charig 1972, which does include a photograph of the pelvis and femur of *Sarcosaurus woodi*, but it shows little (if anything) of the alleged trochanteric shelf. In any case, Gauthier states that "The presence of the trochanteric shelf in only some ceratosaur specimens is perplexing";

further that "Dimorphism in femoral form ... have [*sic*] been attributed to sexual dimorphism." We think it safe, at least for the time being, to ignore the trochanteric shelf as a potential ceratosaur autapomorphy.

The only other alleged ceratosaur autapomorphies mentioned by Gauthier are:

1. The supracetabular shelf is prominent and of a characteristic shape. *Baryonyx* certainly possessed a prominent supracetabular crest (which is the same thing as a supracetabular shelf). This, however, seems to be a plesiomorphous character for all dinosaurs. Nor do we know whether it is of the same shape as that observed by Rowe, Gauthier having failed to specify that shape.

2. The pubis is narrower than in other theropods, aside from birds. No information is available yet on *Baryonyx*, although it may be forthcoming.

3. There is a prominent groove on the ventrolateral side of the fibular condyle of the femur, but this is absent in *Baryonyx*.

4. There is a groove on the proximal end of the medial side of the fibula. (Gauthier actually stated (1984, p. 220; 1986, p. 9) that the groove is on the lateral side, but we have subsequently learned from Gauthier himself and T. Rowe (pers. comm, to R. E. Molnar and thence to us) that this was an error and that the sulcus is on the medial side.) This groove is present in *Baryonyx*.

5. Distal tarsals 2 and 3 are fused to their respective metatarsals. There is no information yet on *Baryonyx*.

(The observation of most of these is attributed by Gauthier to discussions with T. Rowe, who is preparing a more complete discussion of the evidence supporting monophyly of the Ceratosauria.)

Thus it would seem that the evidence for *Baryonyx*'s membership in the Ceratosauria is presently restricted to only three characters, of which no. 1 is indeterminate, no. 3 against, and no. 4 in favor.

The sister-group of the Ceratosauria, according to Gauthier, is his newly named Tetanurae, which includes all other theropod dinosaurs (and the birds). Seventeen autapomorphies are listed for the Tetanurae (Gauthier 1986, pp. 23–26, characters nos. 36–52). We shall cite only six of them here, since for the other ten the relevant parts of *Baryonyx* are as yet unknown.

36. The dentary is without an enlarged fang-like tooth. (Gauthier 1984, p. 256, notes also the absence of a notch between the premaxilla and the maxilla for the reception of an enlarged dentary fang, a character that correlates with the absence of the "fang" itself.) In *Baryonyx* too, there is no such dentary fang. However, we believe that this character is probably plesiomorphous for the Tetanurae, and in any case it is a negative character. We therefore prefer to reject it.

38. The tooth rows (both upper and lower) terminate posteriorly in front of the orbit. They seem to do so too in *Baryonyx*.

39. The axis possesses a spine table, which is absent in *Baryonyx*.

41. The scapula is strap-like, i.e., the distal end of the scapula is not expanded anteroposteriorly. Gauthier, 1984, p. 271, lists "Strap-like scapula" as an autapomorphy (no. 178) for the Coelurosauria. Thus, in his earlier work, he is suggesting that the scapula is not strap-like in the carnosauria, indeed he specificaly states (loc. cit.) that the distal end of the scapula is expanded anteroposteriorly in that group. So how can this character-state be an autapomorphy for the whole of the Tetanurae, linking the Coelurosauria and the Carnosauria together? We prefer to reject it. In any case, it is represented in *Baryonyx* by an intermediate condition, with the upper end of the scapula only slightly flared.

42. The coracoid tapers posteriorly in lateral profile, which is true in *Baryonyx*.

47. The ischium possesses an obturator process, as does *Baryonyx*.

We therefore have a total of six characters that show a different state in Gauthier's Ceratosauria on the one hand and in his Tetanurae on the other, and for which our *Baryonyx* material affords at least some evidence of the relevant character-state in that genus. Three of the characters suggest that *Baryonyx* belongs to the Ceratosauria, and four that it belongs to the Tetanurae. This is hardly conclusive in our attempts to determine positively the phylogenetic relationships of our new genus, and hence its proper position in the classification.

Notwithstanding our failure to achieve any conclusive results by the use of Gauthier's analysis, we have nevertheless thought fit to try it one step further. If we assume, despite the ambivalent indications, that *Baryonyx* is a member of the Tetanurae, we can try to fit the genus into one or other of Gauthier's subdivisions of that taxon. We have performed this seemingly pointless task for two good reasons. First, we wished to give Gauthier's analysis a fair trail by testing *Baryonyx* further against his scheme. Secondly, the two subdivisions in question happen to be the two traditional infraorders of the Theropoda – the Carnosauria and the Coelurosauria – and we were interested in seeing whether *Baryonyx* confirmed the supposed distinction between them.

We shall cite only the following carnosaur autapomorphies listed by Gauthier (1986, p. 10), for the others relate to parts of *Baryonyx* of which we have no knowledge at present.

(Those that Gauthier regards as possibly related to size, perhaps also to macropredacious habits, are marked with an asterisk.)

1. The mandibular symphysis is reduced.
2. The mandibular fenestra is reduced.
3.* The postcranial skeleton is very robust, with stout, relatively thick-walled long bones.
4.* Throughout the vertebral column the neural spines and transverse processes are large.
5.* The cervical and anterior dorsal vertebrae are strongly opisthocoelous.
6. The cervical and dorsal vertebrae are short and stoutly constructed.
7. The ilium is expanded anterodorsally.

His 1984 dissertation (pp. 221–223) includes the additional autapomorphies:

8.* The skull is very large, broad, and high, with a long deep muzzle.

9.   The braincase is short and deep.

10.   The parietals and squamosals are broadly exposed on the posterior surface of the skull.

11.*   The teeth are enlarged.

12.*   The neck is short.

Of these, nos. 1, 3, 5 (cervicals at any rate), 7, and 9 are present in *Baryonyx;* nos. 2, 4, 6 (not short, at any rate), 8, 10, 11, and 12 are not, the opposite condition obtaining in that genus. Thus 5 characters suggest that *Baryonyx* is a carnosaur and seven say that it is not.

For the Coelurosauria, the alleged sister-group of the Carnosauria, Gauthier lists the following relevant autapomorphies (1984, pp. 265–281, character nos. 173–188; 1986, pp. 26–30, characters nos. 53–67).

55.   The cervical ribs are fused to the centra in adult specimens.

56.   The surfaces of the cervical zygapophyses are flexed.

57.   The anterior faces of the anterior cervical vertebrae are broader than deep. The articular surfaces are kidney-shaped, and taller laterally than on the midline.

60, 182.   The forelimb is more than half as long as the hind limb and/or the presacral vertebral column.

63, 185.   The fourth trochanter is feebly developed or absent.

64, 186.   The greater trochanter (= posterior trochanter of Ostrom 1976) is mound-like.

178.   The scapula is strap-like.

179.   The coracoid has a process directed posteroventomedially.

Comparisons of *Baryonyx* are more equivocal here. None of these supposed autoapomorphies is certainly present in the Surrey theropod; nos. 55, 56, and 60 are certainly absent, nos. 178 and 179 show an intermediate condition, and for the rest our present knowledge of *Baryonyx* is insufficient for us to tell.

Thus our comparison of *Baryonyx* with Gauthier's list of coelurosaur autapomorphies tells us little, while our comparison of the genus with his list of carnosaur autapomorphies has produced roughly equal numbers of similarities and dissimilarities. Admittedly our knowledge of *Baryonyx* is far from complete. But even so, we already know more about it than we do about most other carnivorous dinosaurs. Once again the results of our analysis are ambiguous.

Paul (1988) has recently published a more positive view of the systematic position of *Baryonyx*, referring that genus to the family Spinosauridae and placing the family in the suborder Ceratosauria. The book in question, however, introduces a wholly new, idiosyncratic classification that does not even pretend to be based on any phylogenetic framework and often relies on guesses at unknown characters rather than on estab-

Figure 9.7. Restorations of *Baryonyx* in various poses. The discovery of one isolated neural spine, suggesting that the animal possessed a ridge along the centre of its back, was made too late to be considered in preparing these restorations. In any case, the presence of such a ridge is not absolutely certain. (Drawings by John Holmes)

lished fact. Crompton and Gatesy wrote (1989, p. 94) that Paul's modern assessment is almost uninterpretable, lacking, as it does, any reflection of phylogeny.

Conclusions

The only valid conclusions that we feel justified to draw from our investigation are:

1. *Baryonyx* is certainly a theropod.
2. *Baryonyx* seems not to fit into Gauthier's system of classification, which therefore seems not to apply to all theropods.
3. The individual characters found in *Baryonyx* (Fig. 9.7), some of them unique, and its particular combination of characters, indicate unequivocally that it cannot be assigned to any known family of theropods; more specifically, it is not a spinosaurid. We therefore believe that we were justified in placing it in a new theropod family of its own (Baryonychidae Charig and Milner, 1986).

Other material referred to the Baryonychidae

With regard to the two fragmentary snouts from the Aptian of Niger, described by Taquet (1984) as the mandibular symphyses of a spinosaurid, here we can only reiterate our own statement (Charig and Milner 1986) that they are almost identical to the conjoined premaxillae of *Baryonyx* and should therefore be referred to the Baryonychidae rather than to the Spinosauridae.

The maxillary frament from southern Morocco described by Buffetaut (1989) also shows marked similarities to the maxilla of *Baryonyx*, although it is nearly twice as big as the latter. Its general shape is strongly reminiscent of that seen in the Surrey dinosaur. Particular resemblances include the near-horizontal premaxillary-maxillary suture, the peg-like anterior extremity of the bone (which, in *Baryonyx*, fits into a deep notch under the premaxilla), the shape of the dental margin ventrally, the shape of the preserved anteroventral portion of the rim of the external naris (mistakenly referred to by Buffetaut on p. 85 as the antorbital fenestra), the posterior location of the external naris, and all the features of the alveoli and teeth mentioned above except for the absence on the Moroccan teeth of the fine serrations that characterize the keels on the teeth of *Baryonyx*. These similarities leave us in no doubt that Buffetaut's maxilla should be referred to the Baryonychidae as baryonychid gen. et sp. indet.

The material from Gadoufaoua in Niger, from which Taquet (1984) described the two "spinosaurid" jaw fragments, includes undescribed isolated theropod teeth, presently housed in Paris. Examination by one of us (A. C. M.) shows that they possess both serrations and fluting, and there seems to be a strong probability that they too have baryonychid affinities.

If our conclusions are correct, then the known range of the family begins in the basal Barremian of England and extends right up through the Aptian and the Albian and perhaps even into the basal Cenomanian of western and northwestern Africa.

Acknowledgments

Our special thanks are due to Mr. William Walker for finding the original claw-bone and for so kindly donating it to the Museum. We are likewise grateful to Peter Bruckmann, Esq., Chairman of the Board of the Ockley Brick Company Limited, and to his fellow-directors for their interest, for giving the rest of the dinosaur to the Museum and for the gift of a Vacu-blast shot-blasting machine to expedite its preparation. We thank them and their staff (in particular Messrs. Frank Datson and Derek Sturt) for the assistance they gave us in collecting the specimen and for their cheerful sufferance of the inconvenience we must have caused them. We should like to express our gratitude to the dozens of people, both Museum staff and others, who helped us collect the skeleton, to the Museum photographers who recorded the operation and prepared the photographs for this article, to Philip Palmer, who made a detailed study (not yet published) of the Wealden strata at Ockley, and, most of all, to Ronald Croucher and his colleagues in the Museum's Palaeontological Laboratory (William Lindsay, Lorraine Cornish, Gillian Comerford, Adrian Doyle, and David Gray), who are devoting years of skillful work to the development of the *Baryonyx*. Finally we should like to acknowledge the generous assistance given us by Dr. Philip Currie (Tyrrell Museum, Drumheller, Alberta), Dr. Christopher McGowan (Royal Ontario Museum, Toronto), Dr. Philippe Taquet (Museum National d'Histoire Naturelle, Paris) and Dr. Samuel Welles (Museum of Paleontology, Berkeley, California) who let us make comparative studies in their respective institutions; by Dr. Richard Estes, Dr. Ralph Molnar, and Gregory Paul, who made many helpful comments and suggestions; and by John Holmes, who drew the restorations of *Baryonyx* in life. We would not wish to imply, however, that any of these people who have so kindly helped us would necessarily share any of the opinions expressed in this article.

References

Bakker, R. T., and Galton, P. M. 1974. Dinosaur monophyly and a new class of vertebrates. *Nature,* London 248:168–172.

Barsbold, R. 1983 (Published earlier, in Mongolian, in 1982). Khishchnye dinozavri Mela Mongolii (Carnivorous dinosaurs from the Cretaceous of Mongolia). *Trudy sovmestnaya Sovets-Mongol'skaya Paleontologicheskaya Ekspeditsiya* 19:1–120. [in Russian]

Bouaziz, S., Buffetaut, E., Ghanmi, M., Jaeger, J.-J., Martin, M., Mazin, J. M., and Tong, H. 1988. Nouvelles decouvertes de vertebres fossiles dans l'Albien du Sud tunisien. *Bulletin de la Société géologique de France* (8)4, no. 2:335–339.

Buckland, W. 1824. Notice on the *Megalosaurus* or great fossil lizard of Stonesfield. *Transactions of the Geological Society of London* (2) 1 (art. 21):390–396.

Buffetaut, E. 1989. New remains of the enigmatic dinosaur *Spinosaurus* from the Cretaceous of Morocco and the affinities between *Spinosaurus* and *Baryonyx*. *Neues Jahrbuch für Geologie und Paläontologie, Monatshefte* 1989(2):79–87.

Carroll, R. L. 1987. *Vertebrate Paleontology and Evolution* (New York: W. H. Freeman and Co).

Charig, A. J. 1965. Stance and gait in the archosaur reptiles. *Liaison Report, Commonwealth Geology Liaison Office* 6:18–19. [abstract]

1966. Stance and gait in the archosaur reptiles. *Advancement of Science, London* 22:537. [abstract]

1972. The evolution of the archosaur pelvis and hind-limb: an explanation in functional terms. *In* Joysey, K. A., and Kemp, T. S. (eds.), *Studies in Vertebrate Evolution: Essays Presented to Dr. F. R. Parrington, F.R.S.* (Edinburgh: Oliver and Boyd), pp. 121–155.

1976. "Dinosaur monophyly and a new class of vertebrates": a critical review. *In* Bellairs, A. d'A., and Cox, C. B. (eds.), *Morphology and Biology of Reptiles* (Linnean Society Symposium, Series 3) (London: Academic Press), pp. 65–104.

1981. Cladistics: a different point of view. *Biologist* 28:19–20.

1982. Systematics in biology: a fundamental comparison of some major schools of thought. *In* Joysey, K. A., and Friday, A. E. (eds.), *Problems of phylogenetic reconstruction* (Systematics Association, Special Volume 21) (London and New York: Academic Press), pp. 363–440.

Charig, A. J., and Milner, A. C. 1986. *Baryonyx,* a remarkable new theropod dinosaur. *Nature,* London 324:359–361.

Colbert, E. H., and Russell, D. A. 1969. The small Cretaceous dinosaur *Dromaeosaurus. American Museum of Natural History Novitates* 2380:1–49.

Crompton, A. W. and Gatesy, S. M. 1989. A cold-eyed look at a treatise on warm-blooded dinosaurs (review of G. S. Paul's *Predatory Dinosaurs of the World,* 1988). *Scientific American* 260:92–95.

Gauthier, J. A. 1984. *A Cladistic Analysis of the Higher Systematic Categories of the Diapsida.* Ph.D. dissertation, University of California; published 1986 by University Microfilms International, Ann Arbor, Michigan, pp. vii + 564.

1986. Saurischian monophyly and the origin of birds. *In* Padian, K. (ed.), *The Origin of Birds and the Evolution of Flight, California Academy of Sciences,* Memoir 8, pp. 1–55.

Gilmore, C. W. 1920. Osteology of the carnivorous Dinosauria in the United States National Museum, with special reference to the genera *Antrodemus (Allosaurus)* and *Ceratosaurus. Bulletin of the U.S. National Museum* 110:1–159.

Madsen, J. H., Jr. 1976. *Allosaurus fragilis*: a revised osteology. *Utah Geological and Mineralogical Survey Bulletin* 109:i–xii, 1–163.

Milner, A. C., Croucher, R., and Charig, A. J. 1987. *"Claws". The Story (so far) of a Great British dinosaur* Baryonyx walkeri. [London: British Museum (Natural History)].

Ostrom, J. H. 1976. On a new specimen of the Lower Cretaceous theropod dinosaur *Deinonychus antirrhopus. Breviora* 439:1–21.

Owen, R. 1855. The fossil Reptilia of the Wealden and Purbeck Formations. Part II: Dinosauria (*Iguanodon*). (Wealden). *Palaeontographical Society of London, Monograph* 1854:1–54.

1857. The fossil Reptilia of the Wealden and Purbeck Formations. Part III: Dinosauria (*Megalosaurus*). (Wealden). *Palaeontographical Society of London, Monograph* 1855:1–26.

Panchen, A. L. and Smithson, T. R. 1987. Character diagnosis, fossils and the origin of tetrapods. *Biological Reviews* 62:341–438.

Paul, G. S. 1988. *Predatory Dinosaurs of the World* (New York: Simon and Schuster), pp. 1-464.

Romer, A. S. 1956. *Osteology of the Reptiles* (Chicago and London: University of Chicago Press).

1966. *Vertebrate Paleontology,* 3rd ed. (Chicago: University of Chicago Press).

Steel, R. 1970. Saurischia. *In* Kuhn, O. (ed.) *Handbuch der Paläoherpetologie,* (Gustav Fisher Verlag: Stuttgart and Portland, U.S.A.), 14:i–v, 1–87.

Stromer, E. 1915. Ergebnisse der Forschungsreisen Prof. E. Stromers in den Wusten Aegyptens. II:Wirbeltierreste der Baharije-Stufe (unterstes Cenoman). 3: Das Original des Theropoden *Spinosaurus aegyptiacus* nov. gen., nov. spec. *Abhandlungen der Königlischen bayerischen Akademie der Wissenschafte* 3:1–32.

1934. Ergebnisse der Forschungsreisen Prof. E. Stromers in den Wusten Aegyptens. II: Wirbeltierreste der Baharije-Stufe (unterstes Cenoman). 13: Dinosauria. *Abhandlungen der Königlischen bayerischen Akademie der Wissenschafte* (n.f.) 22:1–79.

Taquet, P. 1984. Une curieuse specialisation du crane de certains Dinosaures carnivores du Cretace: le museau long et etroit des Spinosaurides. *Comptes rendus Hebdomadaires des séances de l'Academie des Sciences* 299(2) no. 5:217–222.

Walker, A. D. 1964. Triassic reptiles from the Elgin area: *Ornithosuchus* and the origin of carnosaurs. *Philosophical Transactions of the Royal Society (B)* 248:53–134.

Welles, S. P. 1954. New Jurassic dinosaur from the Kayenta Formation of Arizona. *Geological Society of America Bulletin* 65:591–598.

1984. *Dilophosaurus wetherilli* (Dinosauria, Theropoda): osteology and comparisons. *Palaeontographica A* 185:85–180.

10 Variation in Tyrannosaurus rex

KENNETH CARPENTER

Abstract

Individual variation for the large theropod *Tyranno-saurus rex* may be seen in the maxilla, dentary and ischium. The maxilla is variable in its depth, the size and shape of the maxillary and antorbital fenestrae, and the size and shape of the lacrimal and jugal processes. Even the left and right maxillae of the same skull show variation. Sexual dimorphism is suggested by the presence of two morphs, one more robust than the other. The angle between the ischia and caudals of the robust morph is greater than in the slender morph, and would provide ample space for the passage of eggs. On this basis, the robust morph is considered the female.

Introduction

The large theropod, *Tyrannosaurus rex*, was named by Osborn in 1905 on the basis of a partial skull and skeleton from the Hell Creek Formation of eastern Montana. The holotype (AMNH 973) was later transferred to the Carnegie Museum of Natural History (CM 9380) where it is presently on display. A second specimen (AMNH 5866) from the Lance Formation of eastern Wyoming, was named *Dynamosaurus imperiosus* by Osborn (1905) but later synonymized with *T. rex* (Osborn, 1906). This specimen [BM(NH) R7994 and R7995] is now mounted at the British Museum (Natural History). Numerous additional specimens have since been recovered from the Scollard and Willow Creek formations of Alberta, the Hell Creek Formation of North Dakota, South Dakota, and Montana, and the Laramie Formation of Colorado. These specimens are presently under study by Robert Bakker, Philip Currie, Ralph Molnar, Greg Paul, and myself. As will be demonstrated below, it is doubtful that the specimen reported by Lawson (1976) from the Tornillo Formation is *Tyrannosaurus rex*.

A detailed osteology of *T. rex* has not yet been presented, although Osborn did have a monograph in

In Dinosaur Systematics: Perspectives and Approaches, *Kenneth Carpenter and Philip J. Currie, eds. Copyright © Cambridge University Press, 1990.*

preparation. Preliminary results of Osborn's study were given in a series of short papers (Osborn 1905, 1906, 1912, 1913, 1917). Subsequent papers on *Tyrannosaurus*, in whole or in part, include Romer (1923), Newman (1970), Russell (1970), Molnar (1973), Tarsitano (1983), Bakker et al. (1988), Paul (1988), and Carpenter (in press). Several popular articles have also appeared (Anonymous 1910; Brown 1915; Hill 1983). Except for a brief discussion by Carpenter (in press), none of these papers and articles have examined variation and sexual dimorphism in *Tyrannosaurus*.

Skulls

Six partial and complete skulls are now known for *Tyrannosaurus rex*, including the holotype CM 9380 (formerly AMNH 973), AMNH 5027, LACM 23844, MOR 009, SDSM 12047, and TMP 81.6.1. Two additional specimens, AMNH 5029 and AMNH 5117, are braincases (Osborn 1912). Five of the skulls are shown in Figure 10.1. As may be seen, there is a considerable amount of variation in the size and shape of all the cranial openings (e.g., orbit and lateral temporal fenestra), as well as all the individual elements, including the lacrimal, postorbital, quadratojugal, jugal, and surangular. In fact, no two specimens are identical.

Seven maxillae were available for comparison. Five of these were overlain in Figure 10.2A using the anterior-most margins of the maxilla and maxillary fenestra as standard lengths. Most of the maxillae resemble one another, except for TMM 41436-1, the specimen reported from the Tornillo Formation by Lawson (1976).

Variation can be seen in the depth of the maxilla, size and shape of the maxillary and antorbital fenestrae, position of the lacrimal processes, and the position and shape of the jugal process (Fig. 10.2A). Differences in the depth of the maxilla affect the position of the lacrimal process, which in turn influences the height and shape of the antorbital fenestra. There is also a considerable amount of variation in position of the jugal process that forms the lower rim of the antorbital fenestra, but this does not seem to correlate with the size of the antorbital process. The shape of the maxillary fenestra ranges from almost oval (CM 9380) to subtriangular (LACM 23844 and TMP 81.6.1) to almost square (AMNH 5012). There is little or no difference in the position of the largest teeth and therefore this feature is not affected by the depth of the maxilla or the shape of the antorbital fenestra. Even the left and right maxilla of the same skull show variation (Figs. 10.2B,C).

The maxilla from the Tornillo Formation lacks the anterior-most margin and was scaled to the other maxilla using height and the anterior margin of the maxillary fenestra (Fig. 10.2A). The nasal, or dorsal, margin of the maxilla arcs sharply down suggesting a face considerably shorter than in *T. rex*. Other differences include a slightly larger maxillary fenestra in proportion to the size of the maxilla and a much deeper jugal process. The antorbital fenestra appears to be smaller, but the lacrimal process is too incomplete to be certain. These differences are great enough to suggest that TMM 41436-1 falls out-

Figure 10.1. A comparison of four *Tyrannosaurus rex* skulls. **A**, CM 9380 (formerly AMNH 973); **B**, AMNH 5027; **C**, LACM 23844; **D**, TMP 81.6.1; **E**, SDSM 12047. The reconstruction of CM 9380 differs from that given by Osborn (1906) in the orbital and postorbital regions. Osborn had erroneously used the right ectoptygoid as a right postorbital in the reconstructed skull at the Carnegie Museum. As may be seen in the figure, the squamosal no longer extends into the orbit. TMP 81.6.1 is more complete than shown, but preparation was not complete at the time this figure was made. SDSM 12047 is crushed dorsolaterally, which has greatly affected the snout, no attempt has been made to compensate for this. (Scale bar = 10 cm.)

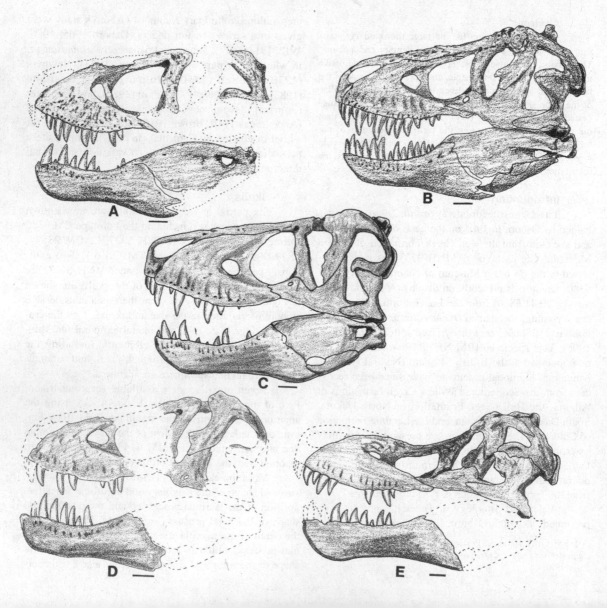

Figure 10.2. **A**, an overlay of maxillae: CM 9380, AMNH 5027, LACM 23844, TMP 81.6.1, and TMM 41436-1. Teeth not shown. **B**, right (reversed) and left maxillae of LACM 23844 and **C**, AMNH 5027.

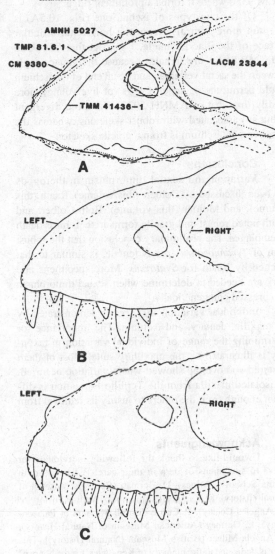

side the range of individual variation for *Tyrannosaurus rex*. It may belong to a new genus, but no name should be proposed until additional material is available. Because it is unusual, it is excluded from discussions on variation.

Variation in the dentary is demonstrated by six specimens (Fig. 10.3). The distance between the anterior tip and the highest part of the dentary were used for the standard length. The greatest difference is the depth of the posteroventral margin of the dentary. It is relatively shallowest in SDSM 12047 and deepest in AMNH 5027 (Fig. 10.3). However, AMNH 5027 also has a tooth row that curves more ventrally than the others and this may have had an influence on the posteroventral margin. The posterior margin of the dentary is thin and easily damaged, although three of the dentaries have complete margins (CM 9380, AMNH 5027, and TMP 81.6.1). The shape of the anterior portion of the external mandibular fossa (expressed as a notch at the lower edge of the posterior margin, see Fig. 10.1) is variable. The teeth seem to show more variation in positional size, but this may be due to how the dentaries were overlain.

Cervicals

Complete cervical series are known for only two specimens, AMNH 5027 and BM(NH) R7994. There is a considerable amount of variation in the shape of the neural spines. Similar variation appears to be present in the tyrannosaur *Albertosaurus libratus*, so the use of neural spine shape in the diagnosis of the tyrannosaur *Daspletosaurus torosus* (Russell 1970) is suspect.

The neck is more robust in BM(NH) R7994 than in AMNH 5027 (Fig. 10.4). This is especially evident in the atlas and its intercentrum, and in the neural spines of cervicals two and three.

Asymmetrical co-ossification of the last cervical and first dorsal in AMNH 5027 has not been observed in any other specimen, and is probably pathological.

Figure 10.3. An overlay of the dentaries of CM 9380, AMNH 5027, BM(NH) R7994 (formerly AMNH 5866), LACM 23844, SDSM 12047, and TMP 81.6.1. Teeth not shown.

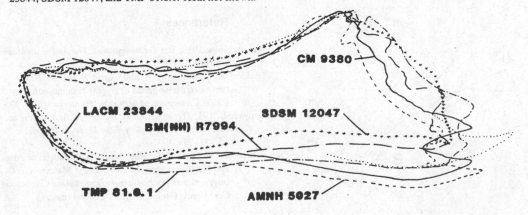

Ischiae

Three ischia were available for comparison. Length was standardized and the articular surface of the iliac peduncle was used to determine orientation in Fig. 10.5. Points of variation include the relative size of the iliac peduncle, the relative size and position of the pubic

Figure 10.4. A comparison of the cervicals of **A**, AMNH 5027 and **B**, BM(NH) R7994. Scale = 10 cm.

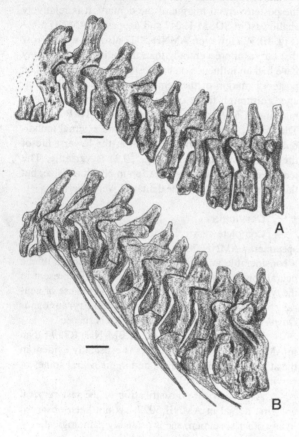

Figure 10.5. An overlay of the ischia of **A**, CM 9380; **B**, TMP 81.61; and **C**, AMNH 5027.

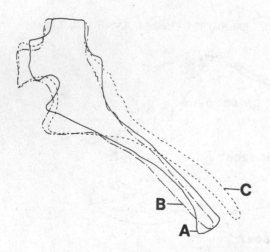

peduncle, the position and size of the obturator process, and the development of the insertion scar for the M. flexor tibialis internus part 3. This scar is best developed in CM 9380 where it forms a prominent ridge.

Of the two types of ischia, one (Fig. 10.5A) is oriented more ventrally from the horizontal articular surface of the iliac peduncle. I suspect that CM 9380 and TMP 81.6.1 are females because the greater angle between the sacral vertebrae and distal end of the ischium would permit the passage of eggs (or live young) more readily than that of AMNH 5027. The more divergent ischia are associated with robust skeletons, whereas the less divergent ischium is from a gracile skeleton.

Conclusions

Variation and sexual dimorphism in theropods has been discussed by Colbert (this volume), Raath (this volume), and Molnar (this volume). Both Colbert and Raath noted robust and gracile forms in their mass death assemblages. The preliminary conclusion that the robust form of *Tyrannosaurus rex* is female, is similar to that reached by Raath for *Syntarsus*. More specimens and work are needed to determine when sexual dimorphism is expressed ontogenetically.

Individual variation in *T. rex* is demonstrated by the maxilla, dentary, and ischium. The importance for determining the range of individual variation in taxonomy is illustrated by the maxillary suite. Most of them clustered and overlap showed where variation occurred. An isolated maxilla from the Tornillo Formation is different enough from the others to justify its removal from *T. rex*.

Acknowledgments

I would like to thank the following individuals for access to specimens or data in their care: P. Bjork (South Dakota School of Mines), D. Berman (Carnegie Museum of Natural History), K. Campbell (Museum of Natural History of Los Angeles County), P. Currie (Tyrrell Museum of Palaeontology), E. Gaffney (American Museum of Natural History), and Angela Milner [British Museum (Natural History)]. This paper is dedicated to the memory of Ken Sauer, for the hours of discussions we had on large theropods, variation, and life.

References

Anonymous. 1910. The *Tyrannosaurus. The American Museum Journal* 10:3–8.

Bakker, R. T., Williams, M. and Currie, P. J. 1988. *Nanotyrannus*, a new genus of pygmy tyrannosaur, from the Latest Cretaceous of Montana. *Hunteria* 1(5):1–30.

Brown, B. 1915. *Tyrannosaurus*, the largest flesh-eating animal that ever lived. *The American Museum Journal* 15:271–279.

Carpenter, K. in press. Tyrannosaurids (Dinosauria) of Asia and North America. *In* Shen, Y. and Mateer, N. (eds.), *International Symposium on Non-marine Cretaceous Correlation* (Beijing: China Ocean Press).

Hill, P. 1983. Haystack Butte surrenders terrible lizard. *American West* 20(2):22–29.

Lawson, D. 1976. *Tyrannosaurus* and *Torosaurus*, Maestrichtian dinosaurs from Trans-Pecos, Texas. *Journal of Paleontology* 50:158–164.

Molnar, R. 1973. The cranial morphology and mechanics of *Tyrannosaurus rex* (Reptilia: Saurischia). Unpublished Ph.D. thesis (Los Angeles: University of California.)

Newman, B. 1970. Stance and gait in the fleshing-eating dinosaur *Tyrannosaurus*. *Biological Journal of the Linnean Society* 2:119–123.

Osborn, H. 1905. *Tyrannosaurus* and other Cretaceous carnivorous dinosaurs. *American Museum of Natural History Bulletin* 21:259–265.

1906. *Tyrannosaurus*, Upper Cretaceous carnivorous dinosaur. *American Museum of Natural History Bulletin* 22:281–296.

1912. Crania of *Tyrannosaurus* and *Allosaurus*. *American Museum of Natural History Memoirs*, new series 1:1–30.

1913. *Tyrannosaurus*, restoration and model of the skeleton. *American Museum of Natural History Bulletin* 32:91–92.

1917. Skeletal adaptations of *Ornitholestes, Struthiosaurus, Tyrannosaurus*. *American Museum of Natural History Bulletin* 35:733–771.

Paul, G. 1988. *Predatory Dinosaurs of the World* (New York: Simon and Schuster).

Romer, A. 1923. The pelvic musculature of saurischian dinosaurs. *American Museum of Natural History Bulletin* 48:605–617.

Russell, D. 1970. Tyrannosaurs from the Late Cretaceous of western Canada. *National Museums of Canada, National Museum of Natural History, Publications in Palaeontology* 1:1–34.

Tarsitano, S. 1983. Stance and gait in theropod dinosaurs. *Acta Palaeontologica Polonica* 28:251–264.

IV Ornithopoda

11 A review of *Vectisaurus valdensis*, with comments on the family Iguanodontidae

DAVID B. NORMAN

Abstract

Vectisaurus valdensis is considered to be the imperfectly preserved remains of juvenile *Iguanodon*. Based on juvenile material of *Iguanodon* recovered from Nehden (Federal Republic of Germany), *Vectisaurus* is most probably referrable to *I. atherfieldensis*.

Ornithopod relationships have been evaluated by reference to the detailed recent systematic reviews of the Ornithischia. Results from this study suggest that the family Iguanodontidae is valid, if more restricted than previously argued; that *Tenontosaurus tilletti* is more closely related to hypsilophodontians than iguanodontians; and, that *Probactrosaurus gobiensis* is the sister-taxon of the Hadrosauridae. A revised cladogram of advanced ornithopod dinosaurs is proposed.

Introduction

In 1879, John Wittaker Hulke described the partial, associated skeleton of a small ornithopod that had been recovered from the Wealden Marls near Brixton (Brighstone) Chine (Isle of Wight, southern England). In the description of this material (Figs. 11.1, 11.2, 11.3), which included five dorsal vertebrae, an anterior caudal centrum, and a partial ilium, Hulke recognised several characters that merited the erection of a new genus and species distinct from the then known ornithopods from the Isle of Wight *(Hypsilophodon foxii* and *Iguanodon mantelli)*. The principal characters by which Hulke distinguished this new taxon were: opisthocoelous dorsal centra with a nearly flat anterior articular surface, a heart-shaped posterior articular surface, a blunt ventral keel on the centra, and the quadrangular appearance of the anterior caudal centrum. Based on the preacetabular process of the ilium, Hulke referred this species to the family Iguanodontidae.

Since Hulke's original article, *Vectisaurus* has received, until relatively recently, little attention in the

In Dinosaur Systematics: Perspectives and Approaches, *Kenneth Carpenter and Philip J. Currie, eds. Copyright © Cambridge University Press, 1990.*

literature relating to the fauna of the Wealden supergroup. Watson (1930) and Swinton (1936) referred a small skull roof fragment found at Yaverland Point (Isle of Wight) to *Vectisaurus*, but Galton (1971) subsequently recognised this as a fragment of a pachycephalosaurian skull, which he named *Yaverlandia bitholus*. Steel (1969) followed Hulke and listed *Vectisaurus* as an iguanodontid, and Ostrom (1970) included it as a valid species in his review of the Wealden fauna. Somewhat later, Galton (1975) discussed this species in a review of British hypsilophodontid dinosaurs and briefly mentioned that another specimen was also referrable to *Vectisaurus valdensis*. This material was later described in some detail by Galton (1976). In recent years, more material has been referred to this species by Huckriede (1982) on the basis of small ornithopod femora and vertebrae collected from a Late Barremian–Early Aptian clay fissure at Nehden (FRG); the latter has been re-identified as juvenile *Iguanodon* sp. (Norman 1987).

Since its description, *Vectisaurus* has consistently been referred to the ornithopod family Iguanodontidae. However, the advent of more rigorous systematic procedures (cladistics) has resulted in a variety of competing character-state trees (cladograms), which affect the classification of ornithopod dinosaurs (Milner and Norman 1984; Norman 1984a,b; Sereno 1984, 1986). The most recent review (Sereno 1986) has provided an explicit statement of both character-states and phylogeny of the Ornithischia, and more particularly of the forms related to *Iguanodon* (Iguanodontia *sensu* Sereno). Sereno proposes that the Iguanodontidae is not monophyletic, and its constituent taxa form a serially derived set of ornithopods that can be informally referred to as "iguanodonts." These occupy a systematic position between the monophyletic Hypsilophodontia and Hadrosauridae.

Recent work on Wealden ornithopods (Galton 1974a, 1975, 1976; Norman 1977, 1980, 1986, 1987) has resulted in a far clearer understanding of the anatomy of these dinosaurs, a factor that impaired the work of Hulke, and permits a long overdue review of the status of *Vectisaurus valdensis*. This paper addresses three specific aims. First, to review the status of the type material *Vectisaurus*. Second, to re-evaluate material that has been referred to *Vectisaurus* more recently, in the light of the status of the type material. Third, to review the status of the family Iguanodontidae, to which *Vectisaurus* has previously been referred.

I. Description of Vectisaurus valdenis Hulke 1879

Ornithischia Seeley 1887
Ornithopoda Marsh 1881
Iguanodontidae Cope 1869
Type species. Vectisaurus valdensis Hulke 1879
Type material. BM(NH) R. 2494–R. 2500.
BM(NH) R. 2494 – dorsal centrum with imperfect neural arch that lacks most of the neural spine (Fig 11.1A);
BM(NH) R. 2495 – dorsal centrum, with eroded neural

Figure 11.1. *Vectisaurus valdensis*, holotype. Dorsal vertebra (BMNH R.2494) in **A**, lateral view; **B**, dorsal view; **C**, anterior view. Dorsal vertebra (BMNH R.2495) in **D**, lateral view; **E**, ventral view; **F**, posterior view. Dorsal neural arch (BMNH R.2498) in **G**, lateral view; **H**, dorsal view. Dorsal neural arch fragment (BMNH R.2499) in **I**, lateral view; **J**, posterior view. Dorsal vertebra (BMNH R.2500) in **K**, lateral view; **L**, ventral view; **M**, posterior view. Caudal centrum (BMNH R.2496) in **N**, lateral view; **O**, posterior view; **P**, anterior view; **Q**, ventral view. (Dotted lines indicate reconstructed areas, cross-hatching indicates broken areas.)

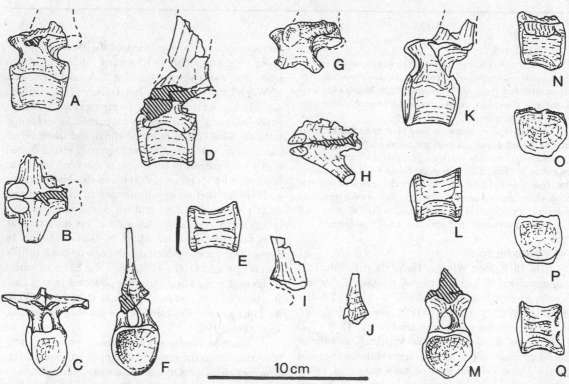

Figure 11.2. *Vectisaurus valdensis*. Type specimen. Ilium (BMNH R.2497) in **A**, lateral; **B**, medial; **C**, dorsal; and **D**, ventral views. (Dotted lines indicate reconstructed areas, cross-hatching indicates broken surfaces.)

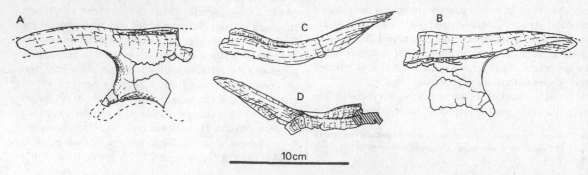

arch including base of the neural spine (Fig. 11.1B); BM(NH) R. 2496 – an isolated caudal centrum (Fig. 11.1F); BM(NH) R. 2497 – the anterior portion of a left ilium, lacking the posterior blade and much of the acetabular margin (Fig. 11.2, 11.3); BM(NH) R. 2498 – isolated base of a dorsal neural arch, lacking spine (Fig. 11.1C); BM(NH) R. 2499 – small fragment of the posterior portion of a neural spine (Fig. 11.1D); BM(NH) R. 2500 – dorsal centrum and base of the neural arch, lacking the major part of the neural spine (Fig. 11.1E).

Type locality. 300 yards east of the flagstaff at Brixton Chine, Isle of Wight, southern England.

Horizon. Wealden Marls (Late Barremian-Early Aptian).

Descriptive comments

All of the type material has been illustrated (Figs. 11.1, 11.2, 11.3) and is briefly redescribed for the purposes of comparison. The dorsal vertebrae are small (40–50 mm long) and seem to come from the middle-posterior part of the series based on the position of the rib-head facets high on the side of the neural arch and near the base of the transverse process. The centra are amphiplatyan, with a slightly greater concavity on their posterior surfaces. The most posterior of the dorsals [BM(NH) R. 2500, identified as such because of its greater width and shorter anteroposterior dimensions] has a well developed, heart-shaped, posterior articular surface that was commented upon by Hulke. However, this feature is not nearly so well seen on the other two centra. The articular margins of the centra are everted, and there is a distinct pinching in of the ventral surface to produce a rounded keel.

The ilium, which lacks much of the posterior portion of the blade and much of the acetabulum, is similarly small (the preserved portion is approximately 190 mm long). The anterior process is long, curved, and quite robust, being reinforced by a ledge of bone on its

Figure 11.3. Lateral views of the ilia of **A**, *Iguanodon atherfieldensis* BMNH R.6462; and **B**, of the type specimen of *Vectisaurus valdensis* BMNH R.2497. (Scale bar = 2 cm.)

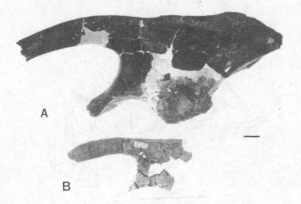

medial surface; the latter is an extension of the area for the attachment of the sacral ribs. The anterior process curves laterally along its length. The dorsal margin of the iliac blade above the acetabulum is rugose and thick, and has a slight inflexion marking the junction between the base of the anterior process and the margin of the main portion of the iliac blade.

The caudal centrum lacks both caudal ribs and the neural arch, which is a consequence of the lack of fusion with the centrum. In lateral view, the centrum as preserved is somewhat angular-sided, with a ventral sulcus separating the anterior and posterior chevron facets, and is somewhat inclined forwards. This centrum appears to be an anterior caudal, judging from the angular nature of the centrum and its anterior inclination. The latter is a feature that has been noted in a variety of ornithopods (Galton 1974a, *Hypsilophodon*; Norman 1980, 1986, *Iguanodon*) and seems to reflect the slope of the tail vertebrae away from the sacrum.

Comparative anatomy

i. Hypsilophodon foxii. The type material of *Vectisaurus valdensis* can readily be distinguished from comparable material of *Hypsilophodon foxii* on the basis of size (*Vectisaurus* is approximately twice the size of the latter species) and on the basis of absolute differences in shape and proportions. The dorsal vertebrae of *Hypsilophodon* lack the prominent ventral keel seen in *Vectisaurus*, have much more prominent zygapophyses, and have significantly smaller neural spines that are broader based and not so posteriorly displaced or inclined. The ilium is also different in form; the anterior process of the ilium of *Hypsilophodon* is long, somewhat variably curved, and narrow (lacking the prominent medial ridge seen in *Vectisaurus*). The dorsal margin of the ilium lacks the prominent longitudinal thickening seen in that of *Vectisaurus*. The angulation of the sides of the caudal centrum and ventral sulcus differ from that seen in *Hypsilophodon*.

ii. Valdosaurus canaliculatus. This is a poorly known species from the Wealden of the Isle of Wight (Galton 1975; Galton and Taquet 1982), but seems to be closely related to the Jurassic genus *Dryosaurus*. It has been characterised by the structure of the femur, which is obviously of no use for comparison with *Vectisaurus*. Referred material, however, includes an ilium [BM(NH) R. 2150; Galton and Taquet 1982], the anterior of which has a concave, longitudinally oriented groove running down its medial surface and therefore is totally different from that of *Vectisaurus*.

iii. Iguanodon spp. Both of the contemporary European species (*I. atherfieldensis* and *I. bernissartensis*) are well known (Norman 1980, 1986, 1987) and can be readily compared to *V. valdensis*.

The dorsal vertebrae of *Vectisaurus* differ in no significant way from those that have been described for small specimens of either species of *Iguanodon*. The ventral keel is a normal feature of dorsals of the latter species. There is also a tendency for the articular surfaces of the centrum to be amphiplatyan with a slightly greater concavity on the posterior surface in *Iguanodon*. In this respect, the slightly greater concavity of the posterior articular surface of the centrum in *Vectisaurus* is interpreted to be due to the relative immaturity of this specimen (i.e., the incomplete ossification of the articular surface). Similarly, the heart-shaped articular surface of the centrum is again interpreted as an immature character, reflecting the incomplete formation of the medial portions of the neural arch pedicles where they attach to the dorsal surface of the centrum. In the midline of the dorsal margin of the vertebral centrum, the lack of contact between adjacent pedicles leaves a "notch" that produces the characteristic heart-shaped profile in end view. The latter feature is also seen in undoubtedly juvenile specimens of *I. bernissartensis* (Norman 1987, Fig. 16).

Comparison between the remains of juvenile *Iguanodon* material from Nehden and *V. valdensis* indicates that the type material of *Vectisaurus* is referrable to *I. atherfieldensis* on the basis of the more slender proportions of the centra. In contrast, the caudal centra of *I. bernissartensis* are more anteroposteriorly compressed and broader, especially in more posterior members of the dorsal series.

The ilium fragment is extremely similar to those of *Iguanodon atherfieldensis* (Norman 1986, Fig. 54). The form of the anterior process (including its lateral curvature and the medial ridge) and base of the main portion of the iliac blade are identical to those of *I. atherfieldensis* (Fig. 11.3).

The anterior caudal centrum is difficult to interpret accurately, because so many of the important structures are missing. However, from what remains, it does not differ significantly from that which might be expected in the caudal of an immature anterior caudal of *Iguanodon atherfieldensis*.

Conclusion

The type material of *Vectissaurus valdensis* Hulke 1879, which has been considered on previous occasions to be a valid genus and species of ornithopod dinosaur, is here regarded as, in all probability, juvenile material of *Iguanodon*, and a junior subjective synonym of *Iguanodon atherfieldensis*.

II. The status of referred material

Apart from the newer material from Nehden that was referred to *Vectisaurus* by Huckriede (1982), Galton (1976) referred to this taxon another partial skeleton [BM(NH) R. 8649] that was both larger and more complete than the type.

Figure 11.5. *Vectisaurus valdensis*. Referred specimen (BMNH R.8649) in **A**, ventral view showing the attached left ilium, sacral ribs and sacral vertebrae (dorsal vertebral centrum "sacrodorsal" later removed when the "Fibronyl" adhesive was dissolved). **B**, Dorsal view of the ilium and sacral ribs of the right side. **C**, Dorsal view of the ilium and sacral vertebrae. (Cross-hatching indicates broken areas, stippled area indicates matrix.)

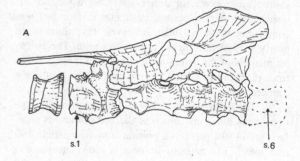

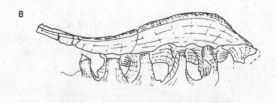

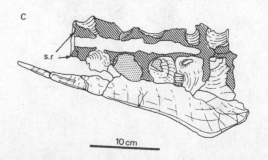

Figure 11.4. *Vectisaurus valdensis*. Referred specimen. Ilium and sacrum (BMNH R.8649), in **A**, left lateral view; and **B**, right lateral view. (Cross-hatching indicates broken areas.)

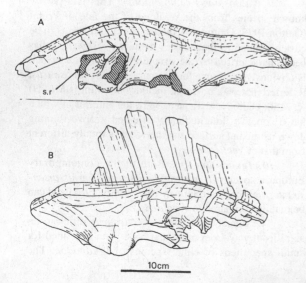

Locality. 100 yards west of Chilton Chine, Isle of Wight, England. Collected by Reginald Walter Hooley in 1916.

Horizon. Wealden Marls (Late Barremian).

Material. Comprises 13 dorsal vertebrae, the sacrum with both ilia (Figs. 11.4, 11.5), portions of the acetabular margins of the pubis and ischium, as well as a variety of broken fragments (mostly ribs and neural spines).

Descriptive comments and discussion

Despite Galton's recent description of this material, it is considered necessary to partially redescribe some of the more relevant parts of the skeleton in order to facilitate comparison and evaluation of the taxonomic position of the specimen.

The associated series of dorsal vertebrae are quite well preserved, although all are somewhat crushed and broken. One of the principal characters used by Hulke (1879) to erect *Vectisaurus* is the unusually deep cupping of the posterior articular surfaces of the centra. This feature is also seen in the vertebrae of BM(NH) R. 8649, although to a lesser degree than in the type specimen. The neural spines of the dorsals are moderately well preserved and tall. In no respects do these characters differ from those seen in the dorsal vertebrae of *Iguanodon*. The relatively tall neural spines suggest that this skeleton may be attributable to *I. atherfieldensis*, because this is a feature that has been noted in the latter species (Norman 1986, 1987).

Both ilia are preserved, but they have been somewhat distorted by vertical crushing (Figs. 11.4, 11.5). The principal features of these ilia are: the anterior process is elongate, laterally flattened, and twisted outward along its length; the medial surface of this process bears a low ridge that arises from a prominent facet for the attachment of the transverse process of the first sacral. The main body of the iliac blade, immediately behind the anterior process, bears a transversely rounded dorsal edge that becomes thicker and more prominently everted posteriorly where it overhangs the ischiadic peduncle. The external surface of the blade is concave vertically, beneath the dorsal edge, although in the preserved specimens this feature has been exaggerated by crushing. The dorsal edge, after becoming thicker, gradually thins and descends to the bluntly pointed hind end. The pubic peduncle is virtually missing in both ilia, being represented only by a broken stump. The acetabulum and the ischiadic tuber are poorly preserved, but were evidently similar in overall shape to those of *Iguanodon*. The posteroventral edge of the blade is inflected posteromedially to produce a narrow brevis shelf beneath the posterior portion of the blade. The internal surface of the ilium bears a prominent dorsomedial ridge on which are preserved the facets for the attachment of six sacral ribs.

When comparison is made between the referred specimen and the type specimen of *Vectisaurus*, the only demonstrable difference in the form of the ilium is in the degree of development of the medial ridge on the anterior process. In specimen BM(NH) R. 2497 (Fig. 11.2) this ridge is proportionately larger. This character, however, is variable within known material of *Iguanodon*, and is therefore not considered to have any taxonomic significance.

Five sacral vertebrae are preserved (Figs. 11.4, 11.5), but one vertebra is missing from the posterior end of the series [there is a sacral rib at the junction between the last preserved sacral and the (lost) last sacral]. Thus, the sacrum appears to have possessed at least six vertebrae, compared with *I. atherfieldensis*, which has seven (including the sacrodorsal) and *I. bernissartensis* with eight (including the sacrodorsal).

There is doubt about the arrangement of the anterior sacral vertebrae in the referred material. The centrum of the dorsal vertebra immediately anterior to the sacrum was artificially cemented to the pedicels of a neural arch, the latter being preserved in natural articulation with the succeeding neural spines of the sacrum. However the centrum of this apparent last dorsal was cemented in place using fibronyl, a heavy jute-alvar paste. Once the cement had been removed, it could be seen that there was no correspondence between the broken surfaces of the centrum and its supposed neural arch. In addition, the supposed first sacral is actually the first true sacral, or the second vertebra of the sacrum following the co-ossified sacrodorsal. Although broken, the anterodorsal margins of the "sacrodorsal" evidently possessed a prominent bony area (Figs. 11.4, 11.5) that coincides with the area for attachment of a sacral rib. This type of arrangement suggests that a sacral rib was attached to the anterodorsal margin of the "sacrodorsal." This implies the presence of a true sacrodorsal vertebra anchored in position by the first sacral rib. It is quite probable that the absence of the sacrodorsal vertebra and the last sacral may be linked to the immaturity of this specimen, because incomplete fusion of the sacrum is seen in even larger individuals of *I. atherfieldensis* (Norman 1986).

Other material attributed to *Vectisaurus* by Galton (1976) included another imperfect sacrum [BM(NH) R. 8650] found on the beach between Barnes High and Cowleaze Chine by R. W. Hooley. This specimen differs in no significant way from the sacra of *Iguanodon*. A jaw fragment [BM(NH) R. 180] is most probably referable to the dryosaurid genus *Valdosaurus* (Norman 1977; Galton and Taquet 1982).

Conclusion

The material that has been subsequently referred to *Vectisaurus* is mostly attributable to *Iguanodon*. The only material clearly not referrable to the latter species is the small jaw [BM(NH) R. 180], which probably belongs to the dryosaurid ornithopod *Valdosaurus canaliculatus*.

III. Comments upon iguanodontian systematics

Sereno (1986) has provided the most recent and comprehensive review of ornithopod systematics and phylogeny. The relationships of the taxa traditionally included within the Iguanodontidae can be summarized briefly. The suborder Ornithopoda (which is determined to exclude both fabrosaurs and heterodontosaurs) consists of two infraorders, the Hypsilophodontia and the Iguanodontia. The former taxonomic group includes traditional hypsilophodonts [*Hypsilophodon, Thescelosaurus, Othnielia, Zephyrosaurus,* and *Yandusaurus* (= *Xiaosaurus*)]. However, *Tenontosaurus,* which has been placed previously within the Hypsilophodontidae (Dodson 1980; Weishampel and Weishampel 1983; Milner and Norman 1984; Norman 1984a,b; Weishampel 1984), is transferred to the Iguanodontia. Ostrom (1970) and Galton (1974a,b) also regarded *Tenontosaurus* as an iguanodontid.

The Iguanodontia is considered a serially derived set of sister taxa to the Hadrosauridae (Fig. 11.6), rather than being divided into a mixture of isolated taxa and discrete families (e.g., Dryosauridae and Iguanodontidae) as has been favored by Norman (1984a), and Milner and Norman (1984).

Certain modifications to the scheme suggested by Sereno (Fig. 11.6) are proposed below and are summarized in Figure 11.7.

Node 1. Dryosauridae + "higher" ornithopods

1.   Contact between premaxilla and lacrimal excludes external contact between nasal and maxilla (independently acquired derived character in heterodontosaurids).

2.   Quadratojugal reduced in size.

3.   Small embayment, or notch, in the quadrate forming the posterior margin of the quadrate foramen.

4.   Prominent primary ridge on lateral surface of maxillary teeth.

5.   Humerus elongate, relatively narrow and slender, with a nonprominent deltopectoral crest.

6.   Anterior ramus of pubis laterally flattened (independently acquired derived character in *Tenontosaurus*).

7.   Shaft of the ischium curved and caudodorsally convex (independently acquired derived character in the hypsilophodontian *Rhabdodon*).

8.   Ischium with expanded (i.e., footed) distal end.

Characters that are not fully substantiated include:

a.   Well developed anterior intercondylar groove on distal end of femur.

b.   Proximally positioned obturator process (possibly primitive for all ornithischians based on material referred to *Lesothosaurus* by Santa Luca 1984).

c.   Occiput narrower than the orbital width of the skull (possibly primitive for a wide range of ornithopods).

d.   Prominent and transversely thick lesser (anterior) trochanter of the femur.

e.   Posterior dentary alveoli medial to the coronoid process of the mandible.

(?) Node 2. *Camptosaurus* + "higher" ornithopods

As indicated in the diagram (Fig. 11.7), the status of this node is dubious, and is probably more accurately portrayed as an unresolved trichotomy between the Dryosauridae, *Camptosaurus,* and "higher" ornithopods. Under such a scheme, all of the characters of Node 2 would be included in Node 1.

1.   Dentary tooth crowns, with a medial (lingual) surface that has a distally offset and low primary ridge separated by

Figure 11.6. Cladogram of iguanodontian ornithopods abstracted from Sereno (1986). Nodes identified in the text.

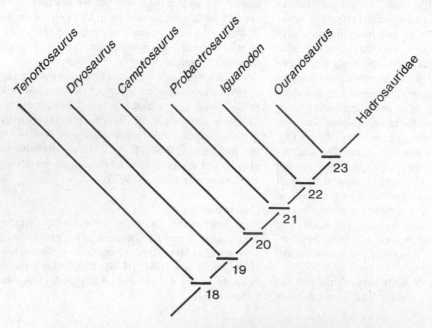

a shallow vertical trough from a low, and broad, secondary ridge. A small number of tertiary ridges developed from the bases of the marginal denticles [also partially developed in *Dryosaurus* (Galton 1983)].

2. Partial fusion of carpals into two blocks, one associated with the distal end of the radius, the other with the distal end of the ulna, in order to facilitate quadrupedal locomotion (condition unknown in *Dryosaurus*).

3. Modification of the manus digit I into a defensive spur by incorporation of the metacarpal into the carpus, the shortening of the first phalanx, and conversion of the ungual into a spur-like structure (condition unknown in *Dryosaurus*).

4. Transversely flattened and deep prepubic process with no distal expansion (parallel development also found in *Tenontosaurus* and partially developed in *Dryosaurus*).

Characters that are not fully substantiated include:

a. Small size of antorbital fenestra (possibly affected by scaling).

b. Development of opisthocoely in cervical and anterior dorsal vertebrae (possibly affected by scaling).

Node 3. Iguanodontidae + Probactrosaurus + Hadrosauridae

1. Supraoccipital excluded from the dorsal margin of the foramen magnum.

2. Rod-shaped caudolaterally directed processes on the sternal bones.

3. Metacarpals II–IV closely appressed, with metacarpals III and IV subequal in length and metacarpal II shorter. Phalanges of digits II–IV shorter than phalanx 1, unguals modified as hooves for locomotion. (It is of course possible to list each item as individual characters to support this node. However, I regard them as an integral component of a character complex.)

4. Distal expansion of the anterior ramus of the pubis.

5. Reduction in length of posterior ramus of the pubis.

(Parallel development occurred in *Tenontosaurus*. There is also the distinct possibility that this character is size-related and reflects the migration of abdominal as well as femoral adductor and retractor musculature from the pubis to the adjacent ischial shaft.)

Characters that are not fully substantiated include:

a. Strong opisthocoely in the cervical and anterior dorsal vertebrae. This feature is probably size-related, because this would appear to be a mechanical modification to support the vertebral column in larger bipeds. This trend is also seen in large theropods.

b. Rhomboidal lattice of ossified tendons (size-related).

Node 4. Probactrosaurus + Hadrosauridae

1. More than one replacement tooth in each of the maxillary and dentary vertical tooth rows.

2. Symmetrical ridging on either side of a median primary ridge in the dentary tooth crowns (possible reversion to the situation seen to some extent in dryosaurs and hypsilophodontians).

3. Straight ischium (possible reversion to the condition seen in hypsilophodontians, and also seen to some extent in *Ouranosaurus*).

Character that has not been fully substantiated:

a. Loss of metatarsal I.

Node 5. Hadrosauridae

The character listing for the Hadrosauridae is not complete, but includes the following: skulls with a modified pleurokinetic skull (pre-orbital portion of the joint between the jugal and lacrimal, instead of the lacrimal and prefrontal as seen in hypsilophodontians, dryosaurs, camptosaurs, and iguanodontids); tooth replacement with multiple replacement teeth in each alveolus and crowns firmly cemented together to form a rigid battery; metacarpal I and associated digit absent;

Figure 11.7. Cladogram of iguanodontian ornithopods indicating the pattern of relationships discussed in the text of this article.

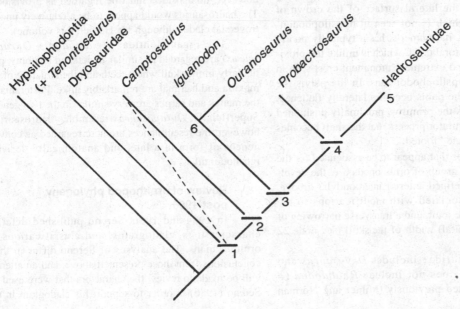

carpals reduced to two small rounded ossifications; humerus sigmoidally curved and with a prominent deltopectoral crest (reminiscent of hypsilophodontians); antitrochanter on ilium; fully enclosed anterior intercondylar groove on femur.

Node 6. Iguanodontidae

1. Maxilla with a well developed jugal process that projects obliquely caudoventrally from the lateral body of the maxilla.
2. Prominent lacrimal process of the maxilla forms the rostral margin of the antorbital fenestra.
3. Lacrimal block-like and bearing a distinctive pattern of depressions on its internal surface, and forming the lateral wall of the tubular passage that leads to the antorbital fenestra.
4. Quadratojugal small, overlain by the jugal rostrally, and embayed caudally where it forms the rostral margin of the quadrate foramen. [Also present in *Dryosaurus* (Galton 1983); position in *Camptosaurus* not clear.]
5. Enlarged conical pollex ungual with first phalanx reduced to a flattened disc.

Ornithopod phylogeny summarized

Among ornithopods the hypsilophodontians and iguanodontians are closely related, but distinct, groups. Considering iguanodontians alone, which comprise the dryosaurids, camptosaurs, iguanodontids, and hadrosaurids, several anatomical changes unify them. These include: modifications to the cheek region of the skull through the development of a contact between premaxilla and lacrimal, associated with the simplification of the pleurokinetic hinge system (Norman 1984b), and an associated reduction in the size and mechanical importance of the quadrotojugal in the coupling across the jugal arch, with concommitant posterior migration of the quadrate foramen into a notch in the jugal wing of the quadrate. There is also the development of a strong primary ridge on the lateral surface of the crown of maxillary teeth, which is not seen in hypsilophodontians. Postcranially, the humerus has a typically poorly developed deltopectoral crest, which is unlike the proximally positioned and extremely prominent crest seen in the majority of hypsilophodontians. In the pelvis the anterior ramus of the pubis becomes laterally flattened. The ischium retains the primitive proximally positioned and leaf-shaped obturator process, but the shaft becomes noticeably curved and "footed."

Other features that appear to be associated for the most part with this group of ornithopods are: the development of a well-defined anterior intercondylar groove, which may be associated with modifications to the structure of the knee joint, and a transverse narrowing of the posterior (occipital) width of the skull (see node 22 below).

The Dryosauridae includes *Dryosaurus* and *Valdosaurus*, but does not include *Rhabdodon* (= *Mochlodon*) as stated previously (Milner and Norman

1984), which appears to be a derived hypsilophodontian.

The genus *Camptosaurus* was previously included in the family Iguanodontidae (Milner and Norman 1984), but has now been excluded and occupies an uncertain position either as the sister taxon of the Iguanodontidae + *Probactrosaurus* + Hadrosauridae (the favored position in this account), or as a basal iguanodontian with the dryosaurs. The skull of *Camptosaurus* is relatively long and low when compared with that of dryosaurids, and the teeth closely resemble those of iguanodontids. The changes in the construction of the manus and carpus, which are associated with defence and locomotion, also resemble iguanodontids. Unfortunately, the detailed structure of the manus and carpus are not known in dryosaurids (Galton 1981). The anterior ramus of the pubis is deeper than in dryosaurids, although the posterior ramus is unshortened.

Iguanodontids, *Probactrosaurus*, and hadrosaurids show a reinforcement of the occipital region of the skull through the exclusion of the supraoccipital from the foramen magnum. Postcranially, the sternal bones develop rod-shaped posterior processes, and the manus becomes narrower, with the metacarpals of the middle three digits (II–IV) becoming bound together to form a weight supporting unit (paralleling the structure of the pes), and the digits terminating in flattened hooves. In the pelvis, the anterior pubic ramus develops an expansion distally, while the posterior ramus is significantly shorter than the ischium.

The critical feature that ties *Probactrosaurus* to hadrosaurids is the development of an incipiently hadrosaurid-like tooth replacement pattern, and modification of the crowns of the teeth of maxilla and dentary into similarly shaped, symmetrical, narrow, diamond-shaped forms. Until further work is done on *Probactrosaurus*, however, this position must be regarded as provisional. The hadrosaurids would appear to be a relatively uncontroversial clade (although see Horner, this volume).

The Iguanodontidae (*Iguanodon* + *Ouranosaurus*) are regarded as similar in detailed anatomy, particularly in the skull where teeth and the structure of the maxilla and lacrimal are remarkably alike. Postcranially, the manus and carpus are very similar in the two genera. Superficially, *Ouranosaurus* resembles hadrosaurids; however, the resemblances are here regarded as convergences of form in a late, and anatomically derived, iguanodontid.

Review of ornithopod phylogeny, post-1986

In 1984 and 1986, Sereno published detailed character lists, cladograms, and classifications of ornithischians. The analyses of Sereno differ in their conclusions from those presented above, and an attempt will be made to review the characters that were used by Sereno (1986) in order to support his cladogram in the light of the revision proposed above. A small section of the original cladogram produced by Sereno is reproduced here (Fig. 11.6) and the characters used to support each of his numbered Nodes are discussed briefly.

Node 18 (Fig. 11.6). Iguanodontia
Tenontosaurus + *all higher*
(non-hypsilophodontian) ornithopods

i. "PM teeth absent." The absence of premaxillary teeth is consistent within the iguanodonts here considered, and premaxillary teeth are present, at least primitively, within hypsilophodontians. However, it should be noted that premaxillary teeth are lost within disparate groups of ornithischians (loss is seen in ceratopsians, ankylosaurs, and pachycephalosaurs), as well as within the hypsilophodontians (including *Rhabdodon* and *Tenontosaurus*). Therefore, the loss of premaxillary teeth cannot be considered to be a compelling unifying character.

ii. "Leaf-shaped denticles." This character refers to the irregular margin that can be observed upon the denticles on mesial and distal edges of tooth crowns. I have doubts about my interpretation of this character because it appears in Node 20 iii. Small crenellations can be seen upon the marginal denticles in large dentary or maxillary crowns, while they are absent from small crowns in *Rhabdodon*.

iii. "Strong primary ridge on the medial side of the D crowns." A strong primary ridge is developed upon the crowns of dentary teeth in a variety of hypsilophodontians: *Hypsilophodon*, *Parksosaurus*, and *Thescelosaurus* (Weishampel pers. comm.), as well as *Tenontosaurus* (Ostrom 1970) and *Rhabdodon*. This character is not therefore of use in defining a more exclusive group of ornithopods as proposed.

iv. "Enamel restricted to the distal half of the crown on the medial side of M teeth and the lateral side of the D teeth." Enamel is found *predominantly* on the medial side of dentary and lateral side of maxillary teeth in the majority of ornithopods (heterodontosaurs, *Hypsilophodon*, *Tenontosaurus*) as well as higher ornithopods, but is retained as a thin veneer on the lateral surface of the dentary and medial surface of maxillary crowns in all but hadrosaurids. (Character repeat: Node 19).

v. "Eversion of ventral PM margin." This is not proven in *Dryosaurus* (Galton 1983), but is exhibited in *Tenontosaurus*, which is classified as an hypsilophodontian (Dodson 1980; Weishampel and Weishampel 1983; Milner and Norman 1984; Norman 1984b; this paper). I would argue that the eversion of the ventral margin of the premaxilla is more likely to be a function of size than of phylogeny: a larger herbivore (such as *Tenontosaurus* when compared to most other hypsilophodontians) is likely to develop a larger cropping device, and would have therefore a broader beak.

vi. "External opening of the antorbital fossa is relatively small or entirely absent." The hypsilophodontian *Parksosaurus* shows a small opening, as does *Hypsilophodon* when compared to forms such as *Heterodontosaurus* and *Lesothosaurus*. (Character repeat: Node 20)

vii. "External nares enlarged relative to the orbit." The small *Dryosaurus* retains a small narial opening.

However, of greater importance is the fact that this character is likely to scale with the relative size of the animal. It is an allometric fact that larger skulls have relatively smaller orbits. Factors governing the size of nares vary, but the net effect will be for large skulls to display relatively larger nares.

viii. "M with paired anterior prs.: primitive anteromedial pr. and new anteroventral pr. that laps the PM palate ventrally." Neither *Tenontosaurus* (Weishampel pers. comm.) nor *Dryosaurus* (Galton 1983) possess a paired anterior process on the maxilla. This character may be of use for higher ornithopods (although see Horner this volume).

ix. "QJ reduced in size relative to the Q." *Tenontosaurus* has a large quadratojugal and is herein regarded as an hypsilophodontian.

x. "D with parallel dorsal and ventral borders." Operationally, this is difficult to justify as a usable character. For example Galton (1983, Fig. 11.5) illustrates two skulls of species of *Dryosaurus* one of which apparently has parallel margins to the dentary, the other has not. The configuration of the margins of the dentary are more likely to be governed by biomechanical factors than phylogenetic ones.

xi. "PD with paired ventral prs." Consistent with the taxa, but this feature is present in *Rhabdodon*, as well as *Tenontosaurus*, which are here regarded as hypsilophodontians, but by Sereno (1986) as basal iguanodonts.

xii. "Denticulate PD bill margin." Consistent with the taxa, but *Tenontosaurus* is here regarded as an hypsilophodontian.

xiii. "Manus digit III with only three phalanges (one phalanx absent)." In those forms that have well preserved manuses, this holds true.

xiv. "Femur with a weak anterior intercondylar groove and a deep posterior intercondylar groove." This character appears to define the form of the distal end of the femur in hypsilophodontians generally [including *Tenontosaurus* (Dodson 1980)].

Node 19 (Fig. 11.6). Dryomorpha
(Dryosaurus + *all other ornithopods*)

i. "M crown narrower anteroposteriorly than the opposing D crown." This is a marginal decision in the case of both dryosaurids (Galton 1983) and *Tenontosaurus* (pers. obs.), and only becomes clear in *Camptosaurus* and *Iguanodon*.

ii. "Lateral M primary ridge stronger than the medial D primary ridge." Consistent for all higher (non-hadrosaurid) ornithopods.

iii. "Diamond shaped M and D tooth crowns; M crowns with rounded anterior and posterior corners." The character is difficult to assess in practice, even though it may sound straightforward as stated. The teeth of *Tenontosaurus* seem to show the characters described here, as do those of several other forms including *Hypsilophodon*.

iv. "Enamel absent from the medial side of the M teeth and the lateral side of the D teeth." Enamel is not completely localized until the hadrosaurid level of organization, where adjacent and successive crowns are cemented together (repetition of Node 18iv).

v. "Space separating the ventral margin of the QJ from the jaw articulation." The quadratojugal is clearly separated from the jaw articulation in *Hypsilophodon* and *Tenontosaurus* among other non-dryomorphan ornithopods (Weishampel 1984, pers. comm.).

vi. "Ischial shaft round in cross section; transversely compressed distally (dorsoventrally compressed distal blade absent)." This complex of characters do not seem particularly characteristic of any particular group of ornithopods. A rounded cross-section of the ischial shaft is seen in *Tenontosaurus*, as well as in such forms as *Lesothosaurus* and *Scutellosaurus*. A straight and laterally compressed distal shaft is also seen in *Hypsilophodon*, *Othnielia*, and *Thescelosaurus* among hypsilophodontians generally.

vii. "Distal shaft with a moderate foot." This is consistent with other taxa (Norman 1984).

viii. "More proximally positioned obturator process." This is consistent with other taxa (Norman 1984). *Lesothosaurus* has been reported as having a proximally positioned obturator process (Santa Luca 1984).

Node 20 (Fig. 11.6). Ankylopollexia

i. "Close packing along the tooth row and in the replacement series eliminating spaces between the bases of the crowns of adjacent functional teeth." This feature would appear to have been illustrated by Galton (1983) in *Dryosaurus*.

ii. "Prominent primary ridge on the lateral side of the M crown." This feature is already present in *Dryosaurus* (Galton 1983) and is in essence a repetition of Node 19ii.

iii. "Ornamentation of the apical margin of individual denticles." These are already present on the teeth of *Tenontosaurus* and *Rhabdodon* (pers. obs.), which are here regarded as hypsilophodontians, and would appear to be a size-related character.

iv. "External opening of antorbital fossa relatively very small or absent." This is a repetition of Node 18vi, with the addition of the word "very," and is difficult to employ because it is defined in such a way that it permits a variety of anatomical conditions.

v. "Cervical neural spines very weak or absent." This condition is seen in *Hypsilophodon*, *Tenontosaurus*, and a wide range of ornithischians.

vi. "Robust, arching cervical postzygapophyses posterior to the axis." Difficult to apply in practice, this character could, for example, be applied to the cervicals of *Hypsilophodon* (Galton 1974a).

vii. "Moderate opisthocoely in cervical vertebrae 4 to 9; slight opisthocoely in dorsal vertebrae 1 to 2." Moderate opisthocoely is reported in *Tenontosaurus* (Forster 1985, in press). The appearance of this type of vertebral modification in large hypsilophodontians as well as in large iguanodonts is suggestive of a biomechanical function (convexoconcave articulations serve to increase the load-bearing capacity of the vertebral column).

viii. "Partial fusion of carpals into two blocks; block 1 consisting of the radiale, intermedium, distal carpals 1 and 3, and metacarpal 1; block 2 consisting of the ulnare and distal carpals 4 and 5." See below.

ix. "Metacarpal 1 inset into carpus and fused to distal carpal 1 and the radiale." See below.

x. "Metacarpal 1 shorter relative to other metacarpals." See below.

xi. "40° to 50° angle of divergence of manus digit 1 from axis of the forearm." See below.

xii. "Subconical manus digit 1 ungual." The array of characters (viii–xii) should not be listed piecemeal because they are without doubt part of a character complex. Taking a biological perspective, it is intuitively obvious that the manus in these forms is specialized for two primary functions: walking and defence. The fusion of the carpals into two blocks is undoubtedly a specialization associated with walking (Norman 1980), while reduction of metacarpal 1 and its incorporation into the carpus, and the conversion of the ungual into a divergent spike can be correlated with the defensive function. Thus, it seems reasonable to make the five characters of the list above into two: fusion of the carpalia into two blocks and the conversion of metacarpal 1 and its ungual into a defensive spike.

xiii. "Stout phalanx 1 of manus digit 1; wider transversely than long anteroposteriorly." This appears to define *Camptosaurus* alone.

xiv. "Pes digit 1 relatively shorter and less robust." This is not a usable statement of character because it conflicts with the absence of a first digit in *Dryosaurus*.

xv. "Metatarsal 1 markedly less robust relative to the other metatarsals." This is in essence a repetition of the previous character, and could equally well be applied to the description of the pes of *Hypsilophodon* or *Thescelosaurus*.

Node 21 (Fig. 11.6). Styracosterna

i. "At least 25 vertical columns in the M and D tooth rows." Rozhdestvensky (1966) indicates that there are " ... 23 or a few more [tooth positions] in the maxilla, 21–23 in dentary bone" in *Probactrosaurus gobiensis*. In *Iguanodon bernissartensis* there are 29 maxillary and 25 dentary tooth positions, while in *Iguanodon atherfieldensis* there are 23 maxillary and 21 dentary tooth positions (Norman 1980, 1986).

ii. "Lanceolate-shaped M crowns." *Camptosaurus* differs little in terms of its detailed tooth morphology from know species of *Iguanodon* and the tooth morphology of *Probactrosaurus* is as yet undescribed (Norman and Kurzanov in preparation).

iii. "Postdentary elements of the lower jaw are positioned posterior to the vertical midline of the coronoid pr." It is well known that some of the postdentary bones of the lower jaw (splenial, prearticular angular, and coronoid) all have portions that lie anterior to the midline of the coronoid process.

iv. "Strong opisthocoely in cervical vertebrae, beginning with the third cervical." This is consistent with the taxa concerned, but there is a suspicion that this feature may be a size related trend, and indeed a continuation of the character used at Node 20vii.

v. "Sternal ventrolateral pr." This is consistent with the taxa in the cladogram.

vi. "Humerus with proximally and posteriorly prominent head." This feature is also observed in *Camptosaurus*

(Gilmore 1909) and *Tenontosaurus* (Dodson 1980; Forster 1985, in press).

vii. "Shafts of metacarpals 2 to 4 closely appressed." This is consistent with the taxa in the cladogram.

viii. "Metacarpal 4 subequal in length to metacarpal 3." This is consistent, but linked with vii, above.

ix. "Distal end of the prepubic pr. moderately expanded dorsoventrally." This is consistent with taxa in the cladogram.

x. "Pubis with distinct, stout iliac peduncle." There is no significant difference in the structure of the iliac peduncle in *Dryosaurus*, *Camptosaurus*, or even *Tenontosaurus*.

xi. "Postpubic pr. consisting of a tapering rod approximately half the length of the ischium." This feature is exhibited by *Tenontosaurus* (Dodson 1980; Forster 1985, in press).

xii. "Femur with deep intercondylar groove." This feature is visible in *Camptosaurus* (Gilmore 1909) and is quite a prominent feature in the dryosaurid *Valdosaurus* (Galton 1980).

Node 22 (Fig. 11.6). Iguanodontoidea

i. "Space between the first D tooth and the PD." A small diastema is present in *Tenontosaurus* (Weishampel 1984) and *Camptosaurus*, but is reduced or absent in the hadrosaurid *Telmatosaurus* (Weishampel pers. comm.). The existence of a diastema is more likely to be related to size and jaw function than phylogeny in the same way as it is with mammals (Turnbull 1970).

ii. "External nares enlarged." Repetition of Node 18vii, and difficult to interpret unambiguously.

iii. "At least a slight narrowing of the cranium from postorbital region posteriorly in dorsal view." This character is also discernable in *Dryosaurus* (Galton 1983).

iv. "Paroccipital pr. relatively broader proximally and narrower distally." This character is also well developed in *Camptosaurus* (isolated cranial fragments from the Cleveland Lloyd quarry, courtesy James Madsen).

v. "Postpalatine foramen absent." This is true as far as can be examined, but few dinosaurs have sufficiently well preserved palates.

vi. "Complete fusion of the radiale, distal carpal 1, and metacarpal 1." Repetition of Node 20ix.

vii. "Metacarpal 1 relatively very short or absent." Repetition of Node 20x.

viii. "Phalanx 1 of manus digit I represented by a flattened disk or absent altogether." Because two character states are included at this point, it is difficult to assess its value.

ix. "Phalanx 2 of manus digits II and IV very short relative to phalanx 1 of the respective digit." This character may be associated with Node 20viii–xii and the development of a competent weight-bearing hand.

x. "Ungual of manus digit II with transversely narrower proportions than the ungual of manus digit III." This may be true for *Iguanodon*, but it is marginal in hadrosaurids, is unknown in *Ouranosaurus*, and has not yet been made known in *Probactrosaurus*.

xi. "Manus digit V with at least three phalanges (one phalanx added)." The distribution of this character is not adequately known among the taxa considered in this cladogram because there are so few well preserved manuses. This is particularly true with respect to *Probactrosaurus* which has not been adequately described to date, and *Ouranosaurus* (Taquet 1976).

xii. "Hoof-shaped unguals on manus digits II and III." This is consistent with the taxa described (possibly linked developmentally with point xiv).

xiii. "Metatarsal 1 represented by a transversely thin, short splint." There is no mention of a first metatarsal in *Probactrosaurus* (Rozhdestvensky 1966), the situation is unknown in *Ouranosaurus* (Taquet 1976), and the first metatarsal of *Dryosaurus lettowvorbecki* and *D. altus* is identical to the condition described above (Galton 1983).

xiv. "Hoof-shaped unguals on pes digits II and IV." This is consistent with the taxa listed, but is possibly linked to point xii.

xv. "Pes digit V absent." This character appears to be variable in a variety of ornithopods, including *Dryosaurus* (Galton 1981), *Tenontosaurus* (only a small percentage of the population appears to possess a fifth metatarsal according to Forster 1985), and *Camptosaurus* (Gilmore 1909). A fifth metatarsal is also absent in *Heterodontosaurus* (Santa Luca 1980).

xvi. "Double layer lattice of ossified tendons from the posterior cervical-anterior dorsal region to the midcaudal region of the vertebral column; tendons of both deep and superficial layers insert high on the lateral side of the neural spines, the former coursing anteroventrally and the latter posteroventrally across adjacent neural spines." The establishment of a lattice of tendons will depend upon the existence of tall neural spines (to provide sufficient area for such an arrangement) as is the case in *Iguanodon* (Norman 1980, 1986). Galton and Powell (1980) describe a lattice of tendons in *Camptosaurus*, but it has not been described in *Ouranosaurus*. Information on the tendon arrangement in *Probactrosaurus* is not available.

Node 23 (Fig. 11.6). Hadrosauroidea

i. "Relatively greater space between the first D tooth and the PD." *Telmatosaurus* has little if any diastema (Weishampel pers. comm.) and, as observed above (Node 22i), this feature may be of greater importance as a biomechanical indicator than as a phylogenetically important character.

ii. "Anterior end of the premaxillary snout expanded transversely; narial fossa lengthened anteroposteriorly and defined laterally by a reflected rim." The first part of this character state is in effect a repetition of Node 18v. The lengthening of the narial fossa is evident in *Iguanodon* sp., and reflection of the premaxillary rim is not seen in lambeosaurine hadrosaurids (see also Horner this volume).

iii. "Anterior end of jugal expanded dorsoventrally in front of the orbit." The expansion of the anterior end of the jugal in *Ouranosaurus* is a marginal feature at best, and there are absolute differences in form between the jugals of hadrosaurids (Lull and Wright 1942) and those of *Iguanodon* and *Ouranosaurus* (compare Norman 1986, Fig. 15 with Taquet 1976, Fig. 19).

iv. "Distinct transverse narrowing of the cranium from postorbital region posteriorly in dorsal view." This is a repetition of Node 22iii.

v. "SQ approach the midline of the posterior skull roof, separated by only a narrow band of the P." This is no different from the situation in *Iguanodon* (Norman 1980, 1986).

vi. "Distal end of the paroccipital and accompanying SQ pr. curve anteriorly." This is consistent with the taxa discussed.

vii. "Caudal neural spines exceed their respective chevrons in length." This is true, although it is a marginal character. The lengths of the neural spines of *Iguanodon atherfieldensis* are close to those of their chevrons.

viii. "Scapular blade with a convex dorsal margin." Whichever way the scapular blade is orientated for descriptive purposes, this feature does not ally *Ouranosaurus* with hadrosaurids (Norman 1986).

ix. "Phalanx I of manus digit 1 absent." This is not testable given the currently available material, and this is repetitive of Node 22viii.

x. "Pubic peduncle of ilium relatively small; articulates against the prominent dorsally directed iliac peduncle of the pubis." This is a repetition in substance of Node 21x.

xi. "Iliac antitrochanter present." There is no hadrosaurid-like antitrochanter on the ilium of *Ouranosaurus*. In fact, the ilium of the latter differs relatively little from that of *Iguanodon* species.

xii. "Iliac preacetabular pr. relatively longer." This is a subjective statement. The ilium of *Iguanodon atherfieldensis* (Norman 1986) is long, and overall is more similar to that of *Ouranosaurus* than hadrosaurids.

xiii. "Marked dorsoventral expansion of the distal prepubic blade." This is a partial repetition of Node 21ix. The pubis of *I. atherfieldensis* exhibits an expanded distal prepubic process (Norman 1986).

xiv. "Proximally positioned obturator process." This feature appears in *Dryosaurus* (Galton 1981), *Camptosaurus* (Gilmore 1909), and *Iguanodon* (Norman 1980, 1986).

xv. "Deep obturator notch between the obturator pr. and the pubic peduncle of the ischium." This feature is widely distributed in ornithopods with a proximally positioned obturator process (see above).

xvi. "Distal tarsals 2 and 3 absent." It is difficult to use this character because these elements are infrequently fossilized, and there is the problem of distinguishing between genuine loss, as opposed to fusion with the proximal ends of the metatarsal.

xvii. "Pes Digit I absent." This digit is lost in *Dryosaurus* (Galton 1983), as it is in *Iguanodon*, where only a metatarsal splint remains. The loss of this digit and its metatarsal in *Ouranosaurus* is conjectural at present (Taquet 1976).

Discussion

As can be seen, Sereno (1986) has provided an extremely useful cladogram of ornithischian dinosaurs that is supported by detailed documentation of the individual taxa concerned. This cladogram is derived from that published earlier (Sereno 1984), with some fundamental alterations to the relations of the major taxonomic groups, and, to some extent, converges on the tentative proposals of Norman (1984a). Such publications will inevitably attract considerable attention and debate among research workers in this area. The net result of such debate should be the development of a measure of unanimity over the issue of ornithischian relationships. In this review ornithopod relationships are discussed.

Although superficially the pattern of relationships proposed here appears comparable to that advocated by Sereno, there are subtle and fundamental differences both in terms of the taxa, how they are supposedly related, and the value attached to the characters that support the positions of taxa within the cladogram. The character lists provided by Sereno have been discussed above, and this has led to the development of an alternative set of characters and cladogram. The merits of these new proposals will undoubtedly be subjected to similar scrutiny by workers in the future. However, I would suggest that the position of three taxa are crucial to the relationships advocated here, and also to the interpretation of Sereno (1986): *Tenontosaurus*, *Ouranosaurus*, and *Probactrosaurus*.

i. Tenontosaurus tilletti

According to the analysis above, there are at least eight anatomical features that distinguish hypsilophodontians and *Tenontosaurus* from *Dryosaurus* and related ornithopods. In contrast, Sereno (1986) placed the former taxon as the sister-taxon of all iguanodontians, which in turn differed considerably from Sereno (1984) where *Tenontosaurus* was placed between *Camptosaurus* and *Iguanodon*. As has been demonstrated earlier, the value of many of the 14 characters provided for Node 19 (Fig. 11.6) are subject to some doubt. Those which are consistent with Sereno's cladogram (denticulate premaxillary margin, paired posteroventral processes on the predentary, and, possibly, the eversion of the occlusal margin of the premaxilla) require the inclusion of *Tenontosaurus* in the Iguanodontia, instead of the Hypsilophodontia as is regarded here.

The single outstanding character that unites *Tenontosaurus* with other iguanodontians is the phalangeal reduction of digit III of the manus. One additional character regarded by Sereno as particularly important is the possession of "leaf-shaped denticles." I must confess to not understanding this character in detail, and I am, therefore, unable to interpret it. In the case of the manus character, it is unfortunate that articulated material of the manus of ornithopods, and particularly that of *Dryosaurus* (Galton 1981), is not more completely known. Following Dodson (1980), Weishampel and Weishampel (1983), Weishampel (1984), Norman (1984a), Milner and Norman (1984), Forster (1985 in press), and Sues and Norman (in press),

Tenontosaurus is referred to the Hypsilophodontia. The revised diagnosis is:

1. Foramen fully within the body of the quadratojugal. This condition is seen in *Hypsilophodon* and *Tenontosaurus*, and may be a derived feature associated with these two taxa alone. In the latter taxon the foramen is large and appears to be bounded by the caudal margin of the jugal, although the foramen lies completely within the body of the quadratojugal (Weishampel 1985, pers. comm. 1987).

2. Dentary teeth with a radiating pattern of ridges developed down the surface of the crown. In several taxa, *Hypsilophodon*, *Tenontosaurus*, and *Rhabdodon*, a more derived condition is found whereby a prominent primary ridge is developed in a medial position on the lateral surface of the crown – reminiscent of dentary crowns of hadrosaurids.

3. Maxillary teeth with an array of subsidiary ridges and lacking a prominent primary ridge. The prominent primary ridge is consistently found in the maxillary teeth of *Dryosaurus* and more derived iguanodonts, but not in hypsilophodontians. *Muttaburrasaurus* (Bartholomai and Molnar 1981) may be a hypsilophodontian for this reason.

4. Length of scapula equal to or shorter than the length of the humerus. (This condition is found in a derived condition in *Tenontosaurus*, but it is found primitively within hypsilophodontians. Even in *Tenontosaurus* the scapula is no more than 5–10% longer than the humerus.)

5. Ossified hypaxial tendons in the tail [only so far noted in *Hysilophodon* (Galton 1974a), *Thescelosaurus* (Galton 1974b), *Parksosaurus* (Parks 1926) and *Tenontosaurus* (Ostrom 1970)].

6. Ischium shaft straight, or nearly straight, with laterally compressed and broadening shaft distal to the obturator process (seen in *Hypsilophodon*, *Othnielia*, *Thescelosaurus*, *Parksosaurus*, *Yandusaurus*, and *Tenontosaurus*).

7. Obturator process positioned between one-third and one-half of the length of the shaft away from the acetabular margin (seen in *Tenontosaurus*, *Hypsilophodon*, *Othnielia*, *Yandusaurus*, *Thescelosaurus*, and *Parksosaurus*).

ii. Ouranosaurus nigeriensis

This taxon was referred to as the sister-taxon of the Hadrosauridae (Sereno 1984, 1986), but the list of characters provided has not proved convincing. As in the case of *Tenontosaurus* (which is here regarded as a large hypsilophodontian that mimics some features seen in larger iguanodontians), evolutionary parallelism or convergence of anatomical features seems to play a major role in the differing opinions over the relationships of this species. In some respects, *Ouranosaurus* is much like a primitive hadrosaur: it has the broad flattened snout, and some elaboration to the nasal region [although it must be noted that nasal elaboration is evident in *Iguanodon orientalis* (Norman 1985) and *Muttaburrasaurus* (Bartholomai and Molnar 1981)]. I favor the view that this form is a "progressive" iguanodontid that has developed features paralleling those seen in hadrosaurs. The detailed anatomy of *Ourano-saurus* resembles that of *Iguanodon*, which supports the inclusion of *Ouranosaurus* in the family Iguanodontidae.

iii. Probactrosaurus gobiensis

Regarded here as the sister-taxon of hadrosaurids, this form was positioned as the sister-taxon of *Iguanodon* + *Ouranosaurus* + Hadrosauridae by Sereno (1986 only). Unfortunately, the anatomy of this ornithopod is not known in detail and awaits review (Norman and Kurzanov in preparation). A number of features reported (Rozhdestvensky 1966) indicate that it may be the primitive sister-taxon of all hadrosaurids. There is no evidence of flattening of the muzzle, or of a cranial crest seen in many later hadrosaurs (also absent in some primitive incontrovertible hadrosaurids like *Telmatosaurus*). The most important features for systematic purposes are the tooth replacement pattern (a second replacement crown is in each vertical tooth position) and the architecture of the enamelled surface of the tooth crowns (the crown is symmetrical on either side of the primary ridge in both maxillary and dentary teeth). Both of these conditions presage the elaborate dental batteries seen uniquely in hadrosaurids, and are not found in more distantly related forms such as dryosaurids and iguanodontids.

Summary

Vectisaurus valdensis, which has on previous occasions been referred to as a valid species of iguanodontid ornithopod, is regarded as a junior subjective synonym of *Iguanodon atherfieldensis*.

Referred material of *V. valdensis* is largely attributable to *Iguanodon* spp., although some has also been referred to the poorly known genus *Yaverlandia* and to the dryosaur *Valdosaurus*.

The status of the family Iguanodontidae and taxa within the Ornithopoda are the subject of some discussion following the publication of cladistic systematic analyses. It is concluded that the case for disbanding the family is not convincing because the existing character-state analyses are not sufficiently rigorous. In an attempt to clarify the views, an effort has been made to comment upon, and revise, a recent cladogram. The most important results of this are that the Hypsilophodontia has been modified to include *Tenontosaurus*, that the family Iguanodontidae has been re-instated and re-defined to include *Ouranosaurus*, and *Probactrosaurus* is regarded as the sister-taxon of the Hadrosauridae.

Acknowledgments

This work would not have been possible without the support and facilities provided by the Nature Conservancy Council, and the Department of Zoology (University Museum), University of Oxford. Financial support was provided by the Royal Society and by NATO scientific affairs division.

I am indebted to various people for assistance with this work. For permission to work with their collections Dr. Angela Milner [British Museum (Natural History)], Dr. N. Hotton III

(U.S. National Museum, Washington), Dr. Michael Novacek (American Museum of Natural History, New York) and Pierre Bultynk (Royal Institute of Natural Sciences, Brussels). The manuscript has also benefitted from discussions or comment from Dr. Peter Dodson (University of Pennsylvania), Dr. David Weishampel (Johns Hopkins University, Baltimore), Dr. Andrew Milner (Birkbeck College, University of London), Kenneth Carpenter (University of Pennsylvania) and Dr. Paul Sereno (University of Chicago).

References

Bartholomai, A. and Molnar, R. 1981. *Muttaburrasaurus*, a new iguanodontid (Ornithischia: Ornithopoda) dinosaur from the Lower Cretaceous of Queensland. *Memoirs of the Queensland Museum* 20:319–349.

Cope, E. D. 1869. Synopsis of the extinct Batrachia, Reptilia and Aves of North America. *American Philosophical Society Transactions* 14:1–252.

Dodson, P. 1980. Comparative osteology of the American ornithopods *Camptosaurus* and *Tenontosaurus*. *Memoires de la Societe Geologiques de France*, ns 1980, no. 139:81–85.

Forster, C. A. 1985. A description of the postcranial skeleton of the early Cretaceous ornithopod *Tenontosaurus tilletti*, Cloverly Formation, Montana and Wyoming. Unpublished Masters thesis. Philadelphia: University of Pennsylvania.

in press. The postcranial skeleton of the ornithopod dinosaur *Tenontosaurus tilletti*. *Journal of Vertebrate Paleontology*.

Galton, P. M. 1971. A primitive dome-headed dinosaur (Ornithischia: Pachycephalosauridae) from the Lower Cretaceous of England and the function of the dome of pachycephalosaurids. *Journal of Paleontology* 45:40–47.

1974a. The ornithischian dinosaur *Hypsilophodon* from the Wealden of the Isle of Wight. *British Museum (Natural History) Bulletin, Geology* 25:1–152.

1974b. Notes on *Thescelosaurus*, a conservative ornithopod dinosaur from the Upper Cretaceous of North America, with comments on ornithopod classification. *Journal of Paleontology* 48:1048–1067.

1975. English hypsilophodontid dinosaurs (Reptilia: Ornithischia). *Palaeontology* 18:741–752.

1976. The dinosaur *Vectisaurus valdensis* (Ornithischia: Iguanodontidae) from the Lower Cretaceous of England. *Journal of Paleontology* 50:976–984.

1980. *Dryosaurus* and *Camptosaurus*, intercontinental genera of Upper Jurassic ornithopod dinosaurs. *Memoires de la Societe Geologique de France*, ns 1980:103–108.

1981. *Dryosaurus*, a hypsilophodontid dinosaur from the Upper Jurassic of North America and Africa. Postcranial skeleton. *Palaontologische Zeitschrift* 55:271–312.

1983. The cranial anatomy of *Dryosaurus*, a hypsilophodontid dinosaur from the Upper Jurassic of North America and East Africa, with a review of the hypsilophodontids from the Upper Jurassic of North America. *Geologica et Palaeontologica* 17:207–243.

Galton, P. M. and Powell, H. P. 1980. The ornithischian dinosaur *Camptosaurus prestwichii* from the Upper Jurassic of England. *Paleontology* 23:411–443.

Galton, P. M. and Taquet, P. 1982. *Valdosaurus*, a hypsilophodontid dinosaur from the Lower Cretaceous of Europe and Africa. *Geobios* 15:147–159.

Gilmore, C. 1909. Osteology of the Jurassic reptile *Camptosaurus*, with a revision of the genus and description of two new species. *United States National Museum, Proceedings* 36:197–332.

Huckriede, R. 1982. Die unterkretazische Karsthohlen-Fullung von Nehden im Sauerland. 1. Geologische, palaozoologische und palaobotanische Befunde und Datierung. *Geologica et Palaeontologica* 16:183–242.

Hulke, J. W. 1879. *Vectisaurus valdensis*, a new Wealden dinosaur. *Quarterly Journal of the Geological Society of London* 35:421–424.

Lull, R. S. and Wright, N. E. 1942. Hadrosaurian dinosaurs of North America. *Geological Society of America Special Paper* 40:1–133.

Marsh, O. C. 1881. Principal characters of the American Jurassic Dinosaurs. *American Journal of Science* 21:417–423.

Milner, A. R. and Norman, D. B. 1984. The biogeography of advanced ornithopod dinosaurs (Archosauria: Ornithischia) – a cladistic-vicariance model. *In* Reif, W. E. and Westphal, F. (eds.), *Third Symposium on Mesozoic Terrestrial Ecosystems*, short papers (Tübingen: Attempto Verlag), pp. 145–151.

Norman, D. B. 1977. On the anatomy of the ornithischian dinosaur *Iguanodon*. Ph.D. thesis. London: University of London, King's College.

1980. On the ornithischian dinosaur *Iguanodon bernissartensis* of Bernissart, Belgium. *Memoires de l'Institut Royal des Sciences Naturelles de Belgique* 178:1–105.

1984a. A systematic reappraisal of the reptile Order Ornithischia. *In* Reif, W. E. and Westphal, F. (eds.), *Third Symposium on Mesozoic Terrestrial Ecosystems*, short papers (Tübingen: Attempto Verlag), pp. 157–162.

1984b. On the cranial morphology and evolution of ornithopod dinosaurs. *Symposium of the Zoological Society of London* 52:521–547.

1985. *The Illustrated Encyclopedia of Dinosaurs*. London: Salamander Books.

1986. On the anatomy of *Iguanodon atherfieldensis* (Ornithischia: Ornithopoda). *Bulletin de l'Institut Royal des Sciences Naturelles de Belgique (Sciences de la Terre)* 56:281–372.

1987. A mass-accumulation of vertebrates from the Lower Cretaceous of Nehden (Sauerland), West Germany. *Proceedings of the Royal Society of London*, series B, 230:215–255.

Ostrom, J. H. 1970. Stratigraphy and paleontology of the Cloverly Formation (Lower Cretaceous) of the Big Horn Basin of Wyoming and Montana. *Bulletin of the Peabody Museum of Natural History* 35:1–234.

Parks, W. A. 1926. *Thescelosaurus [Parksosaurus] warreni*, a new species of orthopodous dinosaur from the Edmonton Formation of Alberta. *University of Toronto Studies, Geological Series* 21:1–42.

Rozhdestvensky, A. K. 1966. Novyyeiguaondonty iz

Tsentral'noy Azii. Filogeneticheskiye i taksonomisheskiye zvaimootnosheniya pozdnikh Iguanodonitdae ranikh Hadrosauridae. *Paleontologischeski Zhurnal* 1966:103–116.

Santa Luca, A. P. 1980. The postcranial skeleton of *Heterodontosaurus tucki* (Reptilia: Ornithischia) from the Stormberg of South Africa. *Annals of the South African Museum* 79:159–211.

—— 1984. Postcranial remains of Fabrosauridae (Reptilia: Ornithischia) from the Stormberg of southern Africa. *Palaeontologica Africana* 25:151–180.

Seeley, H. G. 1887. On the classification of the fossil animals commonly named Dinosauria. *Proceedings of the Royal Society of London* 43:165–171.

Sereno, P. C. 1984. The phylogeny of the Ornithischia: a reappraisal. *In* Reif, W. E. and Westphal, F. (eds.), *Third Symposium on Mesozoic Terrestrial Ecosystems* (Tübingen: Attempto Verlag), pp. 219–226.

—— 1986. Phylogeny of the bird-hipped dinosaurs (Order Ornithischia). *National Geographic Research* 2:234–256.

Steel, R. 1969. Ornithischia. *In* Kuhn, O. (ed.), *Handbuch der Paläoherpetologie*, part 15, pp. 1–84. (Stuttgart: Gustav Fischer Verlag).

Sues, H.-D. and Norman, D. B. In press. Hypsilophodontidae, *Tenontosaurus* and Dryosauridae. *In* Weishampel, D. B., Dodson, P., and Osmólska, H. (eds.), *The Dinosauria* (Los Angeles: University of California Press).

Swinton, W. E. 1936. The dinosaurs of the Isle of Wight. *Geological Association Proceedings* 47:204–220.

Taquet, P. 1976. *Géologie et paléontologie du gisement de Gadoufaoua (Aptien du Niger).* (Paris: Centre National de la Recherche Scientifique), pp. 1–191.

Turnbull, W. D. 1970. Mammalian masticatory apparatus. *Fieldiana Geology* 18:153–356.

Watson, D. M. S. 1930. [No title] *Isle of Wight Natural History Society Proceedings* 2:60.

Weishampel, D. B. 1984. Evolution of jaw mechanisms in ornithopod dinosaurs. *Advances in Anatomy, Embryology and Cell Biology* 87:110.

Weishampel, D. B. and Weishampel, J. B. 1983. Annotated localities of ornithopod dinosaurs: implications to Mesozoic paleobiogeography. *Mosasaur* 1:43–87.

12 Morphometric observations on hadrosaurid ornithopods

RALPH E. CHAPMAN AND
MICHAEL K. BRETT-SURMAN

Abstract

Results are presented of preliminary morphometric analyses on hadrosaurs using the landmark shape analysis method Resistant-Fit Theta-Rho-Analysis (RFTRA). The analyses were performed on both cranial and postcranial material. They show this approach to be useful for the analysis of hadrosaur morphology and provide insight into how this morphology varies within the context of the phylogenetic structure of the family. Further, the patterns are related to two other groups of Euornithopods, the iguanodontids and camptosaurids. The results highlight the distinct morphology of the lambeosaurine hadrosaurs, confirm that most of the significant morphological variation in hadrosaur crania is concentrated in the muzzle and narial regions, and indicate that pelvic element shape should be useful for taxonomic identification and discrimination. In general cranial shape, the lambeosaurines are shown to be most closely related to the hadrosaurines, supporting a monophyletic Hadrosauridae.

Introduction

Hadrosaurs have one of the most complex taxonomic histories of all the dinosaurs; over 100 species representing 44 genera have been named. This unusually high taxonomic diversity is a consequence of the interplay between the taxonomic philosophies of the many researchers studying hadrosaurs, the high level of real taxonomic diversity, the unusually abundant material available, and the high degree of morphological variability within populations and between age groups. The latter is the result of allometric and ontogenetic effects over a wide range of sizes (see Dodson 1975; Hopson 1975; Molnar 1977). Herein, we will present the results of a series of preliminary shape analyses of hadrosaur crania and pelves, and discuss these in the context of hadrosaur taxonomy, phylogeny, and identification.

Hadrosaurs are unusual among the dinosaurs

because they are represented by large numbers of well-documented specimens with both cranial and postcranial material. Contrast this with the pachycephalosaurs, for example, for which little postcranial material is available, and most taxa and specimens are represented by only incomplete crania (Maryańska and Osmólska 1974; Sues and Galton 1987; Goodwin this volume). In fact, hadrosaur material can be so abundant that it is often left uncollected when resources restrict the number of specimens that can be removed during a field season (P. Currie pers. comm. 1986).

In addition to this abundance, hadrosaurs exhibit a high degree of morphological variability, especially in cranial structures. These are thought by some to play a role in social behavior (Dodson 1975; Hopson 1975; Molnar 1977). Dodson (1975), for example, analyzed morphometric data for the crania of 36 specimens of lambeosaurine hadrosaurs referable to three genera (*Corythosaurus*, *Lambeosaurus*, and "*Procheneosaurus*") and 12 species. The approaches used included standard bivariate allometric analyses and principal coordinates analysis. Dodson concluded that the taxon "*Procheneosaurus*" includes only juvenile forms of other taxa, and that only one species of *Corythosaurus* and two of *Lambeosaurus* were represented, as well as both sexes. Clearly, the wide range of body size, the great diversity of display structures, and sexual dimorphism combined to produce a great deal of morphological variation among individuals of a single species.

The taxonomy of hadrosaurids has been the subject of discussion recently. Sereno (1986), in his cladistic analysis of the ornithischians, retained the hadrosaurs as a monophyletic taxon, whereas Horner (this volume) has suggested that the group is diphyletic.

Dinosaur morphometric analyses have been sporadic and limited by the small numbers of specimens available for most groups (Chapman this volume). The most comprehensive studies to date are Dodson's (1975)

In Dinosaur Systematics: Perspectives and Approaches, *Kenneth Carpenter and Philip J. Currie, eds. Copyright © Cambridge University Press, 1990.*

work on lambeosaurine hadrosaurs, Dodson's (1976) allometric analysis of growth and sexual dimorphism in *Protoceratops*, the study of Chapman et al. (1981) on cranial allometry and sexual dimorphism in the pachy-cephalosaurian *Stegoceras*, the study of variation in *Plateosaurus* femora by Weishampel and Chapman (this volume), and the review by Chapman (this volume) demonstrating the application of shape analysis methods, specifically Resistant-Fit Theta-Rho-Analysis, to general problems of dinosaur paleobiology.

Generally, morphometric methods can provide important insights into the morphology of dinosaurs, and the implication of this morphology for interpreting phylogeny, ontogeny, paleoecology, and taphonomy. Morphometric methods already have provided important information on lambeosaurine variability (Dodson 1975). Here, we apply it more generally to hadrosaur morphology and to the interpretation of hadrosaur taxonomic structure.

Materials and methods

Resistant-Fit Theta-Rho-Analysis (RFTRA) is a form of landmark shape analysis that provides superimposed figures representing the fit of one specimen onto another after size and positional differences are removed, and gives distance values representing an estimate of the "goodness of fit." The fit is made using the relative positions of groups of landmarks (homologous or geometrically equivalent points) and the original geometry of the landmarks is maintained without the specimens being distorted during the analysis (Benson et al. 1982).

Landmark shape analysis methods are derived from the transformation grids developed by D'Arcy Thompson (1942), and include tensor methods (see Bookstein et al. 1985, and references therein) and vector methods (Sneath 1967). RFTRA was developed by Siegel and Benson (1982) and Benson et al. (1982), who modified Sneath's least-squares approach by applying more robust statistical methods (Siegel and Benson 1982). The RFTRA algorithm is a major improvement on the least-squares method (referred herein as LSTRA) because it allows localized change to be identified.

Chapman (this volume) presents a detailed discussion on the philosophy and mechanics of the method. In this study, a series of photographs/illustrations representing the same view was obtained for the specimens, and the questions to be addressed by this analysis were developed. Next, a series of homologous landmarks or equivalent points were located on each illustration. The resulting constellation of landmark points then was used to provide the fit of one specimen onto the base specimen. Each landmark had to be found on all specimens. A polygonal or "skeletal" diagram was developed that connected specified pairs of these landmarks representing functional units or individual skeletal elements. The RFTRA programs were then run on a computer to provide the necessary calculations, graphical output, and distance

coefficients (the estimate of the closeness of fit between the two specimens based on the superimposition).

The morphometric approaches used here were applied to provide insight into the morphology of the hadrosaurs and how the different hadrosaur morphologies interrelate. Iguanodontids were used as the primary outgroup to provide examples of the most closely related group to the hadrosaurs, and two specimens of camptosaurids also were used as the second outgroup to give an indication of the morphological trajectory of the euornithopods (*sensu* Sereno 1986). A specimen of "*Procheneosaurus*," considered by us to be a juvenile lambeosaurine (fide Dodson 1975), was included to show how an immature form would compare with the adults, and whether it would cluster with the lambeosaurines.

One characteristic that morphometric analyses share with modern phylogenetic analyses is that they are both dialetic in nature: the results of an analysis are not considered to be the final answer but, instead, indicate the direction that further analyses should take. Because of this, we recognize these results as a first approach in the ongoing analysis of both cranial and postcranial elements of hadrosaurs and other euornithopods. The number of complete and articulated specimens available is exceedingly small. However, the results do indicate where expanded studies should concentrate.

The results of RFTRA within this context do provide information relevant to the interpretation of taxonomic and phylogenetic structure. If the results of a morphometric analysis do not agree with conventional phylogenetic reconstructions, then they raise questions that must be addressed before those phylogenies can be accepted. Often the differences can be recognized as convergence, providing information that may be relevant to the functional morphology or ecology of the taxon. Where convergence is not apparent, however, the characters used in the phylogenetic analysis should be reconsidered and additional morphometric analyses developed to try to reconcile the differences. Agreement between the two provides support in much the same way that the addition of new characters strengthens a phylogenetic analysis. It is more parsimonious to accept morphological similarity between the members of two groups as the result of recent ancestry rather than just a chance convergence, if other factors independently suggest the connection. In this way, morphometric analyses can help in the choice between two phylogenies developed using more conventional approaches.

In vertebrate paleontology, reconstructions such as those used for the analyses here represent the original skeletal material interpreted by paleontologists during the process of reconstruction. The results of morphometric analyses then provide a way to evaluate these interpretations within the context of those available for other related forms. As the resulting patterns are interpreted, they can suggest information that is relevant to the taxonomy and biology of the original animals, or they may suggest where the reconstruction process needs to be reconsidered (see example with *Protoceratops* in Chapman, this volume). In either case, the information is useful.

Two groups of analyses were done on specimens of hadrosaurids, including both hadrosaurines and lambeosaurines (Appendix 2), iguanodontids and *Camptosaurus*. The first analyses concentrated on illustrations of skulls, and used sutural connections of cranial bones, fenestrae, and geometrical points as landmarks. The specimens used and the source of the illustration are given in Table 12.1. Figure 12.1 shows representatives of the hadrosaurines and Figure 12.2 a lambeosaurine for comparison. An example illustrating the landmarks used and the resulting polygonal figure is shown in Figure 12.3.

The following 20 cranial landmarks were chosen as the most representative set to delineate the basic features of hadrosaur cranial morphology (see Figure 12.3). All points are those seen in lateral view.

1. The lateral interior inflection point of the premaxillary "lip."
2. The lateral exterior inflection point of the premaxillary "lip."
3. The medial exterior inflection point of the premaxillary "lip."
4. The medial interior inflection point of the premaxillary "lip."
5. The contact between the dorsal ramus of the premaxilla and the nasal.
6. The posterior limit of the true external narial opening.
7. The most posterior extent of the lower ramus of the premaxilla.
8. The posterior limit of the nasal (nasal/frontal contact in most cases).
9. Frontal/parietal contact.
10. The most dorsal extent of the lateral temporal fenestra.
11. The most anteroventral extent (inflection point) of the lateral temporal fenestra.
12. The most dorsal extent of the quadrate.
13. The ventral limit of the quadrate.
14. The superior jugal/quadratojugal/quadrate contact.
15. The inferior jugal/quadratojugal contact.
16. The posterior end of the maxillary dental battery.
17. The lacrimal/jugal/orbit contact.
19. The maxilla-lacrimal contact.
20. The anterior end of the maxillary dental battery.

Table 12.1. *Specimens/illustrations used for analysis of hadrosaur crania*

No.[a]	Taxon [Group][b]	Source[c]
1	*Camptosaurus depressus* [C]	GAL, p. 82, Fig. 7A
2	*Iguanodon bernissartensis* [I]	OWN, Pl. 9, Fig. 1
3	*Ouranosaurus nigeriensis* [I]	TAQ, p. 62, Fig. 10A
4	*Edmontosaurus regalis* [H]	L&W, p. 152, Fig. 52
5	*Anatotitan copei* [H]	L&W, p. 158, Fig. 54
6	*Hadrosaurus notabilis* [H]	L&W, p. 167, Fig. 59
7	*Brachylophosaurus canadensis* [H]	HOLO., NMC 8893
8	*Maiasaura peeblesorum* [H]	HOR, p. 82
9	*Prosaurolophus maximus* [H]	L&W, p. 173, Fig. 63
10	*Saurolophus osborni* [H]	L&W, p. 176, Fig. 65
11	"*Procheneosaurus praeceps*" [L]	L&W, p. 181, Fig. 67
12	*Lambeosaurus lambei* [L]	L&W, p. 189, Fig. 72
13	*Lambeosaurus magnicristatus* [L]	L&W, p. 194, Fig. 76
14	*Corythosaurus casuarius* [L]	L&W, p. 196, Fig. 77
15	*Hypacrosaurus altispinus* [L]	L&W, p. 206, Fig. 86
16	*Parasaurolophus walkeri* [L]	L&W, p. 211, Fig. 89
17	*Corythosaurus excavatus* [L]	L&W, p. 198, Fig. 80
18	*Lambeosaurus lambei* [L]	L&W, p. 190, Fig. 73
19	*Corythosaurus intermedius* [L]	L&W, p. 200, Fig. 81
20	*Iguanodon atherfieldensis* [I]	ROM, p. 148, Fig. 79B

Note: [a]For RFTRA specimen file numbers add 2000 to each number indicated. For example, cranial data for *Parasaurolophus walkeri*, #16, are contained in file #2016.

[b][C] = camptosaurid; [H] = hadrosaurine hadrosaur; [I] = iguanodontid; [L] = lambeosaurine hadrosaur

[c]Fig. = Figure; GAL = Galton 1980; HOLO. = Photograph of holotype; HOR = Horner and Gorman 1988; L&W = Lull and Wright 1942; NMC = National Museum of Canada; OWN = Owen 1855; p. = Page; Pl. = Plate; ROM = Romer 1956; TAQ = Taquet 1976.

The dentary and predentary were not used in this pre-liminary analysis because the latter is frequently lost. Many landmarks in an articulated dentary are hidden in lateral view.

These points were designed to highlight the four basic "taxonomic" skull sections when seen in lateral view. Each section displays its own unique attributes that have previously proven to contain most of the major morphological features used to delineate genera. Section 1 contains the reflected margin of the premaxilla, which in hadrosaurines forms the "lips." This area is especially important in the edmontosaur clade (Brett-Surman 1988) that contains *Edmontosaurus* (= *Anato-*

saurus), *Shantungosaurus*, and *Anatotitan* (Appendices 1 and 2). This area is represented by landmarks one through four. Section 2 is the most important area of analysis for hadrosaurines, and contains the external nares and the external narial pockets that figure promi-nently in hadrosaurines such as *Edmontosaurus* and *Hadrosaurus* (= *Kritosaurus* fide Horner). The section is bordered anteriorly by the reflected premaxillary lips and posteriorly by the closure of the external narial opening. Section 3 includes the nasal-frontal complex and represents the most diagnostic area for all lambeo-saurines. Section 4, containing the quadrate and tempo-ral fenestrae, shows the least amount of evolutionary change from the standpoint of shape analysis.

A second group of analyses used individual pelvic elements. True homologous landmarks were more diffi-cult to find on these elements and, as a result, we relied more on geometrically analogous points, reflecting inflection points that control the major elements of shape for each element (see Bookstein et al. 1985). For this reason, the results should be used more as an indicator of how well these elements can be used for taxonomic identification, which points are most relevant in shape-change, and the potential for developing further studies that may provide more direct phylogenetic input. The

Figure 12.1. A comparison of two hadrosaurines: **A**, *Edmontosaurus regalis* (type) and **B**, *Anatotitan copei* (type), drawn to the same quadrate height to minimize size differences. Drawings by Gregory S. Paul.

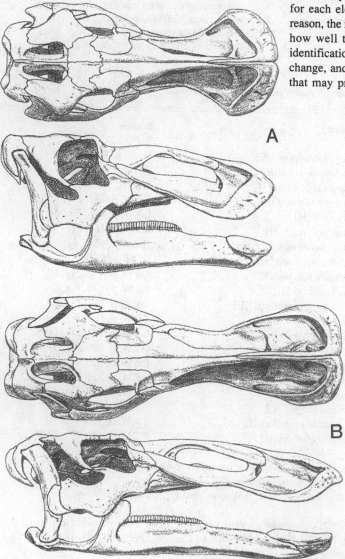

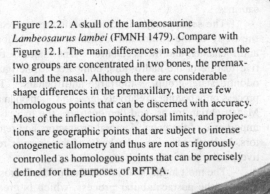

Figure 12.2. A skull of the lambeosaurine *Lambeosaurus lambei* (FMNH 1479). Compare with Figure 12.1. The main differences in shape between the two groups are concentrated in two bones, the premaxilla and the nasal. Although there are considerable shape differences in the premaxillary, there are few homologous points that can be discerned with accuracy. Most of the inflection points, dorsal limits, and projections are geographic points that are subject to intense ontogenetic allometry and thus are not as rigorously controlled as homologous points that can be precisely defined for the purposes of RFTRA.

Figure 12.3. Landmarks and polygonal diagrams used for Resistant-Fit Theta-Rho-Analysis of cranial illustrations for hadrosaurids and advanced ornithopods. **A**, shows the illustration of *Edmontosaurus* from Figure 12.1 with the landmarks indicated by the black circles. **B**, gives an sample polygonal diagram of that specimen for use with RFTRA.

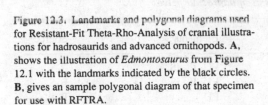

A

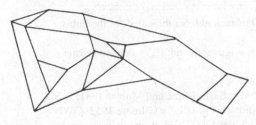

B

specimens used are given in Table 12.2, along with their current taxonomic assignment. Analyses were performed for specimens using the pubis, ilium, and ischium, the most diagnostic postcranial elements.

The landmarks used for the analysis of the pubis are illustrated in Figure 12.4A. The pubis is divided into three functional areas, the postpubis and acetabulum, the prepubic neck, and the expanded prepubic blade (Brett-Surman 1975, 1988). Each of these regions is represented by one or more polygons in the analyses. The true postpubis is vestigial and the ischial peduncle is relatively reduced. In each of the five main lineages of hadrosaurs (Brett-Surman 1979), the blade is the most important feature and is consistent in overall shape through time. In the transition from the hadrosaurine condition to that of the most advanced lambeosaurines (parasaurolophs), there is a consistent shortening of the prepubic neck and an expansion (dorsoventrally) of the neck and blade.

The landmarks used for analysis of the ilium (Fig. 12.4C) were chosen to highlight the three functional/taxonomic sections of the hadrosaur ilium, each delimited by one or more polygons in the graphical output. They are:

1. The anterior limit of the preacetabular process.
2. The point halfway on the dorsal rim of the preacetabular process. The length of the preacetabular process is measured along the midline between perpendicular lines through the anterior tip of the process and the inflection point where the process meets the body of the ilium.

3. The most dorsal aspect of the iliac body and inflection point of the iliac rim above the pubic peduncle.
4. The anterior limit of the antitrochanter.
5. The most lateral extent and inflection point of the antitrochanter.
6. The posterior end of the antitrochanter.
7. The inflection point on the dorsal rim of the postacetabular process where it meets the antitrochanter.
8. The most posterior extension of the postacetabular process on its midline.
9. The ventral inflection point of the postacetabular process where it meets the body of the ilium.
10. The posterior node of the ischial peduncle.
11. The anterior node of the ischial peduncle.
12. The midline or inflection point of the acetabulum.
13. The node of the pubic peduncle.
14. The inflection point between the iliac body and the preacetabular process.
15. The ventral midline point of the preacetabular process.

The hadrosaur ilium is divided into three sections, each represented by one or more polygons in the graphics. The preacetabular process is, in most cases, a simple vertical bar that forms a shallow angle with the dorsal rim of the iliac body. In primitive hadrosaurs, this process is mostly straight and slightly deflected ventrally.

In advanced hadrosaurs, this process becomes strongly deflected, especially in *Hadrosaurus*. In old hadrosaurs that display hyperostosis, this process becomes T-shaped in cross-section and deepens. Lambeosaurine ilia are, in general, thicker and more robust than those in hadrosaurines.

The second major iliac section is represented by the body of the ilium, and contains the peduncles, antitrochanter, and the acetabulum. One major distinctive feature of the hadrosaurs, compared to the iguanodontid sister-group, is the large antitrochanter. This process only becomes relatively large in Campanian and Maastrichtian forms, and demonstrates the increasing importance of the pelvic-femoral protractors and retractors. In lambeosaurines, the body of the ilium is relatively taller than in the hadrosaurines.

The most highly evolved section of the hadrosaur ilium is the postacetabular process, which is greatly increased in length compared with iguanodontids. In lambeosaurines, this process is not as lengthy as in the hadrosaurines, but is much thicker and taller. In pre-Santonian hadrosaurs (*Secernosaurus* and *Gilmoreosaurus*), this process is dorso-medially twisted and asym-

Table 12.2. *Specimens/illustrations used for analysis of hadrosaur pelves*

No.[a]	Taxon [Group][b]	Elements[c]	Source[d]
1	*Parasaurolophus* [L]	PU, IL, IS	B-S
2	*Lambeosaurus* [L]	PU, IL, IS	B-S
3	*Saurolophus* [H]	PU	B-S
4	*Hadrosaurus* [H]	PU, IL, IS	B-S
5	*Edmontosaurus* [H]	PU, IL	B-S
6	*Camptosaurus* [C]	PU, IL, IS	B-S
7	*Ouranosaurus* [I]	PU, IL, IS	TAQ
8	*Hypacrosaurus* [L]	PU, IL, IS	BRO
9	*Hadrosaurus* (2)? [H]	PU, IL, IS	PHO1
10	*Camptosaurus* (2) [C]	PU, IL, IS	GIL1
11	*Bactrosaurus* [I]	PU, IL, IS	GIL2
12	*Iguanodon* [I]	PU, IL, IS	OWN
13	*Muttaburrasaurus* [I]	PU, IL, IS	B&M
14	*Anatotitan* [H]	PU, IL, IS	PHO2
15	*Gilmoreosaurus* [H]	IL, IS	B-S
16	*Shantungosaurus* [H]	IS	B-S

Notes: [a]For RFTRA specimen file numbers add 1000 to each number indicated for the pubis, 1100 for the ischium, and 1200 for the ilium.
[b][C] = camptosaurid; [H] = hadrosaurine hadrosaur; [I] = iguanodontid; [L] = lambeosaurine hadrosaur.
[c]IL = ilium; IS = ischium; PU = pubis.
[d]AMNH = American Museum of Natural History; B&M = Bartholomai and Molnar 1981; B-S = Brett-Surman 1975; BRO = Brown 1913; GIL1 = Gilmore 1909; GIL2 = Gilmore 1933; OWN = Owen 1855; PHO1 = Photograph of specimen AMNH 5465; PHO2 = Photograph of specimen AMNH 5730; TAQ = Taquet 1976.

metrical in shape. In later hadrosaurs it becomes more vertical and more symmetric in shape.

The landmarks used for the analysis of the ischium are illustrated in Figure 12.4B. The ischium is divided into three sections, the head, shaft, and foot, all represented by one or more polygons in the graphics. There are three types of footed ischia in hadrosaurs, not two as commonly believed (Lull and Wright 1942). The first is the lambeosaurine condition with a fully formed, distally expanded foot. The second is the bulbous or partially formed foot found in the earliest hadrosaurs, such as *Gilmoreosaurus* and *Bactrosaurus*. This is also seen in the iguanodontids, and is the ancestral condition for

Figure 12.4. Drawings of the type pelvis of *Parasaurolophus cyrtocristatus* made from original photos. Included are **A**, the pubis; **B**, the ischium; and **C**, the ilium. Note the damaged acetabulum on the ilium. Black circles delineate the landmark points used in the RFTRA technique. Landmark point number is not included but can be deduced from the discussion in the text. Labels refer to major sections of the elements. In the top figure (pubis): 1. acetabulum; 2. iliac peduncle; 3. pre-pubic neck; and 4. pre-pubic blade. In the middle figure (ischium): 1. acetabulum; 2. iliac peduncle; 3. obturator notch; 4. shaft; 5. toe of the foot; 6. body of the foot; and 7. heel of the foot. In the bottom figure (ilium): 1. preacetabular process; 2. iliac body; 3. acetabulum; 4. antitrochanter; and 5. postacetabular process.

hadrosaurs. The last, and most derived type, is the complete absence of a distal expansion, found only in the hadrosaurines.

Results of preliminary morphometric analyses
Morphometric analysis of crania

The results of the Resistant-Fit Theta-Rho-Analysis (RFTRA) of the cranial material listed in Table 12.1 are summarized in the dendrogram in Figure 12.5, and an example comparing two specimens is presented in Figure 12.6. The dendrogram exhibits two major clusters, one for the Lambeosaurinae and the other for the hadrosaurine, camptosaurid, and iguanodontid taxa. Within the latter cluster two subgroups are evident, the first including both *Iguanodon* specimens and *Camptosaurus*, and the other including all the hadrosaurines and *Ouranosaurus*.

In initial analyses, *Maiasaura*, based on the original reconstruction published in Horner (1983), clustered problematically with *Camptosaurus*. The cause of this is apparent upon examination of the type material for *Maiasaura*, which shows that reconstruction to be misleading due to the limited quality and amount of material available at that time. Using a new reconstruction published in Horner and Gorman (1988), based on additional material, *Maiasaura* clusters with the other hadrosaurines, as would be expected. This illustrates the need to evaluate the results of any quantitative analysis for possible outliers, and the capability of RFTRA to indi-

Figure 12.5. The dendrogam resulting from UPGMA cluster analysis of RFTRA distance coefficients for crania.

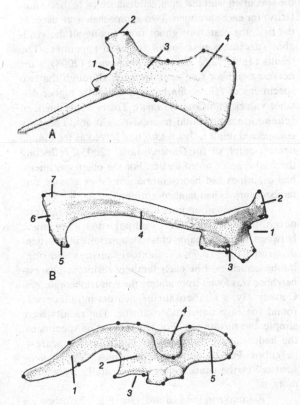

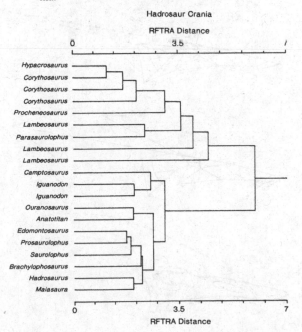

Hadrosaur Crania

RFTRA Distance

Hypacrosaurus
Corythosaurus
Corythosaurus
Corythosaurus
Procheneosaurus
Lambeosaurus
Parasaurolophus
Lambeosaurus
Lambeosaurus
Camptosaurus
Iguanodon
Iguanodon
Ouranosaurus
Anatotitan
Edomontosaurus
Prosaurolophus
Saurolophus
Brachylophosaurus
Hadrosaurus
Maiasaura

RFTRA Distance

cate where possible problems occur in the data. The *Maiasaura* specimen does tend to have the highest similarity among the hadrosaurines with the iguanodontids and *Camptosaurus*, reflecting Horner and Gorman's (1988) suggestion that *Maiasaura* is an evolutionarily conservative "generalized hadrosaur" with close affinities to the iguanodontids.

The pairing of *Anatotitan* with *Ouranosaurus* is considered to be convergence. Both taxa have evolved independently towards an extremely elongate muzzle with an unexpanded narial opening.

The structure of the RFTRA distance matrix can be examined further by observing patterns in the average distance values within and among the major groups. The results are summarized in Table 12.3 for each comparison possible except for the camptosaurid-

Figure 12.6. Resistant-Fit Theta-Rho-Analysis of hadrosaurid crania. The top illustration provides superimposed polygonal diagrams, the middle figure a vector diagram of changes from the base specimen to the other, and the bottom figure superimposed outlines. Specimens 2005 (base) and 2016, D = 13.2418. Method = RFTRA, Skeleton = hadrosaur skull lateral. Specimen 2005 = *Anatotitan copei*. Specimen 2016 = *Parasaurolophus walkeri*. Figure 12.6 shows a comparison of a hadrosaurine (*Anatotitan*) to a lambeosaurine (*Corythosaurus casuarius*). In actuality, the hadrosaurine skull is much longer than the lambeosaurine skull. RFTRA, however, shows that when absolute size is eliminated, the major differences are concentrated in the elaboration of the muzzle in hadrosaurines and the elaboration of the narial apparatus in lambeosaurines. Changes in the postorbital area are relatively minor.

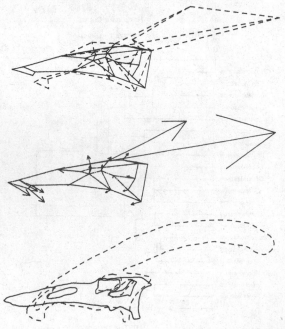

camptosaurid pairing because there was only one specimen in that group. The data presented include mean values, standard deviations, and the number of total comparisons.

The results show that *Camptosaurus* has the highest similarity (= lowest RFTRA distance coefficients) with the iguanodontids and progressively lower similarities with the hadrosaurines and lambeosaurines. The lambeosaurines exhibit the opposite trend with the highest similarity with the hadrosaurines. The iguanodontids and hadrosaurines show a close affinity. Within-group comparisons show the highest average similarity within the hadrosaurines and the lowest in the lambeosaurines.

The largest RFTRA distance coefficient value, 8.129, is between *Lambeosaurus lambei* (reference #2018) and *Camptosaurus depressus* (2001). The lowest distance, 1.154, is between *Hypacrosaurus altispinus* (2015) and *Corythosaurus "intermedius"* (2019). Table 12.4 gives the mean RFTRA coefficient values and standard deviations for each specimen. The mean values range from low values of 3.803 for *Saurolophus* (2010) and 3.832 for *"Procheneosaurus"* (2011), to high values of 5.881, 5.680, and 5.270 for the specimens of *Lambeosaurus* (2018, 2013, and 2012) and 5.380 for *Parasaurolophus* (2016).

The second method for elucidating matrix structure is to perform a nearest-neighbor analysis, finding the specimen with the smallest distance (= highest similarity) for each specimen. Two approaches were used. In the first, the nearest-neighbor from among all the available specimens was found for each specimen. The results (Table 12.5) show that *Maiasaura* (2008) is the nearest-neighbor to *Camptosaurus*, although the two specimens of *Iguanodon* have just slightly higher distance values with *Camptosaurus*. The two specimens of *Iguanodon* are mutual nearest-neighbors. The only unexpected result is that *Anatotitan* served as the mutual nearest-neighbor for *Ouranosaurus* (2003), reflecting the convergence noted earlier. For the other specimens, hadrosaurines had hadrosaurine nearest-neighbors, and lambeosaurines had lambeosaurines.

The second approach used a modified nearest-neighbor analysis. Here, comparisons were made between the two major clusters apparent in the dendrogram (Fig. 12.7), the lambeosaurines and non-lambeosaurines. For each lambeosaurine, a nearest-neighbor was found from among the non-lambeosaurines. Conversely, a lambeosaurine nearest-neighbor was found for each non-lambeosaurine. The results were simple and unanimous. For lambeosaurine specimens, the hadrosaurine *Saurolophus* is always the nearest-neighbor. For all non-lambeosaurines, *"Procheneosaurus"* is the nearest-neighbor, usually by a wide margin.

A sample individual analysis (Fig. 12.6) shows a comparison of a hadrosaurine (*Anatotitan*) to a lambeosaurine (*Corythosaurus casuarius*). In reality, the hadrosaurine skull is absolutely much longer than the lambeosaurine skull. RFTRA, however, shows that when absolute size is eliminated, the major differences are in the muzzle in the narial apparatus. Changes in the postorbital area are relatively minor.

Morphometric analysis of the pubis

The dendrogram resulting from the UPGMA cluster analysis of RFTRA distance coefficients (Fig. 12.7) shows overall pubic shape to be quite useful for delimiting major taxa within the advanced ornithopods, and should be useful for identifying isolated elements to

Table 12.3. *Resistant-Fit Theta-Rho-Analysis distance coefficient values for major taxa using cranial data*

Groups used for comparison	Mean	S.D.	N
Camptosaurs-iguanodonts	2.710	0.220	3
Camptosaurs-hadrosaurines	3.231	0.493	7
Camptosaurs-lambeosaurines	6.750	0.913	9
Iguanodonts-iguanodonts	3.041	0.798	3
Iguanodonts-hadrosaurines	2.824	0.432	21
Iguanodonts-lambeosaurines	6.097	0.921	27
Hadrosaurines-hadrosaurines	2.282	0.347	21
Hadrosaurines-lambeosaurines	5.858	0.952	63
Lambeosaurines-lambeosaurines	3.492	1.046	36

Note: N = number of comparisons used for calculation of mean and standard deviation; S.D. = standard deviation.

Table 12.4. *Resistant-Fit Theta-Rho-Analysis distance coefficient data for specimens used for cranial analysis*

No.	Taxon [Group][a]	Mean	S.D.
1	*Camptosaurus* [C]	4.816	2.010
2	*Iguanodon* [I]	4.395	1.863
3	*Ouranosaurus* [I]	4.253	1.624
4	*Edmontosaurus* [H]	4.092	1.954
5	*Anatotitan* [H]	4.532	2.108
6	*Hadrosaurus* [H]	4.158	1.922
7	*Brachylophosaurus* [H]	4.064	1.660
8	*Maiasaura* [H]	3.934	1.844
9	*Prosaurolophus* [H]	4.198	2.009
10	*Saurolophus* [H]	3.803	1.386
11	"*Procheneosaurus*" [L]	3.832	0.676
12	*Lambeosaurus* [L]	5.270	1.678
13	*Lambeosaurus* [L]	5.680	1.439
14	*Corythosaurus* [L]	4.850	1.693
15	*Hypacrosaurus* [L]	4.598	1.646
16	*Parasaurolophus* [L]	5.380	1.436
17	*Corythosaurus* [L]	4.284	1.364
18	*Lambeosaurus* [L]	5.881	1.459
19	*Corythosaurus* [L]	4.744	1.815
20	*Iguanodon* [I]	4.527	1.943

Note: S.D. = standard deviation.
[a][C] = camptosaurid; [H] = hadrosaurine hadrosaur; [I] = iguanodontid; [L] = lambeosaurine hadrosaur.

Table 12.5. *Nearest-neighbor data for cranial analyses using Resistant-Fit Theta-Rho-Analysis distance coefficients*

Original specimen		Nearest neighbor		
No.	Taxon [Group][a]	No.	Taxon [Group][a]	Value
1	*Camptosaurus* [C]	8	*Maiasaurua* [H]	2.513
2	*Iguanodon* [I]	20	*Iguanodon* [I]	2.051
3	*Ouranosaurus* [I]	5	*Anatotitan* [H]	2.020
4	*Edmontosaurus* [H]	6	*Prosaurolophus* [H]	1.801
5	*Anatotitan* [H]	3	*Ouranosaurus* [I]	2.020
6	*Hadrosaurus* [H]	4	*Edmontosaurus* [H]	1.919
7	*Brachylophosaurus* [H]	10	*Saurolophus* [H]	1.943
8	*Maiasaura* [H]	6	*Hadrosaurus* [H]	2.009
9	*Prosaurolophus* [H]	4	*Edmontosaurus* [H]	1.801
10	*Saurolophus* [H]	4	*Edmontosaurus* [H]	1.892
11	"*Procheneosaurus*" [L]	17	*Corythosaurus* [L]	2.532
12	*Lambeosaurus* [L]	16	*Parasaurolophus* [L]	2.422
13	*Lambeosaurus* [L]	12	*Lambeosaurus* [L]	3.747
14	*Corythosaurus* [L]	19	*Corythosaurus* [L]	1.522
15	*Hypacrosaurus* [L]	19	*Corythosaurus* [L]	1.154
16	*Parasaurolophus* [L]	12	*Lambeosaurus* [L]	2.422
17	*Corythosaurus* [L]	19	*Corythosaurus* [L]	2.011
18	*Lambeosaurus* [L]	19	*Corythosaurus* [L]	3.421
19	*Corythosaurus* [L]	15	*Hypacrosaurus* [L]	1.154
20	*Iguanodon* [I]	2	*Iguanodon* [I]	2.051

Note: [a][C] = camptosaurid; [H] = hadrosaurine hadrosaur; [I] = iguanodontid; [L] = lambeosaurine hadrosaur.

Figure 12.7. A dendrogram resulting from the UPGMA cluster analysis of RFTRA distance coefficients for the pubis.

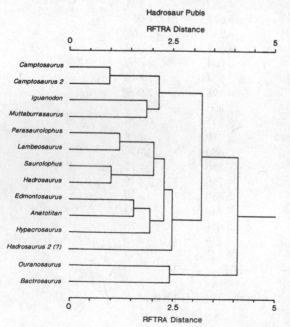

their correct taxon. The dendrogram shows two problematical outliers, *Ouranosaurus* and *Bactrosaurus*, that cluster together at a low level of similarity, suggesting unusual shapes for each and a small degree of convergence between the two taxa.

The other taxa fit nicely into the conventional phylogenetic structure accepted for advanced ornithopods. One major grouping clusters the hadrosaurids, and another the non-hadrosaurid taxa. Within the latter, the two specimens of *Camptosaurus* cluster together at a high level and the two remaining iguanodontids cluster together at a lower level, reflecting family level similarities. The hadrosaurid group includes a subcluster of two of the lambeosaurines, and others that give pairings of *Hadrosaurus* (= *Kritosaurus* following the classification given in Appendix 2) with *Saurolophus*, and *Edmontosaurus* with *Anatotitan*.

An example of the results from the individual comparisons is illustrated in Figure 12.8. Figures 12.8A-C demonstrate the shape changes from an iguanodontid (*Iguanodon*) to a hadrosaurid (*Parasaurolophus*). The results show that the acetabulum is enlarged, the prepubic neck and blade are enlarged, and that most of the changes are concentrated in the enlargement of the dorsal region of the blade.

Figures 12.8D-F demonstrates the changes of the pubis from a hadrosaurine (*Edmontosaurus*) to a lambeosaurine (*Parasaurolophus*). With the exception of the acetabulum, the same types of changes are seen as those from an iguanodontid to a hadrosaur, only to a smaller degree. In actuality, the hadrosaurine prepubis is absolutely much longer but the RFTRA technique reduced the elements to the same "best fit" to show true shape differences.

Morphometric analysis of the ilia

The dendrogram from analysis of the ilia is shown in Figure 12.9. The two specimens of *Camptosaurus* cluster together and with the iguanodontid *Muttaburrasaurus* at a low level of similarity, suggesting a high degree of variability in *Camptosaurus* and an ilium shape in those specimens distinct from that seen for the other taxa. Within the major cluster, including the iguanodontids and hadrosaurids, little structure of taxonomic interest is apparent, although *Edmontosaurus* and *Anatotitan* are joined with *Iguanodon*. In general, the dendrogram demonstrates the distinctiveness of the camptosaurids from the iguanodontids and hadrosaurids.

Figures 12.10A, B, C portray, as an example, the changes from an iguanodontid (*Iguanodon*) ilium to a hadrosaurid (*Parasaurolophus*) ilium. In hadrosaurs, the preacetabular process becomes longer and thicker, the acetabulum becomes deeper, the pubic peduncle is relatively reduced in size, and the postacetabular process becomes more delineated from the body of the ilium as a separate feature (rather than an elongated posterior extension of the iliac body as in iguanodontids). This may be due to the increasing size and importance of the caudifemoralis musculature. The two largest changes are the increased height of the ilium and the tremendous increase in the size of the antitrochanter.

Figures 12.10D, E, F shows the shape change between a hadrosaurine (*Edmontosaurus*) and a lambeosaurine (*Parasaurolophus*) ilium. As before, the two major changes are the height of the iliac body and the relative increase of the antitrochanter.

Morphometric analysis of the ischia

The analysis of the ischia (Fig. 12.11) are the least useful taxonomically of all those presented, and suggest that either the ischium is of little use for taxonomic discrimination at the generic level, or that new analyses should be run using different landmarks. The connections appear to cut across recognized taxonomic lines except for the close pairing of the two specimens of *Camptosaurus*. Figure 12.12 compares a hadrosaurine (*Hadrosaurus*) ischium to a lambeosaurine (*Parasaurolophus*) ischium. In lambeosaurines, the iliac peduncle is enlarged and widened, and the ischial head is relatively larger. The greatest change is in the ischial foot. The increased thickness of the shaft is not shown here due to the lack of landmarks in this area.

Figure 12.8. Results of a Resistant-Fit Theta-Rho-Analysis of the pubis for hadrosaurids and *Iguanodon*. Illustrations are as in Figure 12.6. **A**, **B**, **C**, demonstrate the shape changes from an iguanodontid (*Iguanodon*) to a hadrosaurid (*Parasaurolophus*). The RFTRA technique matches overall shape to a "best fit", consequently elements that are of different size are relatively reduced for the most accurate comparison. **D**, **E**, **F**, demonstrate the changes from a hadrosaurine (*Edmontosaurus*) to a lambeosaurine (*Parasaurolophus*).

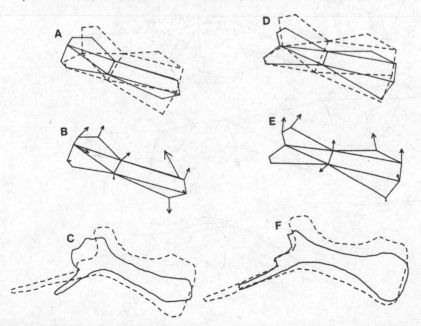

Interpretation of morphometric analyses

The results of the morphometric analyses provide important insight into the morphological and possible phylogenetic relationships among the taxa studied. As

Figure 12.9. A dendrogram resulting from the UPGMA cluster analysis of RFTRA distance coefficients from the ilia.

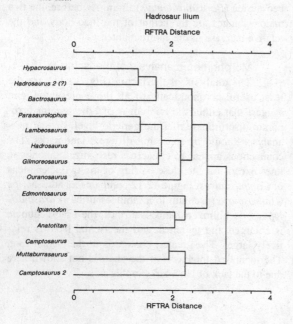

Hadrosaur Ilium

RFTRA Distance

expected, cranial analyses provide the most insight, although analyses of the pelvic elements suggest that they have an excellent potential for the identification of suprageneric taxa among advanced ornithopods. At present, the types of landmarks used in the studies of the pelvic elements preclude making convincing statements in evaluating phylogenies except to support evidence derived from other analyses. However, the results suggest that useful insights may be obtained within this context if homologous points can be found on these elements and new RFTRA analyses run. This is an area of ongoing investigation (Brett-Surman and Chapman in progress).

The results of the cranial analysis (Fig. 12.5) strongly demonstrate the distinct nature of the lambeosaurines, especially considering the conservative nature of the landmarks chosen. Examination of individual comparisons confirms that the major differences between the hadrosaurines and the lambeosaurines are the result of changes mostly in the premaxillae and nasals.

The intergroup relationships among the non-lambeosaurines show that the groups are less distinct morphologically than are the lambeosaurines, but differences clearly are evident. The non-lambeosaurine subclusters separate the hadrosaurines from the iguanodontids and *Camptosaurus*, with the exception of the convergence of *Ouranosaurus* with the hadrosaurine *Anatotitan*.

The results of the cranial analysis suggest that the landmarks used are quite conservative among the

Figure 12.10. Results of a Resistant-Fit Theta-Rho-Analysis of ilia. Illustrations are as in Figure 12.6. **A, B, C**, portray the changes from an iguanodontid (*Iguanodon*) ilium to a hadrosaurid (*Parasaurolophus*) ilium. **D, E, F**, show the shape change between a hadrosaurine (*Edmontosaurus*) and a lambeosaurine (*Parasaurolophus*).

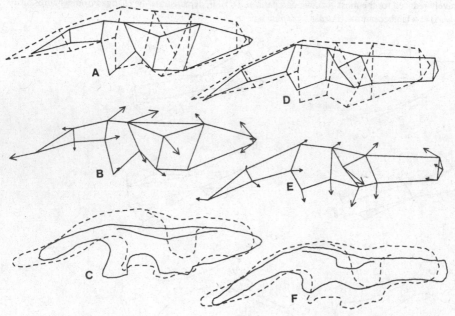

advanced ornithopods with moderate to small changes in position common even in the more distant pairings (e.g., *Camptosaurus* and *Parasaurolophus*). This is evidence for overall similarities in cranial morphology within the Euornithopoda. As would be expected from conventional taxonomic studies, the landmarks demonstrating the greatest changes, and thereby providing the greatest discrimination among taxa, are concentrated in the premaxillary lips, nasal area, and skull roof. The posteroventral landmarks tend to be far more conservative. The greater average distances among the lambeosaurines (Table 12.3) is to be expected due to high variability in forms thought to exhibit strong social behavior and sexual dimorphism (see Dodson 1975; Hopson 1975; Molnar 1977; Chapman et al. 1981).

The analyses run using pelvic elements, especially for the pubis, did provide additional insight into the relationships of the taxa studied. In general, the distinctness of the pelvic morphology of iguanodontids was clearly evident in all analyses. A close relationship between the lambeosaurines and hadrosaurines is evident in the analysis of the pubis, due to the presence only in hadrosaurids of a prepubic blade distinct from the prepubic neck.

Documenting whether the Hadrosauridae, including both the lambeosaurines and hadrosaurines, is monophyletic, as is accepted in most studies, or diphyletic, as suggested by Horner (this volume), is less straightforward. However, the analyses do provide important insights.

The average between-group RFTRA distance coefficients (Table 12.3) show that the lambeosaurines have the greatest average similarity with the hadrosaurines, the second greatest with the iguanodontids, and the least with *Camptosaurus*. This is as would be expected from published discussions (Lull and Wright 1942; Brett-Surman 1979; Sereno 1986) that assume monophyly. The relatively higher similarities among the non-lambeosaurines may appear to contradict this, but, instead, only indicate the high degree of morphological evolution within the lambeosaurine line. More relevant information is contained in the similarities of the lambeosaurines with all the other groups.

Powerful and supporting evidence comes from the second nearest-neighbor analysis between the lambeosaurine and non-lambeosaurine groups. If the lambeosaurines are derived from the iguanodontids independently from the hadrosaurines, then this analysis should have shown the lambeosaurines connecting preferentially to iguanodontids or, at the least, showing a

Figure 12.11. A dendrogram resulting from the UPGMA cluster analysis of RFTRA distance coefficients for the ischia.

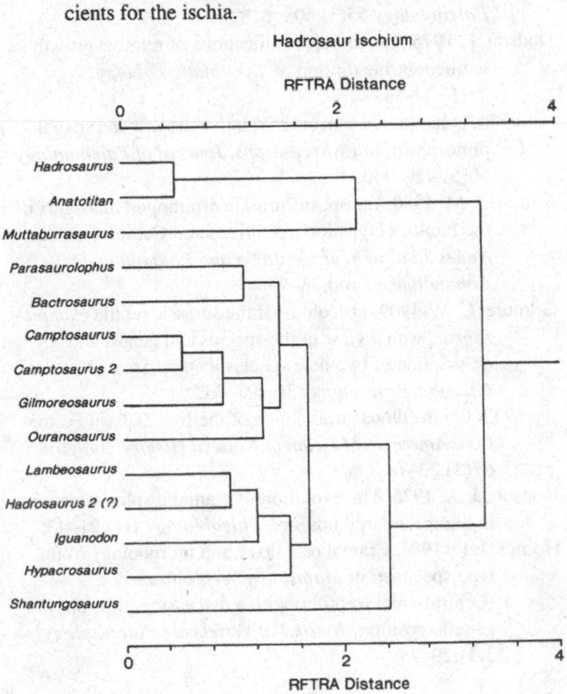

Figure 12.12. Results of a Resistant-Fit Theta-Rho-Analysis of the ischium. Illustrations are as in Figure 12.6. Figure 12.12 compares a hadrosaurine (*Hadrosaurus*) ischium to a lambeosaurine (*Parasaurolophus*) ischium. In lambeosaurines, the iliac peduncle is enlarged and widened, and the ischial head is relatively larger. The increased thickness of the shaft is not shown here due to the lack of identifiable landmarks in this area. The greatest change is in the ischial foot. *Iguanodon*, *Bactrosaurus*, and *Gilmoreosaurus* all have an expanded bulb at the end of the ischium. Lambeosaurines have elaborated this condition in adults while juveniles still retain the plesiomorphic bulb. The synapomorphic condition is the loss of the distal ischial expansion found in all hadrosaurines.

Hadrosaur Ischium

RFTRA Distance

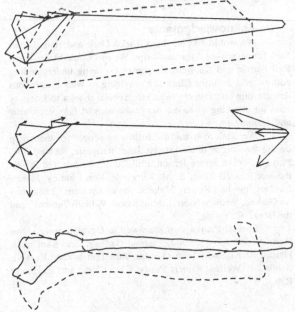

wide variation. Instead, the lambeosaurines connect unanimously to the hadrosaurine *Saurolophus* and tend to have their lowest three to four distances with hadrosaurines. This, combined with recent phylogenetic analyses (e.g., Sereno 1986), strongly supports a close lambeosaurine-hadrosaurine link and argues convincingly for the Hadrosauridae as monophyletic. The less rigorously defined data for the pubis also supports a close lambeosaurine-hadrosaurine link, although far less convincingly.

Summary and conclusions

The hadrosaurids are a distinctive group of advanced ornithopods. The results of the morphometric analyses using the landmark shape analysis method Resistant-Fit Theta-Rho-Analysis (RFTRA) support a monophyletic interpretation for the Hadrosauridae.

Application of RFTRA to the crania of hadrosaurids, iguanodontids, and *Camptosaurus* provided important insights into morphological variability and taxonomic structure, and supports conventional phylogenetic interpretations of ornithopod phylogeny (e.g., Sereno 1986). Among hadrosaurids, most of the important morphological variability or evolution occurs in the muzzle and narial regions, providing the distinct cranial morphologies characterizing the hadrosaurid subfamilies (Figs. 12.1, 12.2).

Morphometric analyses using the pelvic elements were less successful because of the lack of homologous points and the use of geometrically defined points. However, the data did suggest strongly that element shape can be useful for the identification of taxa using isolated elements. For this study, the pubis provided the most information and the ilium, less. The ischium provided the least information, suggesting that it is not as diagnostic and that additional analyses need to be run (work in progress by the authors).

Acknowledgments

We would like to thank Linda Deck and Hans-Dieter Sues for reviewing our manuscript. We also would like to thank Phil Currie and Kenneth Carpenter for being understanding editors, and Jennifer Clark for providing us with illustrations despite our last minute requests. Special thanks to Gregory Paul for allowing us to use his restorations of *Edmontosaurus* and *Anatotitan*.

We wish to thank the following people for their help during the course of this study: Jose Bonaparte, Richard Fox, Phil Gingerich, James Jensen, Phil Currie, the late Ted White, the late Russell King, S. M. Kurzanov, Don Lindsey, Joanne Lindsey, the late Robert Makela, James Madsen, Christopher McGowan, William Morris, John Storer, William Turnbull, and the late C. C. Young.

Special thanks also are owed to David Berman, John Bolt, A. Gordon Edmund, Eugene Gaffney, Erle Kauffman, Harold E. Koerner, John Ostrom, Jane Danis, Dale Russell, Samuel P. Welles, Robert Purdy, Chip Clark, and Raymond Rye.

We are particularly indebted to the following individuals for their generous time and effort in aiding us during this study by providing data, equipment, photographs, helpful discussions, and active support: Don Baird, Alan Charig, John Horner, Joseph Gregory, Nicholas Hotton III, Douglas A. Lawson, Robert Long, David Norman, George Olshevsky, Halska Osmólska, Gregory S. Paul, Phillipe Taquet, Peter Galton, and John S. McIntosh.

References

Bartholomai, A. and Molnar, R. E. 1981. *Muttaburrasaurus*, a new iguanodontid (Ornithischia: Ornithopoda) dinosaur from the Lower Cretaceous of Queensland. *Queensland Museum, Memoirs* 20(2):319–349.

Benson, R. H., Chapman, R. E. and Siegel, A. F. 1982. On the measurement of morphology and its change. *Paleobiology* 8(4):328–339.

Bookstein, F. L., Chernoff, B., Elder, R., Humphries, D., Smith, G. and Strauss, R. 1985. Morphometrics in evolutionary biology. *Academy of Natural Sciences, Philadelphia, Special Publications* no. 15:1–277.

Brett-Surman, M. K. 1975. The appendicular anatomy of hadrosaurian dinosaurs. M.Sc. thesis (Berkeley: University of California).

1979. Phylogeny and paleobiogeography of Hadrosaurian dinosaurs. *Nature* 277(5697):560–562.

1988. Revision of the Hadrosauridae. Ph.D. thesis (Washington, D. C.: George Washington University).

Brown, B. 1913. A new trachodont dinosaur *Hypacrosaurus* from the Edmonton Cretaceous of Alberta, *American Museum of Natural History Bulletin* 32(20):395–406.

Chapman, R. E., Galton, P. M., Sepkoski, J. J. Jr., and Wall, W. P. 1981. A morphometric study of the cranium of the pachycephalosaurid dinosaur *Stegoceras*. *Journal of Paleontology* 55(3):608–618.

Dodson, P. 1975. Taxonomic implications of relative growth in lambeosaurine dinosaurs. *Systematic Zoology* 24(1):37–54.

1976. Quantitative aspects of relative growth and sexual dimorphism in *Protoceratops*. *Journal of Paleontology* 50(5):929–940.

Galton, P. M. 1980. European Jurassic ornithopod dinosaurs of the families Hypsilophodontidae and Camptosauridae. *Neues Jahrbuch für Geologie und Paläontologie, Abhandlungen* 160:73–95.

Gilmore, C. W. 1909. Osteology of the Jurassic reptile *Camptosaurus*, with a view of the species and genus, and description of two new species. *United States National Museum, Proceedings* 36:197–332.

1933. On the dinosaurian fauna of the Iren Dabasu Formation. *American Museum of Natural History, Bulletin* 67(2):23–78.

Hopson, J. A. 1975. The evolution of cranial display structures in hadrosaurian dinosaurs. *Paleobiology* 1(1):21–43.

Horner, J. R. 1983. Cranial osteology and morphology of the type specimen of *Maiasaura peeblesorum* (Ornithischia: Reptilia) with a discussion of its phylogenetic position. *Journal of Vertebrate Paleontology* 3(1):29–38.

Horner, J. R. and Gorman, J. 1988. *Digging Dinosaurs* (New York: Workman Publishing).

Lull, R. S. and Wright, N. E. 1942. Hadrosaurian dinosaurs of North America. *Geological Society of America, Special Paper* no. 40:1–242.

Maryańska, T. and Osmólska, H. 1974. Pachycephalosauria, a new suborder of Ornithischian dinosaurs. *Palaeontologia Polonica* 30:45–102.

Molnar, R. E. 1977. Analogies in the evolution of combat and display structures in ornithopods and ungulates. *Evolutionary Theory* 3(3):165–190.

Owen, R. 1855. Monograph on the fossil Reptilia of the Wealdon and Purbeck Formations. Part II. Dinosauria (*Iguanodon*) (Wealdon). *Palaeontographical Society, Monographs* no. 8:1–54.

Romer, A. S. 1956. *Osteology of the Reptiles* (Chicago: University of Chicago Press).

Sereno, P. C. 1986. Phylogeny of the bird-hipped dinosaurs (order Ornithischia). *National Geographic Research* 2(2):234–256.

Siegel, A. F., and Benson, R. H. 1982. A robust comparison of biological shapes. *Biometrics* 38:341–350.

Sneath, P. H. A. 1967. Trend-surface analysis of transformation grids. *Journal of Zoology* 151:65–122.

Sues, H.-D. and Galton, P. M. 1987. Anatomy and classification of the North American Pachycephalosauria (Dinosauria: Ornithischia). *Palaeontographica A*, 198:1–40.

Taquet, P. 1976. *Géologie et Paléontologie du Gisement de Gadoufaoua (Aptien du Niger)* (Paris: Centre National de la Recherche Scientifique), pp. 1–191.

Thompson, D. W. 1942. *On Growth and Form* (2nd edn.), vol. 1, 2 (Cambridge: Cambridge University Press).

Appendix 1

The diagnosis for *Anatotitan* is given below.
Family Hadrosauridae
   Subfamily Hadrosaurinae
   Anatotitan Brett-Surman, new genus
   A. copei (Lull and Wright 1942) new combination
   Holotype: AMNH 5730
   Referred: AMNH 5886, 5887, CM 16520, and a mounted specimen in the Ekalaka Museum, Montana.
   Type locality: Near Moreau River, Black Hills, S. Dakota
   Age: Late Cretaceous, Maastrichtian
   Horizon: Hell Creek Formation
   Etymology: *Anas* (Latin: duck) and *Titan* (Greek: "large")

Diagnosis: Skull longer and lower than in any other hadrosaur, muzzle wider than in any other hadrosaur, quadrate/mandible ratio the smallest of all hadrosaurs, edentulous portion of the mandible relatively longer than in any hadrosaur, appendicular elements relatively longer and more gracile than in any hadrosaur of the same quadrate height, limb elements up to 10% longer than in an *Edmontosaurus* of the same quadrate height, neck of prepubis relatively longer and shallower than in any hadrosaur, postacetabular process more dorso-medially twisted and relatively shorter than in an *Edmontosaurus* of the same size.

Appendix 2

The following classification was used in this paper. Clade definitions and discussion are given in Brett-Surman (1988).

   Class Reptilia
   Subclass Archosauria
   Order Ornithischia
      Suborder Ornithopoda
         Infraorder Iguanodontia
            Family Hadrosauridae
Subfamily Hadrosaurinae
   (edmontosaur clade) *Anatotitan*, *Edmontosaurus*, *Shantungosaurus*, *Tanius*, *Telmatosaurus* (= *Orthomerus*)
   (hadrosaur clade) *Aralosaurus*, *Brachylophosaurus*, *Hadrosaurus* (= *Kritosaurus* fide Horner, and *Gryposaurus*)
   (sauroloph clade) *Lophorhothon*, *Maiasaura*, *Prosaurolophus*, *Saurolophus*
   Hadrosaurinae *incertae sedis*: *Secernosaurus*, *Gilmoreosaurus*

Subfamily Lambeosaurinae
   (corythosaur clade) *Corythosaurus*, *Hypacrosaurus*, *Lambeosaurus*, *Nipponosaurus* (specimens assigned to "*Procheneosaurus*" are assumed to be juvenile members of this clade)
   (parasauroloph clade) *Bactrosaurus*, *Parasaurolophus*, *Tsintaosaurus*
   Lambeosaurinae *incertae sedis*: *Barsboldia*, *Jaxartosaurus*

   Hadrosauridae *incertae sedis*: *Mandschurosaurus*, *Cionodon*, *Hypsibema*, *Microhadrosaurus*, *Ornithotarsus*, *Pneumatoarthrus*

   Ceratopsia occasionally referred to the Hadrosauridae: "*Agathaumus*," *Claorhynchus*, (?)*Notoceratops*, (?)*Arstanosaurus*

13 Evidence of diphyletic origination of the hadrosaurian (Reptilia: Ornithischia) dinosaurs

JOHN R. HORNER

Abstract

The family Hadrosauridae as originally named by E. D. Cope in 1869 is reevaluated using shared derived characters, and found to exhibit a dichotomy suggesting a diphyletic origin. The newly-defined Hadrosauridae is found to share a number of derived characters with the genus *Iguanodon*, whereas the newly defined Lambeosauridae shares a number of derived characters with the genus *Ouranosaurus*. The newly defined families are defined exclusively with shared derived characters.

Introduction

In 1869, Edward Drinker Cope proposed the family name Hadrosauridae for the herbivorous bipedal dinosaurs having several vertical rows of teeth forming a dental battery. Although there was some discussion in later years as to whether or not the family should be called the Trachodontidae (Lydekker 1888; Marsh 1890), the Hadrosauridae has been retained. For nearly a century, all of the "duckbilled" dinosaurs have been considered a single family, broken down into a variety of subfamilies (Brown 1914; Lambe 1918; Parks 1923; Lull and Wright 1942; Sternberg 1954). Lull and Wright (1942) give the family characters of the Hadrosauridae as follows:

Bipedal, unarmored dinosaurs, with a complicated dental battery consisting of many rows of teeth whose enamelled crowns, at least in the dentary, form a tessellated pavement. Individual teeth enamelled on one face only. Premaxillae edentulous and, together with the predentary, forming a more or less broadened, duck-like beak. Rear of skull either flat, or crested in varying degree.

Cervical vertebrae opisthocoelous, about 15 in number, presacrals about 30 to 34, average 32, sacrals usually 8. Tail long, laterally compressed for swimming. Femur longer than tibia, hind limb about twice the length of the fore limb (average ratio

In Dinosaur Systematics: Perspectives and Approaches, *Kenneth Carpenter and Philip J. Currie, eds. Copyright © Cambridge University Press, 1990.*

is 1:1.568). Manus with four digits, of which but two as a rule bear unguals, pes three toed, digitigrade in posture. Subaquatic in manner of life.

The Hadrosauridae are distinguished from the Camptosauridae and Iguanodontidae by the reduced number of digits in manus and pes, the presence of the antitrochanter on the ilium, and the reduction of the post-pubis, also by the distribution of these dinosaurs in time, the hadrosaurs being entirely confined to the Upper Cretaceous, and by their evident adaptation to an amphibious if not aquatic mode of life.

Most important of these characters, and those which not only appear to unite the group, but set it aside from the Iguanodontidae, include the presence of a dental battery that has more than one replacement tooth per tooth position, large antitrochanter on the ilium, and reduction of the digits on the manus. Additional characters include an extended diastema between the predentary and the anterior-most dentary tooth position, squamosals separated from one another by a thin parietal process, reduction of the oblique, caudoventral jugal process of the maxilla (see Norman this volume), reduction or loss of the quadrate foramen, angular on medial side of surangular, and wedge-shaped phalanges of the manus.

In the following, the primitive and derived characters of the two different hadrosaur groups (hadrosaurids and lambeosaurids after Sternberg 1954) are discussed, followed by a discussion of the similarities and dissimilarities of two iguanodontid genera. A reclassification of the "hadrosaurs" is suggested. Character polarity has been determined on the basis of *Camptosaurus*, *Dryosaurus*, *Hypsilophodon*, *Orodromeus*, *Tenontosaurus*, and *Zephyrosaurus*.

Hadrosaurid characters

The flat-headed hadrosaurs (Hadrosaurinae, of Sternberg 1954 and Horner 1983) can be characterized as having a long, low skull, with or without a solid nasal/frontal crest (Fig. 13.1A), widely expanded duck-like, premaxillary bill with anterior and laterally raised "lip," large, open external naris located at the anterior end of the snout, nasal extremely near or forming a portion of the orbit, an antero-dorsal maxillary process (anterior maxillary process of Sternberg 1953 and Heaton 1972) separated from an antero-ventral maxillary process by an anterior maxillary notch (Figs. 13.2A, B,C), an angle of 120° to 140° between the root and crown of all dentary teeth (Figs. 13.3A,B; also see Sternberg 1936; Lull and Wright 1942; Langston 1960), bifurcated foramina penetrating the anterior end of premaxillary beak (Sternberg 1953; Horner 1983), circumnarial depression (Hopson 1975), parietal with posterior squamosal process, supraoccipital process of the squamosal infolded, vertebral neural spines relatively low, scapular blade with near parallel dorsal and ventral borders, humerus with a relatively narrow deltopectoral crest, sacrum with a median longitudinal furrow on the ventral surface (Young 1958), a long, dorsoventrally restricted prepubic shaft, and an ischium with a straight narrow distal shaft.

Of these, the long low skull, open external naris, low neural spines, narrow deltopectoral crest of the humerus, and the long, dorso-ventrally restricted prepubic shaft are primitive characters.

Lambeosaurid characters

The "crested hadrosaurs" are characterized by a short, high skull with some sort of enclosed narial tract (Fig. 13.1B) bounded by the premaxillae and nasals (see Lull and Wright 1942, and Young 1958, for variations), external nares surrounded by the premaxillae, extreme excavation of the nasals, narrow premaxillary bill with a deflected anterior and antero-lateral edge, anterior maxillary shelf (Figs. 13.2D,E,F; Heaton 1972), ventrally deflected narrow dentary, dentary teeth with crown and root at an angle greater than 145° (Fig. 13.3C; also see Sternberg 1936; Langston 1960), supraoccipital with squamosal bosses (Gilmore 1937), vertebral neural spines relatively high, scapular blade expands posteriorly, humerus with a massive and extended deltopectoral crest, sacrum with a longitudinal median keel on its ventral surface (Young 1958), a short, dorsoventrally expanded prepubic shaft, and an ischium with an expanded distal process.

Of these, the narrow premaxillary bill, large crown to root angle, posteriorly expanded scapular blade, sacrum with longitudinal keel on the ventral surface, and ischium with expanded distal process are primitive characters.

Figure 13.1. **A**, Left lateral view of *Prosaurolophus* sp. (MOR 454) cut away anteriorly for view of antero-dorsal process of the maxilla; **B**, left lateral view of *Corythosaurus casuarius* (AMNH 5240) cut away anteriorly for view of antero-dorsal maxillary flange; **C**, left lateral view of *Iguanodon bernissartensis* (after Norman 1980) cut away anteriorly for view of antero-dorsal process of the maxilla (reconstructed from *I. atherfieldensis* Norman 1986); **D**, left lateral view of *Ouranosaurus nigeriensis* (after Taquet 1976) cut away anteriorly for view of antero-dorsal flange.

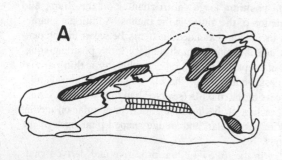

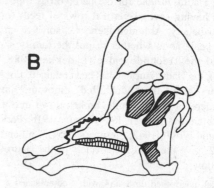

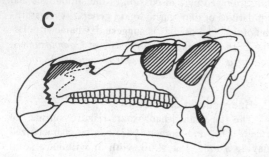

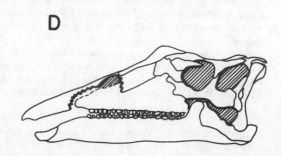

Descriptive discussion

Architecture of the premaxillary nasal apparatus is not only important in identification of the two hadrosaur groups, but is also commonly used to determine generic identity. In all hadrosaurids including babies of *Maiasaura peeblesorum* Horner and Makela (1979), the external naris is located at the anterior end of the snout and overlies part of the anterior end of the maxilla. It is bordered (externally) anteriorly and ventrally by the premaxilla, and above and posteriorly by the nasal. As a result, a process of the nasal extends forward dorsomedially to the anterior end of the narial opening. In all lambeosaurids, including embryonic individuals, the nasals reside well posterior of the narial opening. In this situation, the narial openings are bound entirely, including dorso-medially, by the premaxillae (see Ostrom 1961). In *Tenontosaurus* (Ostrom 1970) and *Camptosaurus* (Gilmore 1909), the nasal forms the posterior wall of the open nares, whereas in *Hypsilophodon* (Galton 1974) the anterior process of the nasal extends about halfway over the open nares.

The structure of the premaxilla-maxilla joint, or contact surface, appears to be significant. In hadrosaurids the antero-dorsal process of the maxilla fits into a groove on the dorsal surface of the lower premaxillary process, forming a portion of the ventro-medial border of the naris (Sternberg 1953; Horner 1983). The antero-ventral process of the maxilla extends forward, beneath the premaxilla for support. In *Hypsilophodon* (Galton

1974), *Dryosaurus* (Galton 1983), *Camptosaurus* (Galton and Powell 1980), and *Orodromeus* (pers. observ.), the anterior or antero-dorsal process of the maxilla fits into a conical pit on the medial side of the premaxilla. The antero-dorsal maxillary processes in these genera do not border any part of the external naris. The antero-ventral process of the maxilla is not present in *Hypsilophodon* (Galton 1974), *Dryosaurus* (Galton 1983), *Tenontosaurus* (see Norman this volume), or *Orodromeus*. In lambeosaurids there exists only a maxillary shelf (?antero-ventral maxillary process) and medial flange upon which the premaxilla rests. No maxillary processes exist to contact the dorsal surface of any portion of the premaxilla. The naris of lambeosaurids, therefore, has no restriction as to its position, other than an entrance into the skull anterior to the cerebral cavity.

On the dorsal surface of the hadrosaurid premaxilla, at the base of the dorsal process, is a large foramen that extends down into the body of the beak (Horner 1983). Interiorly, the foramen opens into a chamber that exits on the antero-dorsal surface and on the ventral surface. A similar chamber is present in *Hypsilophodon* (Galton 1974, Fig. 13.5A), although the medial wall is absent allowing the right and left chambers to unite. Other hypsilophodonts such as *Zephyrosaurus* (Sues 1980) and *Orodromeus* also have similar premaxillary foramenae, as apparently does *Camptosaurus* (Gilmore 1909, Fig. 6). In the lambeosaurid premaxilla there are no penetrating foramina or internal chambers. The ante-

Figure 13.2. **A-C**: Anterior end of maxilla belonging to *Brachylophosaurus goodwini* (Horner 1988). **A**, medial view; **B**, anterior view; **C**, lateral view. **D-F**: Anterior end of maxilla belonging to *Corythosaurus* sp. (TMP 79.8.220). **D**, medial view; **E**, anterior view; **F**, lateral view. (Scale bar = 5 cm.)

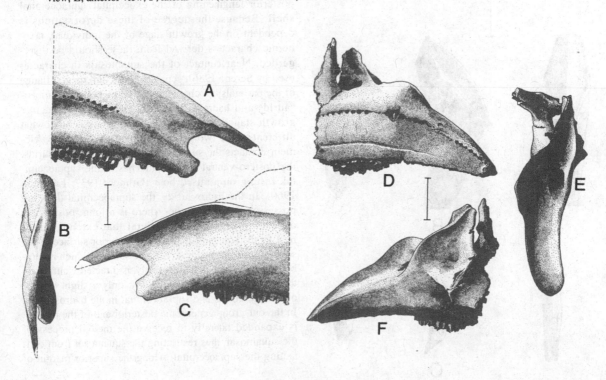

rior and antero-lateral lip on the edge of the dorsal surface of the premaxilla of hadrosaurids is in contrast to the deflected anterior edge of the lambeosaurids. A slightly raised premaxillary lip is present in *Tenontosaurus* and *Camptosaurus*, but is absent on the premaxillae of *Hypsilophodon* and *Orodromeus*.

Another major difference between the two hadrosaur groups is the overall construction of the dentary teeth. All hadrosaurid teeth have diamond-shaped crowns, whereas lambeosaurid teeth are more elongate (Figs. 13.4A–H). Angulation between crown and root is significant in that there are no exceptions in either group. Primitive hadrosaurid dentary teeth such as those of *Claosaurus agilis* Marsh (1872) and the undescribed hadrosaur AMNH 5465 have short, wide crowns (Fig. 13.4E). Furthermore, based on AMNH 5465, they have the lowest angulation of any known hadrosaur (120°).

Figure 13.3. Dentary teeth of hadrosaurs. **A**, posterior view of right dentary tooth of *Prosaurolophus* sp. (MOR 454); **B**, posterior view of right dentary tooth of "*Kritosaurus*" *incurvimanus* (UM 5204); **C**, posterior view of right dentary tooth of *Hypacrosaurus* sp. (MOR 355).

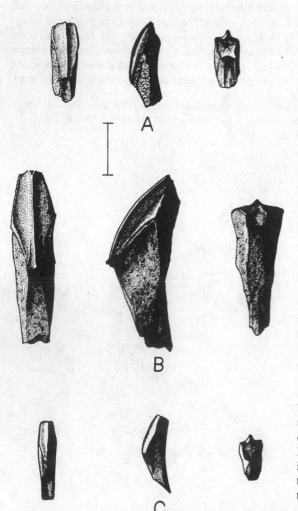

The lowest crown-fang angulation known for a lambeosaurid is 145°, measured from dentary teeth (AMNH 6553) referable (Weishampel and Horner 1986) to *Bactrosaurus johnsoni* Gilmore (1933). *Bactrosaurus*, thought to represent a primitive lambeosaur, has tooth crowns that are more elongate (Fig. 13.4A) than any known primitive hadrosaurid. The lambeosaurids packed a greater number of teeth in a shorter (antero-posterior) and thinner (labial-lingual) area than do the hadrosaurids. The dentary teeth of *Hypsilophodon*, *Tenontosaurus* (Ostrom 1970), and *Camptosaurus* have crown to root angles most similar to the lambeosaurids.

Overall skull shape of the two hadrosaur groups was apparently influenced by premaxillary-nasal architecture and tooth construction. Hadrosaurids possessed relatively long, low skulls because of limitations to the posterior migration of the external nares. The posterior nasal migration in lambeosaurids coincided with deformations of nearly all cranial elements, the orbits, and fenestrae. The upturned antero-dorsal process of the jugal is one example, effectively dragged up the face in response to the posterior migration and vertical vaulting of the nasals and premaxillaries. In either hadrosaurid (i.e., *Maiasaura peeblesorum*, PU 23430) or lambeosaurid (*Hypacrosaurus* sp., MOR 434) embryonic individuals, the antero-dorsal process of the jugal raises very slightly.

In juvenile or adult hadrosaurids or lambeosaurids posterior nasal position correlates with the degree of upturning of the antero-dorsal process of the jugal, foreshortening of the lateral and temporal fenestrae, transverse narrowing of the skull roof, and antero-posterior lengthening of the exoccipital-supraoccipital shelf. Because the degree of these deformations is dependent on the growth stage of the individual, taxonomic characters derived from these should be disregarded. Near contact of the squamosals (a character used by Sereno 1986), separated by a thin medial flange of the parietals is a character shared by both the lambeosaurids and hadrosaurids throughout their respective growth stages. Shape of the squamosals is somewhat different in the two groups, particularly where the element contacts the supraoccipital. In the lambeosaurids, the postero-ventral surface attaches to the supraoccipital, resting on a raised boss (Gilmore 1937; Langston 1960). In the hadrosaurids, the supraoccipital does not possess a boss, but rather, there is a long, posteriorly directed process of the parietal that extends out to receive the squamosal. Also, the posterior surface of the squamosal folds under so that the surface, which in lambeosaurids is directed posteriorly, is directed ventrally in the hadrosaurids. There is, if any, only a slight contact of the squamosal and supraoccipital in the hadrosaurids. In the out-group genera, the posterior end of the parietal is expanded laterally to receive the medial process of the squamosal, thus restricting the squamosal from contacting the supraoccipital. Along the superior margin of the orbit in hadrosaurids, the surface is extremely rough and rugose, and the prefrontal is penetrated by one to three foramina interpreted as remnants of a supraorbital fenestra (Horner 1983). The rugose surface has been shown (Maryańska and Osmólska 1979; Horner 1983) to be fused palpebrals. In lambeosaurids, the superior edge of the orbit is smooth or only slightly rough, and there is no indication, even in embryonic or juvenile specimens, to indicate fusion of palpebral elements. No foramina penetrate the prefrontal. In *Hypsilophodon* and *Orodromeus* the palpebral attaches to the prefrontal and lacrymal, and extends posteriorly into the orbital fenestra (Galton 1974; Coombs 1972). In *Zephyrosaurus* the palpebral is short, rugose, and attaches to the prefrontal along the superior border of the orbit (Sues 1980). In *Dryosaurus* the palpebral appears to cover the prefrontal and extend posteriorly along the superior border of the orbit attaching to the postorbital (Coombs 1972). A large fenestra medial to the palpebral excludes the frontal from reaching the palpebral. The *Camptosaurus* palpebral is similar to those of *Hypsilophodon* and *Orodromeus* (Gilmore 1909; Coombs 1972).

In the postcranium of the hadrosaurids and lambeosaurids, the shape of the scapular blade, construction of the ventral surface of the sacrum, and construction of the distal end of the ischium appear to be of greatest use for taxonomic differentiation. The neck of the lambeosaurid scapular blade is restricted and the blade fans out distally. The neck of the hadrosaurid scapula is relatively broader than that of the lambeosaurid and expands little before both the dorsal and ventral edges become parallel. Each of the out-group genera have scapulae that fan out posterior from a short, restricted shaft. The sacral vertebrae of the hadrosaurids and lambeosaurids are similar except along the ventral surface of the posterior centrae. In hadrosaurids, the posterior four or five centra possess a longitudinal median furrow, whereas in the lambeosaurids this area is keeled. In *Hypsilophodon*, *Orodromeus*, and *Camptosaurus*, the ventral surfaces of the posterior sacrals are either rounded or keeled.

The shaft of the hadrosaurid ischium narrows distally from the obturator process to a small, often slightly inflated termination. In lambeosaurids, the shaft begins to expand just distal to the obturator process and contin-

Figure 13.4. Crown view of lambeosaurid and hadrosaurid dentary teeth. **A**, *Bactrosaurus johnsoni* (AMNH 6553); **B**, *Corythosaurus casuarius* (USNM 10309); **C**, *Hypacrosaurus* sp. (USNM 11950); **D**, *Parasaurolophus* sp. (KU 17800); **E**, undescribed gryposaur-like hadrosaurid (AMNH 5465); **F**, *Gryposaurus notabilis* (AMNH 5350); **G**, *Saurolophus osborni* (AMNH 5221); **H**, ?"*Anatosaurus*" *copei* (AMNH 1811). (Scale bar = 2 cm.)

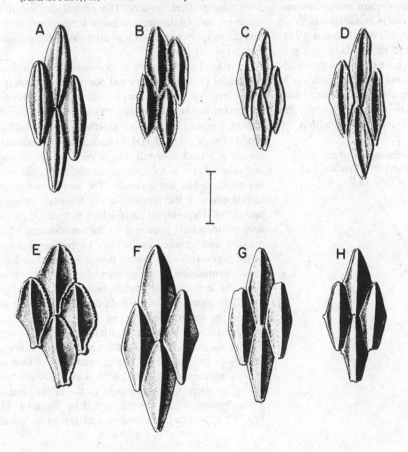

ues into a footed process. In each of the out-group genera, the shaft of the ischium expands posterior from the obturator process.

Preliminary indications are that the neural spines of hadrosaurids are relatively shorter than those of the lambeosaurids, and the deltopectoral crest of the hadrosaurid humerus is relatively less robust than the corresponding crest of the lambeosaurids. The prepubic blades of the hadrosaurids are generally longer and less inflated than those of the lambeosaurids. In the outgroup genera the neural spines are relatively short; the deltopectoral crest of the humerus is narrow in *Hypsilophodon*, *Dryosaurus*, *Orodromeus*, and *Camptosaurus*, but relatively wide and robust in *Tenontosaurus* (Forster 1985), and the prepubic blade is not, or only slightly, expanded anteriorly.

Hadrosaur ancestry: the iguanodonts

As stated by Sereno (1986), the genera *Iguanodon* and *Ouranosaurus* appear to be more closely related to the hadrosaurs than any other iguanodont-like ornithopods. Sereno also suggests that *Iguanodon* and *Ouranosaurus* are paraphyletic and are part of a series of sister-taxa of the monophyletic family Hadrosauridae.

Based on the descriptions of *Iguanodon bernissartensis* by Norman (1980), and *I. atherfieldensis* by Hooley (1925) and Norman (1986), the genus *Iguanodon* appears to share more characters with the hadrosaurids (flat-headed hadrosaurs) than with *Ouranosaurus*. The maxilla of *Iguanodon* possesses both an antero-dorsal maxillary process (Weishampel 1984; Norman 1986, this volume) and an antero-ventral process, each similar in shape to those of the hadrosaurids (Fig. 13.1C). Although the antero-dorsal process is incomplete in *Iguanodon bernissartensis* (Norman 1980) and *Iguanodon atherfieldensis* (Norman 1986), the antero-lateral portion of the maxilla, upon which the

Figure 13.5. Phylogenetic relationships of the Iguanodontoidea (after Sereno 1986). Characters in support of nodes are listed in Table 13.1.

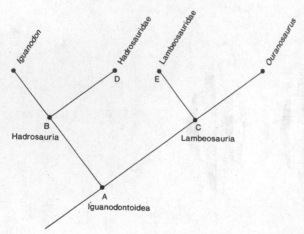

premaxilla rests, has a steeply inclined surface that corresponds with the relatively steep, laterally inclined premaxillary lower limbs. The anterior end of the maxilla reaches a point ventral to the open external naris. The antero-dorsal bar or limb of the nasal extends anteriorly, lateral to the upper limb of the premaxilla, and assists in forming a portion of the superior, external border of the external naris as in all hadrosaurs. It is unclear whether foramina penetrate the premaxillae because they are not mentioned by Norman (1980, 1986), yet appear in a figure by Casier (1960, pl. XIII, Fig. 1). These foramina may be crushed closed or unprepared as is the case in numerous hadrosaurid skulls (Horner 1983). The anterior and antero-lateral borders of the premaxilla have a raised rim similar to that seen in the hadrosaurids. The supraoccipital of *Iguanodon* (Norman 1980, 1986) is bound by the parietals, squamosals, and exoccipitals (including the opisthotics). The squamosal appears to meet the supraoccipital laterally, as the squamosal is slightly infolded as in hadrosaurids. The dentary of *Iguanodon* is massive and relatively wide (labial-lingual). The teeth (based on a cast of a dentary tooth originally described by Mantell 1825, and a tooth illustrated by Norman 1980, Figure 18) have a crown-fang angulation of 120° to 130°, which agrees closely with the hadrosaurids.

The neural spines of *Iguanodon bernissartensis* are relatively short, and the scapula possesses a blade with near parallel borders. The ventral surface of the posterior half of the sacrum has a longitudinal groove, and the ischium possesses a slender shaft with a slightly expanded distal end.

Based on the study of *Ouranosaurus nigeriensis* by Taquet (1976) and personal observation of the type, it is clear that this ornithopod possesses a number of important lambeosaurid-like characters. The maxilla, although incomplete, has an antero-maxillary shelf and medial flange (Fig. 13.1D). In dorsal view the premaxillae show a duck-like bill that is very similar in shape and relative size to the lambeosaurids, and is definitely not penetrated by any foramina. The anterior and antero-lateral edges of the premaxilla do, however, possess a raised rim. The external naris, which is retracted posteriorly, resides well posterior to the anterior end of the maxilla, and is roofed exclusively by the dorsal process of the premaxilla. The antero-dorsal limb of the nasal is short, terminating at the posterior end of the external naris. As in the lambeosaurids, lack of an antero-dorsal maxillary process, which would fit onto the dorsal surface of the ventro-lateral premaxillary process, permitted unrestricted posterior migration of the external naris. The nasals are excavated on the ventral surface directly underlying the nasal crest. The supraoccipital of *Ouranosaurus* appears to be wide, with dorso-lateral bosses nearly identical with those seen in the lambeosaurids. These bosses, which are illustrated by Taquet (1976, Fig. 13.13) as being processes of the exoccipitals,

Table 13.1. *Characters in support of the phylogenetic relationships shown in Figure 13.5*

A. IGUANODONTOIDEA (see Sereno 1986 for list of characters)
   Additional characters:
   1. Squamosal reaches supraoccipital
   2. Large ascending process of astragalus

B. HADROSAURIA
   1. Antero-dorsal process of maxilla extends into and forms medial floor of external naris
   2. Extension of antero-dorsal process of nasal forms supero-lateral border of external naris
   3. Near contact or fusion of supraorbitals to post-orbital, frontal and prefrontal
   4. Postero-ventral surface of squamosal infolded
   5. Crown-root angulation of dentary teeth less than 130°
   6. Grooved ventral surface on posterior sacral centra
   7. Parallel borders of scapular blade
   8. Long narrow prepubic shaft with expanded terminal blade
   9. Ischium with shaft that narrows posterior to obturator process
   10. Premaxillary beak with anterior and lateral lip
   11. Basioccipital nearly or completely excluded from foramen magnum
   12. Large, long paraoccipital process

C. LAMBEOSAURIA
   1. Anterior maxillary shelf for reception of premaxilla
   2. Absence of premaxillary foramen
   3. Ventrally excavated nasals
   4. Dorso-lateral bosses on supraoccipital for reception of squamosals
   5. Nasal does not meet maxilla
   6. Very short parietal and supratemporal fenestra
   7. Tall neural spines
   8. Short prepubic shaft

D. HADROSAURIDAE
   1. Circumnarial depression
   2. Greater lateral expansion of premaxillary beak
   3. Greater anterior elongation of antero-dorsal process of nasal
   4. Expansion of narial opening
   5. Nasal near or forming part of interior orbital wall
   6. Parietal with posterior squamosal process
   7. Ischial shaft slender and nearly straight

E. LAMBEOSAURIDAE
   1. Nasal cavity enclosed by folded premaxillae and nasals
   2. Extreme excavation of nasals
   3. External naris surrounded entirely by premaxilla
   4. Crown width of dentary teeth greater than labio-lingual breadth
   5. ?Absence of palpebrals
   6. Deltopectoral crest of humerus long and large

appear rather to be derived from the supraoccipital (personal observation). As in the lambeosaurids, the squamosals unite with these bosses with no evidence of infolding of the postero-ventral process of the squamosal. The jaw is thin (labial-lingual), and the dentary teeth have a crown-root angulation of 135°.

As in the lambeosaurids, the neural spines of *Ouranosaurus* are tall, the borders of the scapular blade diverge, and the shaft of the ischium is massive and terminates in an expanded, foot-like process. A longitudinal ridge resides on the ventral surface of all sacral centra.

Discussion

The most important derived characters of *Iguanodon* + the hadrosaurids are (Fig. 13.5; Table 13.1): the apparent presence of an antero-dorsal process of the maxilla that extends into, and forms the floor of the external naris, extension of the antero-dorsal bar of the nasal over the external naris, near contact or fusion of the supraorbitals to the postorbital, frontal, and prefrontal, infolding of the postero-ventral surface of the squamosal, crown-root angulation of the dentary teeth less than 130°, a grooved ventral surface on the posterior sacral centra, parallel borders of the scapular blade, a long narrow prepubic shaft with expanded terminal blade, and an ischium with a shaft that narrows posterior to the obturator process.

The most important derived characters of *Ouranosaurus* + the lambeosaurids include the anterior maxillary shelf for reception of the premaxilla, absence of premaxillary foramina, ventrally excavated nasals, dorso-lateral bosses on the supraoccipital for reception of the squamosals, tall neural spines, and a pubis with a short shaft and greatly expanded (dorso-ventral) prepubic blade.

Because of the limited number and preservation of specimens of the sister-taxa *Iguanodon bernissartensis*, *Iguanodon atherfieldensis*, and *Ouranosaurus nigeriensis*, it is difficult to ascertain the importance of certain characteristics. Another difficulty lies in the fact that the ornithopods *Iguanodon orientalis* Rozhdestvenski (1952) and *Probactrosaurus gobiensis* Rozhdestvenski (1966), which may be important to this reevaluation, have only received preliminary description.

From the above, it is clear that *Iguanodon* shares more features with the hadrosaurids than with the lambeosaurids, and that *Ouranosaurus* shares more features with the lambeosaurids than with the hadrosaurids. There is, however, no conclusive evidence that either genus was the immediate ancestor to either "hadrosaur" group. The evidence does indicate that important hadrosaurid and lambeosaurid characters appeared among the "iguanodonts" and therefore suggests that the hadrosaurs and lambeosaurs are diphyletic. If Sereno is correct and the genera *Iguanodon* and *Ouranosaurus* were paraphyletic, then the divergence of the pre-hadrosaurid and pre-lambeosaurid ancestors must be much earlier than previously suspected. Similarity of the hadrosaurs and lambeosaurs most likely represents a case of convergent evolution and long-time, shared derived characters. On the basis of the characters described above, I reaffirm (Horner 1988) the classification of the "duck-billed" dinosaurs into two families, the Hadrosauridae and Lambeosauridae.

Acknowledgments

I wish to extend my thanks to Dr. Philippe Taquet (Museum of Natural History, Paris), Dr. Alan Charig [British Museum (Natural History), London], Dr. Eugene Gaffney (American Museum of Natural History, New York), Dr. William Clemens (University of California, Berkeley), Dr. Philip Currie (Tyrrell Museum of Palaeontology, Drumheller, Alberta), and Dr. Nicholas Hotton III (United States National Museum, Washington, D.C.) for permission to examine specimens in their care. I also much appreciate the advice and comments of Dr. David Weishampel, Dr. Paul Sereno, Dr. Dale Russell, and Mr. Kenneth Carpenter. I also thank Kris Ellingsen for her drawings. Support for this project was derived from the Museum of the Rockies, a grant from the National Science Foundation (EAR-8507031), and a fellowship from the John D. and Catherine T. MacArthur Foundation.

References

Brown, B. 1914. *Corythosaurus casuarius* a new crested dinosaur from the Belly River Cretaceous, with provisional classification of the family Trachodontidae. *American Museum of Natural History Bulletin* 33(35):559–565.

Casier, E. 1960. *Les iguanodonts de Bernissart* (Bruxelles: Instit royal des Sciences naturelles de Belgique), pp. 1–134.

Coombs, W. P. 1972. The bony eyelids of *Euoplocephalus* (Reptilia: Ornithischia). *Journal of Paleontology* 46(5):637–650.

Cope, E. D. 1869. Synopsis of the Batrachia, Reptilia and Aves of North America. *American Philosophical Society Transactions* 14:1–252.

Forster, C. A. 1985. A description of the postcranial skeleton of the early Cretaceous ornithopod *Tenontosaurus tilletti*, Cloverly Formation, Montana and Wyoming. M.S. thesis, University of Pennsylvania.

Galton, P. M. 1974. The ornithischian dinosaur *Hypsilophodon* from the Wealdon of the Isle of Wight. *British Museum (Natural History) Bulletin, Geology* 25(1):1–152.

1983. The cranial anatomy of *Dryosaurus*, a hypsilophodontid dinosaur from the Upper Jurassic of North America and East Africa, with a review of hypsilophodontids from the Upper Jurassic of North America. *Geologica et Palaeontologica* 17:207–243.

Galton, P. M. and Powell, H. P. 1980. The ornithischian dinosaur *Camptosaurus preswichii* from the Upper Jurassic of England. *Palaeontology* 23(2):411–443.

Gilmore, C. W. 1909. Osteology of the Jurassic reptile *Camptosaurus*, with a revision of the species of the genus, and descriptions of two new species. *U.S. National Museum Proceedings* 36:197–332.

1933. On the dinosaurian fauna of the Iren Dabasu Formation. *American Museum of Natural History Bulletin* 67:23–78.

1937. On the detailed skull structure of a crested hadrosaurian dinosaur. *U.S. National Museum Proceedings* 84:481–491.

Heaton, M. J. 1972. The palatal structure of some Canadian Hadrosauridae (Reptilia: Ornithischia). *Canadian Journal of Earth Sciences* 9:185–205.

Hooley, R. W. 1925. On the skeleton of *Iguanodon atherfieldensis* sp. nov., from the Wealdon Shales of Atherfield (Isle of Wight). *Quarterly Journal of the Geological Society of London* 81:1–61.

Hopson, J. A. 1975. The evolution of cranial display structures in hadrosaurian dinosaurs. *Paleobiology* 1(1):21–43.

Horner, J. R. 1983. Cranial osteology and morphology of the type specimen of *Maiasaura peeblesorum* (Ornithischia: Hadrosauridae), with discussion of its phylogenetic position. *Journal of Vertebrate Paleontology* 3(1):29–38.

1988. A new hadrosaur (Reptilia: Ornithischia) from the Upper Cretaceous Judith River Formation of Montana. *Journal of Vertebrate Paleontology* 8(3):314–321.

Horner, J. R. and Makela, R. 1979. Nest of juveniles provides evidence of family structure among dinosaurs. *Nature* 282:296–298.

Lambe, L. M. 1918. On the genus *Trachodon* of Leidy. *Ottawa Naturalist* 31:135–139.

Langston, W. Jr. 1960. The vertebrate fauna of the Selma Formation of Alabama, part VI, the dinosaurs. *Fieldiana: Geology Memoirs* 3(6):315–363.

Lull, R. S. and Wright, N. E. 1942. Hadrosaurian dinosaurs of North America. *Geological Society of America, Special Papers* 40:1–242.

Lydekker, R. A. 1888. *Catalogue of fossil Reptilia and Amphibia in the British Museum*, Pt. 1 [London: British Museum (Natural History)].

Mantell, G. A. 1825. Notice on the *Iguanodon*, a newly discovered fossil reptile, from the sandstone of Tilgate Forest, in Sussex. *Philosophical Transactions of the Royal Society of London* 65:179–186.

Marsh, O. C. 1872. Notice of a new species of *Hadrosaurus*. *American Journal of Science* 3(3):301.

1890. Additional characters of the Ceratopsidae, with notice of new Cretaceous dinosaurs. *American Journal of Science* 3(39):418–426.

Maryańska, T. and Osmólska, H. 1979. Aspects of hadrosaurian cranial anatomy. *Lethaia* 12:265–273.

Norman, D. B. 1980. On the ornithischian dinosaur *Iguanodon bernissartensis* from the Lower Cretaceous of Bernissart (Belgium). *Institut royal des Sciences naturelles de Belgique, Memoire* 178:1–104.

1986. On the anatomy of *Iguanodon atherfieldensis* (Ornithischia: Ornithopoda). *Bulletin de L'institut royal des Sciences naturelles de Belgique: Sciences de la Terre Aardwetenschappen* 56:281–372.

Ostrom, J. H. 1961. Cranial morphology of the hadrosaurian dinosaurs of North America. *American Museum of Natural History Bulletin* 122(2):35–186.

1970. Stratigraphy and paleontology of the Cloverly Formation (Lower Cretaceous) of the Bighorn Basin area, Wyoming and Montana. *Peabody Museum of Natural History Bulletin* 35:1–234.

Parks, W. A. 1923. *Corythosaurus intermedius*, a new species of trachodont dinosaur. *University of Toronto Studies, Geologic Series* 15:1–57.

Rozhdestvenski, A. K. 1952. Otkrytiye iguanodonta v Mongolii (discovery of an iguanodont in Mongolia). *Doklady Akademie nauk SSSR* 84(6):1243–1246.

——— 1966. New iguanodonts from central Asia. *International Geology Review* 9(4):556–566.

Sereno, P. C. 1986. Phylogeny of the bird-hipped dinosaurs (Order Ornithischia). *National Geographic Research* 2(2):234–256.

Sternberg, C. M. 1936. The systematic position of *Trachodon*. *Journal of Paleontology* 10(7):652–655.

——— 1953. A new hadrosaur from the Oldman Formation of Alberta: discussion of nomenclature. *Canada Department of Resources and Development Bulletin* 128:275–286.

——— 1954. Classification of American duck-billed dinosaurs. *Journal of Paleontology* 28:382–383.

Sues, H-D. 1980. Anatomy and relationships of a new hypsilophodontid dinosaur from the Lower Cretaceous of North America. *Palaeontographica A* 169:51–72.

Taquet, P. 1975. Remarques sur l'evolution des iguanodontides et l'origine des hadrosaurides. *Centre National de la Recherche Scientifique Colloquim International* 218(2):503–511.

——— 1976. *Geologie et paléontologie du gisement de Gadoufaoua (Aptian du Niger)* (Paris: Centre National de la Recherche Scientifique), 1–191.

Weishampel, D. B. 1984. Evolution of jaw mechanics in ornithopod dinosaurs. *Advances in Anatomy, Embryology and Cell Biology* 87:1–109.

Weishampel, D. B. and Horner, J. R. 1986. The hadrosaurid dinosaurs from the Iren Dabasu Fauna (People's Republic of China, Late Cretaceous). *Journal of Vertebrate Paleontology* 6(1):38–45.

Young, C. C. 1958. The dinosaurian remains of Laiyang, Shantung. *Palaeontologia Sinica*, series C 16:1–138.

V Pachycephalosauria

14 Morphometric landmarks of pachycephalosaurid cranial material from the Judith River Formation of northcentral Montana

MARK B. GOODWIN

Abstract

A small, morphologically variable collection of pachy-cephalosaur cranial material from the Judith River Formation includes *Stegoceras*, the first record of *Ornatotholus* in Montana, and an unnamed full-domed pachycephalosaur represented by a partial skull.

The wide range of variation among skulls in this assemblage demonstrates the need for a repeatable system of measurements using anatomical landmarks. Toward this end, fifteen measurements are defined as a basis for gathering morphometric data from the pachycephalosaur cranium.

Introduction

The suborder Pachycephalosauria (Maryańska and Osmólska 1974) is characterized by the thickened frontoparietal region of the skull. While most taxonomic studies of pachycephalosaurs are based on isolated skull material, skulls with associated postcranial remains have been described from Alberta (Gilmore 1924) and Mongolia (Maryańska and Osmólska 1974; Perle, Maryańska, and Osmólska 1982). Wall and Galton (1979) and Galton and Sues (1983) reported on this diverse group of ornithischian dinosaurs with emphasis on North American specimens. Sereno (1986) offered a new classification of the pachycephalosaurs in which he arranged several genera into nested groups, from primitive flat-headed to more derived fully-domed forms. Sues and Galton (1987) reviewed North American representatives of the Pachycephalosauria in their recent anatomical and systematic study.

The Pachycephalosauria are divided by some authors (Perle, Maryańska and Osmólska 1982; Sues and Galton 1987) into two families: the domed Pachycephalosauridae Sternberg 1945 and the flat-headed Homalocephalidae Dong 1978. The Pachycephalosauridae include: *Stegoceras* Lambe 1902, *Pachy-*

In Dinosaur Systematics: Perspectives and Approaches, *Kenneth Carpenter and Philip J. Currie, eds. Copyright © Cambridge University Press, 1990.*

cephalosaurus Brown and Schlaikjer 1943, *Yaverlandia* Galton 1971, *Tylocephale* Maryańska and Osmólska 1974, *Prenocephale* Maryańska and Osmólska 1974, *Majungotholus* Sues and Taquet 1979, *Gravitholus* Wall and Galton 1979, *Ornatotholus* Galton and Sues 1983, and *Stygimoloch* Galton and Sues 1983 (=*Stenotholus* Giffin, Gabriel and Johnson 1988). Members of the family Homalocephalidae Dong 1978 include: *Homalocephale* Maryańska and Osmólska 1974, *Wannanosaurus* Hou 1978, *Micropachycephalosaurus* Dong 1978, and *Goyocephale* Perle, Maryańska and Osmólska 1982.

Deposits of the Judith River Formation in Montana continue without interruption into the plains of southern Alberta and contain one of the richest dinosaur faunas in the world (Dodson 1983). Although a large sample of pachycephalosaurid cranial material was described from the Upper Cretaceous Judith River Formation of Alberta, previously known records of pachycephalosaurs from the Judith River Formation of Montana are limited. Except for isolated teeth (Sahni 1972), the first cranial material of *Stegoceras* was a rugose and weathered frontoparietal dome from exposures 80 km south of Chinook, Montana (Wall and Galton 1979). The original specimen is in the Blaine County Museum, Chinook, Montana. Galton and Sues (1983) referred two well preserved fragments of a squamosal shelf (UM 5251) from Kennedy Coulee, Hill County, Montana, to *Stegoceras* sp. Fieldwork conducted by the University of California Museum of Paleontology (UCMP) since 1980 has yielded eight new specimens worthy of description. UCMP parties recovered one skull, two frontoparietal domes, a parietal, and three squamosals from the Upper Cretaceous Judith River Formation, exposed in the steep-walled coulees north of Rudyard, Hill County, Montana. An additional frontoparietal dome was found in the Judith River Formation north of Chester, Liberty County, Montana. John Horner loaned for study an unfused pair of frontals of a juvenile pachycephalosaur from the Judith River Formation of Wheatland County, Montana (Fig. 14.1, index map). This material represents new records of a full-domed and a flat-headed pachycephalosaur from the Judith River Formation of Montana.

Comparative studies and taxonomic assignment of pachycephalosaurid dinosaurs are often based upon lists of measurements taken at arbitrary points on the skull (Brown and Schlaikjer 1943; Maryańska and Osmólska 1974; Wall and Galton 1979) and on morphometric studies using principal component and bivariate analyses (Chapman et al. 1981). Cranial characters are used to distinguish taxa and appear to be highly variable in shape and size. Because these measurements provide the basis for any morphometric study, they must be consistent and repeatable. Measurements need to be applicable to domes of varying size and shape in order to determine where variation in the skull occurs. In attempting to identify and describe the new material from Montana, it became apparent that some of the measurements used in previous analyses of pachycephalosaur skulls in the literature are ambiguous and need more rigorous definition.

In any morphometric analysis, measurement systems should be designed to elucidate patterns of ontog-

Figure 14.1. Index map showing Kennedy Coulee, Turner Ranch, and Museum of the Rockies (MOR) sites where pachycephalosaurids were collected. The type area for the Judith River Formation is about 125 km southeast of Kennedy Coulee.

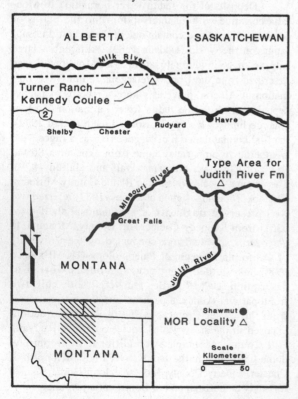

eny, phylogeny, and sexual dimorphism. Based on qualitative assessments and measurements taken at arbitrary points on the dome, several authors have speculated on ontogenetic or sexual characters of pachycephalosaurs (Galton 1970, 1971; Wall and Galton 1979; Chapman et al. 1981; Galton and Sues 1983). The use of inadequately defined measurements resulted in confusion and

Figure 14.2. Landmark points located on the midline of a pachycephalosaurid dome oriented in a defined plane (1–3).

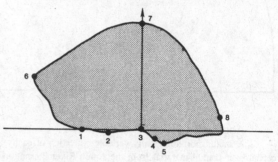

1. anterior margin of olfactory lobes
2. center of olfactory tract
3. frontoparietal suture on roof of braincase
4. depression for cartilaginous portion of the supraoccipital
5. supraoccipital–parietal suture
6. dorsal terminus of the frontal–nasal suture
7. terminal point of dome along a 90 degree projection from point 3
8. projection of anterior-most point of the supratemporal fenestra onto midline

Figure 14.3. Landmark points located along the lateral margin of a pachycephalosaurid dome.

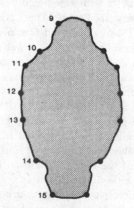

9. frontal – nasal contact
10. prefrontal–supraorbital 1 contact
11. supraorbital 1–supraorbital 2 contact
12. supraorbital 2 – postorbital contact
13. lateral point of the frontoparietal suture
14. anterior-most point of the supratemporal fenestra
15. parietal bridge between the squamosals

lack of uniformity in comparisons. Investigators should replace their use of arbitrary measurements with anatomical landmarks located at homologous points that can be recognized on all the specimens. I discuss below, a suitable measurement system for the pachycephalosaur dome, new pachycephalosaurid cranial material from the Judith River Formation of Montana, and landmark points as a means to compare shape change in pachycephalosaurid domes.

Methods

Fifteen landmark points are identifiable on the pachycephalosaur dome (Figs. 14.2, 14.3). The anterior-posterior orientation of the domes is here defined as a line projected through points 1 and 3. The distance between points 1 and 3 on the braincase remains relatively constant compared with other loci on the frontoparietal dome. These points define a line of reference to communicate the orientation of the dome and are recognizable on most domes. They are only a tool to orient the skull and no biological function is implied by using these two points. The eight points on the perimeter of the dome (Fig. 14.2) are

1. anterior margin of olfactory lobes,
2. center of olfactory tract,
3. frontoparietal suture on the roof of the braincase,
4. depression for cartilaginous portion of the supraoccipital,
5. supraoccipital-parietal suture,
6. dorsal terminus of the frontal-nasal suture,
7. terminal point of dome along a 90 degree vertical projection from point 3,
8. projection of anterior-most point of the dorsal rim of the supratemporal fenestra onto midline.

The seven pairs of points on the external surface used for measurements across a frontoparietal dome (Fig. 14.3) are

9. junctures at frontal-nasal contacts,
10. junctures at prefrontal-supraorbital 1 contacts,
11. junctures at supraorbital 1–supraorbital 2 contacts,
12. junctures at supraorbital 2–postorbital contacts,
13. lateral points of the frontoparietal sutures,
14. anterior-most points of the supratemporal fenestrae,
15. posterior-most points of contact of parietal bridge with the squamosals.

Systematic paleontology

Order Ornithischia Seeley 1887
Suborder Pachycephalosauria Maryańska and Osmólska 1974
Family Pachycephalosauridae Sternberg 1945
Stegoceras Lambe 1902
Type species. *Stegoceras validum* Lambe 1902.
Distribution. Late Cretaceous (Campanian to Maastrichtian) of western North America.
Etymology. Greek *Stego* = covered plus Greek *ceras* = horn.
Diagnosis. [from Sues and Galton 1987, p. 32, based on *Stegoceras validum*] "Small to medium-sized

Pachycephalosauridae. Frontoparietal thickened. Antero-posterior length of dome significantly greater than transverse width. Parietosquamosal shelf developed. Supratemporal fenestrae reduced. Postorbital without prominent tubercles. Diastema between premaxillary and maxillary teeth short."

Stegoceras validum Lambe 1902

Holotype. NMC 515. Lambe (1902), a well preserved frontoparietal dome.
Horizon and Locality. Judith River ("Oldman") Formation, east side of the Red Deer River below the mouth of Berry Creek, Alberta, Canada.
Etymology. Latin *validum* = strong.
Diagnosis. [from Sues and Galton 1987, p. 32] "Nasofrontal boss developed, laterally delimited by a groove on either side."

Referred specimens

UCMP 130048, a well preserved medium-sized frontoparietal dome (Figs. 14.10A–C; Table 14.1) from V82179, Kennedy Coulee, Hill County. Deep lateral grooves are present in the nasal region of the frontals. The anterior region of the dorsal surface of the dome arcs downward. The supratemporal fenestrae are indicated by a groove on the posterolateral edge of the dome. The vertical thickness of the sutural surface for the lateral cranial bones is irregular and the position of the sutural facets for the prefrontal, supraorbitals 1 and 2, and postorbital bones indicate they were not incorporated into the dome. The dorsal surface of the dome is pitted. Nodular or tubercular ornamentation is present only on the lateral margins of the frontal-nasal grooves. Ventrally, the frontoparietal suture is recognizable on the roof of the braincase. Grooves for the olfactory nerves project anteriorly from elongated depressions for the olfactory lobes. The sutural surfaces for endocranial bones show that the olfactory lobes were enclosed by bone. The supraoccipital-parietal suture can be traced directly behind the depression for the cartilaginous portion of the supraoccipital.

UCMP 130050 is a well preserved medium-sized frontoparietal dome (Figs. 14.10D–F; Table 14.1) from V81231, Turner Ranch, Liberty County. Lateral grooves in the nasal region of the frontals are present but shallow. The anterior and posterior dorsal surfaces of the dome slope gently downward. The supratemporal fenestrae are marked by a groove on the posterolateral edge of the dome. The sutural surfaces for the lateral cranial bones are of uniform vertical thickness. The dorsal surface of the dome is lightly sculptured by irregular furrows and there is no nodular or tubercular ornamentation. Ventrally, the frontoparietal suture is recognizable across the roof of the braincase and the sutures for the endocranial bones are also clearly preserved. The depressions for the olfactory lobes are broadened laterally and were enclosed by bone.

UCMP 131679 is a right squamosal of an adult individual (Fig. 14.11L) from V77086, Kennedy Coulee, Hill County. The supratemporal fenestrae are open. Seven primary nodes are preserved in a row on the posterior margin of the squamosal and additional tubercles are present on the dorsal and ventral surfaces.

UCMP 131680 is a right squamosal of a juvenile individual (Fig. 14.11J) from V77083, Kennedy Coulee, Hill County. The posterior edge of the squamosal supports a row of seven primary nodes. The tubercles ornamenting the dorsal and ventral surfaces and the row of seven nodes are smaller than those of UCMP 131679.

UCMP 130296 is a partial right squamosal of a juvenile (Fig. 14.11K) from V78092, Kennedy Coulee, Hill County. A row of four nodes is present on the posterior margin of the squamosal.

MOR 295 is a pair of articulated but unfused, moderately thickened frontals (Figs. 14.11A–C; Table 14.2) collected 16 miles (9.6 km) south of Shawmut, Wheatland County. Lateral grooves in the nasal region of the frontals are absent and the dorsal surface has small pits. The radiating internal trabeculae of the frontals are clearly visible in sagittal view and radiate outward toward the surface of the dome as described by Sues (1978).

Dimensions of these domes and frontals are given in Table 14.1. They are referred to *Stegoceras validum* because of their relative size and amount of thickening of the frontoparietal dome. Only MOR 295 lacks lateral grooves in the anterior frontonasal region. These grooves contribute to the recognition of a frontonasal boss, a character used to separate *Stegoceras validum* from *S. edmontonensis* (Sues and Galton 1987). Variation of the frontoparietal dome in *Stegoceras* was recognized by Gilmore (1924) and analyzed by Chapman et al. (1981). This resulted in the synonymy of *S. breve* Lambe 1918, *S. sternbergi* (Brown and Schlaikjer 1943), and *S. lambei* Sternberg 1945 with *S. validum*. The high variability of the frontoparietal dome remains, in part, the justification for naming new pachycephalosaurid taxa (*Gravitholus* Wall and Galton 1979; *Ornatotholus* Galton and Sues 1983; *Stenotholus* Giffin, Gabriel and Johnson 1988). Gabriel and Berghaus (1988) synonymized *Stenotholus* with *Stygimoloch* on the basis of a partial, narrow-domed pachycephalosaurid skull and associated squamosal displaying the characteristic horns. This variation needs further quantification in order to define the range of variation in the pachycephalosaur cranium and test the validity of the characters chosen.

Table 14.1. *Dimensions (in mm) of new pachycephalosaurs*

| Measurement | Stegoceras validus | | Stegoceras sp. | Pachycephalosaurid gen. undet. |
	UCMP 130048	UCMP 130050	UCMP 130049	UCMP 130051
Length of skull (nasal to squamosal)	—	—	—	186
Frontoparietal length (6–8)	84	91	58	122
Frontoparietal width (13–13)	59.5	71	42	116
Height of dome (at 3)	45	60	13	77e
Anterior braincase length	32	28	22.5	36
Posterior braincase length	13.5	11	9	14.5
Braincase width (across f/p)	22	18	19	19.5

[a]Numbers refer to landmark points defined in Figures 14.5 and 14.6. Anterior braincase length is measured from the anterior margin of the olfactory lobes to the frontoparietal suture (f/p). Posterior braincase length is measured from the frontoparietal suture to the posterior margin of the braincase. e = estimate.

Table 14.2. *Dimensions (in mm) of MOR 295*, Stegoceras validus

Frontal length	51
Frontal width	48.5
Height of dome above f/p	34
Anterior braincase length	29
Braincase width (across f/p)	12.5

Table 14.3. *Dimensions (in mm) of UCMP 130295*, Ornatotholus browni

Parietal length	31.5
Parietal width	45
Thickness at f/p	8
Posterior braincase length	14.5
Braincase width at f/p	18

Stegoceras, species indeterminate

UCMP 130049 is a small, slightly worn fronto-parietal dome (Figs. 14.11D–F; Table 14.1) from V83037, Kennedy Coulee, Hill County. Its relative size, open supratemporal fenestra, and relatively large braincase indicate a young individual. The frontonasal region shows slight development of lateral grooves and the dorsolateral surface is flattened and covered with tubercles. A worn but recognizable vertical series of tubercles occurs posteriorly on the flattened dorsal surface of the parietals. The frontoparietal suture is recognizable on the abraded ventral surface. A portion of the supraoccipital is preserved along the parietal-supraoccipital suture in the posterior region of the braincase.

The small size and low frontal dome of UCMP 130049 are characters also seen in AMNH 5450, identified as *Ornatotholus browni* (Galton 1971). UCMP 130049 does not have the bidomed aspect of the frontals nor the lateral depression separating the frontals from the parietals that characterize AMNH 5450. For the present, this small dome is assigned to *Stegoceras* sp.

Ornatotholus Galton and Sues 1983

Type species. Stegoceras browni Wall and Galton (1979).

Distribution. Upper Cretaceous (Campanian) of western North America.

Etymology. Latin *ornatus* = adorned, decorated plus Latin *tholus* = dome.

Diagnosis. [from Sues and Galton 1987, p. 28] "Frontoparietal dome low (in the holotype divided into a frontal and a parietal dome by a shallow transverse depression). Dorsal surfaces of frontal and parietal covered with prominent tubercles."

Ornatotholus browni (Wall and Galton 1979)

Holotype. AMNH 5450, a frontoparietal.

Horizon and Locality. Judith River ("Oldman") Formation, 21.9 m above the Red Deer River, 1.6 km below Steveville, Alberta, Canada.

Etymology. For Barnum Brown, collector of the type specimen.

Diagnosis. Same as the type and only known species of the genus.

Comments. The holotype, AMNH 5450, was described by Galton (1971) as possibly a female *Stegoceras validum*, but later it was recognized as a new species, *Stegoceras browni*, by Wall and Galton (1979). Subsequently, Galton and Sues (1983) made *Stegoceras browni* the type species of a new genus, *Ornatotholus*.

Referred specimen

UCMP 130295 is an isolated parietal (Figs. 14.11G–I; Table 14.3) from V82218, Kennedy Coulee, Hill County. This specimen is the first record of *Ornato-*

tholus browni from the Judith River Formation of Montana. It has large supratemporal openings and displays a pronounced thickening of the parietal. Numerous tubercles radiate from a longitudinal series along the midline. The ventral surface clearly shows the posterior region of the braincase roof and the suture for the supraoccipital. The sutural surface for the frontal is thickened and slightly wedge-shaped.

Pachycephalosauridae genus undetermined

UCMP 130051 is a partial skull (Fig. 14.12A–C, 14.5; Table 14.1) from V83237, south bank of the Milk River, Super Coulee, Hill County. UCMP 130051 is the best preserved pachycephalosaur skull from the Campanian of the United States. The most prominent feature of UCMP 130051 is its high, robust frontoparietal dome. It is larger than any known skull of *Stegoceras*.

UCMP 130051 shares important characters with the Asian genus *Prenocephale* Maryańska and Osmólska (1974). In both, the dome is fully developed, incorporating almost all of the prefrontal, supraorbital, postorbital, and squamosal elements. As a result, the posterior squamosal and lateral cranial shelves, as seen in *Stegoceras* (UA no. 2, Gilmore 1924, pl. 2), are reduced and the supratemporal fenestrae are closed. These features are listed as shared derived characters for the Domocephalinae by Sereno (1986). Absence of a linear row of primary tubercles on the posterior margin of the squamosal in UCMP 130051 is a primitive character for the Pachycephalosauria (Sereno 1986).

Anteriorly, the frontal contacts the paired nasal bones across a wide suture. The dorsal surface of the dome is relatively smooth with shallow furrows in places, but the frontoparietal suture is not visible on the dorsal surface. The frontal contacts the prefrontal, supraorbitals, and postorbital laterally and the orbitosphenoid and laterosphenoid ventrally.

The paired nasals are not as concave as in the skull of *Stegoceras* (UA no. 2, Gilmore 1924, pl. 1) and descend anteriorly much as in *Prenocephale*. The anterior margin of the prefrontal contacts the nasals. Laterally, the prefrontal contacts the lateral surface of the frontal and the supraorbital, and is suturally distinct from its contact to the frontal. The anterior margin of the prefrontal contacts the nasals. Gilmore (1924) did not recognize a suturally distinct prefrontal in his description of the skull of *Stegoceras* (UA no. 2), but Sereno (pers. comm. June 1986) noted its presence. Sues and Galton (1987, Fig. 1A,C) depict a fused prefrontal and supraorbital for *Stegoceras* (UA no. 2). Maryańska and Osmólska (1974) agreed with Gilmore's interpretation that supraorbital I is fused with the prefrontal, but noted that *Prenocephale* retains a separate prefrontal bone.

A distinct suture separates supraorbital I from supraorbital II (Fig. 14.4). Supraorbital I is bordered by the prefrontal anteriorly and the frontal medially, and contacts supraorbital II posteriorly. Supraorbitals I and II form the anterior portion and part of the posterior region of the orbit respectively. Supraorbital II contacts the frontoparietal dome medially and the postorbital posterolaterally. Supraorbitals I and II and the postorbital are incorporated into the robust dome, a feature shared with *Prenocephale*. This condition is not shared with *Stegoceras validum* (UA no. 2, Maryańska and Osmólska 1974) in which supraorbital II is reduced and contacts the posterior region of the orbit.

The parietal forms a bridge between the squamosals on the posterior margin of the skull and contacts the supraoccipital (Fig. 14.5). The surface of this parietal bridge is furrowed and ornamented with small tubercles. It descends posteriorly and contacts the supraoccipital with a tongue and groove suture resembling a "W," similar to that developed in *Stegoceras*, *Prenocephale*, and *Homalocephale*. Muscle scars are preserved on the surface of the parietal just above the

supraoccipital contact. A raised longitudinal keel serves as the insertion of the nuchal ligament. Accessory scars on the supraoccipital for the rectus capitis posterior muscles are also present (Maryańska and Osmólska 1974).

The thickened squamosal (Fig. 14.5) is incorporated into the dome and does not form a prominent shelf, but overhangs the occipital region by approximately 35 mm. It is furrowed and ornamented with small tubercles, some flattened. The sculpted appearance and abundance of tubercles increases laterally away from the parietal bridge, but there is no evidence of a linear row of pronounced nodes. Because the fine surface details of the parietosquamosal region are preserved, it is unlikely that they have been broken or worn off.

The ventral surface of UCMP 130051 is well preserved but endocranial bones that contribute to the braincase are slightly crushed and the supraoccipital is wedged between the exoccipital and the squamosal. The

Figure 14.4. Right lateral view of pachycephalosaurid skull, UCMP 130051.

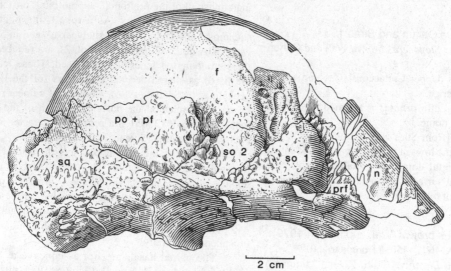

Figure 14.5. Posterior view of pachycephalosaurid skull, UCMP 130051.

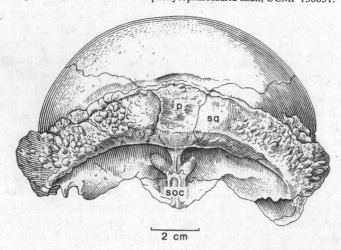

supraoccipital displays the characteristic wing-like structure with a middle tongue that projects dorsally and articulates with the parietal in a "W" shaped suture. It forms the posterior upper border and is not excluded from the margin of the foramen magnum. The sutural contact for the exoccipital is preserved along the ventral surface of the supraoccipital. A fragment of the exoccipital is preserved in articulation with the right squamosal and the sutural surface for the exoccipital can be seen under the wing of the left squamosal.

The anterior region of the basioccipital is crushed. The laterosphenoids are preserved but broken along the suture with the anterior orbitosphenoid. The entire posterior border of the laterosphenoid contacts the parietal.

The frontoparietal suture can be traced across the roof of the braincase and the supraoccipital-parietal suture is slightly open. In the anterior region of the braincase, the orbitosphenoid and parasphenoid are broken and slightly crushed.

UCMP 130051 is more derived than *Stegoceras* in that it incorporates almost all of the prefrontal, supraorbitals, squamosal, and postorbital into the dome. The result is that the lateral shelves are absent, the posterior marginal shell is significantly reduced, and the supratemporal fenestrae are closed. The series of nodes on the squamosal bar is lost. This skull does not share the high vaulting of the frontoparietal dome, well developed ornamentation, and large size of *Pachycephalosaurus*. UCMP 130051 does not have the wide, flattened frontoparietal dome and small braincase that distinguish *Gravitholus*. In the development of the cranium, UCMP 130051 appears similar to *Prenocephale*. I regard this skull as distinct in form, perhaps transitional between *Stegoceras* and *Prenocephale*, but prefer for the present to refer it to cf. Pachycephalosauridae, rather than placing it in yet another taxon.

Results of using specific landmarks

Chapman et al. (1981) were the first to attempt a morphometric study of *Stegoceras*. With principal component and bivariate analyses, they analyzed the frontoparietal domes of 29 specimens referred to *Stegoceras*. They hypothesized two distinct phenons for *S. validum* as sexual dimorphs of a single species and assigned the relatively larger and thicker domes to males. The results of this study are questionable because they did not employ a consistent method of measuring domes using homologous points. Maximum and minimum measurements of the dome may be useful for descriptive purposes, but they do not accurately reveal the locus of morphological difference. Their measurements of the dome are difficult to repeat on domes without a pronounced parietal shelf (Fig. 14.6). Also their measurements fail to reflect the spatial relationship of the dimensions measured because the orientation of the specimen is not explicitly defined. For example, their point 5 (maximum dome thickness) is defined by a ver-

tical line drawn from the frontoparietal suture to the surface of the dome. Dome thickness above the frontoparietal suture can only be measured on specimens oriented in a predefined plane, otherwise this measurement intersects at a different point on the curved surface of the dome. Their points 4 and 6 (posterior and anterior dome thickness) are measured from the frontoparietal suture to the surface of the dome. Again, these measurements are arbitrary because there is no plane of reference. The intersection of the lines on the posterior and anterior "dome margin" do not occur at a recognizable landmark.

Alternative measurements are from the frontoparietal suture to the supratemporal fenestra projected along the midline of the skull (Fig. 14.2, points 3–8) and to the frontal-nasal suture along the midline of the skull (Fig. 14.2, points 3–6). Pachycephalosaur domes are usually broken along distinct sutural surfaces. Additional measurements can be taken from an established plane of reference to any defined point on the margin of the dome if the sutural contacts are not preserved due to damage or abrasion.

The line intersecting point 1 (anterior margin of the olfactory lobes) and point 3 (frontoparietal suture on the roof of the braincase) along the midline forms the plane of reference (Fig. 14.2) when the dome is bal-

Figure 14.6. Measurements of the posterior portion of the skull of *Stegoceras* from Chapman et al. (1981, p. 611, fig. 2). **A**, dorsal, and **B**, ventral views: 1. maximum dome length; 2. maximum dome width; 3. anterior dome width; 4. posterior dome thickness (F is frontoparietal suture); 5. maximum dome thickness; 6. anterior dome thickness; 7. dome length measured on the curve; 8. dome width measured on the curve; 9. anterior dome length; 10. posterior dome length.

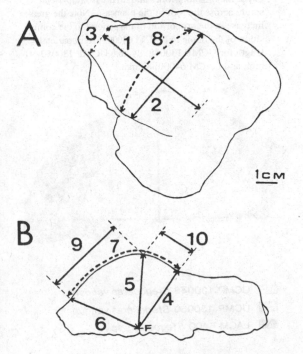

anced symmetrically along the line. In Figure 14.7, three domes of *Stegoceras* are scaled to equal size by superimposing their sagittal outlines at points 1 and 3. Proportional differences in the area of the dome anterior and posterior of the vertical line at point 3 are shown. In two domes of *Stegoceras validum*, UCMP 130048 and UCMP 130050, the frontals and parietals contribute almost equally to the development of the dome. UCMP 130050 is higher and slightly longer posteriorly. There are also minor differences in the orientation of the posteroventral region of the parietal. A significant feature, however, is the amount that the parietal contributes to the dome in LACM 64000, a well preserved frontoparietal of *Stegoceras edmontonensis* (Sues and Galton 1987) from the Hell Creek Formation of Garfield County, Montana. These specimens are nearly the same length from the terminus of the frontal-nasal suture to the projection of the anterior-most point of the dorsal rim of the supratemporal fenestrae.

In Figure 14.8, LACM 64000 and UCMP 130048 (*Stegoceras*) are compared with UCMP 131334, a frontoparietal dome referred to *Pachycephalosaurus wyomingensis*. LACM 64000 and UCMP 13134 share the same proportions in the frontal-parietal development of the dome. This contrasts with the greater percentage of the frontal contribution of the dome in *Stegoceras validum* from the Judith River Formation (Campanian). This is significant as it may indicate a closer affinity of *Stegoceras edmontonensis* with *Pachycephalosaurus*.

Figure 14.7. Scaled diagrammatic outlines of three specimens of *Stegoceras* show the relative proportion of frontal vs. parietal composition of the domes. The domes are scaled to and oriented along (1) anterior margin of the olfactory lobes and (3) the frontoparietal suture across the roof of the braincase. Note the greater increase in the proportion of the parietal to the composition of the dome in LACM 64000. Frontoparietal length for UCMP 130048: 84 mm; UCMP 130050: 91 mm; and LACM 64000: 83 mm.

Another morphotype of *S. validum*, represented by NMC 138, is compared in Figure 14.9 with UCMP 130048 and UCMP 130050. The shapes of the domes of NMC 138 and UCMP 130048 are similar although the

Figure 14.8. Scaled (as in Fig. 14.7) diagrammatic sagittal outlines of *Stegoceras* and *Pachycephalosaurus* show the relative proportion of frontal vs. parietal composition of the domes. The two pachycephalosaurids from the Hell Creek Formation, LACM 6400, *Stegoceras edmontonensis*, and UCMP 131334, *Pachycephalosaurus*, display similar proportions of their domes, despite a difference in frontoparietal length of 83 mm vs. 265 mm respectively.

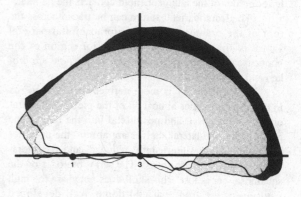

☐ **UCMP 130048** *Stegoceras validum*

▨ **LACM 6400** *Stegoceras* sp.

■ **UCMP 131334** *Pachycephalosaurus wyomingensis*

Figure 14.9. Scaled (as in Fig. 14.7) diagrammatic sagittal outlines of three "morphotypes" of *Stegoceras validus*. The relatively low domes and small size of NMC 138 and UCMP 130048 may indicate these are juvenile pachycephalosaurids.

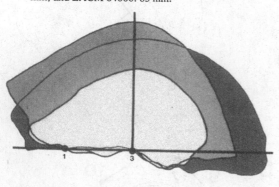

▨ **UCMP 130048** *Stegoceras validum*

▨ **UCMP 130050** *Stegoceras validum*

■ **LACM 6400** *Stegoceras* sp.

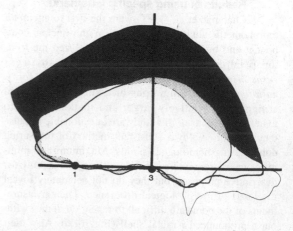

☐ **NMC 138** *Stegoceras validum*

▨ **UCMP 130048** *Stegoceras validum*

■ **UCMP 130050** *Stegoceras validum*

parietal is not completely thickened to form the dome in NMC 138. A hypothesis of increased growth accounting for a thickened parietal that contributes to the dome indicates that NMC 138 may be a subadult. Open supra-temporal fenestra and a visible frontoparietal suture on the dorsal surface of the skull also suggests that these morphotypes represent young individuals. As the dome thickens in growth, the supratemporal fenestrae close

Figure 14.10. Frontoparietal domes of *Stegoceras validum*: **A**, dorsal; **B**, ventral; and **C**, right lateral views of UCMP 130048. **D**, Dorsal; **E**, ventral; and **F**, right lateral views of UCMP 130050.

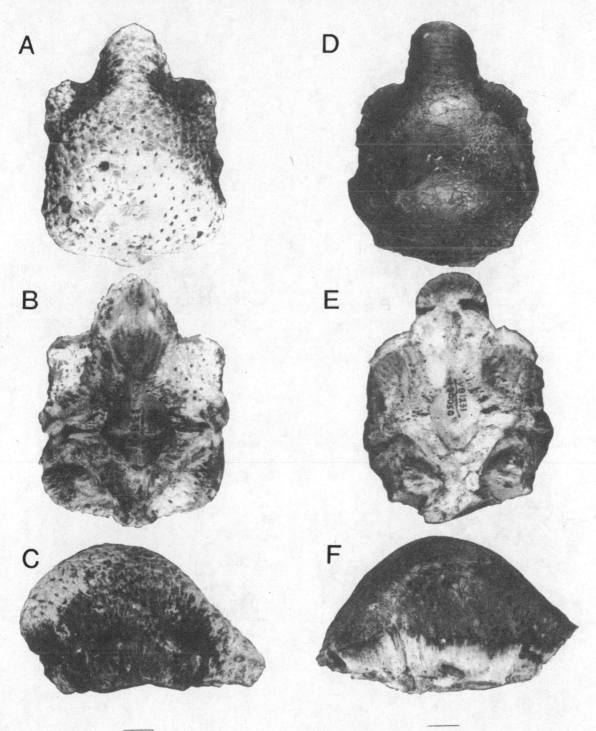

Figure 14.11. Paired frontals of *Stegoceras validum*, **A–C**: **A**, dorsal; **B**, internal view of right frontal showing internal architecture of the dome (note trabeculae radiating outward above braincase roof toward the surface of the dome); and **C**, right lateral view of MOR 295. Frontoparietal dome of *Stegoceras* sp., **D–F**: **D**, dorsal; **E**, ventral; and **F**, right lateral views of UCMP 130049. Parietal of *Ornatotholus browni* **G–I**: **G**, dorsal; **H**, ventral; and **I**, left lateral views of UCMP 130295. Squamosals of *Stegoceras validum* **J–L**: **J**, posterior view of right squamosal, UCMP 131680; **K**, posterior view of partial right squamosal, UCMP 130296; and **L**, dorsal view of right squamosal, UCMP 131679 (cast). Abbreviation: f/p, sutural surface for frontal.

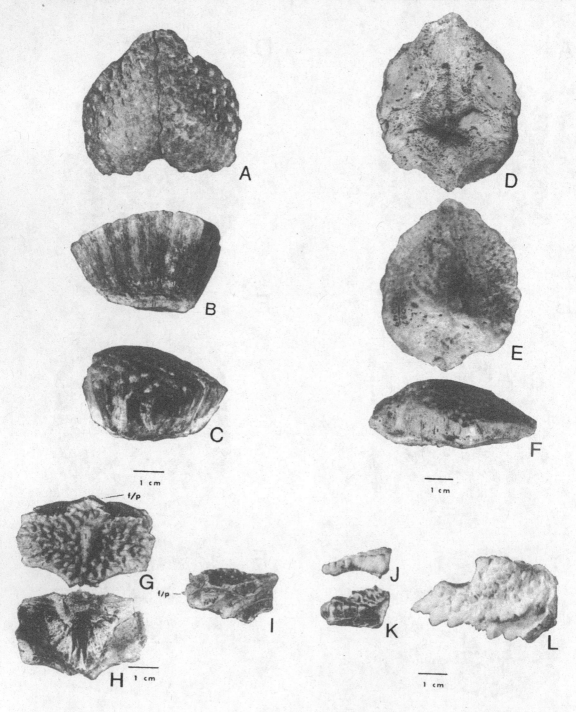

up, the frontoparietal suture becomes fused on the dorsal surface and the parietal thickens as it becomes domed.

Conclusions

Because frontoparietal domes of pachycephalo-

saurs are preserved more often than other parts of the skull or skeleton, variation of the dome is used as a major character in studies of this group. Cranial characters that have been used as taxonomic indicators are relative dome size and shape, size and shape of the cranial

Figure 14.12. "Full-domed" pachycephalosaurid, genus undetermined, A–C: **A**, dorsal; **B**, ventral; and **C**, right lateral views with diagrammatic outlines of UCMP 130051. Abbreviations: eo, exoccipital; f, frontal; lat, laterosphenoid; n, nasal; os, orbitosphenoid; p, parietal; pasp, parasphenoid; pf, postfrontal; prf, prefrontal; so, supraoccipital; so 1, supraoccipital 1; so 2, supraoccipital 2; sq, squamosal.

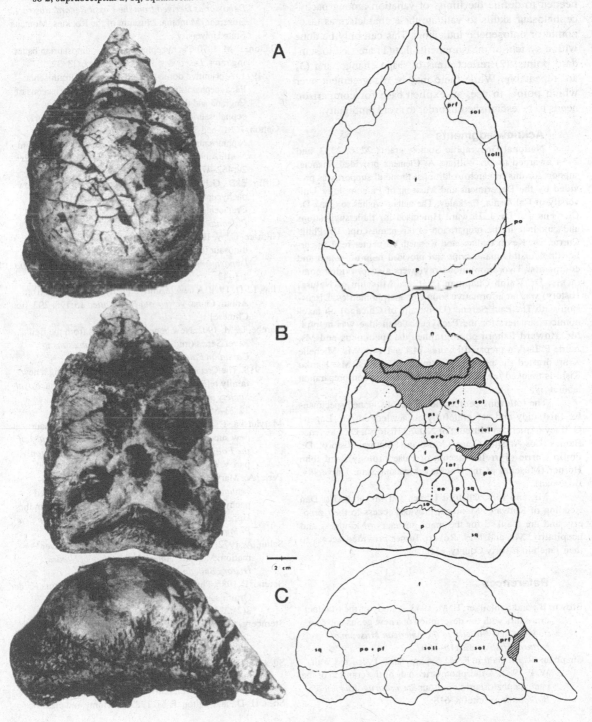

elements that compose the dome, and relative size of the opening of the supratemporal fossa (Sereno 1986). These may be affected by growth, which could be at least partly responsible for the range of variation of the domes. Sues and Galton (1987) recognize the considerable variation in the amount of vaulting and size of the frontoparietal dome, the size of the supratemporal fenestrae, and the development of the parietosquamosal shelf in defining species. Further morphometric analysis is needed to define the limits of variation among pachycephalosaur skulls to validate these characters as taxonomic or ontogenetic indicators. This can only be done with a system of measurements that (1) are tied to standard points, (2) reflect areas of shape change, and (3) are repeatable. While there may be disagreement over which points to use, an explicit base for comparison needs to be established in order to avoid ambiguity.

Acknowledgments

National Geographic Society grants 2226, 2353, and 2454 awarded to Dr. William A. Clemens provided financial support for this research. Additional financial support was provided by the Department and Museum of Paleontology, University of California, Berkeley. The author wishes to thank Dr. Clemens and Dr. J. Howard Hutchison for their suggestions and guidance in the preparation of this manuscript. Dr. Philip Currie, Dr. Kevin Padian, and Kenneth Carpenter read an earlier draft of this manuscript and provided helpful insight and comments. Two anonymous reviewers also provided comments. Dr. Ralph Chapman (National Museum of Natural History) was an informative source on morphometrics. Discussion with Dr. Paul Sereno (University of Chicago) on taxonomic characters for the Pachycephalosauridae was helpful. Mr. Howard Schorn photographed the specimens and Ms. Jaime P. Lufkin prepared Figures 14.4 and 14.5. Ms. Michelle Krup drafted the plates and remaining figures. Ms. Kyoko Kishi is thanked for her assistance in the field and preparation laboratory.

The following people and institutions loaned specimens for this study and are gratefully acknowledged: Dr. Eugene Gaffney (American Museum of Natural History), Dr. Lawrence Barnes (Los Angeles County Museum of Natural History), Dr. Philip Currie (Tyrrell Museum of Palaeontology), and John Horner (Museum of the Rockies, Montana State University, Bozeman).

Mr. and Mrs. Clifford Ulmen and Mr. and Mrs. Dan Redding of Rudyard, Montana, provided access to their property and are thanked for the many summers of kindness and hospitality. Mr. and Mrs. Robert Turner provided access to their ranch in Liberty County, Montana.

References

Brown, B., and Schlaikjer, E. M. 1943. A study of the troodont dinosaurs with the description of a new genus and four new species. *Bulletin of the American Museum of Natural History* 82:115–150.

Chapman, R. E., Galton, P. M., Sepkowski, J. J., Jr., and Wall, W. P. 1981. A morphometric study of the cranium of the pachycephalosaurid dinosaur *Stegosaurus. Journal of Paleontology* 55:608–618.

Dodson, P. 1983. A faunal review of the Judith River (Oldman) Formation, Dinosaur Provincial Park, Alberta. *The Mosasaur* 1:89–118.

Dong, Z.-M. 1978. A new genus of Pachycephalosauria from Laiyang, Shantung. *Vertebrata PalAsiatica* 16:225–228. [in Chinese]

Gabriel, D. L. and Berghaus, C. B. 1988. Three new specimens of *Stygimoloch spinifer* (Ornithischia; Pachycephalosauridae) and behavioral inferences based on cranial morphology. *International Symposium on Vertebrate Behavior as Derived from the Fossil Record, abstract.* Bozeman, Montana: Museum of the Rockies, Montana State University.

Galton, P. M. 1970. Pachycephalosaurids – dinosaurian battering rams. *Discovery, Yale University* 6:23–32.

1971. A primitive dome-headed dinosaur (Ornithischia: Pachycephalosauridae) from the Lower Cretaceous of England and the function of the dome of pachycephalosaurids. *Journal of Paleontology* 45:40–47.

Galton, P. M., and Sues, H.-D. 1983. New data on pachycephalosaurid dinosaurs (Reptilia: Ornithischia) from North America. *Canadian Journal of Earth Sciences* 20:462–472.

Giffin, E. B., Gabriel, D. L., and Johnson, R. E. 1987. A new pachycephalosaurid skull (Ornithischia) from the Cretaceous Hell Creek. *Journal of Vertebrate Paleontology* 7(4):398–407.

Gilmore, C. W. 1924. On *Troodon validum*, an orthopodous dinosaur from the Belly River Cretaceous of Alberta. *Department of Geology Bulletin, University of Alberta* 1:1–43.

Hou, L.-H. 1978. A new primitive Pachycephalosauria from Anhui, China. *Vertebrata PalAsiatica* 15:198–202. [in Chinese]

Lambe, L. M. 1902. New genera and species from the Belly River Series (mid-Cretaceous). *Contributions to Canadian Palaeontology* 3(2):25–81.

1918. The Cretaceous genus *Stegoceras*, typifying a new family referred provisionally to the Stegosauria. *Royal Society of Canada Transactions*, section 4(3), 12:23–56.

Maryańska, T., and Osmólska, H. 1974. Pachycephalosauria, a new suborder of ornithischian dinosaurs. *In Results of the Polish-Mongolian Palaeontological Expeditions*, part V. *Palaeontologia Polonica* 30:45–102.

Perle, A., Maryańska, T., and Osmólska, H. 1982. *Goyocephale lattimorei* gen. et sp. n., a new flat-headed pachycephalosaur (Ornithischia: Dinosauria) from the Upper Cretaceous of Mongolia. *Acta Palaeontologica Polonica* 27(1–4):115–127.

Sahni, A. 1972. The vertebrate fauna of the Judith River Formation, Montana. *American Museum of Natural History Bulletin* 147(6):321–412.

Sereno, P. 1986. Phylogeny of the bird-hipped dinosaurs (Order Ornithischia). *National Geographic Research* 2(2):234–256.

Sternberg, C. M. 1945. Pachycephalosauridae proposed for dome-headed dinosaurs, *Stegoceras lambei*, n. sp., described. *Journal of Paleontology* 19:534–538.

Sues, H.-D. 1978. Functional morphology of the dome in pachycephalosaurid dinosaurs. *Neues Jahrbuch für Geologie und Paläontologie, Monatshefte* 8:459–472.

Sues, H.-D., and Galton, P. M. 1987. Anatomy and classification of the North American Pachycephalosauria (Dinosauria: Ornithischia). *Palaeontographica* A, 198:1–40.

Sues, H.-D., and Taquet, P. 1979. A pachycephalosaurid dinosaur from Madagascar and a Laurasia–Gondwanaland connection in the Cretaceous. *Nature* 279:633–635.

Wall, W. P., and Galton, P. M. 1979. Notes on pachycephalosaurid dinosaurs (Reptilia: Ornithischia) from North America, with comments on their status as ornithopods. *Canadian Journal of Earth Sciences* 16:1176–1186.

VI Ceratopsia

15 New data on parrot-beaked dinosaurs (*Psittacosaurus*)

PAUL C. SERENO

Abstract

Growth characteristics are observed in *Psittacosaurus mongoliensis*. A taxonomic overview of the fossil remains of psittacosaurs suggests that there is one well-established genus, *Psittacosaurus*, with two well-established species, *P. mongoliensis* and *P. sinensis*. Two new species from the Lower Cretaceous of China have been discovered recently and their diagnostic characters are summarized.

Introduction

Parrot-beaked dinosaurs were first discovered in Lower Cretaceous strata of the Gobi Desert in 1922, and represent the first relatively complete skeletons uncovered by the famous Central Asiatic Expeditions of the American Museum of Natural History. The following year, Osborn (1923) established two new species, *Psittacosaurus mongoliensis* and *Protiguanodon mongoliense*, on the basis of two complete skeletons. Despite subsequent recovery of significant new materials pertaining to psittacosaurs and their bearing on the phylogenetic position of the Ceratopsia, no thorough systematic account is currently available.

In anticipation of a formal systematic review, I present a systematic survey of psittacosaur species including a report on the discovery of two new species. I also briefly discuss the phylogenetic affinity of psittacosaurs.

Growth characteristics in *Psittacosaurus mongoliensis*

Several growth stages are available for *Psittacosaurus mongoliensis* that provide important information regarding character change during ontogeny. The smallest individual is a hatchling (AMNH 6535), the cranium of which has received preliminary treatment by Coombs (1980, 1982). Recently, the lower jaws and partial

In Dinosaur Systematics: Perspectives and Approaches, *Kenneth Carpenter and Philip J. Currie, eds. Copyright © Cambridge University Press, 1990.*

postcranium of the same individual have been located and reassociated with the cranium. The entire skeleton has undergone further preparation revealing significant details, such as wear facets and replacement crowns in the dentition, a left stapes, and an articulated sternum. The skull is 2.8 cm in length (Coombs 1980, 1982), which is just slightly shorter than the skulls of the supposed fabrosaur *Tawasaurus minor* (3.0 cm – Young, 1982) and the sauropodomorph *Mussaurus patagonicus* (3.2 cm – Bonaparte and Vince 1979; Charig 1979). The skull of the hatchling is large relative to the partial postcranium, with an estimated body (snout to tail tip) length of 11 to 13 cm, approximately half that estimated by Coombs (1980, 1982).

The left sclerotic ring is preserved in a somewhat larger hatchling skull (AMNH 6536), which measures 4.6 cm in length (4.2 cm given by Coombs 1980, 1982). This skull is comparable in size to the hatchling skull of *Bugaceratops rozhdestvenskyi* (4.7 cm – Maryańska and Osmólska 1975).

The following comments are confined to two aspects of the dentition of particular systematic interest. Five teeth are present in both maxillary and dentary tooth rows in the small hatchling skull of *Psittacosaurus mongoliensis* (AMNH 6535). The larger hatchling (AMNH 6536) has seven maxillary teeth in the complete row, and subadult individuals of the same species have eight to ten maxillary teeth (AMNH 6661, 21740; PIN 2860/1, 3779/18, 3779/19). Ten to twelve teeth appear to be the usual count for the maxillary and dentary tooth rows of an adult individual of *Psittacosaurus mongoliensis* (AMNH 6254, PIN 3779/25). In summary, the tooth count appears to increase from five to approximately ten or twelve during growth in *Psittacosaurus mongoliensis*. Therefore, a slight difference in tooth count from that of a type specimen, obviously, should not be used to distinguish a new species of psittacosaur, particularly in the absence of other age criteria.

The maxillary and dentary crowns in very young individuals are usually characterized by a more subtle, simplified ornamentation with fewer denticles along the apical margin. This juvenile crown morphology is similar to the smallest crowns at each end of the tooth row in adult individuals. Minor variation in the number of denticles or in the prominence of the ornamentation should not be used to distinguish between psittacosaur species.

Systematics

Psittacosaurs have never been subject to detailed systematic evaluation. Original and currently available diagnoses are inadequate and do not distinguish taxa with valid autapomorphies. One limitation in the available literature has been the absence of comparison between type specimens, which in some cases reside in distant collections. The following summary is based on an examination of type and referred materials.

Psittacosaurus mongoliensis
(Figs. 15.1–15.5)

All except one of the characters listed in Osborn's original diagnosis of *Psittacosaurus mongoliensis* occur in other ornithischians that were known at the time of his writing. The single exception – the parrot-like rostrum – is now known to occur in all psittacosaur species, and, therefore, is not appropriate for diagnosis of the species. *Psittacosaurus mongoliensis* can be diagnosed on cranial characteristics, such as the small, triangular antorbital fossa and the upturned lateral margin of the prefrontal.

Protiguanodon mongoliense (Fig. 15.3)

In the same report describing the type skeleton of *Psittacosaurus mongoliensis*, Osborn (1923) named a second genus and species, *Protiguanodon mongoliense*.

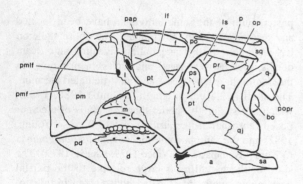

Figure 15.1. Reconstruction of the skull of *Psittacosaurus mongoliensis* in lateral view. Abbreviations for all skull figures: a, angular; ar, articular; bo, basioccipital; d, dentary; ec, ectopterygoid; f, frontal; j, jugal; l, lacrimal; lf, lacrimal foramen; ls, laterosphenoid; m, maxilla; n, nasal; op, opisthotic; p, parietal; pap, palpebral; pd, predentary; pm, premaxilla; pmf, premaxillary foramen; po, postorbital; popr, paroccipital process; pmlf, premaxilla-lacrimal foramen; pr, prootic; pra, prearticular; ps, parasphenoid; pt, pterygoid; q, quadrate; qj, quadratojugal; r, rostral; sa, surangular; so, supraoccipital; sp, splenial; sq, squamosal.

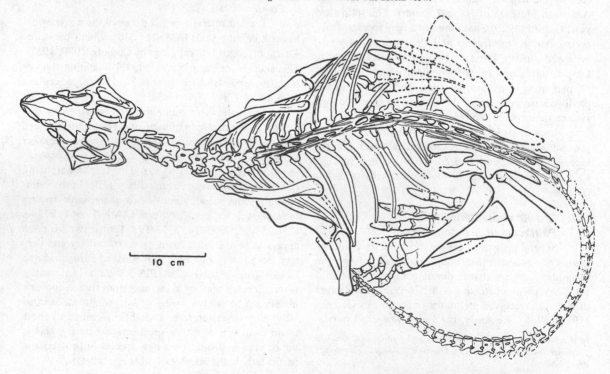

Figure 15.2. The skeleton of *Psittacosaurus mongoliensis* AMNH 6254 in dorsal view.

10 cm

Later Osborn placed this second taxon with *Psittacosaurus mongoliensis* in the Psittacosauridae (Osborn 1924). The supposed differences between these two taxa have never been examined in detail. Young (1958) suggested their generic synonymy and others have argued for specific synonymy (Rozhdestvensky 1955, 1977; Coombs 1982). Reexamination of the type skeletons and referred material supports their detailed similarity. Taxonomic distinction of *Psittacosaurus mongoliensis* and *Protiguanodon mongoliense* can no longer be sustained.

Psittacosaurus osborni
(= *Psittacosaurus tingi*) (Fig. 15.5)

In 1931, Young described two new species, *Psittacosaurus osborni* and *Psittacosaurus tingi*. *Psittacosaurus osborni*, based on a partial disarticulated skull, was distinguished by its small size and by the symmetrical position of the primary ridge in the maxillary and dentary teeth. *P. tingi*, based on two small dentaries, was later reduced to a junior synonym of *P. osborni* (Young 1958).

Figure 15.3. The skeleton of *Psittacosaurus mongoliensis* (*Protiguanodon mongoliense*) AMNH 6253 in dorsal view. Distal tail section figured separately.

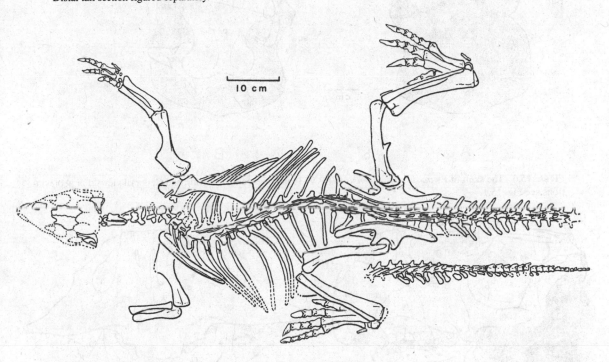

Figure 15.4. Reconstruction of the skeleton of *Psittacosaurus mongoliensis* in lateral view (modified from Osborn).

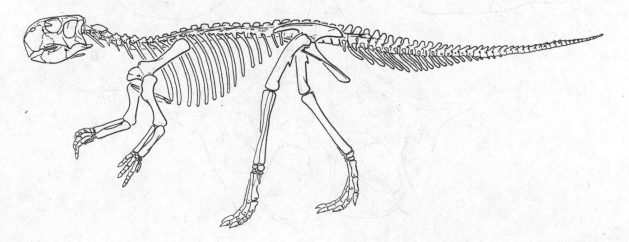

Figure 15.5. The skull of *Psittacosaurus mongoliensis* (*Psittacosaurus osborni*) IVPP RV31039 in **A**, dorsal; and **B**, ventral views. For abbreviations see Fig. 15.1.

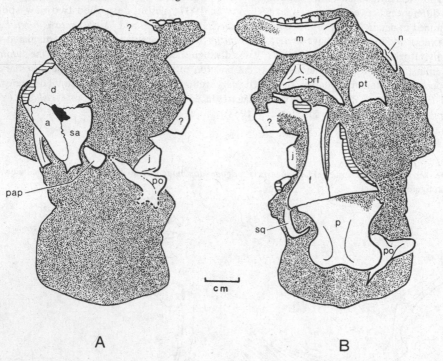

Figure 15.6. The skull of *Psittacosaurus sinensis* IVPP V738 in **A**, lateral; **B**, anterior; and **C**, dorsal views. For abbreviations see Fig. 15.1.

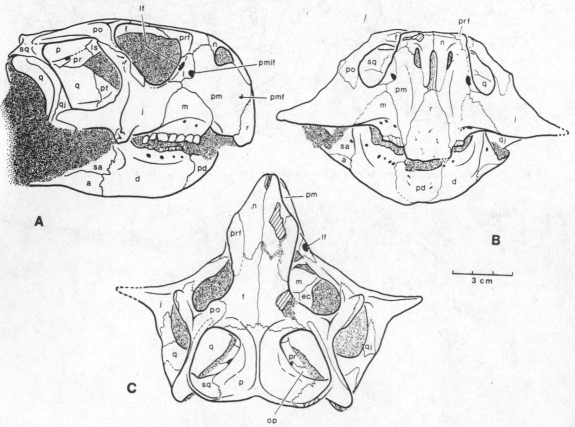

Small size alone is not a satisfactory distinguishing feature for *P. osborni*, particularly given the abundance of juvenile remains at many psittacosaur localities (Coombs 1982). The strong median primary ridge on the dentary teeth of *P. osborni* and *P. tingi* also occurs in the dentary teeth of *P. mongoliensis* and other psittacosaur species. The weathered maxillary crowns of *P. osborni* and *P. tingi* also occurs in the dentary teeth of *P. mongoliensis*, which are now exposed in the type skull. Despite subsequent referral of small dental and postcranial remains to *P. osborni* (Cheng 1983), distinction of this species from *P. mongoliensis* appears unwarranted.

Psittacosaurus guyangensis

Recently, a new species, *Psittacosaurus guyangensis*, has been described on the basis of the anterior portion of a skull from Lower Cretaceous beds in Inner Mongolia, China (IG V.330, Cheng 1983). Postcranial elements of additional individuals from the same locality have been referred to this species (IG V.352). The distinguishing characteristics of the new species are all found in the growth series of *P. mongoliensis*. The size of the skull (intermediate between that of *P. osborni* and *P. mongoliensis*) and the presence of nine maxillary

Figure 15.7. Skull of *Psittacosaurus sinensis* (*Psittacosaurus youngi*) BNHM BPV.149 in **A**, lateral; and **B**, posterior views. For abbreviations see Fig. 15.1.

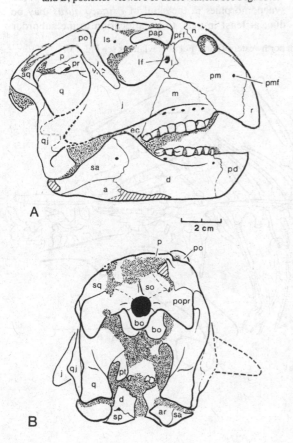

teeth lacking a marked primary ridge does not distinguish this species from *Psittacosaurus mongoliensis*.

Psittacosaurus sinensis
(Figs. 15.6, 15.7)

Young (1958) described abundant psittacosaur remains from Shandong Province, China as a new species, *Psittacosaurus sinensis*. Although his diagnosis of *P. sinensis* was not detailed, Young listed the broadly proportioned skull and lower adult tooth count – valid distinguishing characteristics manifest in all referred material. Many additional features mark this species, such as the sutural contact between the premaxilla and jugal, small jugal-postorbital horn core, absence of the external mandibular fenestra, and laterally convex tooth rows. *P. sinensis* cannot reasonably be considered a junior synonym of *P. mongoliensis*, as suggested by Rozhdestvensky (1955) and Coombs (1982).

Psittacosaurus youngi (Fig. 15.7)

Chao (1962) described a new species, *Psittacosaurus youngi*, based on a partial skeleton with a distorted, but complete skull (BNHM BPV.149). *Psittacosaurus youngi* is similar to *Psittacosaurus sinensis* and was discovered in the same strata at a nearby locality. Its specific distinction from *P. sinensis* can no longer be justified.

In taxonomic overview, there is insufficient evidence for taxonomic distinction between *Psittacosaurus mongoliensis* and *Protiguanodon mongoliense*, *Psittacosaurus osborni*, *Psittacosaurus tingi*, and *Psittacosaurus guyangensis* and between *Psittacosaurus sinensis* and *Psittacosaurus youngi*. The type materials of two psittacosaur species from the published literature, *Psittacosaurus mongoliensis* and *Psittacosaurus sinensis*, exhibit several diagnostic characters.

Psittacosaurus meileyingensis
(Fig 15.8)

The remains of a new species of *Psittacosaurus* were discovered in the Jiufotang Formation near Chaoyoung Liaoning Province in northeastern China (Sereno et al. 1988). The skull, finely preserved in one individual, is proportionately tall but short anteroposteriorly, with a nearly circular profile. Other distinguishing cranial features include a raised rugosity on the quadratojugal and a prominent ventral flange on the dentary. The postcranium, known from a single partial skeleton, is similar to that of *P. mongoliensis*.

Psittacosaurus xinjiangensis
(Fig. 15.9)

A second new species of *Psittacosaurus* was discovered in the Tugulu Group of the Junggar Basin in the Xinjiang Uygur Autonomous Region of western China (Sereno and Chao 1988). An articulated subadult skeleton, the most complete individual, preserves the crushed posterior portion of the skull. Maxillary crowns show a denticulate margin that curves posteromedially onto the side near the base of the crown. The jugal horn is flattened anteriorly, and the postacetabular process is proportionately elongate.

Psittacosaurs as ceratopsians

Recently accepted as matter of fact, the close relationship between psittacosaurs and other ceratop-

Figure 15.8. Lateral view of the skull of *Psittacosaurus meileyingensis* from northeastern China. For abbreviations see Fig. 15.1.

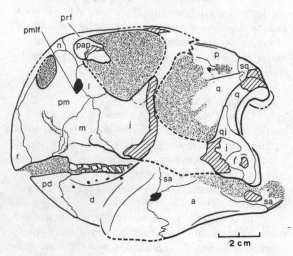

sians went largely unnoticed for a half century primarily for two reasons: (1) the early misidentification of the rostral bone, and (2) the maintenance of the traditional suborder Ornithopoda, a paraphyletic group that includes primitive members of other monophyletic ornithischian clades.

In the initial description of the psittacosaur cranium, Osborn (1923, Fig. 15.2) identified the wedge-shaped element forming the anterior keel of the snout as the premaxilla. Subsequent descriptions of Chinese psittacosaurs followed this identification, despite the fact that the bone in question is clearly an unpaired, median element that did not participate in the border of the external nares (Young 1958, Fig. 51; Chao 1962, Figs. 15.1–3). Romer (1956, pp. 155, 631; 1968, p. 141) first suggested that the element may represent the ceratopsian rostral, and he was followed by subsequent authors (Steel 1969; Maryańska and Osmólska 1975; Coombs 1980, 1982). Romer's reidentification is confirmed by detailed study of the psittacosaur cranium.

Despite longstanding misidentification of the rostral, the similarity in skull shape between psittacosaurs and other ceratopsians was noted soon after the initial description of *Psittacosaurus* (Gregory and Mook 1925; Gregory 1927, 1951; Colbert 1945, 1965; Rozhdestvensky 1955, 1977; Young 1958). No specific anatomical similarities were mentioned, however. The lack of attention to relatively obvious ceratopsian synapomorphies in the skull of *Psittacosaurus* may be due, at least in part, to its allocation to the suborder

Figure 15.9. Skeleton of *Psittacosaurus xinjiangensis* from northwestern China. For abbreviations see Fig. 15.1.

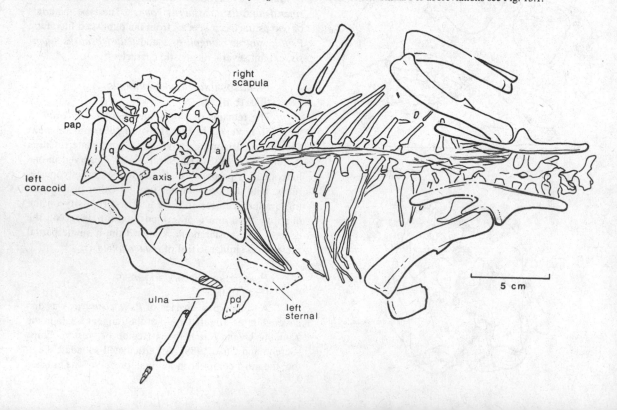

Ornithopoda, a traditional repository for taxa whose affinities (synapomorphies) remain undocumented. Traditionally diagnosed as symplesiomorphy, the Ornithopoda encompasses the "main line" of ornithischian evolution – those ornithischians that lack the conspicuous autapomorphies of other suborders. Symplesiomorphies, however, do not elucidate phylogenetic relationship.

Maryańska and Osmólska (1975, p. 172) were first to list ceratopsian synapomorphies in the skull of *Psittacosaurus,* in addition to the rostral bone mentioned by Romer. One of these characters is considered valid in this consideration.

1. Jugals projecting laterally well beyond the dorsal margin of the orbit.

Other notable ceratopsian characters include the following (Figs. 15.1, 15.6, 15.8; Sereno 1986, 1987):

2. Flat subnarial margin between the narial fossa and ventral margin of the premaxilla.
3. Dorsoventral crest traversing the lateral aspect of the jugal.
4. Jugal infraorbital ramus deeper than the infratemporal ramus.
5. Parietal extends across at least half of the posterior margin of the shelf or frill.
6. Premaxillary palate vaulted.
7. Predentary ventral process is broad, expanding in width towards its proximal end.

Pachycephalosaurs and ornithopods were employed as successive outgroups in the evolution of these characters, although in most instances these characters are derived with respect to any ornithischian outgroup. A monophyletic Ceratopsia, uniting psittacosaurs and neoceratopsians, appears robust.

Conclusions

A growth series for *P. mongoliensis* permits the identification of age-specific characteristics, such as increasing tooth count. Taxonomic overview of psittacosaur remains indicate the presence of one valid genus, *Psittacosaurus,* with four valid species, *P. mongoliensis, P. sinensis,* and two new species from China. A number of cranial synapomorphies diagnose the Ceratopsia, including *Psittacosaurus* and neoceratopsians.

Acknowledgments

I acknowledge the financial support from the National Science Foundation (Doctoral Dissertation Improvement Grant BSR83-05228), National Geographic Society, American Museum of Natural History, Sigma Xi, and the Department of Geological Sciences of Columbia University.

For permission to study material in their care, I am indebted to Eugene Gaffney (American Museum of Natural History), Chang Meemann, Sun Ailing, Dong Zhiming, and Chao Shichin of the Institute of Vertebrate Paleontology and Paleoanthropology (Beijing), Cheng Zhengwu [Institute of Geology (Beijing)], Rinchen Barsbold and Altangerel Perle of the State Museum of the Mongolian People's Republic (Ulan Bator), and Academician L. P. Tatarinov and Sergei Kurzanov [Paleontological Institute (Moscow)]. Special thanks are due to Michelle Koo for her help during assembly of the illustrations.

References

Bonaparte, J. F., and Vince, M. 1979. El hallazgo del primer nido de dinosaurios triasicos (Saurischia, Prosauropoda), Triasico Superior de Patagonia, Argentina. *Ameghiniana* 16(1–2):173–182.

Chao, S. 1962. New species of *Psittacosaurus* from Laiyang, Shantung. *Vertebrata Palasiatica* 6(4):349–360. [in Chinese]

Charig, A. J. 1979. *A New Look at the Dinosaurs* (New York: Mayflower Books).

Cheng, Z. 1983. Reptilia. Part 7. *In* Geological Survey of the Autonomous Region of Inner Mongolia (ed.), *The Mesozoic Stratigraphy and Paleontology of Guyang Coal-Bearing Basin, Nei Monggol Autonomous Region, China* (Beijing: Geology Press). [in Chinese]

Colbert, E. H. 1945. The hyoid bones in *Protoceratops* and *Psittacosaurus. American Museum Novitates* 1301:1–10.

1965. *The Age of Reptiles* (New York: W.W. Norton and Co., Inc.).

Coombs, W. P., Jr. 1980. Juvenile ceratopsians from Mongolia – the smallest known dinosaur specimens. *Nature* 283(5745):380–381.

1982. Juvenile specimens of the ornithischian dinosaur *Psittacosaurus. Palaeontology* 25(1):89–107.

Gregory, W. K. 1927. The Mongolian life record. *Scientific Monthly* 24:169–181.

1951. *Evolution Emerging*, vols. 1–2 (New York: Macmillan Co.).

Gregory, W. K., and Mook, C. C. 1925. On *Protoceratops*, a primitive ceratopsian dinosaur from the Lower Cretaceous of Mongolia. *American Museum Novitates* 156:1–9.

Maryańska, T., and Osmólska, H. 1975. Results of the Polish–Mongolian palaeontological expeditions, part VI, Protoceratopsidae (Dinosauria) of Asia. *Palaeontologica Polonica* 33:133–182.

Osborn, H. F. 1923. Two Lower Cretaceous dinosaurs from Mongolia. *American Museum Novitates* 95:1–10.

1924. *Psittacosaurus* and *Protiguanodon*: two Lower Cretaceous iguanodonts from Mongolia. *American Museum Novitates* 127:1–16.

Romer, A. S. 1956. *Osteology of the Reptiles* (Chicago: University of Chicago Press).

1968. *Notes and Comments on Vertebrate Paleontology* (Chicago: University of Chicago Press).

Rozhdestvensky, A. K. 1955. New data concerning psittacosaurs, Cretaceous ornithopods. *Voprasy Geologii Asia* 2:783–788. [in Russian]

1977. The study of dinosaurs in Asia. *Journal Palaeontological Society of India* 20:102–119.

Sereno, P. C. 1986. Phylogeny of the bird-hipped dinosaurs (Order Ornithischia). *National Geographic Research* 2(2):234–256.

1987. The ornithischian dinosaur *Psittacosaurus* from the Lower Cretaceous of Asia and the relationships of the Ceratopsia. Ph.D. dissertation (New York: Columbia University).

Sereno, P. C., and Chao Shichin. 1988. *Psittacosaurus xinjian-gensis* (Ornithischia: Ceratopsia), a new psittacosaur from the Lower Cretaceous of northwestern China. *Journal of Vertebrate Paleontology* 14:353–365.

Sereno, P. C., Chao Shichin, Cheng Zhengwu, and Rao Chenggang. 1988. *Psittacosaurus meileyingensis* (Ornithischia: Ceratopsia), a new psittacosaur from the Lower Cretaceous of northeastern China. *Journal of Vertebrate Paleontology* 14:366–377.

Steel, R. 1969. Ornithischia. *In* O. Kuhn (ed.), *Handbuch der paleoherpetologie* (Stuttgart: Gustav Fischer Verlag), 15, pp. 1–84.

Young, C. 1931. On some new dinosaurs from western Suiyan, Inner Mongolia. *Bulletin of the Geological Survey of China* 11:159–226.

——— 1958. The dinosaurian remains of Laiyang, Shantung. *Palaeontologica Sinica* whole no. 142, new series C, 16:1–138. [in Chinese with English translation]

——— 1982. A new ornithopod from Lufeng, Yunnan. *In Selected Works of Yang Zhungjian* (Beijing: Science Press), pp. 29–35.

16 The ceratopsian subfamily Chasmosaurinae: sexual dimorphism and systematics

THOMAS M. LEHMAN

Abstract

Advanced ceratopsians (family Ceratopsidae) are divisible into two groups, based primarily on the relative proportions and fenestration of their neck frills. These groups have been informally termed the *Ceratops-Torosaurus* and *Monoclonius-Triceratops* phyla, the "long-frilled" and "short-frilled" ceratopsids, or the "long-faced" and "short-faced" ceratopsids. Although Lawrence Lambe recognized three subfamilies within the Ceratopsidae, subsequent workers have preferred informal groups and have not adopted his classification.

Lambe's subfamily Eoceratopsinae is abandoned because of the placement of *Eoceratops* in synonymy with *Chasmosaurus*, and the inclusion of *Triceratops* within the "long-frilled" ceratopsids. Lambe's subfamilies Chasmosaurinae and Centrosaurinae should be revived, however, for the "long-frilled" and "short-frilled" ceratopsids, respectively. Members of these subfamilies are distinguished by many features of the skull, and it is useful to formally recognize their separation. Although the placement of *Triceratops* has been problematical, its affinities are now believed to lie with the Chasmosaurinae. *Pachyrhinosaurus* represents an aberrant Early Maastrichtian survivor of the Centrosaurinae.

Species-level taxonomy of ceratopsids is complicated by pronounced individual and ontogenetic variability, and sexual dimorphism, in most species. A population sample of *Chasmosaurus* from Texas suggests that orientation of the supraorbital horncores is a useful criterion for separating sexual morphs. Based on this criterion, all chasmosaurine genera contain a species or group of species exhibiting the supposed "female" morph, and one exhibiting the supposed "male" morph. A partial growth series in the Texas *Chasmosaurus* sample illustrates progressive ontogenetic elongation of the facial region and supraorbital horncores, probably typical of most chasmosaurines. The pronounced ontogenetic elongation of the frill, observed in some other chasmosaurines, is not marked in the Texas *Chasmosaurus*. In light of these observations, most ceratopsid genera remain distinct and well-defined, but a revision of all included species is in order. The variation observed among named species within a given ceratopsid genus can in most cases, considering the range of variation observed in the Texas sample, actually be accommodated within a single species.

Introduction

Primitive ceratopsian dinosaurs (Protoceratopsidae) are known from population samples in which the range of intraspecific variation is well-documented (e.g., Brown and Schlaikjer 1940; Dodson 1976). In advanced ceratopsians (Ceratopsidae), however, little or nothing is known in most cases regarding the range of morphological variation within any species. The systematics of ceratopsids is therefore highly conjectural at the species level.

The goals of the present study are threefold. First, in order to formally recognize the long-standing informal arrangement of ceratopsids into two groups, a subfamily division of the Ceratopsidae is presented. The subfamilies, Chasmosaurinae and Centrosaurinae, originally proposed by Lambe (1915), are revived for this purpose. Morphological characters by which the two groups may be distinguished are discussed. Second, in order to document the range of individual, ontogenetic, and sex-associated variation within a single ceratopsid species, a bone-bed accumulation of *Chasmosaurus* from the Upper Aguja Formation (Late Campanian) of west Texas is described. Third, using the range of variation observed in this population sample as a guide, a species-level revision of the long-frilled ceratopsids (Chasmosaurinae) is presented.

Subfamilies of the Ceratopsidae

It became apparent soon after their discovery that advanced ceratopsians (family Ceratopsidae) were divisible into two groups based primarily on the relative proportions and fenestration of their squamosal-parietal frills. John B. Hatcher and Richard S. Lull (in Hatcher

In Dinosaur Systematics: Perspectives and Approaches, *Kenneth Carpenter and Philip J. Currie, eds. Copyright © Cambridge University Press, 1990.*

et al. 1907) informally termed these groups the *Ceratops-Torosaurus* phylum and *Monoclonius-Triceratops* phylum based on presumed phylogenetic relationships. Most later workers, however, preferred the more general terms "long-frilled" and "short-frilled" to describe the same groups (e.g., Lull 1933; Colbert 1948). Wann Langston, Jr. (1967) noted corresponding differences in the facial proportions and narial structure of the two groups, which he instead called the "long-faced" and "short-faced" ceratopsids, respectively. Although Lawrence Lambe had earlier (1915) formally defined and named three subfamilies – Centrosaurinae, Chasmosaurinae, and Eoceratopsinae – subsequent authors have preferred the informal groups and have not adopted Lambe's classification. Langston (1967, p. 184) remarked, however, that it may now be useful to formalize the division of the Ceratopsidae at the subfamily level.

Lambe's (1915) Chasmosaurinae is in most respects comparable to, and is available, for the long-frilled or long-faced group of ceratopsids. Lambe's Centrosaurinae is available for the short-frilled or short-faced group. Although Lambe's (1915) Eoceratopsinae originally contained the genera *Eoceratops* and *Triceratops,* the subfamily should be abandoned for the following reasons. The placement of *Triceratops* within either the short-frilled or long-frilled group has long been a point of contention. Earlier authors followed Hatcher et al. (1907) and referred it to the short-frilled group. However, Sternberg (1949) and Langston (1975) later assigned *Triceratops* to the long-frilled group. Apart from its relatively short, thick, and closed frill, *Triceratops* agrees in all characters with the Chasmosaurinae as defined herein. The modifications of its frill are secondarily derived from the chasmosaurine condition. *Eoceratops* is herein regarded as a junior synonym of *Chasmosaurus*. Hence, no taxa remain within the Eoceratopsinae as originally defined by Lambe. Because the names Chasmosaurinae and Eoceratopsinae were published simultaneously (Lambe 1915) the International Code of Zoological Nomenclature (Arts. 23d, 24a; see Mayr 1969) permits that the name ensuring stability should be chosen to represent the taxon. In view of the fact that neither name has been used for over 70 years, a valid genus (*Chasmosaurus*) is a type of the Chasmosaurinae, and no taxa remain within Lambe's Eoceratopsinae, the name Chasmosaurinae is chosen to represent this subfamily, and the name Eoceratopsinae is abandoned.

The placement of *Pachyrhinosaurus* within either Centrosaurinae or Chasmosaurinae was at first problematic and Sternberg (1950) placed it in its own family. Langston (1967, 1975), however, demonstrated that on the basis of its narial structure and frill *Pachyrhinosaurus* should be placed among the short-frilled group, Centrosaurinae. Sternberg (1938, 1940) showed that *Centrosaurus* is generically distinct from *Monoclonius,* and is not a subgenus or junior synonym. Hence, *Centrosaurus* remains the type genus of the Centrosaurinae.

Considering the preceding remarks, all known genera of ceratopsids may be referred to either Chasmosaurinae or Centrosaurinae as defined below. The following definitions are derived largely from the work of Lambe (1915), Sternberg (1927b, 1938, 1949), and Langston (1967, 1975).

Subfamily Chasmosaurinae Lambe 1915

Definition. Large ceratopsian dinosaurs with long, low facial region (preorbital length/height = 1.4 to 3.0); inter-premaxillary fossae present; premaxilla and predentary with horizontal or poorly developed lateral cutting flange; nasal horncore, if present, small and formed in part by a separate ossification; supraorbital horncores usually large; postfrontal foramen present; postfrontal fontanelle walled mostly by the postorbital bones; frontal bones reduced; prefrontal bones small; cranial frill long (0.94 to 1.70 basal length of skull, except *Triceratops*); triangular squamosal bones (length/height = 2.0 to 3.5); prominent longitudinal channel in ventral surface of sacrum; ulna with large olecranon process; strongly curved ischium.

Type genus. *Chasmosaurus* Lambe.

Included genera. *Chasmosaurus* Lambe, *Pentaceratops* Osborn, *Anchiceratops* Brown, *Arrhinoceratops* Parks, *Torosaurus* Marsh, *Triceratops* Marsh, (*Eoceratops* Lambe is regarded as a junior synonym of *Chasmosaurus* Lambe; *Ceratops* Marsh and *Ugrosaurus* Cobabe and Fastovsky are regarded as *nomina dubia*).

Subfamily Centrosaurinae Lambe 1915

Definition. Large ceratopsian dinosaurs with short, deep facial region (preorbital length/height = 1.2 to 1.4); no inter-premaxillary fossae; finger-like processes of the nasal bones projecting into the narial apertures; wide laterally inclined shearing surfaces on the premaxilla and predentary bones; nasal horncore, if present, large and formed primarily by upgrowth of the nasal bones; poorly developed supraorbital horncores; no postfrontal foramen; postfrontal fontanelle walled primarily by large frontal bones; prefrontal bones large; short cranial frill (0.54 to 1.00 basal length of skull); quadrangular squamosal (length/height about 1.0); first sacral rib expanded ventrally to form triangular web of bone; poorly developed longitudinal channel on ventral surface of sacrum; ulna with reduced olecranon process; relatively straight ischium.

Type genus. *Centrosaurus* Lambe.

Included genera. *Centrosaurus* Lambe, *Monoclonius* Cope, *Styracosaurus* Lambe, *Pachyrhinosaurus* Sternberg. (*Brachyceratops* Gilmore and *Avaceratops* Dodson are regarded as junior synonyms of *Monoclonius* Cope; but see Dodson this volume).

As noted by Sternberg (1938) and Langston (1967), the two subfamilies of ceratopsids are readily separated by characteristics of their facial region (Fig. 16.1). Although the preorbital length of the skull is comparable in both groups, centrosaurines have a much deeper face at a given length. In addition to the relatively lower facial region in chasmosaurines, the portion of their face anterior to the nasal horncore is greater, giving them an even more drawn-out facial appearance. The inter-premaxillary fossae of chasmosaurines is of uncertain function, but may have housed the vomeronasal organs (Lehman 1982). No trace of this feature is apparent in centrosaurines. Likewise, the finger-like processes of the nasal bones that project into the narial apertures in centrosaurines have no counterpart in chasmosaurines. Langston (1975) suggests that these processes may have "supported the vestibulum from below and in some way separated this structure from ramifications of the cavum nasi proprium." The oral surfaces of the premaxillary, rostral, and corresponding predentary are strongly inclined (as much as 60° to the horizontal) in centrosaurines, but are nearly horizontal in chasmosaurines. Perhaps this reflects some subtle difference in diet or mastication processes.

Although all ceratopsids (except *Pachyrhinosaurus*) show some development of both nasal and brow (supraorbital) horncores, chasmosaurines generally possess larger brow horncores and centrosaurines larger nasal horncores. Farlow and Dodson (1975) suggest that this reflects behavioral differences between the two subfamilies. Some evidence suggests that the nasal horncore in centrosaurines is formed predominantly by upward growth of the paired nasal bones (Gilmore 1917 on *Brachyceratops montanensis*). In contrast, the nasal horncore in chasmosaurines was formed, at least in part, by a separate ossification (e.g., *Triceratops flabellatus* Marsh, YPM 1821, and *Triceratops elatus* Marsh, USNM 1201; see Hatcher et al. 1907). Lambe (1915) termed this ossification the epinasal bone. Sternberg (1949) later doubted the morphological significance of such a bone and showed that in some specimens of *Triceratops* the nasal horncore seems to have been wholly comprised of the nasal bones. However, while the paired nasal bones undoubtedly participate in the base of the nasal horncore, the single epinasal bone that formed the body of the horncore is apparently detached and lost prior to preservation in many specimens of chasmosaurines (e.g., *Arrhinoceratops brachyops* Parks 1925 and many specimens of *Triceratops;* see Hatcher et al. 1907). Evidence for the presence and/or importance of the epinasal bone in many chasmosaurines is scant. Supraorbital horncores are formed by outgrowth of the postorbital bones in both chasmosaurines and centrosaurines. Some specimens of centrosaurines (e.g., *Styracosaurus albertensis* Sternberg 1927b) have a pit over the orbit suggesting that a separate ossicle may have

Figure 16.1. Criteria used for differentiation of Chasmosaurinae (typical genus *Chasmosaurus*) and Centrosaurinae (typical genus *Centrosaurus*). Features include: a, pre-orbital length; b, pre-orbital height; c, frill length; d, inter-premaxillary fossa; e, narial process of nasal bone; f, inclined cutting flange on predentary; g, supraorbital horncore; h, nasal horncore; i, squamosal; j, postfrontal foramen; k, frontal.

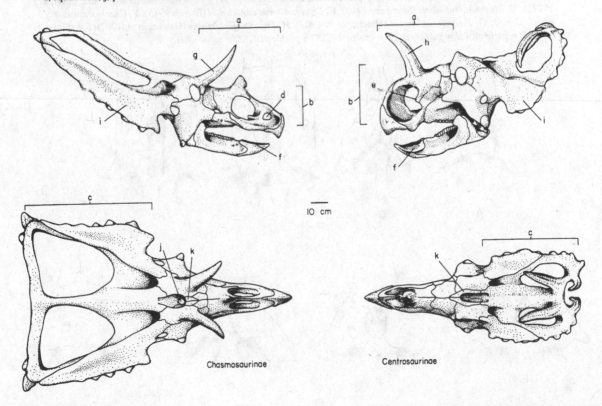

10 cm

Chasmosaurinae

Centrosaurinae

contributed to the formation of the brow horncore. In the advanced chasmosaurines, *Torosaurus* and *Triceratops*, the bases of the supraorbital horncores are hollow; however, in other ceratopsids the structure is solid.

The frontal region of the skull differs markedly between the two subfamilies (Fig. 16.2). In chasmosaurines, the postorbital bones are large and together wall most of the postfrontal fontanelle. (The term "postfrontal fontanelle" refers to this feature's position, not the bones that surround it.) The frontal bones are greatly reduced and participate only in the anterior end of the fontanelle. The prefrontal bones are likewise reduced. A large circular foramen is present where the frontals meet the fused parietals, near the anterior end of the postfrontal fontanelle. Gilmore (1917) termed this feature the "postfrontal foramen." This foramen remains open even in those advanced chasmosaurines in which the skull roof has secondarily roofed over (Fig. 16.3). Its placement suggests that the foramen may not be homologous with either the "parietal foramen" of primitive reptiles, that lies wholly within the fused parietals, or with the "pineal foramen." Hence, the postfrontal foramen appears to be a feature peculiar to chasmosaurines. Its function and derivation is unknown.

By contrast, in centrosaurines the frontal bones remain large and together mostly or entirely enclose the postfrontal fontanelle laterally. The prefrontal bones are also large. The postorbital bones participate little if any in the lateral walls of the postfrontal fontanelle, and no postfrontal foramen is present (at least in adults). Only juvenile specimens of *Monoclonius* (*Brachyceratops montanensis* Gilmore 1917) are known to possess a postfrontal foramen, but this apparently closes with age. A subtle circular depression in the parietal of *Styracosaurus albertensis* Lambe (NMC 344), posterior to the frontal suture, is placed more in accordance with the parietal or pineal foramen and posterior to the postfrontal foramen of chasmosaurines (Fig. 16.3). These differences in the frontal region of the skull probably correlate with the greater development of brow horncores in chasmosaurines. In those centrosaurines with more developed brow horncores (e.g., *Centrosaurus*) the postorbital bones show the greatest development.

The cranial frill in centrosaurines (exclusive of epoccipitals) is relatively short, ranging from 0.54 to 1.00 the basal length of the skull (distance from occipital condyle to tip of premaxilla). In chasmosaurines, except *Triceratops*, the frill is generally longer, ranging from 0.94 to 1.70 the basal length of the skull. In *Triceratops*, the frill is 0.62 to 0.89 the basal length of the skull; however, as noted above, the relative shortness of the frill is considered along with its thickening and the closure of parietal fenestrae, to be secondarily derived from the chasmosaurine condition. Evidence for this is seen in the form of the squamosal, which in *Triceratops* is triangular as in other chasmosaurines, instead of quadrangular as in centrosaurines.

Apart from some subtle differences in the pelvis,

Figure 16.2. Comparison of the dermal skull roof of ceratopsians in dorsal view, not to scale. **A**, *Protoceratops andrewsi* (Brown and Schlaikjer 1940); **B**, *Brachyceratops montanensis* (Gilmore 1917); **C**, *Styracosaurus albertensis* (Sternberg 1927b); **D**, *Monoclonius lowei* (Sternberg 1940); **E**, *Chasmosaurus canadensis* (Lambe 1915); **F**, *Chasmosaurus mariscalensis*; **G**, *Torosaurus gladius* (Hatcher et al. 1907); **H**, *Triceratops serratus* (Hatcher et al. 1907). F = frontal, N = nasal, P = parietal, Pal = palpebral, PF = prefrontal, PO = postorbital, SQ = squamosal.

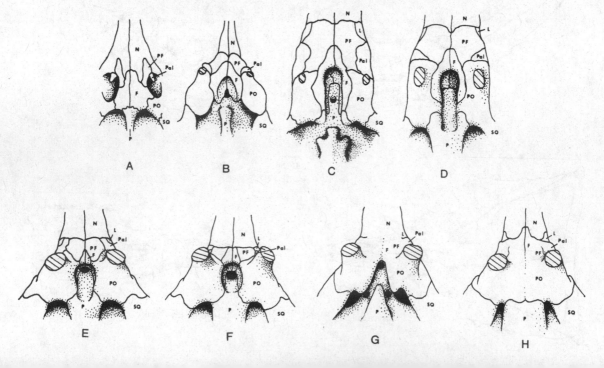

there is little in general to distinguish the postcranial skeletons of chasmosaurines from those of centrosaurines. A longitudinal channel is more deeply impressed in the ventral surface of the sacrum, and the ischium is more strongly curved in chasmosaurines than in centrosaurines. The first sacral rib is expanded ventrally in centrosaurines, but is not expanded in chasmosaurines. The olecranon process of the humerus is more developed in chasmosaurines.

The Chasmosaurinae constitute the most geographically and temporally widespread radiation of ceratopsians in North America, having spread over much of the western shore of the Cretaceous Interior Seaway from southern Canada to southern Texas during Late Campanian to Late Maastrichtian time. In contrast, the Centrosaurinae were a more restricted and short-lived group, never having extended their range farther south than Alberta and Montana (reports of *?Monoclonius* from New Mexico and Mexico are based on inadequate or undocumented specimens). The Centrosaurinae were diverse during the Late Campanian, but survived only through Early Maastrichtian time in the form of the rather aberrant *Pachyrhinosaurus*.

A Texas chasmosaurine

To illustrate the degree of variation present in a single ceratopsid species, a *Chasmosaurus* bone-bed from the Aguja Formation in Texas is described. The Aguja Formation comprises marginal marine, deltaic, and fluvial sediments deposited during Campanian time in the Trans-Pecos region of Texas (Lehman 1982, 1985). These deposits have yielded a diverse dinosaur fauna of Judithian aspect, mostly from the upper part of the formation (upper shale member of Lehman 1985). One locality, herein referred to as "WPA 1," on the northwest flank of Talley Mountain in Brewster County, Texas, yielded abundant ceratopsian remains referable to *Chasmosaurus mariscalensis*. The taphonomy of the WPA 1 bone-bed accumulation was described by Lehman (1982) and Davies and Lehman (1989). The remains were excavated as part of a Works Progress Administration (WPA) project in 1938–1939 under the direction of William S. Strain, and are part of the collections of the Centennial Museum of the University of Texas at El Paso (UTEP). At present, the specimens are housed at the Texas Memorial Museum in Austin, Texas.

Taxonomy

The ceratopsian bones recovered from WPA 1 represent the remains of at least 10 individuals, based on the greatest number of any single skeletal element, 10 left femora. By comparing elements of compatible sizes, as many as 15 individuals may be present. A great deal of variation is present in this sample, and most skeletal elements are represented by graded size series. Because the remains were collected from a single locality and stratigraphic horizon, and because the accumulation probably represents a nontransported assemblage, the variation observed in this sample may represent either the individual variability within a population of a single species, or less likely, between different sympatric species.

I believe the variation in the sample may be individual, ontogenetic, allometric, and secondary sex-associated variation (*sensu* Mayr 1969) within a population of a single ceratopsian species. This species belongs to the genus *Chasmosaurus*, and represents a new species closely related to *C. canadensis* (see below). *Chasmosaurus mariscalensis* is formally described elsewhere (Lehman 1989). The estimated maximum adult

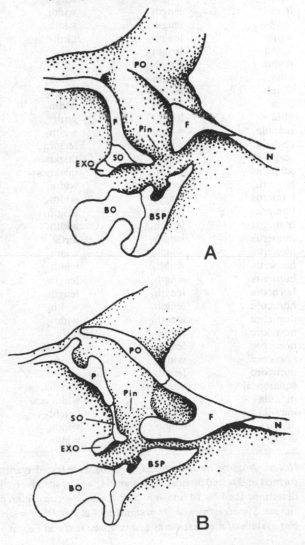

Figure 16.3. Diagrammatic sagittal sections (not to scale) through the skull roof and braincase of **A**, *Chasmosaurus mariscalensis* and **B**, *Triceratops* sp. (after Gilmore 1919), showing the folding of postorbital bones (PO) and invagination of frontal bones (F) with closure of the postfrontal fontanelle. Postfrontal foramen (Pin) remains open. BO = basioccipital, BSP = basisphenoid, EXO = exoccipital, N = nasal, SO = supraoccipital, P = parietal.

size of *Chasmosaurus mariscalensis* is comparable to that in other members of the genus.

In any taxonomic study it is essential to consider the great potential of a species for individual variation, and there are many sources of such variation. The paleontologist-taxonomist can only hope to address some of these and, of course, must gauge for himself the amount of morphological variation that can be accommodated by a single species. Historically, ceratopsian taxonomy has been based almost entirely on cranial features in spite of their high variability, simple structure, and obvious specializations. Hence, it is important to distinguish those morphological characters that are subject to high intraspecific variability from those which indicate taxonomic separation. Information derived from this collection is useful in reevaluating the systematics of advanced ceratopsians, because in the past, few data have been available on the range of intraspecific variation in the Ceratopsidae. In what follows, emphasis is placed on the individual variation observed

Table 16.1. *Results of regression analysis of* Chasmosaurus mariscalensis *sample from quarry WPA 1*

Bone	X	Y	n	r	a	S
ilium	$length_3$	thickness	6	.961	1.20	no
ilium	$length_3$	height	6	.821	.94	no
ilium	$length_3$	width	5	.915	1.49	no
ilium	$length_3$	a	6	.840	.32	**
ilium	$length_3$	b	7	−.819	−.48	**
femur	$length_1$	$length_3$	11	.779	1.38	no
femur	$length_2$	$width_1$	11	.898	1.24	no
femur	$length_2$	$width_2$	12	.979	1.19	no
femur	$length_2$	girth	12	.981	1.13	no
femur	$length_2$	$area_1$	12	.978	2.27	no
femur	$length_2$	a	11	−.898	−.36	**
tibia	length	$width_1$	7	.873	1.06	no
tibia	length	$width_3$	8	.965	1.24	*
tibia	length	$width_2$	9	.920	1.40	no
tibia	length	girth	9	.950	1.43	*
scapula	$width_1$	$width_2$	6	.963	.99	no
scapula	$width_1$	$length_1$	6	.914	.71	no
scapula	$width_1$	$thickness_1$	8	.862	1.30	no
scapula	$width_1$	$thickness_2$	6	.946	1.21	no
humerus	$length_1$	$width_1$	7	.871	.93	no
humerus	$length_1$	$width_2$	8	.947	1.34	no
humerus	$length_1$	$length_2$	8	.989	1.03	no
humerus	$length_1$	$width_3$	8	.942	1.13	no
humerus	$length_1$	girth	8	.944	1.34	no
humerus	$width_1$	$width_2$	7	.906	1.43	no
humerus	$width_4$	$width_5$	7	.909	.66	**
humerus	$length_1$	$length_3$	6	.954	1.51	*
horncore	$length_1$	$length_2$	9	.986	1.12	no
horncore	$length_1$	$width_1$	9	.943	.89	no
horncore	$length_1$	$width_2$	9	.943	.75	*
horncore	$length_1$	girth	9	.926	.66	**
horncore	$length_1$	orbit dia.	6	.971	.77	*
horncore	$width_1$	$width_2$	10	.952	1.01	no
horncore	$length_1$	a	8	.149	.14	—
squamosal	thickness	epo. length	10	.725	1.29	no
maxilla	length	width row	5	.980	2.09	**
maxilla	length	$height_2$	5	.800	1.26	**
dentary	length	$width_4$	5	.587	.54	no
dentary	length	$width_5$	6	.810	1.58	no

Notes: Angular measurements (*a* and *b*) were tested against $H_o = 0$; linear and area measurements were log transformed and tested against $H_o = 1$ and $H_o = 2$, respectively. The method of reduced major axes was employed to determine the best fit line. *n* = sample size, *r* = correlation coefficient, *a* = regression coefficient or allometric coefficient, *S* = *significance*, ** = significant at P = .05, * = significant at P = .1, no = not significant, — = not tested. For system of measurements and data, see text and Lehman (1982, appendix III).

in each element of the skeleton and an attempt is made to differentiate the sources of this variation.

Ontogenetic allometry

The disarticulated condition of the skeletal remains from WPA 1 precludes the possibility of detecting ontogenetic changes in the overall skeletal proportions of *Chasmosaurus mariscalensis*. It is possible, however, to describe the relative growth in one dimension of a particular bone with respect to another dimension of the same bone. The allometric changes for each skeletal element thus detected can be combined to give some idea of relative growth in *Chasmosaurus*.

In spite of the small sample size, I have analyzed the WPA 1 sample biometrically, using the logarithmic expansion of the equation for simple allometry, as described by Gould (1966) and Dodson (1975, 1976). The line-fitting technique of reduced major axes was used to determine the best fit line. Not unexpectedly, this analysis largely failed to demonstrate significant allometry (at a probability level of p = .05) even for

allometric coefficients as low as 0.7 or as high as 1.4 (Table 16.1). The small sample size (n = 5 to n = 12) greatly accentuates individual variation and obscures any real allometric relationship. There also appears to be a high degree of individual variation in *Chasmosaurus mariscalensis*. The sample consists of a few large, presumably mature adults, and many "mediumsized" juveniles or subadults one third to one half the size of the largest adult. Very small individuals are represented only by fragments. This also reduced the possibility of detecting allometric growth by removing the lower end of the size distribution. Moreover, poor preservation of many of the specimens has created additional "non-biologic" variation. Recognizing that these factors have been at work in the sample, it is surprising that any significant results were obtained. Representative results of the regression study, significant and not, are reproduced in Table 16.1, and along with more subjective analysis forms the basis of the following discussion of general ontogenetic trends. The accompanying bivariate plots (Figs. 16.4–16.10) include data for additional

Figure 16.4. Selected bivariate plots of *Chasmosaurus mariscalensis* showing ontogenetic variation in scapula width (above) and humerus girth and width (below). In this and subsequent figures, measurements are in millimeters, and the scale on both axes is logarithmic unless otherwise indicated. Added for comparison are data for *Chasmosaurus canadensis* (solid circles: 1-UA 40, 2-NMC 1254, 3-AMNH 5401) and for *Pentaceratops sternbergii* (open triangles: 1-UNM FKK-081, 2-AMNH 6325, 3-PMU R200). The dashed line represents calculated Reduced Major Axes best fit.

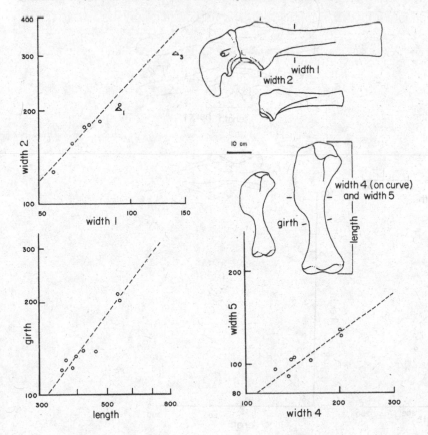

specimens of *Chasmosaurus* and *Pentaceratops* pertinent to the discussion that were not included in the regression analysis.

   While no marked proportional changes are apparent in the axial skeleton of *Chasmosaurus mariscalensis*
with growth, the incorporation of dorsal and caudal vertebrae into the sacrum as well as the serial fusion of the
sacral ribs and expansion of the acetabular bar are notable ontogenetic changes. The largest, presumably most
mature, specimen of *Chasmosaurus mariscalensis* has
four "true" sacral vertebrae, three dorso-sacrals and
three sacro-caudals. All four sacral ribs are fused both to
their corresponding centra and at their distal ends, to
form the acetabular bar. Smaller specimens have one
less dorso-sacral vertebra anteriorly, and exhibit varying
degrees of fusion to the sacral ribs. The first two sacral
ribs are thoroughly fused to their centra and united distally to form the ventro-laterally directed, cup-shaped,
portion of the acetabular bar. The third rib is fused to its
centrum but barely, if at all, united with the acetabular
bar; and the fourth rib is completely free. A disarticulated first sacral rib, about two-thirds the size of the
comparable element in the largest specimen, indicates

that in still smaller individuals the sacral ribs were all
freely articulating. The dorso-sacral and sacro-caudal
vertebral centra become progressively more co-ossified
toward the true sacrals. Hence, the sacrum in *Chasmosaurus mariscalensis* developed ontogenetically by successive addition of dorsal and caudal vertebrae, first by
fusion of their centra, and later by incorporation of their
ribs and transverse processes into the acetabular bar, or as
braces to the anterior and posterior processes of the ilia.

   There are apparently no marked ontogenetic
changes in the pectoral girdle. The distal end of the blade
in a juvenile scapula is irregularly crenulated and concave, indicating that in juveniles the end was incompletely ossified and finished in cartilage (Fig. 16.4).
However, neither the scapula nor the coracoid shows
notable proportional change with growth. More marked
changes are apparent in the pelvis. In juvenile ilia, the
horizontal dorsal blade is much more constricted along
its lateral border opposite the acetabulum, and the postacetabular process is subvertical (Fig. 16.5). In adult
ilia, the dorsal blade widens into a broad horizontal lateral shelf, the pre-acetabular process bends ventrally,
the post-acetabular process twists laterally, and the

Figure 16.5. Selected bivariate plots of *Chasmosaurus mariscalensis* showing ontogenetic variation in the ilium. The scale
for both axes on plots to the right are arithmetic. See caption to Figure 16.4 for additional data.

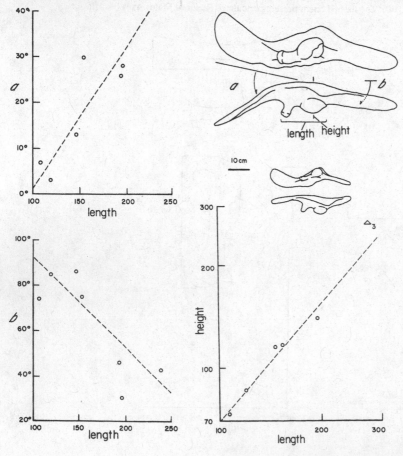

ischial peduncle becomes compressed into the body of the ilium. A pronounced depressed articular facet on the postero-medial surface of the ischial peduncle, seen in adult ilia, is not evident in juveniles. The post-acetabular process is also slightly inclined upward in juveniles, as it is in *Protoceratops* (Brown and Schlaikjer 1940). In many of these features, the ilia of juvenile individuals resemble those of the more primitive proto-ceratopsids, and to some degree centrosaurines. In addition, the acetabular wall of the pubis expands, and the pre-pubic process lengthens relative to the body of the pubis with growth. Too few ischia are preserved in the sample to gauge its relative growth.

In general, the various trochanters and muscle insertions on appendicular bones become more pro-nounced and rugose with growth. All of the limb bones show a slight, though statistically insignificant, increase in girth with increasing size. Among the forelimb bones, the deltopectoral crest of the humerus twists ventrally and extends proximally with growth, and the head extends farther onto the dorsal surface of the bone with growth (Fig. 16.4). In the hind limb, the femur shows a significant medial rotation of the distal condyles with growth. There is a slight increase in the height of the greater trochanter relative to the head of the femur, and a slight descent of the fourth trochanter down the shaft of the femur with growth (Fig. 16.6). Little ontogenetic change is apparent in the radius, ulna, tibia, or fibula. The distal end of the tibia does, however, widen with growth (Fig. 16.7).

Figure 16.6. Selected bivariate plots of *Chasmosaurus mariscalensis* showing ontogenetic variation in the femur. The scale for both cases on the lower left plots are arithmetic. See caption to Figure 16.4 for additional data.

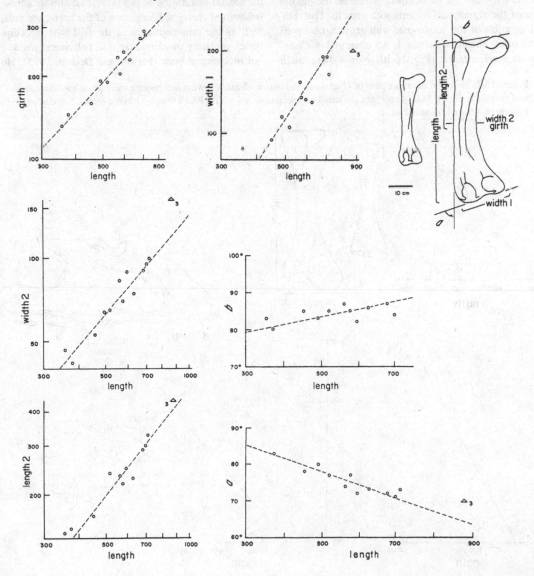

Several ontogenetic changes are evident in the skull of *Chasmosaurus mariscalensis*. There is progressive elongation of the facial region, as evidenced by the greater proportional increase in length, relative to height and width, of the dentary and predentary bones (Figs. 16.7, 16.8). Accompanying this is an increase in the number of tooth positions (from 20 to 28) and an increase in the size of the alveoli and teeth. Although the maxilla does not exhibit a corresponding increase in length, the facial elongation may have here been accommodated by growth of the premaxilla and rostral. It is apparently the edentulous portion of the jaws that experiences the greatest elongation with growth. Too little of the premaxilla and rostral are preserved, however, to determine if this is definitely the case. There is progressive elongation of the brow horncores relative to their basal width and circumference, and to the diameter of the orbit (Fig. 16.9). Epoccipital bones fuse to the margin of the frill, and the paroccipital flange on the medial surface of the squamosal widens with growth. The relative elongation of the squamosal with growth, observed in some other chasmosaurines, is not observed in *Chasmosaurus mariscalensis* (Fig. 16.10). Nor is there much

ontogenetic tendency toward roofing over the postfrontal fontanelle, as there is in some other ceratopsids (e.g., *Torosaurus*, see Fig. 16.2). Interestingly, many juvenile specimens of *Chasmosaurus mariscalensis* show almost complete obliteration of cranial sutures, particularly in the frontal region. A similar phenomenon is observed in juveniles of *Protoceratops* (Brown and Schlaikjer 1940). Hence, apparent closure of the cranial sutures is alone not a safe criterion for age determination. Whether the obscuration of these sutures represents true co-ossification or not is uncertain, though it seems unlikely that it does considering the necessity of further growth.

Changes in the postcranial skeleton with growth in *Chasmosaurus mariscalensis* reflect a wider, heavier body in the adult and the consequent mechanical adjustments necessary to suspend and propel an increased mass. Changes in the skull, however, probably reflect the sexual maturation of the individual and its attendant behavioral changes. Elongation of the horns and frill, as well as the ornamentation of the frill with epoccipital bones probably owes its origin to behavioral needs and not mechanical ones (Farlow and Dodson 1975). How-

Figure 16.7. Selected bivariate plots of *Chasmosaurus mariscalensis* showing ontogenetic variation in the tibia (above) and maxilla (below). Maxilla outlines are based on specimens: a – UTEP P.37.7.086, b – .087, c – .088. See caption to Figure 16.4 for additional data.

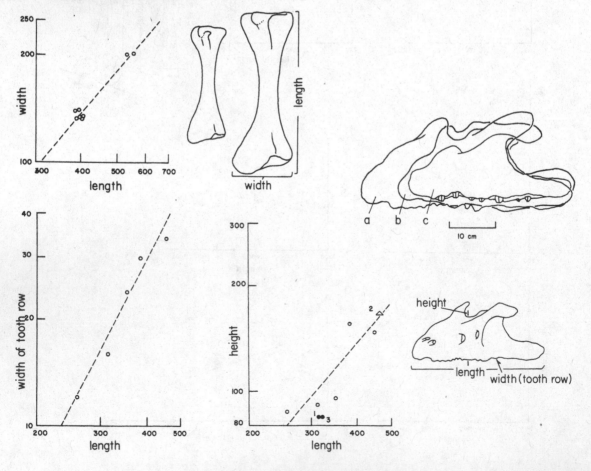

ever, the elongation of the facial region of the skull, with its corresponding increase in the number of tooth rows, may be in response to a need for more surface area or leverage in rending resistant plant food, and thus a reflection of mechanical function (Ostrom 1964, 1966).

Sexual dimorphism

Sexual dimorphism in the primitive ceratopsian *Protoceratops* is well known (Brown and Schlaikjer 1940); and, recently, statistical techniques have been employed to document its sexually dimorphic features (Kurzanov 1972; Dodson 1976). Owing to small sample sizes, such features have not been noted in advanced ceratopsians, though suggestions have been made from time to time that these animals would also prove to be sexually dimorphic.

Authors have drawn an analogy between ceratopsians and modern horned ungulates in relation to their herding behavior, male intrasexual selection, and sexual dimorphism. A death assemblage representing juvenile and adult individuals of the same species, as at WPA 1,

lends support to the opinions expressed by Hatcher et al. (1907), Sternberg (1970), and others that ceratopsians were herding animals. The large ceratopsian bone-beds described by Currie (1981) from the Judith River (Oldman) Formation of Alberta offer even more convincing evidence of this.

In modern ungulates, herding behavior, social structure, and sexual dimorphism are related. A polygamous breeding system forces females to live in groups, which fosters male intrasexual selection, and results in sexual dimorphism (Gingerich 1981). Dodson (1976) and Farlow and Dodson (1975) have argued that the frill and horns in ceratopsians probably served primarily as visual recognition signs for attracting females or driving away rival males or predators. It seems likely therefore, that evidence of sexual dimorphism would be found in the cranial features of advanced ceratopsians.

The variability in cranial morphology shown by *Chasmosaurus mariscalensis*, particularly in its brow horncores, can be resolved into two subtly different morphs that probably represent the two sexes (Fig.

Figure 16.8. Variation in the dentary and predentary of *Chasmosaurus mariscalensis* with growth, showing **A**, elongation of the dentary with respect to width; **B**, diagrammatic reconstruction of the lower jaws in oral view; and **C**, elongation of the edentulous portion of the lower jaw. Inflected ventral border of the dentary is indicated by arrow. Dentary outlines based on specimens: a – UTEP P.37.7.087, b – 073, and c – 095. See caption to Figure 16.4 for additional data.

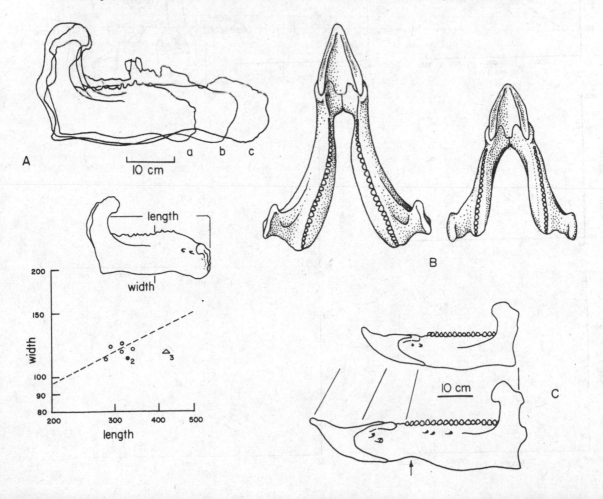

16.11). In one morph, possibly the female, horncores arise from the anterior margin of the orbit, are curved, and are directed antero-laterally and upward from the orbit. Other horncores arise from the dorsal border of the orbit, are more erect, and are directed vertically from the orbit (Fig. 16.11). Because the latter morphology seems better suited for the frontal "shoving matches" postulated by Farlow and Dodson (1975), and because it is shown by the largest individuals, it may represent the male morph. In anterior view, the female horncores present a more widely spread appearance, the males a more narrow one (Fig. 16.11). This dichotomy in horncore orientation is subtle, and difficult to express as linear measurements; hence, it is best shown graphically as the angular measure between the axis of the horncore and the frontal plane of the skull in lateral view (Fig. 16.11). Apart from the orientation of the horncore, there is some variation in the cross-sectional shape of the horncore, and in its basal cross-sectional area that may be correlative with orientation. The orbit appears to be slightly more elliptical in "males," perhaps to strengthen the base of the horn against longitudinal (compressional) stress. However, the angulation of the horn seems to be the best discriminator because it separates the sexes readily, even at juvenile growth stages.

Based on the horncores that can be assigned to one sex or the other, the number of males (4 or 5) and females (3 or 4) is about equal, considering the possibility that some detached horncores may belong to a single individual. There are also several indeterminate individuals in the sample. Two distinct size groups appear to be present in the population, one comprised of juveniles or "subadults" of both sexes, and the other of fully adult animals (Fig. 16.11). Hence, there are three adult females

Figure 16.9. Selected bivariate plots for *Chasmosaurus mariscalensis* showing ontogenetic variation in the supraorbital horncore (postorbital bone). See caption to Figure 16.4 for additional data.

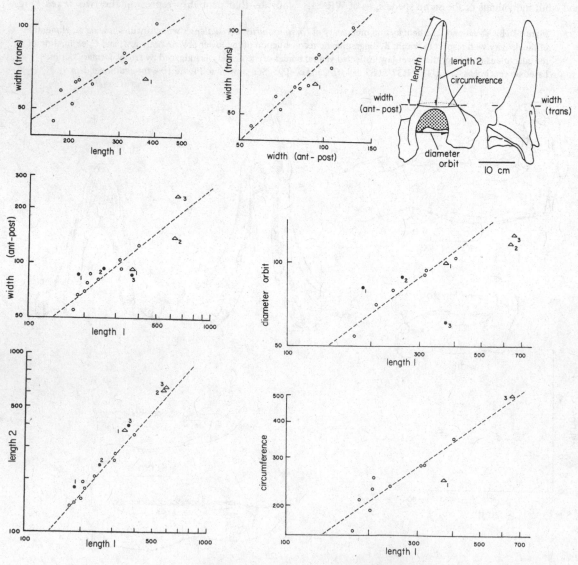

and one adult male. The adult females are slightly smaller than the male, which is the largest individual in the sample.

As Dodson (1976) observed in his study of *Protoceratops*, males and females of *Chasmosaurus mariscalensis* may not be distinguished simply by the presence or absence of osteological characters, but instead by the difference in development of such characters (namely the brow horncores) at a given size. This condition is analogous to that observed in species of African antelopes in which both sexes possess horns (Packer 1983). Hence, at a given size the male horncores appear to have a greater cross-sectional area, but apart from their orientation they do not differ substantially from those of the female. Such features as the degree of expansion and erection of the frill, and the height of the nasal horncore, which exhibit sexual dimorphism in *Protoceratops*, cannot be measured in the *Chasmosaurus mariscalensis* sample owing to incomplete preservation.

Although the form of the squamosal and parietal may differ subtly between the sexes, the frill is apparently so similar in form between males and females of *Chasmosaurus mariscalensis*, and in most ceratopsid species, that its utility for species recognition must have been greater than that for sexual display.

Sexual dimorphism in other chasmosaurines

Chasmosaurus mariscalensis from Texas allows for the first time the description of ontogenetic change and sexual dimorphism in an advanced ceratopsid, and provides a basis for reevaluation of the taxonomy of chasmosaurines. Recognition of the wide range of intraspecific variation present in *Chasmosaurus mariscalensis* necessitates a taxonomic revision of the ceratopsids.

The relatively short squamosal, horncores, and facial region of *Eoceratops canadensis* (NMC 1254 and

Figure 16.10. Ontogenetic variation in the squamosal of *Chasmosaurus mariscalensis* (above), *Chasmosaurus canadensis* (a – NMC 1254, b – TMP 83.25.1, c – AMNH 5401), and *Pentaceratops sternbergii* (d – from Wiman 1930, e – UNM FKK 081, f – MNA 1747, g – AMNH 6325, h – PMU R200). See caption to Figure 16.4 for additional data.

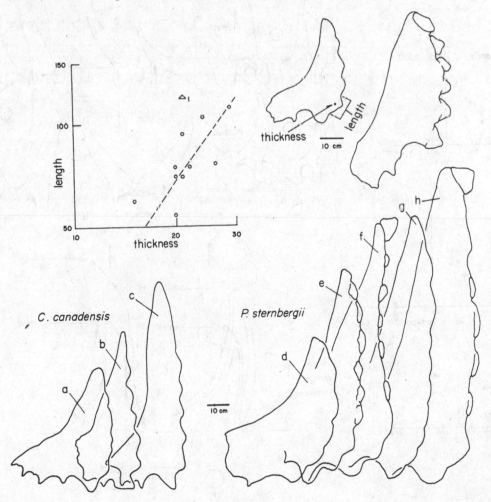

UA 40) are within the realm of variation expected for a juvenile or "subadult" of *Chasmosaurus kaiseni* (AMNH 5401). Thus the two are synonymous and should be called *Chasmosaurus canadensis*. A Tyrrell Museum of Palaeontology specimen on display in the Banff Natural History Museum in Banff, Alberta (identified as *Chasmosaurus* sp.) is probably referable to this species. *Chasmosaurus canadensis* exhibits the same dichotomy in horncore orientation observed in the *Chasmosaurus mariscalensis* sample from Texas. Based on the criteria suggested here, UA 40 (*Eoceratops canadensis* of Gilmore 1923, and Tyson 1977, or *Chasmosaurus* cf. *C. kaiseni* of Lull 1933) and AMNH 5401 (type of *Chasmosaurus kaiseni* Brown 1933) are probably small and large females, respectively; whereas NMC 1254 (type of *Eoceratops canadensis* Lambe 1915) is a small male of the same species.

In other species of *Chasmosaurus*, where the horncores are small, sexual dimorphism is obscure (Fig. 16.12). As originally noted by Sternberg (1927a), described specimens of *Chasmosaurus belli* (Lambe 1902) suggest that the males may have had more pointed epoccipitals, a straighter posterior margin of the parietal and a larger body size than females. The females had rounded, blunt epoccipitals and an indented posterior parietal margin. If Sternberg's (1927a) assertion is correct, then correspondingly, the male specimens have more erect and pointed brow horncores than the females, whose horns were more curved. The variable development of brow horncores in specimens referred to *C. belli*, however, makes this last point tenuous. It is possible that hornless specimens referred to *C. belli* (ROM 5499) belong to *C. russelli* Sternberg (1940). It seems less likely, but possible, that hornless specimens (includ-

Figure 16.11. Sexual dimorphism in the supraorbital horncores of *Chasmosaurus mariscalensis*, showing **A**, variation in inclination of the horn; **B**, variation in spread of the horns; and **C**, separation of individuals with respect to age and sex. Bivariate plots depict separation of sexes and age groups. Specimens used in diagram include: females a – UTEP P.37.7.082, b – 042, c – 079, d – 083; and males a – 086, b – 091, c – 044, d – 043, e – 090.

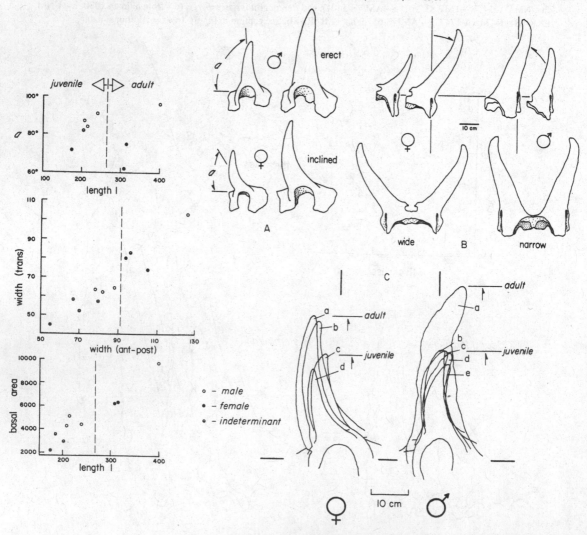

ing *C. russelli*) represent a female morph corresponding to horned male specimens (generally referred to *C. belli*). The former alternative is adopted here, although additional work is needed to establish whether or not horned and hornless morphs can both be present in a single *Chasmosaurus* population. The indented posterior parietal margin, thought to be distinctive of *C. russelli* by Sternberg (1940) is, however, also present in individuals referred to *Chasmosaurus mariscalensis* from Texas, *C. belli*, and *C. canadensis* as diagnosed here (Fig. 16.13). The indented posterior parietal margin is present in all species of *Chasmosaurus*, and is perhaps a sexual feature, though too little is yet known about this character to be certain.

The shortened facial region exhibited by the type specimen of *Chasmosaurus brevirostris* Lull (1933) is, as shown in *Chasmosaurus mariscalensis* from Texas, within the realm of variation expected of *C. belli*. *Chasmosaurus brevirostris* is herein regarded a junior synonym of *C. belli*. Hence, *Chasmosaurus* contains four species: the hornless *C. russelli*, the small-horned *C.*

belli, the large-horned *C. canadensis*, and *Chasmosaurus mariscalensis* from Texas with still larger horns (Lehman 1982).

An examination of other chasmosaurines suggests that sexual dimorphism, displayed in the orientation of the brow horncores, is not only widespread but seen more markedly in other species (Fig. 16.14). *Pentaceratops, Anchiceratops, Torosaurus,* and *Triceratops* all show more pronounced dichotomy in the orientation of their horncores than *Chasmosaurus*.

Pentaceratops fenestratus Wiman (1930) has been regarded as a large pathologic individual of *P. sternbergii* Osborn, and hence should be considered its junior synonym (Lehman 1981). Described specimens of *P. sternbergii* display no marked differences in horncore orientation, and all seem to possess the "male" morphology. An undescribed specimen (UNM FKK-081), referable to *P. sternbergii*, however, possesses a "female" horncore morphology (Fig. 16.14). This specimen is smaller than typical examples of *P. sternbergii*, but on the basis of its fully co-ossified epijugal and

Figure 16.12. Variation in chasmosaurine supraorbital horncores. **A**, *Chasmosaurus belli* (ROM 5499); **B**, *C. belli* (AMNH 5402); **C**, *C. brevirostris* (ROM 5436); **D**, *C. belli* (NMC 2280); **E**, *C. belli* (NMC 2245); **F**, *Ceratops montanus* (USNM 2411); **G**, *Chasmosaurus canadensis* (UA 40); **H**, *C. canadensis* (AMNH 5401); **I**, *C. canadensis* (NMC 1254); **J**, *Chasmosaurus mariscalensis* (UTEP P.37.7.082); **K**, *C.* n.sp. (UTEP P.37.7.086); **L**, *P. sternbergii* (AMNH 6325); **M**, *A. ornatus* (AMNH 5259); **N**, *A. brachyops* (ROM 5135); **O**, *T. callcornis* (USNM 4928); **P**, *T. gladius* (YPM 1831). Proposed synonymies cited in text.

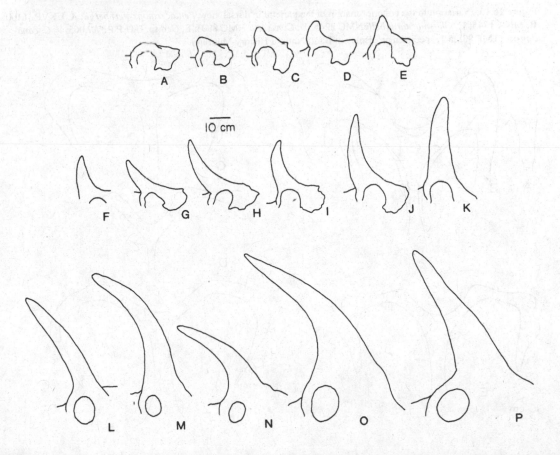

epoccipitals, and its fused scapulocoracoid, appears to represent a fully adult animal. There is thus only one valid species of *Pentaceratops*, namely *P. sternbergii*.

Anchiceratops longirostris Sternberg and *Anchiceratops ornatus* Brown are the female and male morphs, respectively, of a single species that should be designated *A. ornatus*. The elongate rostrum and forwardly inclined brow horncores of *A. longirostris* (NMC 8535) are easily within the realm of variation expected of a single species, when compared to the type and referred specimens of *A. ornatus* (AMNH 5251 and 5259). An undescribed specimen at the Tyrrell Museum of Palaeontology (TMP P83.1.1) represents the most complete male individual of *A. ornatus* known. At present there is only one species of *Anchiceratops*, namely *A. ornatus*.

As originally suggested by Colbert and Bump (1947), *Torosaurus gladius* Marsh should be considered a junior synonym of *Torosaurus latus* Marsh. The type specimen of *T. gladius* (YPM 1831) is herein regarded as a male individual; the type (YPM 1830) and referred specimen (ANSP 15192) of *T. latus* are regarded as females. Specimens referred to *T. utahensis* (Gilmore) reveal identical sexual dimorphism. The type specimen (USNM 15583) is a female individual; the referred specimen USNM 16169 is a male. Gilmore (1946) originally referred this species to *Arrhinoceratops*. However, Lawson (1976) subsequently transferred it to *Torosaurus*. At present, there is no basis for separating *T.*

utahensis from *T. latus*. Thus, *Torosaurus* contains only one species, *T. latus*.

As only the type specimen (ROM 5135) is currently referable to *Arrhinoceratops brachyops* Parks, nothing can be said about its variability (Tyson 1981). The type specimen, however, appears to possess a "male" horncore morphology.

The variation observed in all 14 described species of *Triceratops* can be accommodated within a single, sexually dimorphic, species (Lehman 1982; Ostrom and Wellnhofer this volume). Lull's (1915, 1933) "*T. brevicornus* – *T. prorsus* – *T. horridus* lineage" comprises female individuals while the "*T. calicornis* – *T. elatus* lineage" comprises male individuals of the same species. Lull's "*T. obtusus* – *T. hatcheri* lineage" comprises aged and pathologic male individuals, also referable to the same species. Hence, all of the adequately known species of *Triceratops* represent the varied individuals of a single species, that must be designated *T. horridus* Marsh.

In all of the cases cited above, males apparently attained larger body sizes than females. Thus, small but fully mature skulls often prove to be female (e.g., the type of *Triceratops prorsus*, YPM 1822); while large but immature skulls [e.g., the types of *Triceratops flabellatus* Marsh (YPM 1821) and *Torosaurus gladius* Marsh (YPM 1831)] often prove to be male. It is also interesting that pathologic specimens with perforated squamosals or broken horns generally prove to be male [e.g., the types of

Figure 16.13. Variation in the posterior margin of the parietal in dorsal view. *Pentaceratops sternbergii*: **A**, UKVP 16100; **B**, MNA 1745; **C**, *Chasmosaurus belli* NMC 2245; **D**, *C. russelli* NMC 8803; **E**, *C.* n.sp. UTEP P.37.7.065; **F**, *C. canadensis*, (TMP 83.25.1, specimen on display in Banff Natural History Museum).

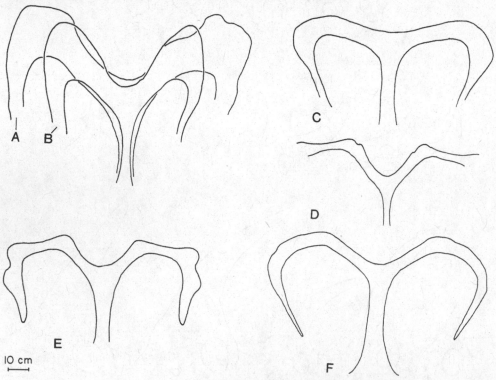

10 cm

Pentaceratops fenestratus (PMU R200) and *Diceratops hatcheri* (USNM 2412)]. This is in accordance with suggestions that the males may have engaged in frontal shoving matches or other combat that might have resulted in injuries to the head (Farlow and Dodson 1975).

Conclusions

Ceratopsids may be separated readily into two groups, herein formally recognized as subfamilies, Chasmosaurinae and Centrosaurinae. These groups are defined on the basis of narial structure, configuration of the frontal region, horncore morphology, and frill proportions. Their distribution varies both geographically and temporally. A population sample of *Chasmosaurus* from Texas illustrates the degree of morphological variation expected within a single chasmosaurine species. Analysis of this variation suggests that the following characters are of little value in distinguishing chasmosaurine species:

1. *Relative elongation of the face.* The facial region, particularly the edentulous part of the jaws, undergoes ontogenetic lengthening and is subject to a wide range of individual variation in adults.
2. *Size and orientation of the supraorbital horncores.* The brow horncores vary in size and orientation within a species, depending on ontogenetic stage and sex, and undergo pronounced ontogenetic elongation. Some evidence suggests that the horncores of adult females are shorter and have a smaller basal area than those of adult males.
3. *Indentation of the posterior parietal margin.* Individuals of a given species may possess either straight or deeply indented posterior parietal margins. This may reflect sex-associated variation.
4. *Length of the squamosal.* The squamosal, and the frill as a whole, undergo ontogenetic elongation in many species, and may be shorter in adult females than adult males. Hence, a wide range of lengths may be characteristic of a single species. The maximum length of the squamosal, however, may be specifically diagnostic.
5. *Presence or size of epoccipitals.* Epoccipital bones fuse to the margins of the squamosal and parietal ontogenetically, beginning at the anterior end of the squamosal. If unfused to the frill, they are often lost in preserved skulls. Where preserved, epoccipitals vary in size and peakedness within a single species.
6. *Presence of nasal horncore.* The nasal horncore was formed partly or entirely by a separate ossification (the epinasal) that is detached and lost in many specimens.
7. *Skull and body size.* Within a single species, males have larger skulls and may likewise have attained a larger body size. The maximum size attained by a given species may, however, be diagnostic.

The foregoing summary leaves us somewhat bereft of characters that may be useful in distinguishing

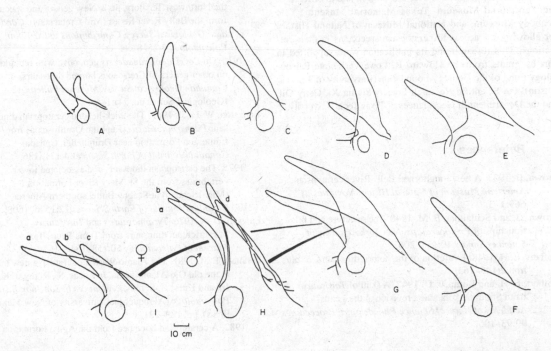

Figure 16.14. Sexual dimorphism in the Chasmosaurinae, as shown by orientation of the supraorbital horncores. Females have forwardly inclined horns, males have erect horns. **A**, *Chasmosaurus canadensis* (male NMC 1254, female AMNH 5401); **B**, *Chasmosaurus mariscalensis* (male UTEP P.37.7.086, female UTEP P.37.7.082), **C**, *Pentaceratops sternbergii* (male AMNH 6325, female UNM FKK-081); **D**, *Anchiceratops ornatus* (male AMNH 5259, female NMC 8535); **E**, *Torosaurus latus* (male YPM 1831, female YPM 1830); **F**, *Torosaurus latus* (male USNM 16169, female USNM 15583: type and referred specimens of *T. utahensis*); **G**, *Triceratops horridus* (male USNM 4928 = type of *T. calicornis*, female USNM 2100 = referred specimen of *T. prorsus*); **H**, variation in males of *T. horridus* (a – USNM 1201 = *T. elatus*, b – USNM 4928 = *T. calicornis*, c – YPM 1821 = *T. flabellatus*, d – USNM 2412 = *T. hatcheri*); **I**, variation in females of *T. horridus* (a – USNM 2100 = *T. prorsus*, b – YPM 1822 = *T. prorsus*, c – YPM 1834 = *T. brevicornus*.)

10 cm

species within a given chasmosaurine genus. In fact, given the range of variation observed in the Texas *Chasmosaurus* sample, all chasmosaurine genera except *Chasmosaurus* contain but a single species. Juvenile individuals of different chasmosaurine species, or even genera, may be virtually impossible to distinguish at current levels of knowledge. In all probability, species may be distinguished only at "adult" body sizes, yet attainment of adult status is difficult to determine. As noted above, apparent closure of cranial sutures alone is not a safe criterion for age determination in ceratopsians; fusion of epoccipital bones to the frill margin may be a more reliable indicator of adult status. Amongst adults, orientation of the supraorbital horn cores is the most useful character for distinguishing the sexual morphs of a given species. Bone-bed samples of other chasmosaurine species must be sought and studied to properly gauge the range of morphological variation present in other species, and to determine whether or not the reduction in number of valid species proposed herein is accurate. A similar evaluation of named species within the Centrosaurinae may likewise reveal similar synonymous relationships (see Dodson this volume).

Acknowledgments

This paper comprises part of my Master's thesis completed under the supervision of Wann Langston, Jr. at the University of Texas at Austin. I thank Dr. Langston for his direction and encouragement while I was a student in Austin, and for reviewing an early version of this paper. The great accomplishments and teachings of Charles M. Sternberg, perhaps the foremost student of the Ceratopsia, inspired most of the ideas presented here. Two anonymous reviewers helped improve the content and presentation of this paper. Thanks are also extended to the officials of Big Bend National Park for their cooperation during fieldwork for this project, and to officials of the Centennial Museum, Texas Memorial Museum, Yale Peabody Museum, and National Museum of Natural History for allowing me access to ceratopsian specimens in their collections. This research and its publication were supported in part by grants from the Howard R. Lowe Vertebrate Paleontology Fund of the Geology Foundation (University of Texas at Austin), the Scientific Research Society Sigma Xi, Getty Oil, and the Department of Geosciences at Texas Tech University.

References

Brown, B. 1933. A new longhorned Belly River ceratopsian. *American Museum of Natural History Novitates* 669:1–3.

Brown, B. and Schlaikjer, E. M. 1940. The structure and relationships of *Protoceratops*. *New York Academy of Science, Annals* 40:133–206.

Colbert, E. H. 1948. Evolution of the horned dinosaurs. *Evolution* 2:145–163.

Colbert, E. H. and Bump, J. D. 1947. A skull of *Torosaurus* from South Dakota and a revision of the genus. *Academy of Natural Science Philadelphia, Proceedings* 99:93–106.

Currie, P. J. 1981. Hunting dinosaurs in Alberta's great bonebed. *Canadian Geographic* 101:34–39.

Davies, K. L. and Lehman, T. M. 1989. The WPA Quarries. *In* Busbey, A. B. and Lehman, T. M. (eds.), *Vertebrate Paleontology, Biostratigraphy, and Depositional Environments, Latest Cretaceous and Tertiary, Big Bend Area, Texas.* Society of Vertebrate Paleontology, 49th Annual Meeting Fieldtrip Guidebook, pp. 32–42.

Dodson, P. 1975. Relative growth in two sympatric species of *Sceloporus*. *American Midland Naturalist* 94:421–450.

1976. Quantitative aspects of relative growth and sexual dimorphism in *Protoceratops*. *Journal of Paleontology* 50:929–940.

Farlow, J. O. and Dodson, P. 1975. The behavioral significance of frill and horn morphology in ceratopsian dinosaurs. *Evolution* 29:353–361.

Gilmore, C. W. 1917. *Brachyceratops*, a ceratopsian dinosaur from the Two Medicine Formation of Montana. *U.S. Geological Survey Professional Paper* 103:1–45.

1919. A new restoration of *Triceratops*, with notes on the osteology of the genus. *Proceedings of the U.S. National Museum* 55:97–112.

1923. A new species of *Corythosaurus* with notes on other Belly River Dinosauria. *Canadian Field-Naturalist* 37:46–52.

1946. Reptilian fauna of the North Horn Formation of central Utah. *U.S. Geological Survey Professional Paper* 210-C:1–52.

Gingerich, P. D. 1981. Variation, sexual dimorphism, and social structure in the early Eocene horse *Hyracotherium* (Mammalia, Perissodactyla). *Paleobiology* 7:443–455.

Gould, S. J. 1966. Allometry and size in ontogeny and phylogeny. *Biology Review* 41:587–640.

Hatcher, J. B., Marsh, O. C., and Lull, R. S. 1907. The Ceratopsia. *U.S. Geological Survey Monograph* 49:1–300.

Kurzanov, S. M. 1972. Sexual dimorphism in protoceratopsians. *Paleontology Journal* 1972:91–97.

Lambe, L. M. 1902. On Vertebrata of the Mid-Cretaceous of the Northwest Territory, pt. 2. New genera and species from the Belly River Series (Mid-Cretaceous). *Canadian Geological Survey, Contributions to Canadian Paleontology* 3:25–81.

1915. On *Eoceratops canadensis* gen. nov., with remarks on other genera of Cretaceous horned dinosaurs. *Canadian Geological Survey, Museum Bulletin* 12 (Geological Series no. 24):1–49.

Langston, W. L., Jr. 1967. The thick-headed ceratopsian dinosaur *Pachyrhinosaurus* (Reptilia: Ornithischia) from the Edmonton Formation near Drumheller, Canada. *Canadian Journal of Earth Sciences* 4:171–186.

1975. The ceratopsian dinosaurs and associated lower vertebrates from the St. Mary River Formation (Maestrichtian) at Scabby Butte, southern Alberta. *Canadian Journal of Earth Sciences* 12:1576–1608.

Lawson, D. A. 1976. *Tyrannosaurus* and *Torosaurus*, Maestrichtian dinosaurs from Trans-Pecos Texas. *Journal of Paleontology* 50:158–164.

Lehman, T. M. 1981. The Alamo Wash local fauna: a new look at the old Ojo Alamo fauna. *In* Lucas, S., Rigby, J. K., Jr., and Kues, B. S. (eds.), *Advances in San Juan Basin Paleontology* (Albuquerque: University of New Mexico Press), pp. 189–221.

1982. A ceratopsian bone bed from the Aguja Formation

(Upper Cretaceous) Big Bend National Park, Texas. Unpublished M.A. thesis (Austin: University of Texas).

1985. Stratigraphy, sedimentology, and paleontology of Upper Cretaceous (Campanian-Maastrichtian) sedimentary rocks in Trans-Pecos Texas. Unpublished Ph.D. dissertation (Austin: University of Texas).

1989. *Chasmosaurus mariscalensis*, sp. nov., a new ceratopsian dinosaur from Texas. *Journal of Vertebrate Paleontology* 9:137–162.

Lull, R. S. 1915. The mammals and horned dinosaurs of the Lance Formation of Niobrara County, Wyoming. *American Journal of Science* 40:319–348.

1933. A revision of the Ceratopsia or horned dinosaurs. *Yale Peabody Museum Memoirs* 3:1–175.

Mayr, E. 1969. *Principles of Systematic Zoology* (New York: McGraw-Hill, Inc).

Ostrom, J. H. 1964. A functional analysis of jaw mechanics in the dinosaur *Triceratops*. *Postilla* 88:1–35.

1966. Functional morphology and evolution of ceratopsian dinosaurs. *Evolution* 20:290–308.

Packer, C. 1983. Sexual dimorphism: the horns of African antelopes. *Science* 221:1191–1193.

Parks, W. A. 1925. *Arrhinoceratops brachyops*, a new genus and species of Ceratopsia from the Edmonton Formation of Alberta. *University of Toronto Studies, Geological Series* 19:5–15.

Sternberg, C. M. 1927a. Horned dinosaur group in the National Museum of Canada. *Canadian Field-Naturalist* 41:67–73.

1927b. Homologies of certain bones of the ceratopsian skull. *Transactions of the Royal Society of Canada* 21:135–143.

1938. *Monoclonius* from southeastern Alberta compared with *Centrosaurus*. *Journal of Paleontology* 12:284–290.

1940. Ceratopsidae from Alberta. *Journal of Paleontology* 14:468–480.

1949. The Edmonton fauna and description of a new *Triceratops* from the upper Edmonton member: phylogeny of the Ceratopsidae. *National Museum of Canada Bulletin* 113:33–46.

1950. *Pachyrhinosaurus canadensis*, representing a new family of Ceratopsia, from southern Alberta. *National Museum of Canada Bulletin* 118:109–120.

1970. Comments on dinosaurian preservation in the Cretaceous of Alberta and Wyoming. *National Museum of Canada, Publications in Paleontology* 4:1–9.

Tyson, H. 1977. Functional craniology of the Ceratopsia (Reptilia: Ornithischia) with special reference to *Eoceratops*. Unpublished M.S. thesis (Edmonton: University of Alberta).

1981. The structure and relationships of the horned dinosaur *Arrhinoceratops* Parks (Ornithischia: Ceratopsidae). *Canadian Journal of Earth Sciences* 18:1241–1247.

Wiman, C. 1930. Ueber Ceratopsia aus der oberen Kreide in New Mexico. *Nova Acta Regiae Societas Scientiarum Upsaliensis*, series 4, 7(2):1–19.

17 On the status of the ceratopsids *Monoclonius* and *Centrosaurus*

PETER DODSON

Abstract

The history of *Monoclonius crassus* Cope 1876 is reviewed. The type is based on a composite of disarticulated and unassociated individuals from the Judith River Formation exposed along the Missouri River in Montana. The parietal (AMNH 3998) is well preserved and diagnostic, and is herein designated as the neotype of *Monoclonius crassus*. Cope named three further species in 1889, all based on fragmentary associations, none of which can be shown to be valid. Lambe named *Centrosaurus apertus* in 1904 on the basis of an isolated parietal from Judith River sediments in southern Alberta. Barnum Brown discovered complete skulls in Alberta, and described them in 1914 and 1917 as new species that he referred to *Monoclonius*, believing *Centrosaurus* to be a synonym. Confusion has persisted to the present day.

In this study, the types of all Judithian centrosaurine ceratopsids are subjected to biometric analysis. It is shown that *Monoclonius* and *Centrosaurus* are distinct from each other, and that *Styracosaurus* is closer to the latter than to the former. Sexual dimorphism is inferred for *Centrosaurus* and *Styracosaurus*. Morphological differences previously regarded as indicative of species-level differentiation are regarded as sexual dimorphism. It is necessary to recognize only single species of *Centrosaurus*, *Styracosaurus*, and *Monoclonius*. *Brachyceratops* is not a juvenile of *Monoclonius*, but is a valid taxon.

Introduction

Horned dinosaurs of the subfamily Centrosaurinae reached the peak of their diversity in the Late Cretaceous Campanian stage, and their remains are well preserved in the Judith River Formation of Montana and particularly of Alberta. *Monoclonius* Cope, 1876 from the classic Judith River exposures along the Missouri River in Montana, was the first described ceratopsid that may still be regarded as valid. However, the type species, and three further species described by Cope (1889), were all based on fragments whose diagnoses have proven

In Dinosaur Systematics: Perspectives and Approaches, *Kenneth Carpenter and Philip J. Currie, eds. Copyright © Cambridge University Press, 1990.*

troublesome. In 1902 Lambe described the first ceratopsids from Alberta, erecting three new species that he initially referred to *Monoclonius*. These too were based on fragmentary associations. Lambe (1904) defined the genus *Centrosaurus* on the basis of a single parietal. The first complete ceratopsid skull described from the Judithian of Alberta was *Styracosaurus* Lambe 1913. Subsequently, many superb centrosaurine skulls have been found, especially at Dinosaur Provincial Park (Dodson 1983), but elsewhere in southern Alberta as well (Sternberg 1938, 1940).

Confusion as to the generic identities of *Monoclonius* and *Centrosaurus* soon arose. Brown (1914, 1917) was convinced the two genera were identical, and used skulls from Dinosaur Provincial Park to emend the diagnosis of *Monoclonius*. Lambe (1915) objected to this, and Sternberg (1938, 1940) used new discoveries from Alberta to make the case for the separate identities of the two genera. No new species have been described since 1940, although skulls continue to be found (P. J. Currie, pers. comm.). Ambiguity still persists, as a survey of recent popular books shows. For instance, Halstead and Halstead (1981) and Benton (1984) regard *Monoclonius* as the senior synonym for *Centrosaurus*, while Sattler (1981), Glut (1982), and Lambert (1983) regard *Centrosaurus* as valid. Another option was followed by Norman (1985), who regarded *Monoclonius* as a *nomen dubium*, and substituted *Centrosaurus* as the legitimate name for all referred species of both genera.

The purpose of this contribution is to review the genera and species relevant to the controversy from both an historical and a modern biometric point of view to provide a rational basis for determining the genera and species of Judithian short-squamosaled ceratopsids. It is an a priori expectation that growth and sexual dimorphism may contribute importantly to the variability among specimens. All type specimens have been examined and measured. The types reside either at the American Museum of Natural History in New York or at the National Museum of Natural Sciences of Canada in Ottawa (formerly designated as the Geological Survey of Canada or as the National Museum of Canada).

Monoclonius crassus

In 1876, Cope visited outcrops of the Judith River Formation on the Missouri River near Cow Creek in what is now Montana. He collected a series of ceratopsian specimens, but none was articulated, and most were not even associated. Moreover, his field records and specimen labelling were inadequate. The first account (Cope 1876) included a brief description, without illustration, of *M. crassus,* and included reference to a sacrum, femur, tibia, three co-ossified "dorsal" vertebrae, "episternum", and teeth. In 1877, Cope published a rather prolix account of further ceratopsian material from the same region. He illustrated a braincase and postorbital horncore, but his descriptions lack insight because a complete ceratopsian skull was then unknown.

In 1888, Marsh described *Ceratops montanus* from the Judith River Formation, based on an occipital condyle and a pair of orbital horn cores. In the following year, Marsh described the first complete ceratopsid skull, that of *Triceratops horridus* from the Lance Formation. Stimulated by Marsh's interest in ceratopsids and in the Judith River Formation, Cope (1889) redescribed materials collected in 1876, and, incidentally, corrected a number of anatomical errors that he had made earlier. He described three new species, and referred further skeletal elements to *M. crassus*. He mentioned for the first time a "frontal" bone bearing a small horn over the orbit, and alludes to a squamosal, but neither bone

was figured. He named three other species, *M. recurvicornis, M. sphenocerus,* and *M. fissus.*

Cope's four species of *Monoclonius* were reviewed by Hatcher (1907), and problems were already apparent. It was clear that the type of *M. crassus* was composite, based on at least two and probably several more individuals. The teeth were hadrosaurian, the "episternal" proved to be a parietal, as Cope himself had realized by 1889, and the fused dorsals turned out to be the fused cervicals so characteristic of ceratopsids. But more important were the problems of association. It is apparent that while ceratopsid material is relatively abundant in Judith River sediments along the Missouri River between Cow Creek and the Judith River (see map in Sahni 1972, Fig. 2), articulated or even associated material is extremely rare. Cope apparently took no pains to establish associations; neither field records nor collection numbers aid in reconstructing such crucial data. Hatcher (1907) specifically disavowed the squamosal and the frontal-postorbital complex as pertaining to the type specimen, he doubted the association of the sacrum, thought that some of the dorsals "may pertain to the same skeleton as the type," referred an ilium to the type "with a query," and could not find a femur or tibia in the collection that agreed with the measurements given by Cope.

What then should the fate of *Monoclonius* be? Is the genus diagnosable or must it be abandoned? Most of Cope's skeletal elements are not diagnostic at the generic or specific level. However, the parietal (Fig. 17.1) is one of the most diagnostic elements of the ceratopsian skeleton (Dodson and Currie 1990). I propose that the parietal described by Cope (1889, Plate XXXIV, Fig. 1; Hatcher, 1907, Fig. 75; Gilmore 1917, Fig. 10), AMNH

Figure 17.1. Representative centrosaurine parietals, not drawn to scale. L refers to sagittal length of the parietal in mm. (Scale bar equals 200 mm.) (After Gilmore and Lambe.) **A**, *M. crassus,* type specimen, L = 495; **B**, *C. apertus,* type specimen, L = 485; **C**, *B. montanensis,* L = 575; **D**, *S. albertensis,* type specimen, L = 520.

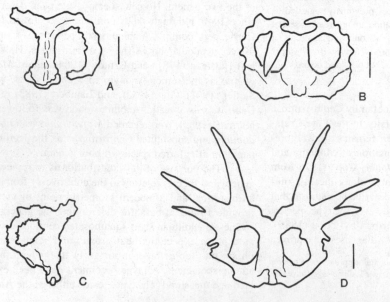

3998, be designated as the neotype of *M. crassus* Cope 1876. Although incomplete, it is well preserved (contrary to the assertions of Brown 1914), contains all essential features of the parietal, and is easily accessible for study.

For diagnosing a species of ceratopsid, it is desirable to have at minimum associated parietals, squamosals, postorbitals, and nasal horncores. Regrettably, no ceratopsid from the Judith River Formation of Montana includes all of these elements. In fact, only a single complete skull is referrable to *Monoclonius*, that named *M. lowei* by Sternberg (1940), who collected it near Manyberries in southern Alberta in 1937. In his description, Sternberg (p. 468) stated, "it is not possible to demonstrate that our new specimen is distinct from *M. crassus*." However, he believed it to be distinct from Cope's other species of *Monoclonius*, which Brown (1914) synonomized with *M. crassus*. One other skull of *Monoclonius* (AMNH 5442) was collected by Brown from the Judith River Formation at what is now Dinosaur Provincial Park, Alberta. It was noted by Russell (1967) but has not been described. It is well preserved but lacks the premaxilla and nasal, including the horncore.

Other species of *Monoclonius*

M. fissus may be dismissed readily as a *nomen nudum*, as it is based only on a partial pterygoid that Cope (1889) took to be a squamosal. *M. recurvicornis* is based on a braincase, a pair of postorbital horncores of moderate size, and a nasal with a damaged horncore. He referred to a squamosal but did not figure it. *M. sphenocerus* is based on a pair of superb nasals bearing a tall, straight, laterally compressed horncore (figured), and a partial left premaxilla. *M. recurvicornis* was found within three kilometers of *M. crassus*, but *M. sphenocerus* was found 50 kilometers from *M. crassus* (Sahni 1972, Fig. 2). If it is to be assumed that there were only one genus of ceratopsid from the Judith River Formation, then assignment of the additional species to the genus *Monoclonius* would be defensible. But this is not a plausible assumption. *Ceratops montanus* Marsh 1888 (also found 50 km from *M. crassus*) is a *nomen dubium*, but the form of its horns strongly suggests that it was a chasmosaurine (see Lehman, Chapter 16, and Dodson and Currie 1990, for the distinction between centrosaurines and chasmosaurines). Other genera of ceratopsids are known from the Judith River Formation farther north in Alberta (Russell 1967, 1984; Béland and Russell 1978; Dodson 1983) and farther south in Montana (Dodson 1986). In retrospect, therefore, it ought not to be assumed that *M. sphenocerus* and *M. recurvicornis* necessarily pertain to *Monoclonius* or even to a centrosaurine at all. The length of the orbital horns of *M. recurvicornis* suggests referral to the Chasmosaurinae. The form of the nasal horn of *M. sphenocerus* is suggestive of that of *C. nasicornus* and *S. albertensis*. Cope's species must be regarded as *nomina dubia* until such time as new material from the type localities clarifies the situation.

Centrosaurus

If *Monoclonius* was born in ambiguity, the discovery of *Centrosaurus* was little more auspicious. In 1902, Lambe described three new species of ceratopsids, the first reported from Canada, from the Judith River (Oldman) Formation along the Red Deer in southern Alberta (now Dinosaur Provincial Park). Included in Lambe's material were two partial skulls and a parietal. The assignment to *Monoclonius* was conservative (Lambe did not even defend the referral), and by 1915 Lambe had erected three new genera, including *Chasmosaurus* (*M. canadensis* = *Eoceratops canadensis* Lambe 1915) to receive his specimens. For present purposes, Lambe's third species, *M. dawsoni*, is crucial. The find included a poorly preserved partial skull, comprising the nasal horn, orbital series, quadrate, occipital condyle, and maxilla. The frill was badly damaged and not collected. However, he referred a *second* frill to *M. dawsoni*. In 1904, Lambe recognized the distinctiveness of the second specimen, and made it the type of a new genus, *Centrosaurus apertus*. He emphasized the "remarkable inwardly directed hook-shaped processes springing from the posterior border of the frill," and also the thickening of the posterior border, reaching a thickness of 6 cm. He possessed a horn-like structure that he took to be a nasal horn, later (Lambe 1910) realizing that it pertained to yet another distinctive feature: horn-like processes projecting forward from the caudal border and overhanging the parietal fenestrae (Fig. 17.1). Lambe clearly regarded *Centrosaurus* as distinct from *Monoclonius*, but again included no reference to *M. crassus*. The type of *Centrosaurus apertus* thus is directly comparable to the type of *M. crassus*; each consists of an incomplete but otherwise well preserved and generically diagnostic parietal. As with the type of *M. crassus*, it is not known precisely what configuration of nasal and orbital horns and what sort of squamosal go with the type of *C. apertus*.

Brown (1914) collected a magnificent complete centrosaurine skull from the Red Deer River in 1912. In his description, he reviewed all relevant taxa from Montana and Alberta and reached some interesting conclusions. He accepted the view of Hatcher (1907) that *M. recurvicornis* should be referred to *Ceratops* (here regarded as an indeterminate chasmosaurine). He believed that his new skull demonstrated that *M. sphenocerus* and *M. crassus* were identical, and that *C. apertus* and *M. dawsoni* were indistinguishable. He prepared a diagnosis of the genus *Monoclonius* based on his new Alberta skull, including the "long curved hook-like processes on the posterior border" of the parietal (a trait unknown in specimens from the Judith River Formation of Montana). Inasmuch as *M. crassus* unequivocally lacks either the caudal hooks or the rostral projections, Brown apparently diagnosed *M. crassus* out of the genus *Monoclonius*! He regarded the parietal of the type of *M. crassus* as that of an old individual "subjected to considerable abrasion during fossilization" and thus not showing the thickness, hooks, and processes of the Alberta specimens. This interpretation is demonstrably erroneous. Finally Brown erected a new species with a rostrally-curving nasal horn, *M. flexus,* distinct from *M. crassus* with an erect horn (as for *M. sphenocerus*), and *M. dawsoni* with a backward-curved horn.

Lambe (1915) rebutted Brown by questioning the validity of the genus *Monoclonius.* He repeated Hatcher's (1907) observations about the composite nature, and errors of anatomy and of association in the type. He also emphasized apparent weathering of the parietal (untrue) and its close resemblance to two separate genera, *Centrosaurus* and *Styracosaurus* Lambe 1913. He recommended "non-employment" of the generic name *Monoclonius.* Brown (1917) was slightly chastened, and thought it advisable to await more complete material from the type locality of *M. crassus* "before referring doubtful Belly River Ceratopsians to Judith River species." He therefore named two new species of *Monoclonius* based on two further Alberta finds! *Monoclonius nasicornus* is based on a complete skull and skeleton. The skull is of typical *Centrosaurus* type, and has a short face, and a tall, erect nasal horn. Brown also named a partial skeleton *M. cutleri,* on account of differences in the form of the ischium and of the relative length of the tibia as compared with *M. nasicornus.* As there is no skull, this species will not be considered further.

Sternberg (1938, 1940) reaffirmed the distinctness of *Monoclonius* and *Centrosaurus* by discovering and describing the only complete skull of *Monoclonius* presently known, *M. lowei.* He clearly demonstrated that a parietal that is thin caudally and lacks hooks or rostral processes is not merely an artifact of old age or erosion. He also found a skull of *Centrosaurus* with a caudally-recurved horn like that of *M. dawsoni.* He therefore transferred *M. dawsoni* to *C. dawsoni.* The nomenclatural history of *Monoclonius* and *Centrosaurus* are summarized in Tables 17.1 and 17.2. Following Lambe (1915) and Sternberg (1940), I designate as *Centrosaurus* those species found in Alberta that have the

thickened caudal parietal margin with hooklike processes projecting caudally and hornlike processes projecting rostrally.

Styracosaurus albertensis is included in the study because of its close resemblance to species of *Centrosaurus,* particularly *C. nasicornus* (Lambe 1913; Brown 1917). Additional species of *Styracosaurus* have been named. Gilmore (1930) named *S. ovatus* on the basis of a fragment of parietal from the Two Medicine Formation of Montana, and Brown and Schlaikjer (1937) named *S. parksi* from the Judith River Formation of the Red Deer River on the basis of a good skeleton but incomplete (and certainly nondiagnostic) skull.

Biometric considerations of *Monoclonius* and *Centrosaurus*
Introduction

The centrosaurine skulls of Alberta demonstrate a wide range of variability of shape and pattern of ornament (Fig. 17.2). Taxonomic practice during the first three decades of this century and earlier generally involved assignment of morphological variants to separate species. Modern practice, however, attempts to define a reasonable range of intraspecific variation in an attempt to reflect a level of speciation and ecological relationships that are consistent with modern ecosystems. It is an a priori expectation that species with elaborate secondary display characters will exhibit morphological variability, and that sexual dimorphism may be an important source of variability (Geist 1966; Dodson 1975, 1976; Packer 1983). So convinced were Ostrom and Wellnhofer (1986, this volume) of the merits of this argument that they reduced all species of *Triceratops* to junior synonymy with *T. horridus* without explicit consideration of the morphologies represented by each named species or referred specimen. Dodson (1975), in a study of lambeosaurine hadrosaurs of the Judith River Formation from Dinosaur Provincial Park, sought to rationalize the taxonomy of three genera and 12 species. He used an explicit null hypothesis that the 12 nominal species rep-

Table 17.1. *The history of* Monoclonius

Species	Reference	Comment
M. crassus	Cope 1876	mixed association of skull and skeletal elements
M. recurvicornis	Cope 1889	incomplete skull: braincase, orbital, and nasal horncores
M. sphenocerus	Cope 1889	complete nasal with horncore, incomplete premaxilla
M. fissus	Cope 1889	incomplete pterygoid
M. dawsoni	Lambe 1902	associated incomplete skull
M. flexus	Brown 1914	complete skull
M. nasicornus	Brown 1917	complete skull and skeleton
M. cutleri	Brown 1917	incomplete skeleton only
M. lowei	Sternberg 1940	complete skull

resented a growth series of a single species. The biometric analysis rejected the null hypothesis, and concluded that there were three sexually dimorphic species.

A desirable approach to the *Centrosaurus* problem involves mapping of the morphological variability of type species and referred specimens in order to determine the range of morphological variability, and to determine appropriate groups independent of historically described species. This study is preliminary, as it includes all types but not all referred specimens. A comprehensive study is underway. Figure 17.2 illustrates and Table 17.3 lists all specimens studied.

Table 17.2. *The history of* Centrosaurus

Species	Reference	Comment
C. apertus	Lambe 1904	incomplete parietal
C. longirostris	Sternberg 1940	complete skull

Method

Each specimen was examined and measured (Fig. 17.3), a suite of approximately 80 measurements being made with dial calipers and steel tape. Standard measurements (denominator) for bivariate analysis included basal skull length (snout to condyle), sagittal length of the parietal (excluding the lengths of spikes, hooks, and processes), and length of the caudal half of the squamosal for complete skulls, specimens consisting only of frills, or single squamosals, respectively. The use of basal skull length instead of total skull length is strongly recommended for ceratopsid morphometrics (Dodson

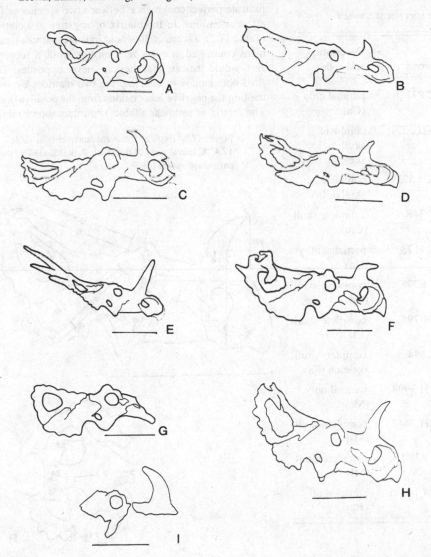

Figure 17.2. Sketches of the centrosaurine skulls utilized in the biometric study. Skulls are not drawn to scale, but are arrayed according to increasing basal length (see Table 17.5). (Scale bar = 500 mm.) Abbreviations as in Table 17.3. (After Brown, Lambe, Sternberg, and original.)

1976), because relative length of the parietal frill is likely to be highly variable and quite possibly sexually dimorphic. Use of total skull length that includes the frill would induce a spurious autocorrelation that would obscure the desired comparison. Similarly, the use of sagittal length of the parietal provides a reasonable basis for comparing the importance of the ornamental processes on the parietal among specimens. In the preliminary analysis, 30 log–log bivariate plots were constructed, using the three standard measures. Variables selected for preliminary analysis (Fig. 17.3, Table 17.4) were known to vary and were postulated to have ontogenetic, taxonomic, and/or behavioral significance (Farlow and Dodson 1975). Table 17.5 lists selected measurements that characterize aspects of the morphological diversity.

There are no small skulls. This is characteristic of ceratopsids generally (Sternberg 1949; Ostrom and Wellnhofer 1986 on *Triceratops*). Small ceratopsid skulls are of exceptional interest (Gilmore 1917 on

Brachyceratops; Dodson 1986 on *Avaceratops;* Dodson and Currie 1988 on cf. *Monoclonius*). The range of total skull length in the study series is from 1,375 mm to 1,840 mm (ROM 767 measures about 1,240 mm), and of basal skull length from 730 mm to 835 mm. The smallest specimen of *Centrosaurus* is nearly four-fifths the size of the largest, in no sense a juvenile. Thus, allometric effects have little explanatory power in this instance.

Each specimen was scored according to its position on the bivariate plot relative to every other specimen (details in Dodson 1976). A score of positive one was awarded for each plot on which a specimen was situated above the line of best fit, a score of zero was recorded when the specimen was on or near the line of best fit, and a score of negative one was assigned for each plot in which a specimen was situated below the line of best fit. Matrices were constructed to compare the scores of each pair of specimens. In the matrix of positive association (Table 17.6), a score of 30 would indicate perfect congruence between two specimens for all relationships. In the matrix of negative association (Table 17.7), a score of 30 would indicate that two specimens contrasted in every relationship, while a score of zero would indicate that they are never opposites. The final operation is to combine the two matrices by subtracting the negative associations from the positive ones. The matrix of residuals (Table 17.8) thus shows robust

Table 17.3. *Centrosaurine specimens studied*

Species	Specimen	Description and code
C. apertus	NMC 971	parietal only (Ca)
C. nasicornus	AMNH 5351	complete skull, skeleton (Cn)
C. flexus	AMNH 5239	complete skull (Cfn)
C. flexus	NMC 348	complete skull (Cfo)
M. dawsoni	NMC 1173	partial skull (Cd)
C. dawsoni	NMC 8798	complete skull (Cd)
C. longirostris	NMC 8795	complete skull (Cl)
S. albertensis	NMC 344	complete skull, skeleton (Sa)
M. crassus	AMNH 3998	parietal only (Mc)
M. sp.	AMNH 5442	complete skull (Msp)
M. lowei	NMC 8790	complete skull (Ml)
Brachyceratops montanus	USNM 14765	partial skull (Bm)

* denotes type specimen.

Figure 17.3. Explanation of measurements in Table 17.4. **A**, lateral view of the skull; **B**, dorsal view of the parietosquamosal frill.

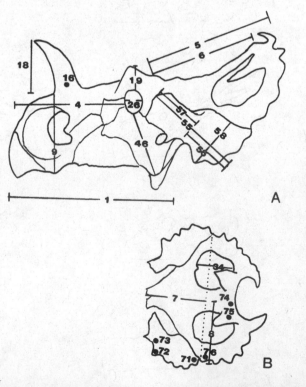

phenetic relationships. Those pairs that have the highest residual scores represent specimens that have the greatest resemblance to each other with the least contradic-

tory evidence. Clusters so obtained offer some reasonable basis for recognizing and defining species.

Results

The matrices of association show both expected and unexpected results. The positive matrix (Table 17.6) confirms a general association of *C. flexus*, *C. longirostris*, and *C. nasicornus*, and less obviously but importantly reveals the association between *S. albertensis*, *C. flexus*, *C. longirostris*, and *C. nasicornus*. It also confirms the relationship between *M. lowei* and the specimen of *Monoclonius* from Dinosaur Provincial Park. Puzzling, however, is a grouping that includes *Monoclonius*, *C. nasicornus*, *C. dawsoni*, and *S. albertensis*. The negative matrix (Table 17.7) confirms the separateness of *Monoclonius* and *Centrosaurus*, and to a lesser extent the distinctiveness of *S. albertensis* and *C. nasicornus* from *C. longirostris* and *C. dawsoni*. The matrix of residuals (Table 17.8) provides the most powerful basis for interpretation of the results, and important clusters are summarized in Table 17.9. The highest residual score for any pair of specimens, 16, linked *S. albertensis* and *C. nasicornus*, revealing an unanticipated but compelling relationship. Also important are the links between *C. dawsoni* and *C. longirostris*, among all three specimens of *Monoclonius*, and among *C. apertus*, *C. longirostris*, *C. nasicornus*, and *C. dawsoni*. An important positive outcome is that, even though there are only five characters available for assigning the type parietals of *M. crassus* and *C. apertus*, each shows unequivocal affinities within the residual matrix. Unexpectedly, the Ottawa and New York specimens of *C. flexus* show no apparent close relationship at all. Indeed, the New York specimen shows some indication of affinity with *C. nasicornus*.

Table 17.4. *List of characters analyzed*

Variable No.	Description[a]
1	snout–condyle L
4	L face orbit–premax
5	parietal maximum L
6	parietal sagittal L
7	squamosal projection on parietal midline L
8	parietal W half
9	face H
10	condyle W
16	nasal horn W base
18	nasal horn H
19	orbital horn H
26	orbit L
34	parietal fenestra L
46	jugal L
55	squamosal L
56	squamosal L caudal half
57	squamosal L rostral half
58	squamosal W maximum
71	squamosal T caudal
72	squamosal T opposite quadrate
73	squamosal T maximum
74	parietal T sagittal
75	parietal T maximum
76	parietal T lateral

[a]L = length; H = height; T = thickness; W = width.

Table 17.5. *Selected measurements (in mm) of centrosaurine skulls*

Specimen code[a]	Skull L		Parietal L 6	Condyle W 10	Nasal horn H 18	Orb horn H 19
	basal 1	max 2				
Cn	730	1410	505	68	470	57
Cd	760	1375	470	—	205	63
Sa	760	1840	520	71	500e	63
Msp	768*	—	570	—	—	66
Md	786*	—	—	57.5	315	62
Ml	805	1575	645	62	190	46
Cl	810	1470	455	72	203	86
Cfo	815	1470	595	78	187	162
Cfn	835	1580	515	74	340	107
Ca	—	—	485	—	—	—
Mc	—	—	530	—	—	—
Bm	—	—	575	—	—	—

[a]Specimen code explained in Table 17.3.

* denotes multiple regression estimate.

e - estimated

Table 17.6. *Matrix of positive associations of centrosaurine specimens based on congruence in pairwise analysis of bivariate plots*

	Cfn	Cfo	Cl	Cn	Cd	Sa	Ml	Md	Msp	Ca	Mc
Cfn	—										
Cfo	9	—									
Cl	13	10	—								
Cn	13	7	7	—							
Cd	7	8	14	7	—						
Sa	12	9	9	17	7	—					
Ml	6	9	7	5	10	8	—				
Md	2	3	1	1	1	2	2	—			
Msp	2	8	5	9	10	9	14	1	—		
Ca	2	3	4	3	3	3	0	—	1	—	
Mc	2	0	2	2	2	1	5	—	4	1	—
Bm	0	0	1	0	1	0	2	—	2	0	2

Note: Maximum number of associations 30 for skulls, 5 for parietals only (Ca, Mc, Bm). See text for details.

Table 17.7. *Matrix of negative associations of centrosaurine specimens based on contrasting position in pairwise analysis of bivariate plots*

	Cfn	Cfo	Cl	Cn	Cd	Sa	Ml	Md	Msp	Ca	Mc
Cfn	—										
Cfo	6	—									
Cl	4	8	—								
Cn	2	6	5	—							
Cd	6	7	1	7	—						
Sa	4	5	7	1	9	—					
Ml	13	6	10	7	3	7	—				
Md	1	0	2	2	0	1	0	—			
Msp	8	3	9	4	4	5	1	0	—		
Ca	0	1	0	0	0	1	2	—	3	—	
Mc	3	3	3	2	0	4	0	—	1	2	—
Bm	1	1	2	1	1	2	0	—	0	2	0

Note: Other details as for Table 17.6.

Table 17.8. *Matrix of residuals formed by subtracting the negative matrix from the positive matrix*

	Cfn	Cfo	Cl	Cn	Cd	Sa	Ml	Md	Msp	Ca	Mc
Cfn	—										
Cfo	3	—									
Cl	9	2	—								
Cn	11	1	2	—							
Cd	1	1	13	0	—						
Sa	8	4	2	16	−2	—					
Ml	−7	3	−3	−2	7	1	—				
Md	1	3	−1	−1	1	1	2	—			
Msp	−6	5	−4	5	6	4	13	1	—		
Ca	2	2	4	3	3	2	−2	—	−2	—	
Mc	−1	−3	−1	0	2	−3	5	—	3	−1	—
Bm	−1	−1	−1	−1	0	−2	2	—	2	−2	2

Interpretation

It is evident that *Monoclonius*, *Styracosaurus*, and *Centrosaurus* are separate from each other, and that *Styracosaurus* and *Centrosaurus* resemble each other more closely than either resembles *Monoclonius*. It is now possible to discriminate *Monoclonius* from *Centrosaurus* morphometrically. *Centrosaurus* is characterized by a parietal that is relatively short and thick along the caudal midline. The squamosal is wider, thicker, and longer relative to the parietal. The jugal is longer, and the occipital condyle is wider. The width of the base of the nasal horn is variable. *Monoclonius* is characterized by a parietal that is long and relatively thin. The squamosal is narrower and thinner, and a little shorter relative to the length of the parietal. The jugal is shorter and the nasal horn has a laterally compressed base. Despite the larger size of the skull of *M. lowei*, its condyle is only 62 mm wide. This width is anomalously low compared with condyles of *Centrosaurus*. Neither AMNH 5442 nor AMNH 3998 has a condyle, but a partial skull of *Monoclonius* in Toronto (ROM 1427) also has a condylar width of 62 mm. The skull of *M. dawsoni* offers too few characters to permit unequivocal assignment to either *Monoclonius* or *Centrosaurus*. However, its condyle is small (57.5 mm), and this suggests that the referral to *Monoclonius* may be correct.

Styracosaurus closely resembles *Centrosaurus* in possessing a thickened parietal, and in lacking the recurved hooks and rostral processes of the parietal (it has instead characteristic thick, caudally-projecting spikes). The nasal horn is tall and straight. *Centrosaurus nasicornus* obviously resembles *S. albertensis* in the nasal horn, and in many other characters as well. I posit that *C. nasicornus* is actually a female of *Styracosaurus*. One subtle character that reveals a relationship between *Centrosaurus* and *Styracosaurus* is shown by several isolated *Styracosaurus* parietals (ROM 1436, TMP P 66.10.28) that show short, abortive rostral processes. These specimens suggest that in *Styracosaurus* the morphogenetic potential to produce *Centrosaurus*-like rostral processes on the parietal was present but "switched off." A weak case may be made that the New York specimen (the type) of *C. flexus* is a large female of the *C. nasicornus* type. If so, then it follows that morphology of the nasal horn is irrelevant in centrosaurine systematics. This may prove to be the case, but this is a conclusion that I am not yet prepared to embrace.

Table 17.9. *Major clusters from matrix of residuals*

Sa	– Cn			
Cd	– Cl			
Ml	– Msp	– Mc		
Cf	– Cl	– Cn	– Sa	
Cl	– Cn	– Cd	– Ca	

Within *Centrosaurus*, there appear to have been two morphological types. One morph is characterized by a longer frill, a deeper face, a nasal horn that is taller and directed forward, relatively strong orbital horns, a jugal that is longer, a parietal that is thicker at the lateral border, and a squamosal that is wider and thicker. In addition, the caudal process of the premaxilla contacts the lacrimal, and separates the nasal from the maxilla. *C. flexus* characterizes this morphological type. The second morph is characterized by a shorter frill, a shallower face, a nasal horn that is lower and variable in orientation, orbital horns that are variable but generally weaker, a shorter jugal, a parietal that is thinner laterally, and a squamosal that is narrower and thinner. The caudal process of the premaxilla is weaker and does not contact the lacrimal. *C. longirostris* and *C. dawsoni* characterize this group. The type of *C. apertus* clusters here as well, making it possible to recognize a single species of *Centrosaurus*, *C. apertus*. The differences are consistent with sexual dimorphism, and it is posited that the first group, the *C. flexus* morphotype, represents males and the second group represents females. Males thus were larger and more robust, females smaller and more gracile. These inferences are summarized in Table 17.10.

Comments on *Brachyceratops*

Brachyceratops Gilmore 1914 from the Two Medicine Formation of Glacier County, Montana, was based on incomplete remains of five immature specimens, from which a composite skeleton 1.6 m long was constructed (Gilmore 1917, 1922). Sternberg (1940, 1949) was convinced that *Brachyceratops* was a juvenile of *Monoclonius*, a view often repeated in popular books (Sattler 1981; Glut 1982; Benton 1984). It is rele-

Table 17.10. *Summary of revised centrosaurine taxa*

	Male	Female
M. crassus		
	?"*M. sphenocerus*"	*M. crassus*
		"*M. lowei*"
C. apertus		
	"*C. flexus*"	*C. apertus*
		"*C. longirostris*"
		"*C. dawsoni*"
S. albertensis		
	S. albertensis	"*C. nasicornus*"
Sex not determinate		
B. montanensis		
A. lammersi		
Nomina dubia		
M. fissus		
M. recurvicornis		
M. cutleri		

vant to consider the status of *Brachyceratops* from a biometric point of view.

The study of *Brachyceratops* presents a number of the same problems that complicate the study of *Monoclonius,* due to the fragmentary and composite nature of the type of *B. montanensis.* A large specimen (USNM 14765), which was collected in 1935, represents an associated partial skull, and provides important insights into the nature of *Brachyceratops* (Gilmore 1939). It is roughly twice the size of the juveniles. While the parietals of the small specimens show relatively smooth margins, the large parietal shows coarse scallops. This parietal (Gilmore 1939, Fig. 10) seemingly argues against the assignment of *Brachyceratops* to *Monoclonius.* In fact, the parietal is quite large, measuring 575 mm along the midline. It is significantly larger than the type specimen of *M. crassus* (530 mm), and is slightly larger than the referred specimen from Dinosaur Provincial Park (570 mm). The question thus is not whether USNM 14765 might plausibly be transformed ontogenetically into *Monoclonius,* but whether it is already referrable to *Monoclonius,* and the answer appears to be "no." It appears to stand apart from the statistical range of variation of both *Centrosaurus* and *Monoclonius.* Whereas *Monoclonius* had large parietal fenestrae, those of *Brachyceratops* were small if they were present at all. (Gilmore assumed parietal fenestrae to have been present, but in all specimens the critical area is incomplete; clearly the parietal of *Brachyceratops* was thin, but not necessarily fenestrated.) Furthermore, whereas the parietal of USNM 14765 was large, the rest of the skull is that of an animal smaller than any other in the *Monoclonius–Centrosaurus* series. Basal skull length was roughly 560 mm, estimated from Gilmore's reconstruction (Gilmore 1939, Fig. 11). Thus the parietal appears to have exceeded the basal skull length, whereas in typical *Centrosaurus* the parietal was only two-thirds of the basal skull length, and in *Monoclonius* it was three-quarters skull length. In *Monoclonius,* the inferred length of the squamosal based on the squamosal facet of the parietal is 62 to 67% of the length of the parietal, while in the adult *Brachyceratops* the inferred length of the squamosal is 75% of the length of the parietal. Finally, the maxilla of *Brachyceratops* was relatively small, and shows no ontogenetic increase in the number of alveoli from small to large size. In both, the number of alveoli was about 20, as compared with 28 to 31 for *Centrosaurus* or *Monoclonius.* I thus conclude that *Brachyceratops montanensis* is a good taxon distinct from other genera.

A further recent complication may be mentioned. John Horner (pers. comm.) collected from the Two Medicine Formation in 1986 and 1987 a series of juvenile and adult centrosaurine skulls. The adult skulls show a peculiar styracosaur morphology but the juveniles seem to show a typical *Brachyceratops* morphology. The locality is close to the type locality of both *Brachycera-*tops montanensis and *Styracosaurus ovatus.* Is it credible that *Brachyceratops* is a juvenile of *Styracosaurus ovatus?* Much study remains to establish this.

Discussion

It is important for several reasons to have a biologically reasonable species-level taxonomy of dinosaurs. It is useful in its own right to understand community-level phenomena such as energy flow (Bakker 1972; Farlow 1976, 1980; Béland and Russell 1979). Species-level taxonomy of dinosaurs also plays a role in larger scale evolutionary questions such as diversity trends across the Cretaceous–Tertiary boundary, and has implications for discerning the mode and tempo of extinction (Russell 1982; Sloan et al. 1986). There are many historical problems that complicate the understanding of the species-level taxonomy of centrosaurine ceratopsids.

Centrosaurine skulls exhibit high levels of morphological variability. The skulls are large, complex, three-dimensional objects that are susceptible to distortion and deformation during fossilization, and to mechanical destruction during erosion. However, there is also intrinsic biological variability expressed in unequal development of left and right orbital horns, parietal processes, and squamosal scallops. No two specimens are alike. It is no wonder that the systematics of centrosaurines has been confusing. This study is an attempt to provide a rational basis for interpreting the systematics of centrosaurines by recognizing that allometry, sexual dimorphism, and individual variation are expected sources of within-species variability. The approach taken is a phenetic one, in which matrices of similarity coefficients are examined to extract clusters of related specimens. Taxonomic assignments are based on ranges of variation. The method has some degree of objectivity and repeatability, but no claim is advanced that the answer obtained is the unique, correct result. No amount of biometric sophistication can replace the taxonomic insight of the investigator. If the measurements taken are inappropriate, no computer manipulation will yield meaningful results. The results presented, in which nine species of *Monoclonius* and two species of *Centrosaurus* are reduced to one sexually dimorphic species of each, seem inherently reasonable. Because Nature is not necessarily so simple, other interpretations are possible. Until a larger sample size is available, a definitive resolution is not yet attainable.

Critical examination of the literature and restudy of the specimens themselves permits a resolution of long-standing problems. The genus *Monoclonius* is well founded and readily distinguished from the related *Centrosaurus.* The type of *M. crassus* is composite, and it is necessary to designate from among those elements collected by Cope in 1876 a single diagnostic element, the parietal (AMNH 3998), as the neotype. Neither nasal or orbital horn cores may be considered generically diagnostic. The gracile *M. lowei* and the referred specimen of *Monoclonius* from Dinosaur Provincial Park form a cohesive group, and may be assigned to *M. crassus*. *M. sphenocerus* consists of a tall nasal horn, reminiscent of that of *C. nasicornus*, but agreeing with *Monoclonius* in its narrow base. It may be tentatively referred to *M. crassus* as a male type, by inferring a range of variation in *M. crassus* similar to that seen in *Centrosaurus*. By this reasoning, the former group, including the type specimen, consists of females. *M. recurvicornis* is probably chasmosaurine, but is indeterminate. Thus, a single species of *Monoclonius* may be recognized, *M. crassus* (Table 17.10).

Skulls of *Centrosaurus* are relatively abundant and well preserved. Apart from *C. nasicornus*, two morphs may be recognized. One morph, typified by *C. flexus*, apparently consists of males, which are relatively large, robust, and extravagant in display structures (horns, frill). The other morph consists of smaller, more gracile, less extravagant females, typified by *C. longirostris* and *C. dawsoni*. The genotype, consisting only of a parietal, clusters with the female assemblage. Here too, a single species suffices, *C. apertus* (Table 17.10).

C. nasicornus presents a special case, as it clusters with *S. albertensis*. It is tentatively posited as a female of *S. albertensis*, and thus shows a connecting link between *Centrosaurus* and *Styracosaurus*, which are acknowledged to be closely related (Lambe 1915; Sternberg 1949). Indeed, a case could be made that *Centrosaurus* and *Styracosaurus* should be regarded as congeneric.

Unfortunately, *M. dawsoni* is too incomplete to permit unambiguous resolution. A weak case may be made for referral to *Monoclonius* on the basis of its small condyle, but for the present it is best regarded as a *nomen dubium*.

A cladistic analysis of centrosaurines based on 15 characters using PAUP (Phylogenetic Analysis Using Parsimony) produced a highly corroborated tree with an excellent consistency index of .947 (Fig. 17.4). *Pachyrhinosaurus canadensis*, the only Maastrichtian centrosaurine, is the most derived centrosaurine (Dodson and Currie in press). It is large (diameter of occipital condyle 96 mm – Langston 1975) and has a massive nasofrontal boss in place of nasal and orbital horns, but has a *Centrosaurus*-like frill. *Styracosaurus albertensis* is the most derived Judithian centrosaurine, with the most highly elaborated display structures. *Centrosaurus apertus* is closely related, but more conservative. *Monoclonius crassus* is less closely related. Apart from the (possible) large nasal horn in males, it is relatively conservative in cranial display structure. The structure of the squamosal is closely similar in *Styracosaurus, Centrosaurus,* and *Monoclonius. Brachyceratops montanensis* too is conservative in cranial display. It is possibly of smaller adult size than the aforementioned species, and the parietal may lack fenestrae. The structure of its squamosal is unknown, but probably is similar to the preceding. In *Avaceratops*, adult size was relatively small (Dodson 1986), the parietal was solid, and the form of the squamosal significantly different, possibly more primitive.

A final question to consider is whether the name *Monoclonius* ought to be used for all Judithian centrosaurines. This is perhaps a matter of taste, and I argue

Figure 17.4. Cladogram of inferred relations of centrosaurine ceratopsids based on PAUP analysis of distributions of 15 characters.

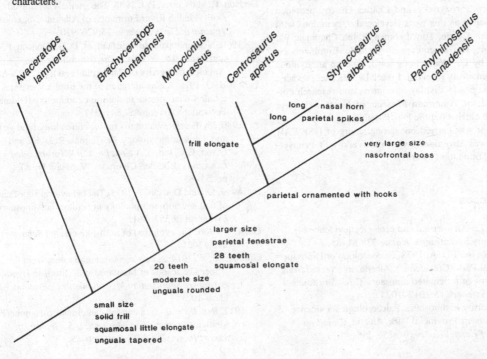

for retaining *Centrosaurus* as in *C. apertus* rather than *M. apertus,* for the following three reasons. First is the historical reason that the foundation of the genus *Monoclonius* is far less secure than that of *Centrosaurus,* as no quality specimens of *Monoclonius* have ever been found in the Judith River Formation of Montana such as to elucidate the type of *M. crassus.* Second, the geometry of the cladogram (Fig. 17.4) would argue against this, unless *Pachyrhinosaurus* be included in *Monoclonius.* Third, it is easier to understand ecological relationships between sympatric, large-bodied herbivores of different genera than of congeneric sympatric species – congeneric sympatry of several-thousand-kilogram herbivores is not seen in the modern world because of the implied close similarity of food requirements (Dodson 1975).

It is trite to suggest the need for further specimens, but this is certainly the case, particularly for *Monoclonius.* Only a single complete skull is known. It would be highly desirable to recover some quality material from Montana that will help to substantiate the range of variability of this genus. The logical extension of this work will be a multivariate analysis of all available specimens of *Centrosaurus.*

Acknowledgments

The bulk of this research was carried out during a sabbatical spent at the Paleobiology Division of the National Museum of Natural Sciences in Ottawa. I am deeply grateful to Drs. C. Richard Harington and D. A. Russell for their hospitality, and to my chairman, Dr. Leon Weiss of the University of Pennsylvania, for financial arrangements that made this all possible. I thank Gail Rice, Richard G. Day, Clayton Kennedy, and Kirin Shephard for their daily aid and cheerful tolerance of the burden of my demands. Dale Russell, David Jarzen, Phil Youngman, Chris Pirozynsky, and Charles Gruchy provided stimulating criticism as this work developed. Discussions with John Horner, Phil Currie, Tom Lehman, Ralph Chapman, Pat Leiggi, and Kenneth Carpenter were valuable. Comments on the manuscript by Dr. Wann Langston Jr. and an anonymous reviewer were gratefully received. I thank Dr. Michael Novacek for arranging access to display specimens and research collections alike at the American Museum of Natural History. Anthony Fiorillo carried out the PAUP analysis.

This work was carried out during tenure of NSF EAR 84-08446, and was also aided by two University of Pennsylvania Research Foundation awards.

References

Bakker, R. T. 1972. Anatomical and ecological evidence of endothermy in dinosaurs. *Nature* 238:81–85.

Béland, P., and Russell, D. A. 1978. Paleoecology of Dinosaur Provincial Park (Cretaceous), Alberta, interpreted from the remains of articulated dinosaurs. *Canadian Journal of Earth Sciences* 15:1012–1024.

1979. Ectothermy in dinosaurs: paleoecological evidence from Dinosaur Provincial Park, Alberta. *Canadian Journal of Earth Sciences* 16:250–255.

Benton, M. J. 1984. *The Dinosaur Encyclopedia* (New York: Wanderer Books, Simon and Schuster).

Brown, B. 1914. A complete skull of *Monoclonius*, from the Belly River Cretaceous of Alberta. *American Museum of Natural History Bulletin* 33:549–558.

1917. A complete skeleton of the horned dinosaur *Monoclonius*, and description of a second skeleton showing skin impressions. *American Museum of Natural History Bulletin* 37:281–306.

Brown, B., and Schlaikjer, E. M. 1937. The skeleton of *Styracosaurus* with a description of a new species. *American Musuem of Natural History Novitates* 955:1–12.

Cope, E. D. 1876. Description of some vertebrate remains from the Fort Union beds of Montana. *Academy of Natural Sciences of Philadelphia, Proceedings* 28:248–261.

1877. Report on the geology of the region of the Judith River, Montana, and on vertebrate fossils obtained on or near the Missouri River. *U.S. Geological and Geographical Survey of the Territories Bulletin* 3:565–597.

1889. The horned Dinosauria of the Laramie. *American Naturalist* 23:715–717.

Currie, P. J. 1981. Hunting dinosaurs in Alberta's huge bonebed. *Canadian Geographical Journal* 101(4):32–39.

Dodson, P. 1975. Taxonomic implications of relative growth in lambeosaurine hadrosaurs. *Systematic Zoology* 24:37–54.

1976. Quantitative aspects of relative growth and sexual dimorphism in *Protoceratops*. *Journal of Paleontology* 50:929–940.

1983. A faunal review of the Judith River (Oldman) Formation, Dinosaur Provincial Park, Alberta. *The Mosasaur* 1:89–118.

1986. *Avaceratops lammersi:* a new ceratopsid from the Judith River Formation of Montana. *Academy of Natural Sciences of Philadelphia, Proceedings* 138:305–317.

Dodson, P., and Currie, P. J. 1988. The smallest ceratopsid skull – Judith River Formation of Alberta. *Canadian Journal of Earth Sciences* 25:926–930.

1990. Neoceratopsia. *In* Weishampel, D. B., Dodson, P., and Osmólska, W. (eds.). *The Dinosauria*. (Los Angeles: University of California Press), pp. 593–618.

Farlow, J. O. 1976. A consideration of the trophic dynamics of a Late Cretaceous large-dinosaur community (Oldman Formation). *Ecology* 57:841–857.

1980. Predator/prey biomass ratios, community food webs and dinosaur physiology. *In* Thomas, R. D. K., and Olson, E. C. (eds.), *A Cold Look at Warm-blooded Dinosaurs* (Boulder, Colorado: Westview Press), pp. 55–83.

Farlow, J. O., and Dodson, P. 1975. The behavioral significance of frill and horn morphology in ceratopsian dinosaurs. *Evolution* 29:353–361.

Geist, V. 1966. The evolution of horn-like organs. *Behavior* 27:173–214.

Gilmore, C. W. 1914. A new ceratopsian dinosaur from the Upper Cretaceous of Montana, with a note on *Hypacrosaurus*. *Smithsonian Miscellaneous Collections* 63 (3):1–10.

1917. *Brachyceratops*, a ceratopsian dinosaur from the Two Medicine Formation of Montana. *U.S. Geological Survey Professional Paper* 103:1–45.

1922. The smallest known horned dinosaur, *Brachyceratops. U.S. National Museum Proceedings* 61(3):1–4.

1930. On dinosaurian reptiles from the Two Medicine Formation of Montana. *U.S. National Museum Proceedings* 77(16):1–39.

1939. Ceratopsian dinosaurs from the Two Medicine Formation, Upper Cretaceous of Montana. *U.S. National Museum Proceedings* 87:1–18.

Glut, D. F. 1982. *The New Dinosaur Dictionary* (Secaucus, N.J.: Citadel Press).

Halstead, L. B., and Halstead, J. 1981. *Dinosaurs* (Poole, Dorset, U.K.: Blandford Press).

Hatcher, J. B. 1907. History of discovery, classification, osteology, and systematic descriptions. *In* Hatcher, J. B., Marsh, O. C., and Lull, R. S. (eds.) The Ceratopsian. *U.S. Geological Survey Monograph* 49:1–160.

Lambe, L. 1902. New genera and species from the Belly River series (Mid Cretaceous). *Contributions to Canadian Palaeontology* 3:25–81.

1904. On the squamoso-parietal crest of two species of horned dinosaurs from the Cretaceous of Alberta. *Ottawa Naturalist* 18:81–84.

1910. Note on the parietal crest of *Centrosaurus apertus* and a proposed new name for *Stereocephalus tutus. Ottawa Naturalist* 24:149–151.

1913. A new genus and species from the Belly River Formation of Alberta. *Ottawa Naturalist* 27:109–116.

1914. On *Gryposaurus notabilis,* a new genus and species of trachodont dinosaur from the Belly River Formation of Alberta, with a description of the skull of *Chasmosaurus belli. Ottawa Naturalist* 27:145–155.

1915. On *Eoceratops canadensis,* gen. nov., with remarks on other genera of Cretaceous horned dinosaurs. *Geological Survey of Canada, Museum Bulletin* 12:1–49.

Lambert, D. 1983. *A Field Guide to the Dinosaurs* (New York: Avon).

Langston, W. L., Jr. 1975. The ceratopsian dinosaurs and associated lower vertebrates from the St. Mary River Formation (Maestrichtian) at Scabby Butte, southern Alberta. *Canadian Journal of Earth Sciences* 12:1576–1608.

Marsh, O. C. 1888. A new family of horned Dinosauria from the Cretaceous. *American Journal of Science* ser. 3, 36:477–478.

Norman, D. B. 1985. *The Illustrated Encyclopedia of Dinosaurs* (New York: Crescent).

Ostrom, J. H., and Wellnhofer, P. 1986. The Munich specimen of *Triceratops* with a revision of the genus. *Zitteliania* 14:111–158.

Packer, C. 1983. Sexual dimorphism: the horns of African antelopes. *Science* 221:1191–1193.

Russell, D. A. 1967. A census of dinosaur specimens collected in western Canada. *National Museum of Canada Natural History Papers* 36:1–13.

1982. The mass extinctions of the late Mesozoic. *Scientific American* 246(1):58–65.

1984. A check list of the families and genera of North American dinosaurs. *Syllogeus* 53:1–35.

Sahni, A. 1972. The vertebrate fauna of the Judith River Formation, Montana. *American Museum of Natural History Bulletin* 147:321–412.

Sattler, H. R. 1981. *Dinosaurs of North America* (New York: Lothrop, Lee and Shepard).

Sloan, R. E., Rigby J. K., Jr., Van Valen, L. M., and Gabriel, D. 1986. Gradual dinosaur extinction and simultaneous ungulate radiation in the Hell Creek Formation. *Science* 232:629–633.

Sternberg, C. M. 1938. *Monoclonius* from southeastern Alberta compared with *Centrosaurus. Journal of Paleontology* 12:284–286.

1940. Ceratopsidae from Alberta. *Journal of Paleontology* 14:468–480.

1949. The Edmonton fauna and description of a new *Triceratops* from the upper Edmonton member; phylogeny of the Ceratopsidae. *National Museum of Canada Bulletin* 113:33–46.

18 Triceratops: an example of flawed systematics

JOHN H. OSTROM AND
PETER WELLNHOFER

Abstract

A total of 16 species have been named or referred to the genus *Triceratops*. Surprisingly, the genus has never been formally diagnosed until recently. The first named species, *T. alticornus*, is not demonstrably referrable to *Triceratops*. Accordingly, the second named species, *T. horridus*, is here designated the type species of the genus. All other named species of *Triceratops* are here considered to be conspecific and thus invalid taxa. The supposed diagnostic differences among them fall well within normal ranges of variation in examples of large and small extant reptilian and mammalian species. Even more conclusive data are the stratigraphic and geographic distribution of the "type specimens" of the 16 "species" of *Triceratops*. Eleven of the sixteen type specimens were collected from the same stratigraphic unit (Lance Formation) and from a small area about one-quarter the size of the state of Rhode Island.

Introduction

The selection of diagnostic anatomical features that enable us to distinguish between different taxa (either species or genera) of related extinct organisms *should* and *must* be a topic of critical importance to all paleobiologists. The criteria by which such diagnostic features are selected often hinge on personal whims and prejudices. Only rarely are they chosen by consensus or based on established rules or procedures. The disturbing impression one receives of paleontologic studies is the apparently pervasive attitude that "if it's different, it *must* be given a new name." Too often, little thought seems to have been given to the character selection process, or to the sources of individual variation. Evaluation of the anatomical (or other) taxonomic differences is a fundamental issue that is rarely discussed and almost never tested. Moreover, the philosophical or systematic theory of a particular author is rarely registered. Is the author a "splitter" or a "lumper"? And how are the resulting binomials (now set in concrete) justified?

Times have changed, though, and there is a better appreciation of population variability and the sources of individual variation. Large samples have been more vigorously sought and reported in recent years (Brown and Schlaikjer 1940; Dodson 1975, 1976; Lehman this volume). We have learned from the sometimes naive conclusions of some of our predecessors, yet new taxa continue to be spawned on the basis of very fragmentary and inadequate material. For example, to cite just a few tenuous recent cases – *Alocodon kuehni, Phyllodon henkeli,* and *Trimucrodon cuneatus,* each of which is based on an assortment of isolated teeth from the Upper Jurassic of Portugal (Thulborn 1973) and attributed to the ornithischian family Hypsilophodontidae. Or consider *Ugrosaurus olsoni,* a new kind of ceratopsian based on a snout fragment, a piece of the dentary, a frill fragment, and a single partial vertebral centrum (Cobabe and Fastovsky 1987).

We do not pretend to have a universal solution to this situation. We merely wish to present an example that we believe has been, and may still be, symptomatic of our discipline – an example that has an important lesson – a well-documented example that is susceptible to ordinary common sense and straightforward logic. We believe this example is typical of paleobiology in general – and dinosaur systematics in particular.

Example

Our example is the taxon *Triceratops,* so well known that it needs no further introduction. As with all paleontologic taxa, the problem with *Triceratops* is an historical consequence of our ever-changing perspectives, the continuous accumulation of new knowledge and new specimens, combined with obsolete concepts and outdated, inflexible systematic practices. Whether comparable evidence was available or not, each new specimen

In Dinosaur Systematics: Perspectives and Approaches, *Kenneth Carpenter and Philip J. Currie, eds. Copyright © Cambridge University Press, 1990.*

was established as a new species – or so it seems. Sixteen separate species of *Triceratops* were christened during the last century (Fig. 18.1). We believe that there is sound rationale for recognizing only one species – *Triceratops horridus*.

Triceratops was established as a formal generic category a century ago in 1889 by O. C. Marsh. It is easy to blame Marsh for the long list of species that were

subsequently attributed to this genus, for Marsh himself was the author of ten of the sixteen species. Under the circumstances that prevailed during the 1890's, none of us should be surprised at the proliferation rate of Marsh's pen. But what is staggering and almost beyond 20th century comprehension is the astonishing productivity of John Bell Hatcher – Marsh's chief collector and field general. Between the spring of 1889 and early

Figure 18.1. Profiles of the *Triceratops* "type" specimen skulls originally assigned to different species, as labelled, but which are judged here to be only variants of a single species *T. horridus*. The profiles are drawn to a common scale to facilitate comparison and are based on the following published illustrations: **A**, *T. prorsus* YPM 1822 from site 3 of Fig. 18.4 (Hatcher et al. 1907, Pl. 34); **B**, *T. horridus* YPM 1820 from site 1 of Fig. 18.4 (Hatcher et al. 1907, Pl. 26); **C**, *T. serratus* YPM 1823 from site 4 of Fig. 18.4 (Hatcher et al. 1907, Pl. 27); **D**, *T. elatus;* **E**, *T. brevicornus*, formerly YPM 1834, now BSP 1964 I 458 from site 33 of Fig. 18.4 (Hatcher et al. 1907, Pl. 41); **F**, *T. hatcheri* USNM 2412 from site 25 of Fig. 18.4 (Hatcher 1905, Pl. 13); **G**, *T. flabellatus* YPM 1821 from site 2 of Fig. 18.4 (Hatcher et al. 1907, Pl. 44); **H**, *T. calicornis* USNM 4928 from site 29 of Fig. 18.4 (Hatcher et al. 1907, Pl. 38); **I**, *T. eurycephalus* MCZ 1102 from Goshen County, Wyoming (Schlaikjer 1935, Fig. 3). From Ostrom and Wellnhofer 1986.

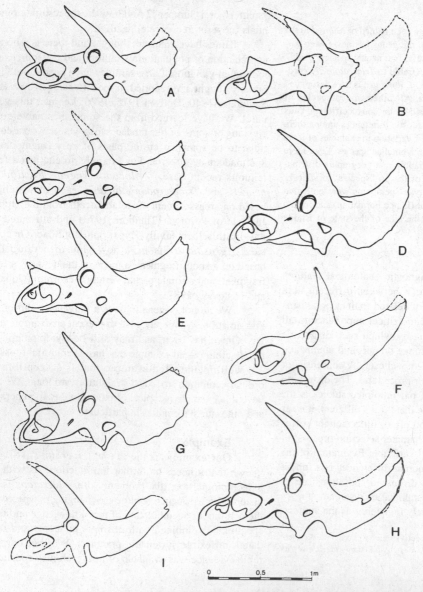

0   0,5   1m

summer of 1892, Hatcher discovered and collected no less than 32 ceratopsian skulls and partial skeletons from eastern Wyoming. Two of these were determined by Marsh to be two separate species that he assigned to a new genus *Torosaurus*. All others were designated new species of *Triceratops* – or assigned to previously named species of that genus.

It may surprise some readers that the genus *Triceratops,* until our recent study (1986), had never been formally defined or diagnosed. It is sufficient here to note that *Triceratops* is the only known three-horned ceratopsian genus with an unfenestrated parietal frill. But please refer to our 1986 study for a more complete diagnosis of *Triceratops*. A brief historical synopsis follows.

History

The first ceratopsian remains discovered and named in publication are as follows:

1. *Agathaumus sylvestris* (a few nondiagnostic vertebrae and a partial sacrum from probable Lance-age strata near Black Butte, Wyoming) established by Cope in 1872;
2. *Polyonax mortuarius* (several vertebrae, supposed limb fragments, and unrecognized horncore pieces from so-called Laramie beds in Colorado) named by Cope in 1874;
3. *Monoclonius crassus* (a parietal-squamosal frill portion, horncores, limb fragments, sacrum, and dorsal vertebrae from the Judith River beds of Montana) established by Cope in 1876;
4. *Bison alticornus* (a pair of brow horncores and a skull roof fragment from the Denver Formation of Colorado) named by Marsh in 1887; and
5. *Ceratops montanus* (a pair of brow horncores and an occipital condyle from the Judith River beds of Montana) described by Marsh in 1888.

Notice that up to this point no evidence had been found that suggested a three-horned ceratopsian.

The discovery and collection of *Ceratops montanus* by J. B. Hatcher in the summer of 1888 might well be described as the beginning of the deluge of ceratopsian discoveries. That fall, Hatcher heard of the discovery of other large fossil horncores to the south near the town of Lusk, Wyoming. The following spring, Hatcher, following up that rumor, was shown a large horncore and then led to the site. There, still in place, was the major part of a skull that turned out to be the first of the many famous horned dinosaurs from Converse County (now Niobrara County), Wyoming (Figs. 18.2, 18.3). This specimen became the holotype of *Triceratops horridus* (YPM 1820). Originally, it was assigned by Marsh to *Ceratops,* but later he placed it in his new genus *Triceratops* because it was obvious that there were three facial horns – something that could not be demonstrated in the *Ceratops* material. YPM 1820 thus is the true genotype of *Triceratops.*

1. T. alticornus. The brow horncores of *Bison alticornus* (USNM 1871E) were first transferred to *Ceratops* by Marsh (1889b), and then to *Triceratops* by Lull in 1907. As Lull later observed in 1933, the speci-

men is inadequate. We agree and consider it to be a *nomen dubium* (a taxon that is not identifiable from the type material). No longer can it be considered the type specimen of *Triceratops.*

2. T. horridus. Based on an incomplete skull (YPM 1820) bearing large brow horncores plus a robust nasal horn. Named by Marsh (1889a). The holotype resides at Yale Peabody Museum.

3. T. flabellatus. Based on a larger skull (YPM 1821) characterized by its long parietal crest ornamented by marginal scallop-shaped epoccipital bones. Designated by Marsh (1889b) on remains from Niobrara County, Wyoming. It is here considered a junior synonym of *T. horridus.*

4. T. galeus. Proposed by Marsh (1889b) for an isolated nasal horncore (USNM 2410) of a much smaller "species" found in Colorado (Denver Formation). We consider this to be a *nomen dubium.*

5. T. serratus. Established by Marsh (1890a) for a well preserved and nearly complete skull (YPM 1823) lacking a nasal horn, but judged by him to be *Triceratops* distinguished by a series of minor mid-line bony bumps on the frill. We believe this to be an individual variation and consider this taxon to be a junior synonym of *T. horridus.*

6. T. prorsus. Based on a nearly complete three-horned skull thought by Marsh (1890a) to be distinctive because of the forwardly directed nasal horn. In our opinion this is a junior synonym of *T. horridus.* The specimen is YPM 1822.

7. T. sulcatus. Established by Marsh (1890b) for an incomplete skull, the most distinctive feature being the presence of prominent grooves on the brow horncores (now known to be highly variable). In our judgment, this is a *nomen dubium,* a specimen that is not specifically identifiable with certainty. The type specimen is USNM 4276.

8. T. elatus. This species was proposed by Marsh (1891) for a three-horned skull that he considered to have a distinctly long and elevated parietal frill. We believe this to be a variable growth feature and consider *elatus* to be a junior synonym of *T. horridus.* The specimen in question is USNM 1201.

9. T. calicornis. Based by Marsh (1898) on a nearly complete skull with a distinctive, posteriorly concave nasal horn that appears to be incompletely developed. We judge this to be a junior synonym of *T. horridus.* The type specimen is USNM 4928.

10. T. obtusus. Erected by Marsh (1898) for a skull collected by Hatcher with a short and blunt nasal horn (which probably is not a horn at all, but more likely is the supporting boss of the nasal bones). In our opinion, this is another example of individual variation and represents one more junior synonym of *T. horridus.* The original type specimen is USNM 4720.

11. T. brevicornus. A nearly complete skull from Niobrara County, Wyoming, distinguished by Hatcher

Figure 18.2. Nearly all of the holotype specimens and most of the other early classic specimens of *Triceratops* were discovered in eastern Wyoming, in Niobrara County. Niobrara County is marked here by the black rectangle to show its size relative to North America. The open triangles indicate the approximate locations of the sites of the type specimens of the few other "species" of *Triceratops* that occur outside of Niobrara County. From south to north, these are the sites of *T. alticornus* (USNM 1871E), *T. galeus* (USNM 2410), *T. eurycephalus* (MCZ 1102), *T. maximus* (AMNH 5040), and *T. albertensis* (NMC 8862). From Ostrom and Wellnhofer 1986.

(1905) on the basis of its relatively short horns. It was this specimen that provoked our inquiry into the validity of all the numerous *Triceratops* species – and especially of all the *type* specimens. We consider this specimen and its label "*brevicornus*" to be a junior synonym of *T. horridus*. The type specimen, originally YPM 1834, is now on exhibit in the Bayerische Staatssummlung für Paläontologie und historische Geologie in Munich. The specimen is now catalogued as BSP 1964 I 458.

12. T. hatcheri. Established by Lull (1905) for a moderately large skull lacking a nasal horn, but with a distinct boss or swelling of the nasal bones. The apparent absence of the nasal horn initially led to recognition of this specimen as a separate genus – *Diceratops*. Lull (1907) transferred it to *Triceratops,* recognizing the anomalous preservation of the nasal horn region. The specimen is clearly a junior synonym of *T. horridus* and is registered as USNM 2412.

13. T. "ingens". A manuscript name mentioned without any description by Lull in 1915. The specimen (YPM 1828) consists of cranial and postcranial material, but remains largely unprepared. Accordingly, it is a *nomen nudum* (a name without definition or description).

14. T. maximus. Created by Barnum Brown (1933) for eight cervical and dorsal vertebrae and cervical ribs (AMNH 5040) from the Hell Creek Formation of Garfield County, Montana. No cranial material is known. For this reason, it cannot be assigned with any certainty to *Triceratops,* and we consider it to be a *nomen dubium* because, except for size, the type material is not distinctive and not identifiable.

15. T. eurycephalus. Established for a nearly complete skull (MCZ 1102) from the Lance Formation of Goshen County, Wyoming, by Erich Schlaikjer (1935). Considered distinctive by Schlaikjer because of the relatively greater width of the frill. In our judgment, this is another example of individual variation and we consider it to be another junior synonym of *T. horridus.*

16. T. albertensis. This final species was proposed by C. M. Sternberg (1949) for an incomplete skull collected from the Scollard Formation at a site not far from the Tyrrell Museum in Drumheller, Alberta (Section 2 of Township 34, Range 22, west of the 4th Meridian). It was believed distinctive because of a "horizontal" frill and "vertical" or backwardly inclined brow horns. We conclude that both of these supposed distinctions are not real, but rather are the result of mis-orientation instead of genuine anatomical differences (see Sternberg 1949). *Triceratops albertensis* (NMC 8862) is a junior synonym of *Triceratops horridus.*

A lesson?

At this point we must repeat our adamant conclusion that the anatomical distinctions that caused Marsh and his associates to create these separate species of *Triceratops* are not acceptable. Again, we refer you to our earlier study (1986) and the supposed distinctions cited therein that led to this plethora of *Triceratops* taxa. Note that of the 16 *type* specimens of *Triceratops* species, eleven came from a narrow stratigraphic range in a single rock formation – the Lance Formation of eastern Wyoming. Even more significantly, those eleven type

Figure 18.3. Location of Niobrara County (previously part of Converse County) and the historic *Triceratops* region in relation to other landmarks within the state of Wyoming. The designated area within Niobrara County includes six townships (Fig. 18.4) covering approximately 216 square miles (560 square kilometers). From Ostrom and Wellnhofer 1986.

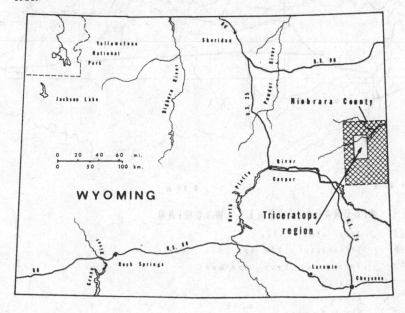

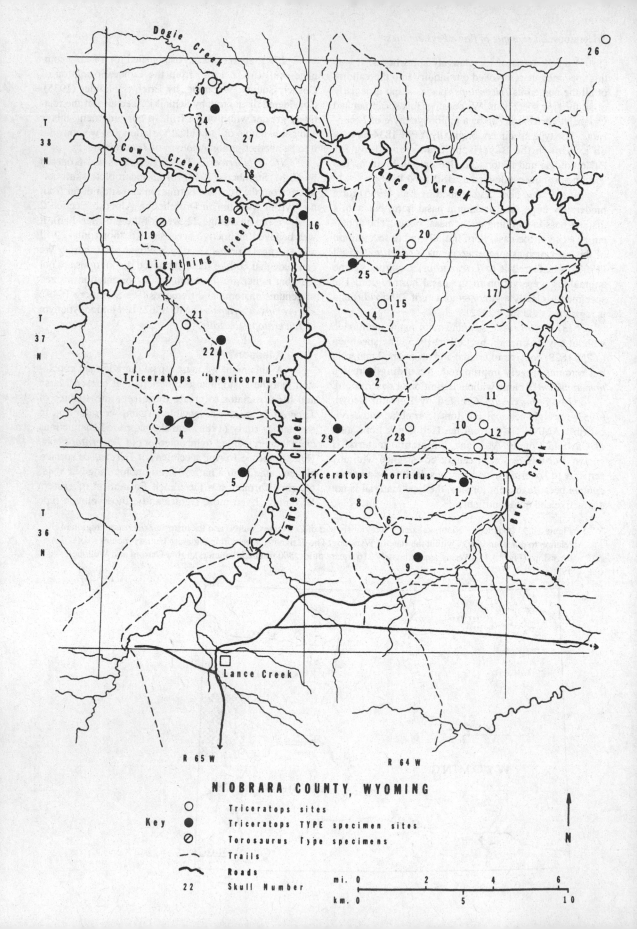

Dogie Creek

30

24

27

Cow Creek

18

T 38 N

Lance Creek

19a

19

16

Lightning Creek

20

23

25

15

17

14

21

T 37 N

22

Triceratops "brevicornus"

2

11

3

10

12

4

28

13

29

Triceratops horridus

1

5

8

6

7

9

Lance Creek

Buck Creek

T 36 N

26

Lance Creek

R 65 W

R 64 W

NIOBRARA COUNTY, WYOMING

Key	○	Triceratops sites
	●	Triceratops TYPE specimen sites
	⊘	Torosaurus Type specimens
	---	Trails
	～	Roads
	22	Skull Number

mi. 0   2   4   6

km. 0   5   10

N

Figure 18.4 (facing). Map of the region just north of the community of Lance Creek in east central Wyoming. It is from this small area that most of the type and other classic specimens of *Triceratops* were recovered by John Bell Hatcher for Yale College Museum during the years 1889 to 1892. The open and solid circles indicate the locations of the original sites of the 32 ceratopsian skulls and partial skeletons collected by Hatcher within this area. Both symbol types are labelled with the skull number, as originally designated by Hatcher in the field. The solid circles mark the sites of specimens that were later designated by Marsh, Hatcher, or Lull as type specimens of "new" species of *Triceratops*. They are: 1. *Triceratops horridus* (YPM 1820); 2. *Triceratops flabellatus* (YPM 1821); 3. *Triceratops prorsus* (YPM 1822); 4. *Triceratops serratus* (YPM 1823); 5. *Triceratops sulcatus* (USNM 4276); 9. *Triceratops obtusus* (USNM 4720); 16. *Triceratops elatus* (USNM 1201); 22. *Triceratops brevicornus* (YPM 1834, *now* BSP 1964 I 458); 24. *Triceratops "ingens"* (YPM 1828); 25. *Triceratops hatcheri* (USNM 2412); 29. *Triceratops calicornis* (USNM 4928).

It is noteworthy that the sites labelled 19 and 19a produced the type specimens of *Torosaurus latus* (YPM 1830) and *Torosaurus gladius* (YPM 1831). (It is entirely possible that the specimens referred to *Torosaurus* actually represent male individuals of *Triceratops*, but we have not attempted to verify that. Certainly, that is just as likely as is the coexistence of two distinct species of such large size.) According to Hatcher's notes and Lull's summary of 1915, all of the sites registered here are within the Lance Formation (locations from Hatcher 1896, Hatcher et al. 1907), the relative stratigraphic positions are recorded in Lull (1915, 1933). This map was drafted from United States Geological Survey map NK 13-2, Newcastle, Wyoming, 1962 edition. Specimen data are from Lull 1915. Abbreviations: R = Range; T = Township. Note: Buck Creek is now called Crazy Woman Creek and Dogie Creek was spelled Doegie Creek in the 1880s. From Ostrom and Wellnhofer 1986.

specimens were recovered from a *very* small area within Niobrara County, Wyoming – an area far smaller (Fig. 18.4) than the state of Rhode Island. Those eleven type specimens were among the 32 specimens collected by Hatcher in that local region from 1889 to 1892.

It is not possible now to reestablish the exact stratigraphic position of each of these type specimens, but the records indicated that all 32 of Hatcher's skulls (including the type specimens of *Torosaurus latus*, YPM 1830, and *Torosaurus gladius*, YPM 1831) came from the upper half of the Lance Formation exposed on both sides of Lance Creek. Hatcher (1893) estimated the thickness of the "*Ceratops* beds" at about 3,000 feet and that the producing horizon "was near the summit." Knowlton (1909) concluded that the bed known as "Laramie" (Lance Formation) could not be more than 2,000 feet thick, and that the fossiliferous level was 100 to 150 feet below the overlying Fort Union Formation. Without quantifying, Stanton (1909) reported that the highest "type" specimen (*T. brevicornus*) was not much higher than the lowest "type specimen" (*T. prorsus*), except for that of *T. horridus*. In other words, all of the "type" specimens of *Triceratops* collected in the Niobrara County area were stratigraphically close together – perhaps less than a few hundred stratigraphic feet apart. It seems most probable to us that these numerous ceratopsian specimens represent a random population sample over a short geological interval within Lancian time.

Even more telling is the restricted geographic distribution of the Niobrara County ceratopsian sample – occurring within an area of approximately 200 square miles. The obvious question cannot be avoided. On the basis of what we now understand about the ecology of large animals (1,000 lb. or 500 kg. or more), is it likely that as many as a dozen closely related species of elephant-sized animals – *Triceratops* – actually coexisted in an area less than one fourth the size of Rhode Island? Is it even reasonable or logical to think that more than one

species the size of *Triceratops* (estimated liveweight of 5 to 8 tons) existed simultaneously in Niobrara County? We think not. Critics may point to the so-called specific differences cited by the original authors. These include horn length, curvature, horn surface, nasal horn size and shape, jugal size and shape, temporal fenestra shape and orientation, and so forth. We think that all these are no more than individual or sexual variations that correspond well with the ranges of variation in modern large mammal species, and we will cite some examples. Some critics might point to the inclusion of the two type skulls of *Torosaurus* (skulls 19 and 19a) among Hatcher's 32 skulls from Niobrara County. They could be claimed as proof that at least two large ceratopsian species did coexist in Niobrara County, but there is a possible alternative explanation. *Torosaurus* may in fact represent a separate taxon, a distinct genus, but it *might also be the male form of Triceratops*.

Modern analogies

We distinguish between permanent horn-like structures of bovids and giraffes and those that are shed and replaced annually, as in cervids. The specimens attributed to *Triceratops* obviously represent animals that possessed permanent brow or frontal horns. No hornless Lance-age ceratopsians are known. On that basis, the bovid/ giraffe cranial protuberances seem to provide the best analogy.

Antlers, as develop in modern cervids, are true bony outgrowths of the frontal bones, nourished by a rich blood supply and sheathed in fur-bearing epidermis during growth. Growth is by an upward and/or outward bone expansion from permanent frontal "pedicels." Antler growth diminishes and ultimately stops with the loss of circulation, the final antlers consisting of necrotic bone, except for the pedicels. Antlers function primarily as sexual ornaments as well as weapons, and are shed annually after the mating season. With each successive year, the new antlers usually are larger and more ornate. Thus, antler morphology or size obviously is not a desirable basis for defining a taxon.

Horns, on the other hand, develop initially as separate structures forming from ossicones (epidermal ossification sites) in the overlying skin and specific skin outgrowths. These are independent bony elements in the skin covering the frontal-parietal region and are not part of the dermal skull bones. In bovids and giraffes these centers become fused to the underlying skull bones shortly after birth. The ossicones grow and become the horncores of bovids and giraffes. They typically are porous or cancellous structures containing large sinuses. The bovid horncore is covered with skin that grows an ever-thickening cover of cornified tissue or horn. It is this overlying cornified sheath that produces the great intraspecies variety of bovid horn shapes and sizes. Horns are not shed, and commonly grow throughout life, although some species may regenerate parts of the horny sheath.

Bruhin (1953) classified horns into three main types:

1. those that are similar in size and shape in both sexes and usually function as defensive weapons against enemies in interspecific combat, e.g., *Bison;*
2. those that appear useless against predators and function only in intraspecific fighting or display, e.g., *Aepyceras* (impala);
3. those that are strictly of ceremonial function and never used as weapons, e.g., *Giraffa.*

We conclude several things from the available material of *Triceratops*. First, we conclude that *both* sexes were equipped with brow horns similar (but possibly distinctive) in size and shape. No hornless ceratopsian specimens are known from Lance-age deposits, so we are forced to conclude that the brow horns were permanent features and used as defensive devices by both sexes. Undoubtedly, there must have been some sexual variation, as well as ontogenetic and individual differences in brow horn shape, orientation, and length within the *Triceratops* population sampled by Hatcher from Niobrara County, Wyoming.

Variation in bovid horn shape and size is evident in virtually all bovid species. That is confirmed by even the most cursory examination of domestic as well as wild species. This reflects age differences and individual variation, as well as sexual differences in particular cases. The *Triceratops* sample cited above seems to us entirely analogous to the bovid example, and we offer two illustrations (Figs. 18.5, 18.6) of living horned species. Our conclusion is that the known sample of *Triceratops* specimens cited here represents a single species – *Triceratops horridus.*

Addendum

Since this preview paper was submitted for the Tyrrell Museum Symposium on Dinosaur Systematics in mid-1986, our monograph (Ostrom and Wellnhofer

1986) on the *Triceratops* issue has been published. To date, we have received no negative reactions to our thesis in that work, yet we hear that some colleagues are sceptical. We emphasize that our study, summarized here, dealt *only* with the *previously designated type specimens* of the several named species of *Triceratops*. Although we did examine other specimens, it was not a comparative analysis of all specimens that have been discovered or referred to that genus. Of course it is entirely possible that more than one species of *Triceratops* existed, but it is most unlikely that the specimens from Niobrara County, Wyoming include two kinds. Nor do the other

Figure 18.5. Examples of horn variation in *Alcelaphus buselaphus* (the hartebeest) of central Africa. The sample series was selected from different locales within the normal range of the species by Ruxton and Schwarz (1929) to demonstrate hybridization between two subspecies of *A. buselaphus* (*A. b. jacksoni* and *A. b. cokii*). **A**, *Alcelaphus buselaphus jacksoni*, Lake Nakuro, Kenya; **B**, *A. b. jacksoni*, Lake Naivasha, Kenya; **C**, *A. b. cokii*, Mlali Plain, Tanganyika; **D**, *A. b. jacksoni*, Lake Nakuro, Kenya; **E**, *A. b. jacksoni*, Guas Ngishu Plateau, Kenya; **F**, *A. b. jacksoni*, Lake Nakuro, Kenya; **G**, *A. b. lelwel*, Bahr et Ghazal, Kenya; **H**, *A. b. jacksoni*, Njoro, Kenya; **I**, *A. b. jacksoni*, Njoro, Kenya. Scale = 400 cm. Redrawn from Ruxton and Schwarz 1929.

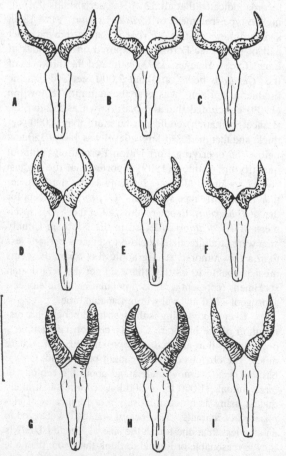

"type" specimens from Colorado, Wyoming, Montana, and Alberta provide persuasive evidence to support separate species designations. Until a thorough multivariate analysis of all *Triceratops* specimens has been performed, the subject will remain something of a moot issue. Sloan (1976, pers. comm.) has observed that *Triceratops* specimens from Montana and Alberta appear to be significantly larger than those from Wyoming (all or just some of them?), but no statistical record or consensus is available. Moreover, size alone is an extremely unreliable basis for taxonomic separation. Do these represent different species?

We stand by our original assessment: the *type* specimens of *Triceratops* do not support the published theories of taxonomic separation expressed in past studies and tradition.

Conclusions

Our conclusion, on stratigraphic and geographic distribution grounds, *and* current understanding of anatomic variation in living populations, is that only a single species of *Triceratops* can be substantiated. All other theories are beyond testing and unsupported by living analogues. The only valid species is *Triceratops horridus.*

Acknowledgments

The authors wish to acknowledge the hospitality and assistance provided by numerous colleagues who permitted access to specimens and data under their charge. Our thanks are extended to John Bolt and Mary Carmen (Field Museum of Natural History, Chicago), Kenneth Carpenter and Peter Robinson (University of Colorado Museum, Boulder), Eugene Gaffney and Barbara Werscheck (American Museum of Natural History, New York City), Nicholas Hotton III and Robert Purdy (National Museum of Natural History, Smithsonian Institution, Washington, D.C.), Wade Miller (Brigham Young University), Farish Jenkins and Charles Schaff (Museum of Comparative Zoology, Harvard University), Mary Ann Turner (Peabody Museum, Yale University), and Gerhard Plodowski (Senckenberg Museum, Frankfurt). The senior author is also grateful to Dietrich Herm, Volker Fahlbusch, and his coauthor, all of the Bayerische Staatssammlung für Paläontologie und historische Geologie, Munich for their hospitality, friendship, and generous assistance. Gratitude must also be expressed to the Alexander von Humboldt Stiftung of West Germany for the honor and financial assistance to J. H. O. that made this collaboration and study possible.

References

Brown, B. 1933. A gigantic ceratopsian dinosaur *Triceratops maximus*, new species. *American Museum of Natural History Novitates* 469:1–9.

Brown, B., and Schlaikjer, E. M. 1940. The structure and relationships of *Protoceratops*. *New York Academy of Science Annals* 40:133–266.

Bruhin, H. 1953. Zur Biologie der Stirnaufsälze bei Huftieren. *Physiologie Comparative et Oecologie* 3:63–127.

Cobabe, E. A., and Fastovsky, D. E. 1987. *Ugrosaurus olsoni*, a new ceratopsian (Reptilia: Ornithischia) from the Hell Creek Formation of eastern Montana. *Journal of Paleontology* 61:148–154.

Cope, E. D. 1872. On the existence of Dinosauria in the transition beds of Wyoming. *American Philosophical Society Proceedings* 12:481–483.

1874. [Footnote.] *United States Geological and Geographic Survey of the Territories Bulletin* 1(1):10.

1876. Descriptions of some vertebrate remains from Fort

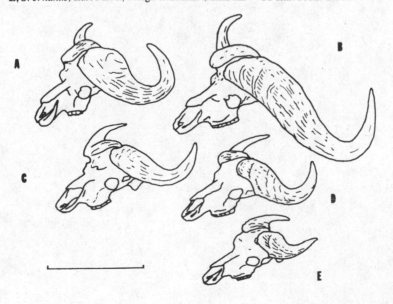

Figure 18.6. Intraspecific variation in the skulls and horns of *Syncerus caffer caffer*, the African forest and savanna buffalo, to show hybridization and "incipient" speciation. **A**, *Syncerus caffer caffer*, Mt. Elgon, Uganda; **B**, *S. c. caffer*, Amala River, Kenya; **C**, *S. c. cottoni*, Kasindi, Lake Edward, Congo Kinshasa; **D**, *S. c. mathewsi*, Mfumbiro, Ruanda; **E**, *S. c. nanus*, Ituri Forest, Congo Kinshasa. (Scale bar = 50 cm.) From Grubb 1972.

Union beds of Montana. *Academy of Natural Sciences of Philadelphia Proceedings* 28:255–256.

Dodson, P. 1975. Taxonomic implications of relative growth in lambeosaurine hadrosaurs. *Systematic Zoology* 24:37–54.

1976. Quantitative aspects of relative growth and sexual dimorphism in *Protoceratops*. *Journal of Paleontology* 50:929–940.

Grubb, P. 1972. Variation and incipient speciation in African buffalo. *Zeitschrift für Säugetierkunde* 37:121–144.

Hatcher, J. B. 1893. *Ceratops* beds of Converse County, Wyoming. *American Journal of Science* 45(3):135–144.

1896. Some localities for Laramie mammals and horned dinosaurs. *American Naturalist* 30:112–120.

1905. Two new Ceratopsia from the Laramie of Converse County, Wyoming. *American Journal of Science* 20(4):413–419.

Hatcher J. B., Marsh, O. C., and Lull, R. S. 1907. The Ceratopsia. *United States Geological Survey Monograph* 49:1–300.

Knowlton, F. H. 1909. The stratigraphic relations and paleontology of the "Hell Creek bed," "*Ceratops* beds," and equivalents, and their reference to the Fort Union. *Washington Academy of Science Proceedings* 11:179–238.

Lull, R. S. 1905. Restoration of the horned dinosaur *Diceratops*. *American Journal of Science* 20(4):420–422.

1907. Phylogeny, taxonomy, habits and environment of the Ceratopsia. *In* The Ceratopsia. Hatcher, J. B., Marsh, O. C., and Lull, R. S. (eds.), *United States Geological Survey Monograph* 49:1–300.

1915. The mammals and horned dinosaurs of the Lance Formation of Niobrara County, Wyoming. *American Journal of Science* 40(4):319–348.

1933. A revision of the Ceratopsia or horned dinosaurs. *Peabody Museum of Natural History, Memoir* 3(3):1–175.

Marsh, O. C. 1887. Notice of new fossil mammals. *American Journal of Science* 34(3):323–331.

1888. A new family of horned dinosaurs from the Cretaceous. *American Journal of Science* 36(3):477–478.

1889a. Notice of new American Dinosauria. *American Journal of Science* 37(3):331–336.

1889b. Notice of gigantic horned Dinosauria from the Cretaceous. *American Journal of Science* 38(3):173–175.

1890a. Description of new dinosaurian reptiles. *American Journal of Science* 39(3):81–86.

1890b. Additional characters of the Ceratopsidae, with notice of new Cretaceous dinosaurs. *American Journal of Science* 3(39):418–46.

1891. Notice of new vertebrate fossils. *American Journal of Science* 3(42):265–269.

1898. New species of Ceratopsia. *American Journal of Science* 6(4):92.

Ostrom, J. H., and Wellnhofer, P. 1986. The Munich specimen of *Triceratops* with a revision of the genus. *Zitteliana Abhandlungen der Bayerischen Staatssammlung für Paläontologie und historische Geologie* 14:111–158.

Ruxton, A. E., and Schwarz, E. 1929. On hybrid hartebeests and on the distribution of the *Alcelaphus buselaphus* group. *Zoological Society of London Proceedings* 38:567–583.

Schlaikjer, E. M. 1935. Contributions to the stratigraphy and paleontology of the Goshen Hole area, Wyoming. II. The Torrington member of the Lance Formation and a study of a new *Triceratops*. *Museum of Comparative Zoology, Harvard University Bulletin* 76:31–68.

Sloan, R. E. 1976. The ecology of dinosaur extinction. *In* Churcher, C. S. (ed.), *Athlon: Essays on Paleontology in Honor of Loris Shano Russell. Royal Ontario Museum Life Science Miscellaneous Publication* 134–154.

Stanton, T. W. 1909. The age and stratigraphic relations of the "*Ceratops* beds" of Wyoming and Montana. *Washington Academy of Science Proceedings* 11:239–293.

Sternberg, C. M. 1949. The Edmonton fauna and description of a new *Triceratops* from the Upper Edmonton member; phylogeny of the Ceratopsidae. *National Museum of Canada Bulletin* 113:33–46.

Thulborn, R. A. 1973. Teeth of ornithischian dinosaurs from the Upper Jurassic of Portugal with a description of a hypsilophodontid (*Phyllodon henkeli* gen. et sp. nov.) from Guimarota Lignite. *Portugal Services Geologicos Memoria* 22:89–134.

VII Stegosauria

19 Stegosaurs of Asia

DONG ZHIMING

Abstract

In the past two decades, specimens of eight genera of stegosaurs from the Early Jurassic to the Late Cretaceous have been discovered in Asia. These include *Tatisaurus*, *Huayangosaurus*, *Chialingosaurus*, *Chungkingosaurus*, *Tuojiangosaurus*, *Monkonosaurus*, *Wuerhosaurus*, and *Dravidosaurus*. This material reveals that stegosaurs arose from a Late Triassic small ornithopod – a fabrosaur or heterodontosaur – and that the birthplace of this group is probably eastern Asia. *Tatisaurus*, originally referred to the Ornithopoda, is transferred to the Stegosauria. *Dravidosaurus* is the youngest stegosaur known. A classification of the stegosaurian dinosaurs is presented, as well as a phylogenetic interpretation.

Introduction

Stegosaurian remains are known from Mesozoic continental deposits of all the continents except Australia, South America, and Antarctica (Fig. 19.1). The first description of a stegosaur was that of *Craterosaurus*, found in the Potton sands (Lower Cretaceous) near Potton, England. It was described by Seeley in 1874, but it was Nopcsa (1912) who recognized it as part of the neural arch of a stegosaurian dinosaur. From the middle of the last century, the study and collection of dinosaurs centered in North America. The first stegosaurs from North America were described by Marsh in 1877, from Colorado and Wyoming. Complete stegosaurian skeletons were collected in Utah, and less complete material in Oklahoma. Outside North America and Asia, stegosaurs have been discovered in Africa and Europe (Hennig 1915; Galton 1985).

Stegosaurs, a group of ornithischian dinosaurs, appeared in the Early Jurassic and died out by the mid-Late Cretaceous, making them one of the first groups of dinosaurs to disappear from the Earth. They seem to have been most successful during the Late Jurassic.

In Dinosaur Systematics: Perspectives and Approaches, *Kenneth Carpenter and Philip J. Currie, eds. Copyright © Cambridge University Press, 1990.*

Stegosaurus, described by Marsh in 1877 and reviewed by Gilmore in 1914, was thought to be the typical stegosaur. *Kentrosaurus* (= *Kentrurosaurus*), found in the Tendaguru Formation of Tanzania, is another stegosaur that is well known. In Europe three genera of stegosaurs have been described: *Dacentrurus*, including two species from the Upper Jurassic (Oxfordian – Kimmeridgian); *Lexovisaurus*, found in the Middle Jurassic (Bathonian and Callovian); and *Craterosaurus* from the Lower Cretaceous (Neocomian). *Scelidosaurus* once was thought to be a stegosaur, but recently was referred to the Ornithopoda by Thulborn (1977), and to the Ankylosauria by Romer (1968).

Hoffstetter suggested that the birthplace of stegosaurs was probably Europe, where *Scelidosaurus* was present in the Early Jurassic (Lower Lias or Sinemurian) and *Lexovisaurus* in the Middle Jurassic. The proposed evolutionary line started with *Scelidosaurus* and progressed through *Lexovisaurus*, *Dacentrurus*, *Kentrosaurus*, and *Stegosaurus*.

Wiman (1929) reported a bony plate from the Upper Jurassic Mengyin Formation. This was the first record of a stegosaur from Asia. On the basis of this specimen, Steel (1969) suggested that the place of origin of stegosaurs was probably Asia. In the past decade, many complete skulls and skeletons of stegosaurs have been found in the Sichuan Basin, the Junggar Basin, and the Qamdo (Chungdu) Basin of China (Fig. 19.1). A small stegosaur, *Dravidosaurus*, has been found in the Upper Cretaceous of central India. These new finds, ranging from the Lower Jurassic to the Upper Cretaceous, extend our knowledge of the stratigraphic range and evolution of this group of dinosaurs. The present work summarizes our understanding of Asian stegosaurs, because most papers on this topic are published in Chinese and are not widely available.

Stegosaurus, with its long, narrow skull and long facial region, has long been thought to be the model stegosaur. The premaxilla is excluded from contact with the lacrimal by the elongate suture between the nasal and maxilla. Three supraorbital elements are present, the orbit extends nearly the full height of the skull, and the superior temporal fenestrae are small with a flat, plate-like parietal between them. The lateral temporal fenestrae are large and placed on the sides of the skull. The quadrate is short and almost vertical. In posterior aspect, the skull is narrow and essentially rectangular, with broad but abbreviated paroccipital processes. The choanae and subtemporal vacuities are elongate and the pterygoid exhibits a rounded flange. In contrast with the earlier stegosaurs, *Stegosaurus* lacks both antorbital and mandibular fenestrae. In the lower jaw, the dentary is relatively long and the retroarticular process poorly developed. The dentition is weak, and the premaxillary teeth were lost as in all stegosaurs except the primitive *Huayangosaurus*. In *Stegosaurus*, 20 to 27 teeth are present in each jaw ramus.

The body of *Stegosaurus* was high and narrow, with 27–29 presacral vertebrae (9–10 cervicals, 17 dorsals and 4–6 sacrals). The centra are platycoelous to amphiplatyan. The cervical neural arches and spines are high. The front legs are shorter than the hind. The olecranal process of the ulna is prominently developed, and the radius is about three-quarters the humeral length. The manus has five short digits, of which the outer two were presumably reduced. The anterior process of the ilium is long, and the posterior short. The hind legs are long, with the tibia about three-quarters the femoral length. In the pes, digit I is smaller than it is in primitive stegosaurs. Digit V is absent or represented only by a metatarsal rudiment. The armor consists mainly of paired dorsal plates of variable shape, and small ossicles on the trunk itself.

Stegosaurs probably arose in the Early Jurassic, 180 million years ago. *Scelidosaurus* from Lower Jurassic rocks was thought to be an ancestor of stegosaurs, but is now referred to the Ornithopoda (Thulborn 1977). Stegosaurs emerged early in the Jurassic from small primitive ornithopods (heterodontosaurian or fabrosaurian dinosaurs) that lived during the Late Triassic. *Tatisaurus*, referred to the Hypsilophodontidae (Simmons 1965) and found in the Lower Lufeng Formation of Early Jurassic age, is probably a stegosaur. Both *Huayangosaurus* and *Lexovisaurus* are primitive stegosaurs of the Middle Jurassic. The latter includes fragmentary material collected in western Europe during the last century. *Huayangosaurus* is known from complete skulls and complete skeletons found in the Sichuan Basin and is important in our understanding of the evolution and origin of the stegosaurs.

In the classification of the stegosaurs, two subfamilies are recognized – the Huayangosaurinae and the Stegosaurinae. The Huayangosaurinae are structurally primitive forms that presumably include the ancestor of the Stegosaurinae. Eight genera of stegosaurs, representing both subfamilies, have been reported from Asia. They are as follows:

Tatisaurus Simmons 1965
Huayangosaurus Dong et al. 1982
Chialingosaurus Young 1959
Chungkingosaurus Dong et al. 1983
Tuojiangosaurus Dong et al. 1977
Monkonosaurus Zhao 1983
Wuerhosaurus Dong 1973
Dravidosaurus Yadagiri and Ayyasami 1979

Figure 19.1. Geographic distribution of stegosaurian dinosaurs.

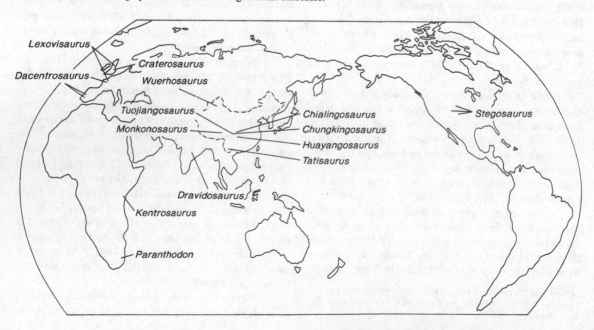

The stegosaurs of Asia

In Asia, the earliest finds of dinosaurs were from the Jabalpur district of India, in the Upper Cretaceous Lameta Beds. This material was described as the sauropod *Titanosaurus indicus* by the British paleontologist Lydekker in 1877. Matley (1923) described a few fragmentary fossils from the Upper Cretaceous of India, consisting of limb and girdle elements and armor. He felt that the large massive bones might belong to the sauropod *Titanosaurus*, while the smaller ones could be stegosaurian. The traditional view that stegosaurs died out during the Lower Cretaceous was thus considered wrong by Matley. One hundred years later, additional stegosaur material was collected from the same area (Yadagiri and Ayyasami 1979).

The first report of a stegosaur from China was in 1929. At that time, the Swedish paleontologist Wiman reported on a caudal spine Zdansky and Tan had collected from the Mengyin Formation. It was then believed to be Lower Cretaceous age, but is now considered to be Upper Jurassic. The sauropod *Euhelopus* was collected from the same site – Ninjiakou village, Mengyin County, Shandong Province. Although the spine is certainly that of a stegosaur (Dong 1977; Galton 1978) this identification of Wiman was not accepted by some paleontologists (Rozhdestvensky 1977). A second bony plate came from the Upper Cretaceous at the Chingkankou site of Laiyang County, Shandong Province (Young 1958).

The first stegosaur named from Asia is *Chialingosaurus*, collected from the Upper Shaximiao Formation of Late Jurassic age in the Sichuan Basin (Young 1959).

From 1963 to 1967, IVPP scientists explored Mesozoic continental deposits in the Junggar basin of northwestern China, and collected many vertebrate fossils from many localities. New faunas of dinosaurs, pterosaurs, and crocodilians of Late Jurassic and Early Cretaceous age were found. The Wuerho site yielded the Pterosaur–*Psittacosaurus* Fauna. An incomplete skeleton of a stegosaur from there was named *Wuerhosaurus homheni* (Dong 1973).

In 1974, members of the Municipal Museum of Chungqing opened a quarry in a suburb of Zigong City, Sichuan Province. They collected 7 tons of vertebrate fossils, including two large sauropod skeletons (*Omeisaurus*), a theropod skeleton (*Szechuanosaurus*), and a nearly complete skeleton of the stegosaur *Tuojiangosaurus* (Dong 1977). An assemblage of Middle Jurassic dinosaur fossils was found in the same area at Dashanpu during the winter of 1979. The most prominent form is *Shunosaurus*, a medium-sized sauropod. The *Shunosaurus* Fauna (Dong 1980) includes abundant sauropods (more than one hundred individuals represented, including 13 complete skeletons), carnosaurs, pterosaurs, crocodilians, and a recently described genus and species of stegosaur, *Huayangosaurus taibaii* (Dong et al. 1982; Zhou 1984).

A complete pelvic girdle and several bony plates were collected from Monkon County, Qamdo District, Tibet, in 1976. A new genus and species of stegosaur, *Monkonosaurus lawulacus*, was erected for this specimen (Zhao 1983).

Systematics

Known stegosaurian dinosaurs may be divided, on the basis of cranial and postcranial elements, into two distinct subfamilies. The first is the relatively unspecialized Huayangosaurinae; the second includes the advanced stegosaurs, the Stegosaurinae.

Order Ornithischia Seeley 1888
Suborder Stegosauria Marsh 1877
Family Stegosauridae Marsh 1880
Subfamily Huayangosaurinae Dong, Tang and Zhou 1982

Diagnosis. Primitive stegosaurs, of small to medium size. Skull high with facial region relatively shorter than in advanced forms. Antorbital fenestra present; jugal unspecialized; 1 – 3 supraorbitals. Slight heterodonty; premaxillary teeth compressed, with deep marginal serrations and enamel on both surfaces. Lower jaw has small mandibular fenestra. There are two genera: *Tatisaurus* and *Huayangosaurus*.

Range. Lower Jurassic to Middle Jurassic.

Tatisaurus oehleri Simmons 1965

Etymology. Tati is a small village situated in the Lufeng Basin where Oehler, a Catholic priest, collected many fossils in 1948 and 1949.

Material. Fragment of left mandible with teeth (CUP 2088 – FMNH).

Locality and horizon. Tati village, Lufeng Basin, Yunnan Province; from the Dark Red Beds of the Lower Lufeng Formation, Lower Jurassic.

Diagnosis. An ornithischian of small size. Mandible low anteriorly, slender and tapered; anterior ventral margin bends medially toward the symphysis. Teeth thecodont, overlapping, relatively simple in form and increasing in size from front to rear. Dentary–predentary junction edentulous. Dentigerous margin sigmoid. Coronoid appears to be absent.

Comments. This small ornithischian was referred to the Hypsilophodontidae by Simmons (1965) when it was compared with the Iguanodontidae, Ankylosauridae, and Acanthopholidae. Simmons initially thought that *Tatisaurus* was ankylosaurian, as many characters were similar to those of nodosaurids. But because the specimen was from the Upper Triassic or Lower Jurassic, and because nodosaurids were then considered strictly Cretaceous, he concluded that the Lufeng form was an ornithopod. However, Galton (1972) suggested that *Tatisaurus oehleri* might not be an ornithischian. In

1985, I had the opportunity to study the specimen (Fig. 19.2). The ramus is low anteriorly and high posteriorly, and broken both in front and in back of the tooth row. Viewed from above, it is slightly sigmoid, with 17 teeth and one vacant alveolus. A piece of attached bone was regarded as a rib by Simmons, but may be a quadratojugal. The teeth increase in size from front to rear, and

Figure 19.2. *Tatisaurus oehleri* Simmons 1965. Lower jaw in **A**, lateral and **B**, dorsal views. (Scale bar = 2 cm.)

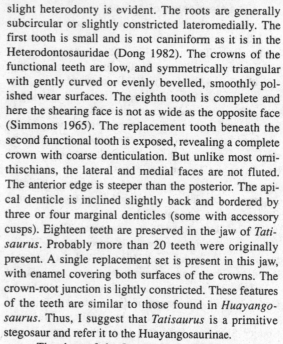

slight heterodonty is evident. The roots are generally subcircular or slightly constricted lateromedially. The first tooth is small and is not caniniform as it is in the Heterodontosauridae (Dong 1982). The crowns of the functional teeth are low, and symmetrically triangular with gently curved or evenly bevelled, smoothly polished wear surfaces. The eighth tooth is complete and here the shearing face is not as wide as the opposite face (Simmons 1965). The replacement tooth beneath the second functional tooth is exposed, revealing a complete crown with coarse denticulation. But unlike most ornithischians, the lateral and medial faces are not fluted. The anterior edge is steeper than the posterior. The apical denticle is inclined slightly back and bordered by three or four marginal denticles (some with accessory cusps). Eighteen teeth are preserved in the jaw of *Tatisaurus*. Probably more than 20 teeth were originally present. A single replacement set is present in this jaw, with enamel covering both surfaces of the crowns. The crown-root junction is lightly constricted. These features of the teeth are similar to those found in *Huayangosaurus*. Thus, I suggest that *Tatisaurus* is a primitive stegosaur and refer it to the Huayangosaurinae.

The date of the Lower Lufeng Formation has been unresolved for a long time. Young (1951) favored a Late Triassic assignment, drawing attention to the number of Lufeng taxa represented by structurally comparable forms in the Stormberg Series of South Africa. Sun et al. (1985), in an analysis of the Lower Lufeng fauna, correlated the upper part (Dark Red Beds) of the Lower Lufeng Formation with the Newark Group (Zone

Figure 19.3. *Huayangosaurus taibaii* Dong et al. 1982. Holotype skull in left lateral view.

3) and the Upper Stormberg (Clarens Formation), which are Early Jurassic.

Huayangosaurus taibaii Dong, Tang and Zhou 1982

Etymology. After a book entitled *Hua Yang Guo Zhi* from the Jin Dynasty (265–317 AD), when Sichuan was called Huayang. Taibai after the famous poet Li Bai, who lived during the Tang Dynasty in Sichuan.

Material. All of the following specimens were collected in Dashanpu, Zigong City, Sichuan Province.

Type. IVPP V6728 – a complete skull (Fig. 19.3), five vertebrae, fragmentary limbs, and three plates.

Referred specimens. ZDM T7001, nearly complete skeleton with complete skull (Figs. 19.4–19.6), 64 vertebrae, limbs, and 12 plates; ZDM T7002, several vertebrae; ZDM T7003, vertebrae and pelvic girdle; ZDM T7004, caudal vertebrae; CV 720, fragmentary skull, 28 vertebrae, and 20 plates; CV 721, seven vertebrae.

Locality and horizon. Dashanpu Quarry, situated 11 km east of Zigong City, 250 km south of Chengdu, Sichuan Province. Lower Shaximiao Formation, Middle Jurassic (Bathonian–Callovian).

Diagnosis. Primitive stegosaur of moderate size. Skull wedge-shaped, with antorbital fenestra, two or three supraorbitals, and elongate jugal. Small mandibular foramen and prominent coronoid process. Premaxillary teeth 6–7 in number, crowns with denticulate margins and weakly developed cingula at base. Long cheek tooth row with as many as 28 maxillary teeth and 27–28 dentary teeth. Vertebrae: 25–27 presacral (including 8 cervical, 17–18 dorsals, 4 co-ossified sacral vertebrae) and 35–42 caudals. Four co-ossified neural spines of sacrum form long, low blade, with three pairs of perforated notches. Processes hamularis of anterior dorsal ribs well developed. Limb bones hollow; femur 113 percent length of humerus; phalangeal formula of pes is 2.3.3.?.?. Pair of large bony plates (parascapular spines) in scapular region resemble sacral-parasacral spines of *Kentrosaurus;* bony plates and spines variable in shape extending along back in two rows.

Comments. The specimens, which include the

Figure 19.4. *Huayangosaurus taibaii* Dong et al. 1982. Skull (T. 7001) in **A**, lateral and **B**, posterior views. (Scale bar = 5 cm.)

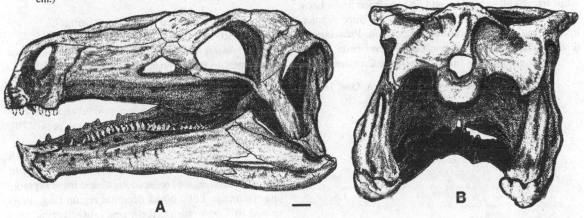

Figure 19.5. *Huayangosaurus taibaii*. Skeletal reconstruction.

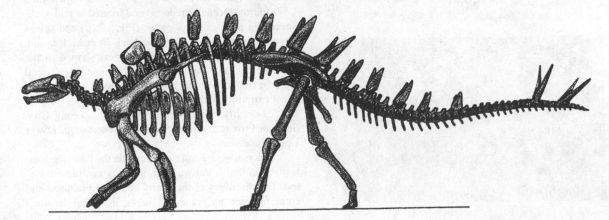

finest stegosaur material known, are considered to represent the most primitive form (Benton 1985). The teeth of the premaxilla indicate a close relationship with those ornithopods with premaxillary teeth (the heterodontosaurs and fabrosaurs of Late Triassic and Early Jurassic age). This suggests that stegosaurian dinosaurs probably arose from a small ornithopod ancestor. The age of this find supports an eastern Asian origin for the group (Dong 1980, 1982).

Until recently, our understanding of stegosaur cranial anatomy was based mostly on a single, relatively complete skull of *Stegosaurus stenops* from Morrison Formation of Colorado. It was first described more than a century ago (Marsh 1877). Therefore, the discovery of the complete skull of *Huayangosaurus taibaii* is particularly noteworthly. The articulated skeletons of *Huayangosaurus* add a new dimension to our understanding of stegosaur morphology and phylogeny. Because of its primitive status among stegosaurs, *Huayangosaurus* carries particular weight in clarifying the definition and relationships of the suborder Stegosauria.

Subfamily Stegosaurinae Nopcsa 1911

Diagnosis. Advanced stegosaurids of medium to large size. Skull rather low and, in proportion to the body, very small, with reduced jugal and two to three supraorbital elements. Antorbital fenestra closed. Premaxillary teeth absent. Coronoid process poorly developed.

Range. Middle Jurassic to Upper Cretaceous.

Figure 19.6. *Huayangosaurus taibaii.* **A**, Skull and **B**, skeleton.

Chialingosaurus kuani Young 1959

Etymology. Named in honor of Kuan Yao Wu, who was surveying the middle reaches of the Chialing River of the Sichuan Basin in 1957.

Material. Holotype, IVPP V2300, six vertebrae, left femur, both humeri, right radius, distal portions of both coracoids, and three dermal spines. Referred specimen, CV 202, a fragmentary skull and lower jaw, several vertebrae, incomplete limb elements, and four dermal plates.

Diagnosis. Medium-sized stegosaur with slender proportions. Skull higher and narrower than *Stegosaurus* and *Tuojiangosaurus*. Quadrate vertical. Teeth fewer in number than in later forms, do not overlap each other, and are in a single row. Lower jaw is thick and deep. Ratio of femoral length to humeral length (1.62) relatively high; femur slender and straight, with fourth trochanter absent. Plate-like spines rather small.

Locality and horizon. Taipingstai, Yunghsing. Lower Upper Jurassic, Upper Shaximiao Formation.

Comments. This was the first stegosaur found and named in China. Dr. Young thought that *Chialingosaurus* was like *Kentrosaurus* because anatomical details appear less specialized than in *Stegosaurus* and *Tuojiangosaurus*.

Tuojiangosaurus multispinus Dong, Li, Zhou, and Chang 1977

Etymology. The Tuojiang River is a tributary of the Yangtzi in the Sichuan Basin. *Multispinus* refers to the 17 pairs of bony plates on the back.

Material. Holotype, CV 209, a skeleton lacking some parts of the skull and lower jaw, cervical and caudal vertebrae, and some limb elements. Paratype, CV210, part of the sacrum.

Diagnosis. Large stegosaur, 7 meters in length (Figs. 19.7, 19.8). Skull similar to *Stegosaurus*, but with elongate facial region, reduced jugal, and three supraorbital elements. Low, broad occipital region (Fig. 19.9) similar to *Kentrosaurus*. Teeth small, overlapping each other in a single functional series; about 27 teeth per jaw. Sacrum consists of five vertebrae (Fig. 19.10), with three small, bean-shaped fenestrae on each side in dorsal aspect. Ratio of femoral length to humeral length is 1.57; fourth trochanter reduced and located at mid-shaft. Seventeen pairs of bony plates (Fig. 19.10) and spines arranged in two rows along the back from neck to tail. Bony plates are symmetrical and pear shaped in the neck region; large, high spines in the lumbar and sacral regions; two to four pairs of large, massive spikes at the distal extremity of the tail.

Locality and horizon. Wujiabai, Zigong City, Sichuan Province. Upper Shaximiao Formation, Lower Upper Jurassic.

Comments. *Tuojiangosaurus* is the first stegosaur from Asia that is known from a nearly complete skeleton. The discovery of the nearly complete braincase and some fragmentary cranial elements, including the maxilla, jugal, quadrate, and pterygoid, and an incomplete lower jaw, is significant because it makes possible detailed description, and comparison with the skulls of other stegosaurs. Unlike *Stegosaurus* itself, where the dorsal plates were arranged in an alternating fashion apparently for purposes of visual display (Gilmore 1914; Galton 1980), here the dorsal plates are symmetrically arranged in opposing pairs. *Tuojiangosaurus* is similar to *Kentrosaurus*, found in the contemporaneous Upper Jurassic Tendaguru Formation of Africa.

Figure 19.7. Skeleton of *Tuojiangosaurus multispinus* Dong et al. 1977.

Figure 19.8. Skeletal reconstruction of *Tuojiangosaurus multispinus*.

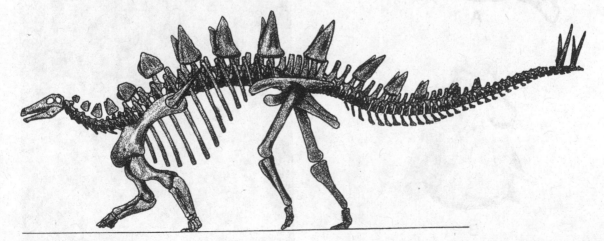

Chungkingosaurus jiangbeiensis Dong, Zhou, and Chang 1983

Etymology. This animal was found near Chungqing (Chungking) in 1977. The specific name is from Jiangbei, the county that yielded the fossils.

Material. Holotype, CV 206, incomplete skeleton, including anterior portion of skull (Fig. 19.11), ten dorsals, complete sacrum, 25 caudals, distal portion of humerus, three metacarpals, pelvic girdle, both femora (complete), a tibia, and five dermal plates.

Diagnosis. Small stegosaur (Fig. 19.12), 3–4 meters in length. Skull rather high and narrow. Lower jaw bears single row of small teeth that do not overlap. Crown of each tooth sharp and compressed. Sacrum consists of five vertebrae with three small fenestrae between sacral ribs (Fig. 19.13A). Ratio of femoral to humeral length ranges from 1.61 to 1.68. Femur straight, round in section, and lacks fourth trochanter. Astragalus fused with tibia. Bony plates large and thick, and have shape intermediate between plates and spines (Fig. 19.13B). Spines present at the end of the tail (Fig. 19.14).

Locality and horizon. Mia-eisni village, Jiangbei County, Chungqing City, Sichuan Province. Upper Shaximiao Formation, Upper Jurassic.

Comments. There are three specimens from the Upper Shaximiao Formation that are referred to *Chungkingosaurus* sp.:

1. A specimen including a complete sacrum and both ischia (CV 207), collected from Oilin Park, Chungqing City. The sacrum, consisting of five vertebrae, is like that of *Chungkingosaurus jiangbeiensis*, but the ischia are different.

2. A specimen including four caudal vertebrae, the right humerus, both femora, and fragmentary bony plates, from the quarry named Hua-ei-pa, Chungqing City. The ratio of the length of the femur to that of the humerus is 1.67, within the range of those of *Chungkingosaurus jiangbeiensis*.

3. This specimen (CV 208) is represented by ten caudal vertebrae from the distal extremity of the tail, together with three pairs of spines. The fossils were collected from Lungshi, Hechuan County, Sichuan Province. Traces of the first pair of caudal spines were found at the site. It is believed that *Chungkingosaurus* had four pairs of spines on the tail.

Wuerhosaurus homheni Dong 1973

Etymology. The fossils were collected close to Wuerho, a small town in Xinjiang. *Homhen* is Latin, meaning wide and flat, in reference to the sacral region of this stegosaur.

Material. Holotype, IVPP V4006, fragmentary and incomplete skeleton, including two vertebrae, com-

Figure 19.9. *Tuojiangosaurus multispinus* skull in **A**, dorsal and **B**, ventral views.

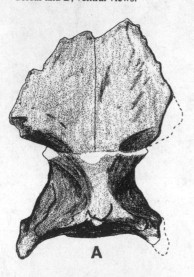

A

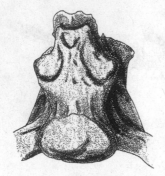

B

Figure 19.10. Two plates of *Tuojiangosaurus multispinus*. (Scale bar = 10 cm.)

plete left scapulocoracoid, both humeri, portion of ulna, pubis, ungual phalanx of pes, and two large bony plates. Paratype, IVPP V4007, three caudal vertebrae.

Diagnosis. Large stegosaur, 7–8 meters in length. Similar to *Stegosaurus* in that neural spine of anterior caudal vertebra has convex dorsal surface and is three times the height of the centrum. Fenestrae between sacral ribs and sacral processes completely closed (Fig. 19.15; 19.16A). Ratio of humeral length to iliac length 1:3 (1:2 in *Stegosaurus*). Differs from *Stegosaurus* in that anterior processes of ilia separated by wide angle. Bony plates large, long, and rather low in profile (Fig. 19.16B).

Locality and horizon. Wuerho, Xinjiang (Sinkiang), Tugulu Group, Lower Cretaceous.

Comments. Wuerhosaurus homheni is definitely a stegosaur. Galton thought that most of the postcranial bones of *Wuerhosaurus* were similar to those of *Kentrosaurus* (Hennig 1924) and *Lexovisaurus* (Nopcsa 1911) as well as to those of *Stegosaurus priscus* (BMNH R 3167, now considered *Lexovisaurus*). Thus, the age of *Wuerhosaurus* was suggested as probably Late Jurassic. This opinion has also been given by some Chinese biostratigraphers. The reptilian fossils of the Tugulu Group include:

Chelonia: *Sinemys wuerhoensis* Yeh

Figure 19.11. Reconstruction of skull of *Chungkingosaurus jiangbiensis*.

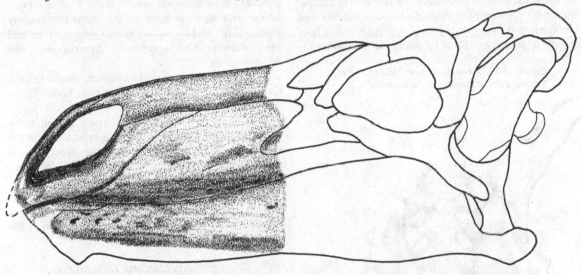

Figure 19.12. Skeleton of *Chungkingosaurus jiangbiensis*.

Pterosauria: *Dsungaripterus weii* Young, *Noripterus complicidens* Young

Crocodilia: *Edentosuchus tienshanensis* Young

Plesiosauria: cf. *Sinopliosaurus wieyuanensis* Young

Saurischia: *Tugulusaurus faciles* Dong, *Phaedrolosaurus ilikensis* Dong, *Kelmayisaurus petrolicus* Dong, *Asiatosaurus* sp.

Ornithischia: *Wuerhosaurus homheni* Dong, *Psittacosaurus xinjiangensis* Sereno and Zhao

This fauna from the Tugulu is named the Pterosaur–*Psittacosaurus* Fauna (Dong 1980), and in general can be correlated with the Morrison fauna of the Rocky Mountain region of western North America. However, the small theropod *Phaedrolosaurus ilikensis* is comparable with *Deinonychus* from the Lower Cretaceous of Montana, and suggests the beds may have had a later origin. A prominent form in this fauna is *Dsungarip-*

Figure 19.13. *Chungkingosaurus jiangbiensis.* **A**, Ventral view of sacrum; **B**, dermal plate.

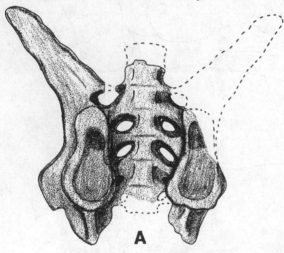

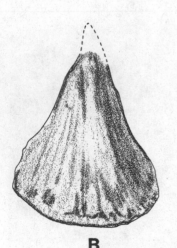

terus, which is a large pterosaur with a well developed medial crest on the skull and lacking anterior teeth. Because of these advanced characters, Young suggested an Early Cretaceous age for it. We therefore tentatively date the Tugulu fauna of Wuerho as Early Cretaceous.

Monkonosaurus lawulacus Zhao 1983

Etymology. Monko is a county in the eastern part of Tibet, where the fossils were collected by a team from the Chinese Academy of Sciences in 1976–1977. *Lawula,* a Tibetan word, is the name of a mountain.

Material. Holotype, two incomplete vertebrae, three plates, and a complete sacrum.

Diagnosis. Stegosaur of medium size, about 5 meters in length. Dermal plates large and thin and similar to those of *Stegosaurus* in shape. Five sacral vertebrae with fenestrae between the sacral ribs entirely closed (Fig. 19.17). Sacral neural spines broad and lower than those of *Stegosaurus, Kentrosaurus,* and *Wuerhosaurus.*

Locality and horizon. Lawulashan, Monko County, Qamdo District, Tibet. Loe-ein Formation, Lower Cretaceous or Upper Jurassic.

Comments. This was the first dinosaur to be found in Tibet. Other vertebrate fossils found in the same beds are similar to those from the Upper Jurassic rocks of the Sichuan Basin. However, the large, thin bony plates and closed sacral fenestrae of *Monkonosaurus* are progressive features that suggest it may be from the Lower Cretaceous. Zhao (1983) published the name only with no drawings or description of the specimen.

Dravidosaurus blanfordi Yadagiri and Ayyasami 1979

Etymology. Dravidosaurus is derived from the word Dravidanadu, commonly used for that part of the Indian peninsula from which this specimen was reported. The species is named after Blanford, who did the pioneering work in the Cretaceous rocks of southern India, and who first reported on dinosaurian fossils from those Cretaceous sediments.

Material. Holotye, GSI SR Pal 1, a partially preserved skull. Referred specimens include a tooth GSI(SR PAL 2), a sacrum GSI(SR PAL 3), an ilium GSI(SR PAL 4), an ischium GSI(SR PAL 5), an armor plate GSI(SR PAL 6), and a spike GSI(SR PAL 7). These are housed in the Palaeontological Laboratory of the Geological Survey of India.

Diagnosis. Small stegosaur. Two supraorbitals, the anterior of which is large; postorbital a thin bone expanded anteriorly and with a straight margin; postfrontal absent. Tooth small, with three crenulations. Four co-ossified, amphiplatyan sacral centra with fused neural arches; neural canal expanded, sometimes excavated into the centrum. Sternum plate-like, with a proximal expansion. Iliac blade curved posteriorly, and ischial peduncle is narrow and weak. Armor plates generally thin (10 mm) with stout bases, and triangular outline; spike curved with medial expansion.

Locality and horizon. Siranattam village (lat. 11°10′N; long. 79°01′E), Tiruchirapalli District, India. Trichinopoly Group, Upper Cretaceous (Coniacian Age).

Comments. This material was found in the *Kosmaticeras theobaldinum* Zone (Coniacian, middle Upper Cretaceous) of the Trichinopoly Group. It is the first and only definite stegosaurian dinosaur from the Upper Cretaceous (Galton 1981). The tooth (Yadagiri and Ayyasami 1979, Plate 1, 3, Fig. 19.4) was obtained while washing the matrix prepared away from the skull, and is only 3 mm in length with a crown width of 1.1 mm. This is much too small for a stegosaurian skull 220 mm long, and it may turn out to be from an ankylosaur. The tooth has three simple, small crenulations on the crown, similar to the teeth of Late Cretaceous ankylosaurs.

The origin of Stegosauria

Late Jurassic and Early Cretaceous stegosaurs are represented by well preserved specimens from around the world. The record of stegosaurian dinosaurs from the Upper Cretaceous is based on fragmentary specimens from India (Yadagiri and Ayyasami 1979) and Madagascar (Charig 1973). Olshevsky (1978) reported that ten genera of stegosaurs had been described, but

four newly discovered Asian forms, and the recognition of the small ornithopod-like *Tatisaurus* as a stegosaur, brings the number of known genera to 15. All of the material found in Asia seems to represent the main line of stegosaurian evolution.

The origin of the stegosaurs has been discussed by Dollo (1911), Gilmore (1914), Hoffstetter (1957), Thulborn (1977), Galton (1980), and Dong (1983) on the basis of material from Europe and Africa. New material found in Asia during the past decade appears to fill some of the gaps in the record of stegosaurs.

For a long time it was generally believed that *Scelidosaurus* was a primitive form of stegosaur (Swinton 1934; Romer 1956; Hoffstetter 1957), but Romer (1968) concluded that it is a primitive ankylosaur. Dollo (1911) suggested that the stegosaurs had arisen, during the Late Triassic, from a small bipedal ornithischian that bore small bony plates on its back. Gilmore (1914) thought that the ancestral stegosaur was similar to *Scelidosaurus* in many characters. Thulborn (1977), in his comparative study of the type specimen of *Scelidosaurus*, regarded it as an ornithopod, but still thought that the origin of the stegosaurs was from ornithopod stock. Although many paleontologists have considered stegosaurs to have arisen from a primitive ornithopod, the origin was shrouded in mystery because of the lack of Early and Middle Jurassic stegosaurs.

Hoffstetter (1957), who studied *Lexovisaurus* from the Middle Jurassic of Europe, suggested that the

Figure 19.14. Caudal vertebrae and spines of *Chungkingosaurus* sp. (CV 208).

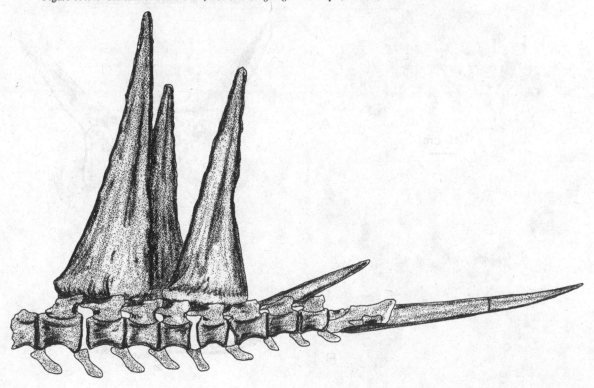

birthplace of the stegosaurs was probably Europe, and proposed a lineage from *Scelidosaurus* through *Lexovisaurus* to *Dacentrurus, Kentrosaurus,* and *Stegosaurus*.

In 1969, Steel suggested that *Chialingosaurus* was a primitive stegosaur, possibly an ancestral form, and that the birthplace of the stegosaurs may have been Asia. Dong (1983) suggested that stegosaurs arose from a small and primitive fabrosaurid-like ornithopod from the Upper Triassic of Lufeng Basin, and that their origins were therefore in southeast Asia.

Mesozoic continental deposits occur extensively in southern China (Yunnan, Guizhou, and Sichuan Provinces). These fluviolacustrine sediments have a thickness of more than three thousand meters, and represent continuous deposition from the Late Triassic to early Early Cretaceous. Five stages in the evolution of the stegosaurs can be seen in this stratigraphical succession, represented by *Tatisaurus* from the Lower Lufeng Formation (transitional between the fabrosaurid-like ornithopods and the Stegosauria), *Huayangosaurus* from the Lower Shaximiao Formation of the Middle Jurassic (Bathonian–Callovian), *Chialingosaurus* and *Tuojiangosaurus* from the lower Upper Jurassic Upper

Shaximiao Formation of Oxfordian age, *Monkonosaurus* from the Loe-ein Formation (Kimmeridgian or Neocomian), and *Wuerhosaurus* from the Lower Cretaceous Tugulu Group of Neocomian age.

The first trend is in the shape and structure of the skull. Primitively, there is (are) one (or two) supraorbital(s), the skull is high, the facial region short and narrow, both antorbital and mandibular fenestrae are present, the jugal and coronoid processes are developed, the quadrate slopes forward from its dorsal end, the retroarticular process is small, there are premaxillary teeth, and there is a slight degree of heterodonty. In the more advanced forms, there are first two supraorbitals (*Huayangosaurus*), and later three (*Tuojiangosaurus* and *Stegosaurus*). The premaxillary teeth are lost in *Tuojiangosaurus* and *Wuerhosaurus* respectively, and the antorbital and mandibular fenestrae are closed in *Wuerhosaurus*. *Stegosaurus* has only a vestigial antorbital notch (Sereno pers. comm.) and a small mandibular foramen (USNM 4934). The retroarticular process progressively enlarges and the skull becomes lower with a longer face.

The second trend is seen in changes in the structure of the sacrum. The anterior process of the ilium is more or less parallel to the longitudinal axis in primitive forms, but diverges progressively more in the later Asian genera. In *Huayangosaurus* the angle is about 30 degrees, and it increases to about 65 in *Wuerhosaurus*.

Figure 19.15. *Wuerhosaurus homheni* Dong 1973. Sacrum in **A**, ventral; and **B**, posterior views.

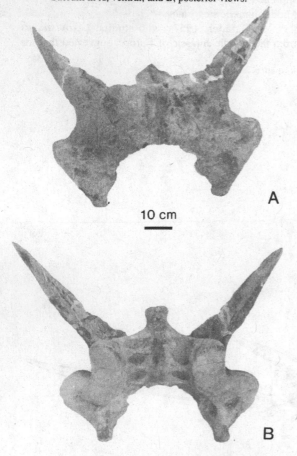

10 cm

A

B

Figure 19.16. *Wuerhosaurus homheni* Dong 1973. **A**, Ventral view of sacrum; and **B**, dorsal plate.

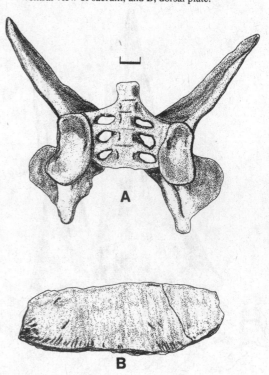

A

B

Stegosaurus deviates from this trend in that it has straightened the anterior process of the ilium (Ostrom and McIntosh 1966). The anterior process also increases steadily in length. The openings between the sacral ribs are large in *Huayangosaurus*, small in *Chialingosaurus* and *Tuojiangosaurus*, and closed in the later forms such as *Wuerhosaurus*.

The spines and plates on the back show several trends. Primitively there are many small, narrow, low plates and spines arranged in symmetrical pairs. The later forms, *Stegosaurus* and *Wuerhosaurus*, appear to be the only stegosaurs with alternating plates. The more advanced forms have progressively fewer, larger, thicker oval plates in the cervical region, and broader, higher plates in the dorsal region. There are many pairs of tail spines in primitive forms, and many small lateral dermal ossicles, as in the earliest ornithischians such as *Scutellosaurus*.

At present, the evidence suggests that China may have been the center of evolution for the Stegosauria. *Tatisaurus* shares a number of derived characteristics with the primitive stegosaur *Huayangosaurus*, and is considered here to be a suitable ancestral morphotype. All the stages of the evolution are found in China in a continuous geologic succession that correlates well with evolutionary change.

Acknowledgments

The author would like to thank the following people for useful discussions on the stegosaurs of Asia: R. Sloan of the University of Minnesota, D. A. Russell of the National Museum of Canada, R. E. Molnar of the Queensland Museum, P. J. Currie of the Tyrrell Museum of Palaeontology, and P. M. Galton of the University of Bridgeport. Special thanks are due to S. Chatterjee of Texas Tech University for providing access to the type of *Tatisaurus* (on loan from the Field Museum of Natural History, Chicago), to P. J. Currie and R. E. Molnar for correcting the English, and to Rebecca Kowalchuk (Tyrrell Museum) for typing several drafts of this paper. Figures 19.4, 19.5, 19.8, 19.9, 19.11, 19.13, 19.14, and 19.16 were drawn by Kenneth Carpenter.

Figure 19.17. *Monkonosaurus lawulacus* Zhao 1983. Sacrum in dorsal view. (Scale bar = 25 cm.)

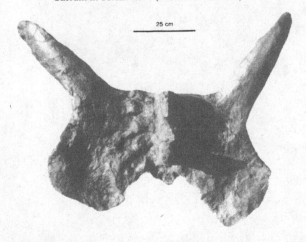

25 cm

References

Benton, M. J. 1985. Dinosaurs that fill the gap. *Nature* 317(6034):199.

Charig, A. J. 1973. Jurassic and Cretaceous dinosaurs. *In* Hallam, A. (ed.), *Atlas of Palaeobiogeography*. (Amsterdam: Elsevier) pp. 339–352.

Dollo, L. 1911. Reptiles. *In* Mourlon, M. (ed.), Gembloux. *Service géologique de Belgique. Texte explicatif du leve géologique de la planchette.*

Dong, Z. M. 1973. Dinosaurs from Wuerho. *Institute of Vertebrate Paleontology and Paleoanthropology Memoirs A* 10:45–52.

1977. On the dinosaurian remains from Turpan, Xingiang. *Vertebrata PalAsiatica* 16(4):225–228.

1980. The dinosaur faunas and their stratigraphic position in China. *Journal of Stratigraphy* 4:256–263. [in Chinese]

1983. [untitled] Kagaku 53(5):315–322. [in Chinese]

Dong, Z. M., Li, X. M., Zhou, S. W., and Chang, Y. H. 1977. On the stegosaurian remains from Zigong (Tzekung) Sichuan Province. *Vertebrata PalAsiatic* 15:307–312. [in Chinese]

Dong, Z. M., Tang, Z. L., and Zhou, S. W. 1982. Note on the new Mid-Jurassic stegosaur from Sichuan Basin, China. *Vertebrata PalAsiatica* 20:83–87.

Dong, Z. M., Zhou, S. W., and Chang, Y. H. 1983. The dinosaurian remains from Sichuan Basin, China. *Palaeontologia Sinica* 162 series C, 23:1–166.

Galton, P. M. 1972. Classification and evolution of ornithopod dinosaurs. *Nature* 239:464–466.

1978. Fabrosauridae, the basal family of ornithischian dinosaurs (Reptilia: Ornithischia). *Paläontologische Zeitschrift* 52:138–159.

1980. Armoured dinosaurs (Ornithischia: Ankylosauria) from the Middle and Upper Jurassic of England. *Geobios* 13:825–837.

1981. *Craterosaurus pottonensis* Seeley, a stegosaurian dinosaur from the Lower Cretaceous of England, and a review of Cretaceous stegosaurs. *Neues Jahrbuch für Geologie und Paläontologie* 161:28–46.

1985. British plated dinosaurs (Ornithischia, Stegosauridae). *Journal of Vertebrate Paleontology* 5:211–254.

Gilmore, C. W. 1914. Osteology of the armored dinosaurs in the United States National Museum, with special reference to the genus *Stegosaurus*. *Bulletin of the U.S. National Museum* 89:1–136.

Hennig, E. 1915. *Kentrosaurus aethiopicus*, der Stegosauride des Tendaguru. *Sitzungsberichte der Gesellschaft naturforschender Freunde, Berlin* 1915(6):219–247.

1924. *Kentrurosaurus aethiopicus*, die Stegosaurier-funde von Tendaguru, Deutsch-Ostafrika. *Palaeontographica*, Supplement 7 (1,1):103–254.

Hoffstetter, R. 1957. Quelques observations sur les Stégosaurinés. *Bulletin du Museum National d'Histoire Naturelle, Paris* 2(29):537–547.

Lydekker, R. 1877. Notices of new and other Vertebrata from Indian Tertiary and Secondary rocks. *Records of the Geological Survey of India* 10:30–43.

Marsh, O. C. 1877. New order of extinct Reptilia (Stegosauria) from the Jurassic of the Rocky Mountains. *American Journal of Science* (3)14:513–514.

Matley, C. A. 1923. Note on an armoured dinosaur from the Lameta Beds of Jubbulpore. *Records of the Geological Survey of India* 55:105–109.

Nopcsa, F. 1911. Notes on British dinosaurs, part IV. *Stegosaurus priscus* sp. nov. *Geological Magazine, London,* 5:109–115.

　　1912. Notes on British dinosaurs, part V: *Craterosaurus* (Seeley). *Geological Magazine* (5)9:481–483.

Olshevsky, G. 1978. The archosaurian taxa (excluding the Crocodilia). *Mesozoic Meanderings* 1:1–50.

Ostrom, J. H. and McIntosh, J. S. 1966. *Marsh's Dinosaurs, the Collection from Como Bluff* (New Haven: Yale University Press).

Romer, A. S. 1956. *Osteology of the Reptiles* (Chicago: University of Chicago Press).

　　1968. *Notes and Comments on Vertebrate Paleontology* (Chicago: University of Chicago Press).

Rozhdestvensky, A. K. 1977. The study of dinosaurs in Asia. *Palaeontological Society of India, Journal* 20:102–119.

Seeley, H. 1874. On the base of a large lacertian cranium from the Potton Sands, presumably dinosaurian. *Quarterly Journal of the Geological Society* 30:690–692.

　　1888. The classification of the Dinosauria. *British Association for the Advancement of Science, Manchester Meeting Report for 1887* 57:698–699.

Simmons, D. J. 1965. The non-therapsid reptiles of the Lufeng Basin, Yunnan, China. *Fieldiana, Geology* 15:1–93.

Steel, R. 1969. *Handbuch der Paläoherpetologie*. Part 15, *Ornithischia* (Jena: Gustav Fischer Verlag).

Sun, A. L., Cui G. H., Li Y. H. and Wu X. C. 1985. Composition and preliminary subdivision of the Lufeng saurian fauna. *Vertebrata PalAsiatica* 23:1–2. [in Chinese]

Swinton, W. E. 1934. *A Guide to the Fossil Birds, Reptiles and Amphibians in the Department of Geology and Palaeontology in the British Museum (Natural History)* (London: Oxford University Press).

Thulborn, R. A. 1977. Relationships of the Lower Jurassic dinosaur *Scelidosaurus harrisonii. Journal of Paleontology* 51:725–739.

Wiman, C. 1929. Die Kreid-dinosaurier aus Shantung. *Palaeontologia Sinica* (c) 6:1–67.

Yadagiri, P. and Ayyasami, K. 1979. A new stegosaurian dinosaur from Upper Cretaceous sediments of south India. *Geological Society of India Journal* 20:521–530.

Young, C. C. 1958. The dinosaurian remains of Laiyang, Shantung. *Palaeontologia Sinica* (c) 16:1–138.

　　1959. On a new Stegosauria from Szechuan, China. *Vertebrata PalAsiatic* 3(1):1–6.

Zhao, X. J. 1983. Phylogeny and evolutionary stages of Dinosauria. *Acta Palaeontologica Polonica* 28:295–306.

Zhou, S. W. 1984. *The Jurassic Dinosaurian Fauna from Dashanpu, Zigong, Sichuan*. Volume II, *Stegosaurs* (Chongqing: Sichuan Scientific and Technical Publishing House).

VIII Ankylosauria

20 Teeth and taxonomy in ankylosaurs

WALTER P. COOMBS, JR.

Abstract

Five sources of variation in dinosaur teeth – positional, ontogenetic, intraspecific, taxonomic, and chimeric – are rarely analyzed in sufficient detail to justify using dental characters to define a taxon. *Palaeoscincus costatus* is a *nomen dubium* because the single holotype tooth falls within the range of variation in teeth of two species that can be distinguished by nondental characters. Two specimens of *Ankylosaurus* have teeth subtly different from the holotype of *A. magniventris*, but the range of variation is incompletely documented, and dental characters are inadequate to establish a second species.

Introduction

Isolated dinosaur teeth have been made the types of several familiar taxa (e.g., *Iguanodon mantelli*, *Palaeoscincus costatus*, *Trachodon mirabilis*, *Troodon formosus*). The maintenance of old taxa based on isolated teeth and the temptation to define new taxa using only subtle differences in single teeth continue without regard to the variability present in dinosaur teeth. This paper explores potential sources of variation in teeth of two ankylosaurs: *Palaeoscincus costatus*, an old taxon based on an isolated tooth, and *Ankylosaurus magniventris*, the three known skulls of which have teeth of slightly different morphology.

Sources of tooth variability

There are five possible sources of variation in tooth morphology: positional, ontogenetic, intraspecific, taxonomic, and chimeric. Positional variation encompasses changes in teeth along a dental row, including size, number of cusps, and ornamentation (grooves, ridges, serrations, and other irregularities on the enamel). Ornithischian teeth, including those of ankylosaurs, commonly increase in size posteriorly along the tooth row. The largest teeth are in the posterior third of the dental bat-

In Dinosaur Systematics: Perspectives and Approaches, *Kenneth Carpenter and Philip J. Currie, eds. Copyright © Cambridge University Press, 1990.*

tery (Fig. 20.1). In some ankylosaur species, the cusp count remains constant along the tooth row, but in other species, larger teeth have higher cusp counts (Figs. 20.5A,D). Larger teeth also commonly have more complex ornamentation, but there are exceptions (Figs. 20.4F,F'). Reptilian maxillary and mandibular teeth are commonly very similar, and for ankylosaurs there appears to be no way to distinguish upper teeth from lowers. Premaxillary teeth may be substantially different from maxillary or dentary teeth (e.g., *Heterodontosaurus tucki*, Crompton and Charig 1962, and *Protoceratops andrewsi*, Brown and Schlaikjer 1940; Dodson 1976). Among ankylosaurs, *Silvisaurus condrayi*, *Sauropelta edwardsi*, and possibly *Struthiosaurus transilvanicus* have premaxillary teeth that are generally similar to the maxillary/dentary teeth, but do not fall within the variation range of the latter (Nopcsa 1929; Eaton 1960; Ostrom 1970; Coombs 1971).

Ontogenetic variation commonly extends the pattern of positional variation into smaller/simpler (younger) or larger/more complex (older) morphologies, but documentation requires complete dentitions of several age categories. Ornithischian teeth presumably change ontogenetically until they reach an adult morphological plateau. Lull (1904) speculated that dinosaurs had a fixed upper body size because they were warm-blooded. At least in the nodosaurid *Edmontonia*, teeth increase in size as skull size increases (Fig. 20.2).

Intraspecific variation exists between same size (and age) individuals within a single species. Clinal variation may exist in species having a large geographic range. Sexual dimorphism in teeth used for display and/or intraspecific combat is likely to be discontinuous and therefore conspicuous, and is unlikely to pass unnoticed (Dodson 1976). Such nonmasticatory function is suspected for premaxillary teeth of *Protoceratops andrewsi* and *Heterodontosaurus tucki*, but there is no viable evidence for such teeth in ankylosaurs.

Taxonomic variation (interspecific variation) should be analyzed at the species level, whether comparing species of one genus or species of different genera. There are three possibilities for morphologic variation in teeth of two species. First, there may be complete coincidence in variation such that all tooth morphologies of each species are present in the other. Thus, there is no possibility of defining the species by tooth morphology. The second possibility is partial overlap in the ranges of variation. This is probably a common phenomenon; juvenile teeth are a major source of morphologic overlap. In this case, definition of species by tooth morphology, taxonomic assignment of isolated teeth, and single tooth types are all possible. Inadequacies in describing the variation creates opportunities for error in direct proportion to the incompleteness of the descriptions. The third possibility is complete separation of the ranges in variation. In this instance, any tooth can be unambiguously identified and single tooth types are possible.

Chimeric teeth may be produced by anomalous ontogeny or mechanical malformation of an enamel organ such that some teeth fall entirely outside the normal range variation. Such cases, although possible, are virtually unknown, or unrecognized, and will hereafter be ignored.

Under ideal circumstances a taxon will be represented by an ontogenetic series of skulls and mandibles containing complete dental batteries that accurately document positional, ontogenetic, and intraspecific variation. If two species are represented by such suites of material, it is possible to test the three hypotheses of variation in dental morphology: coincident overlap, partial overlap, and non-overlap. Examples of such collections (e.g., for *Protoceratops andrewsi* and *Maiasaura peeblesorum* – Brown and Schlaikjer 1940; Dodson 1976; Horner and Makela 1979; Horner 1983) are familiar because they are so rare.

Most dinosaur species are represented by a few adult skulls, and the dentitions may be incompletely preserved. Historically, juveniles have been assigned to their own taxa (Dodson 1975), have only hesitatingly been placed into synonymy of taxa based on adult specimens, and even thereafter have had their status questioned. In the absence of an extensive sympatric (single quarry) collection, any assignment of small and large individuals to a single species is a theory open to

Figure 20.1. Medial (lingual) view of the relatively complete left maxillary dentition of ROM 1215. The graph shows tooth size along this dentition. Graph numbers correspond to numbered teeth in the photograph. Size index = the maximum tooth length measured anteroposteriorly along the base of the crown plus maximum tooth width measured labiolingually along the base of the crown. Abbreviations: A = empty alveolis; R = root only. (Scale in millimeters.)

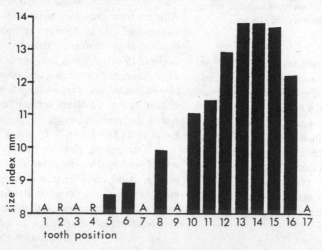

debate. Thus, for the first three sources of tooth variability, positional is moderately well known for some species, but not always well-described and illustrated; intraspecific variation is incompletely known; and ontogenetic variation is poorly known except for exceptional cases.

Ankylosaur teeth, general description

Among ornithischians, the ankylosaurs, stegosaurs, and some pachycephalosaurs have the simplest and, inferentially, the most primitive tooth morphology and packing arrangement. Crowns of ankylosaur teeth are labio-lingually compressed and phylliform, with an apical cusp and a series of secondary cusps along the

Figure 20.2. Graph of size ranges for teeth derived from four nodosaurid skulls: NMC 8531, holotype of *Edmontonia longiceps;* ROM 1215, referred to *E. longiceps;* AMNH 5381 and AMNH 5665, both referred to *E. rugosidens.* Length = maximum anteroposterior tooth measurement at the crown base. Width = maximum labiolingual tooth measurement at the crown base. The circled intersection of lines is the mean value for these two measurements. The relative size of the skulls from which these teeth are derived, as determined by maximum skull length, is as follows: ROM 1215 < AMNH 5381 < NMC 8531. It was not possible to enter the display case to measure AMNH 5665, but it appears to be larger than NMC 8531. The two measurements for ANSP 9263 are marked X.

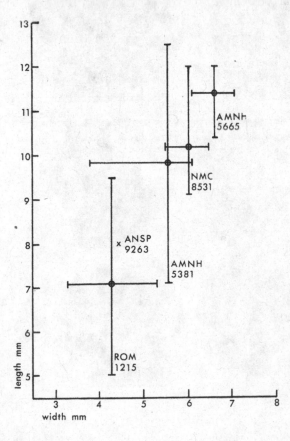

edge of the crown. The apical cusp is larger than secondary cusps in some species (Figs. 20.3, 20.4, 20.5, 20.6) and is commonly offset posteriorly (Figs. 20.5C, C′,D,D′,E,E′), or the entire crown may curve posteriorly (Figs. 20.6B,B′), but there are exceptions (Figs. 20.6G,G′). There may be equal numbers of secondary cusps anterior and posterior to the apical cusp, but usually there are one or two more cusps anteriorly (Figs. 20.5A,A′). In the family Nodosauridae, the flanks of the teeth commonly have fluting that is coincident with the notches between marginal cusps (Figs. 20.3, 20.4, 20.5). In the family Ankylosauridae, teeth may be smooth or fluted, or may differ on the labial and lingual surfaces (Figs. 20.6B,B′). Teeth of the ankylosaurid *Euoplocephalus tutus* have ridges and grooves that have no regular relationship to the marginal cusps (Figs. 20.4F,F′).

A basal cingulum is more commonly present on nodosaurid than ankylosaurid teeth. The cingulum may be smooth or ornamented in various ways (Figs. 20.3C′, 20.4A′,B,E), and the ornamentation is variable among teeth of a single individual (Figs. 20.6E,F). The cingulum is different on the occlusal and non-occlusal crown faces in one of three basic patterns:

1. almost straight and perpendicular to the tooth axis on the occlusal face, apically arched on the non-occlusal face (Figs. 20.5A,A′,B,B′);
2. almost straight and angled to the tooth axis on the occlusal face, straight and perpendicular to the tooth axis on the non-occlusal face (Figs. 20.5E,E′,F,F′); and
3. arched toward the tooth on the occlusal face, and almost straight or apically arched on the non-occlusal face (Figs. 20.3A,A′,B,B′).

Additional variation in cingulum shape blurs the boundaries of these three categories, and several different patterns may be present in teeth of a single individual (Fig. 20.5).

In addition to asymmetry of the cingulum, ornamentation may be different on labial and lingual crown faces. However, enamel distribution (thickness) is symmetrical. The orientation of wear surfaces is variable among teeth of a single individual and can be different on adjacent teeth (Fig. 20.1, teeth 6, 8, 12, and 14; Figs. 20.6G,H). Roots of ankylosaur teeth are long, straight, simple pegs that may be inflated near midlength. Roots and crowns of most teeth are vertically aligned in both lateral and anterior views, but some teeth have a slight bend at the crown–root junction (Figs. 20.6B,B′,C,C′).

Tooth morphology is very conservative within the family Nodosauridae. Teeth of *Sauropelta edwardsi* Ostrom (1970) from the Lower Cretaceous Cloverly Formation differ only in minute details from teeth of *Edmontonia longiceps* and *E. rugosidens* from the Upper Cretaceous Judith River formation (compare Figs. 20.3F, F′,G,G′ with other teeth in Figs. 20.1, 20.3, 20.4, and 20.5, especially Figs. 20.5C,C′). If *Sauropelta edwardsi* teeth were found in the Judith River Formation, they would be assigned to a species of *Edmontonia* or *Palaeoscincus.*

Palaeoscincus
Historical review

Palaeoscincus costatus Leidy (1856), established on the basis of an isolated, worn, stream-abraded tooth from the Judith River Formation, Montana, is the oldest name for a North American ankylosaur (holotype: ANSP 9263). The apical cusp, most of the secondary cusps, and most ornamentation of the crown flanks are not preserved (Figs. 20.3E,E'). Despite the poor quality of the specimen, it is readily identified as an ankylosaur of the family Nodosauridae (*sensu* Coombs 1971, 1978). Only one additional specimen has been referred in print to *P. costatus* (AMNH 5665, Matthew 1922; teeth illustrated in Figs. 20.3A–D). This specimen is one of the

Figure 20.3. **A–D**, teeth of AMNH 5665, referred to *Edmontonia rugosidens*. One badly worn and poorly preserved tooth is not figured. **E**, ANSP 9263, holotype of *Palaeoscincus costatus*. **F** and **G**, teeth of AMNH 3016, referred to *Sauropelta edwardsi*. One tooth of this specimen is not figured. Pairs of letters (A, A', B, B', etc.) are opposite crown views of the same tooth. (Scale in millimeters.)

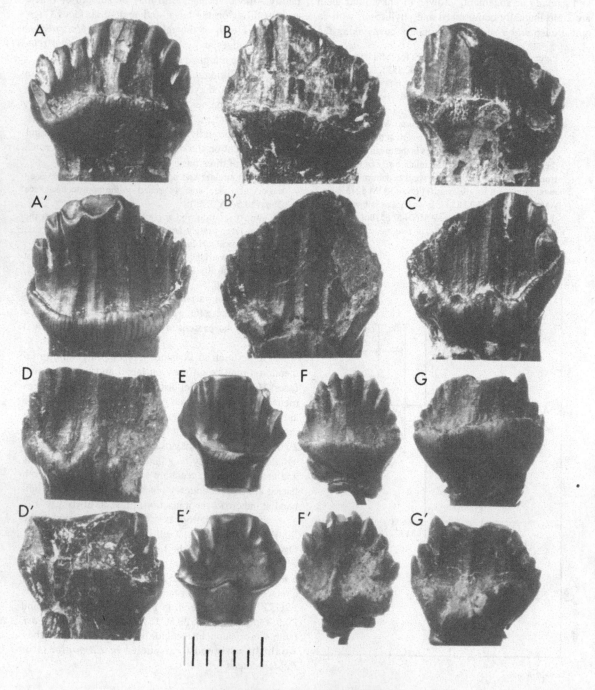

few nodosaurids that has a substantial part of the armor in place and that has been displayed as a free-standing mount. Consequently it has served as a model for several restorations, making *Palaeoscincus* one of the most commonly mentioned ankylosaur genera. Several museums have large numbers of undescribed, isolated teeth from the Judith River Formation catalogued as *Palaeoscincus*. The AMNH has scores of such teeth.

Palaeoscincus latus Marsh (1892) from the Lance Formation, Peterson's Quarry, Niobrara County, Wyoming, is based on a single tooth that is the only specimen ever to be assigned to the species (holotype: YPM 4810).

Figure 20.4. **A–D**, teeth of NMC 8531, holotype of *Edmontonia longiceps*. D is in the maxilla and could not be photographed from the opposite side. Several other teeth in the maxillae are not figured. E, lateral (labial) view of left maxillary teeth of ROM 1215, referred to *Edmontonia longiceps;* medial views are given in Figure 20.1. F, two views of NMC 1349, *Euoplocephalus tutus* (holotype of *Palaeoscincus asper*). Pairs of letters (A, A′, B, B′, etc.) are opposite crown views of the same tooth. (Scale in millimeters.)

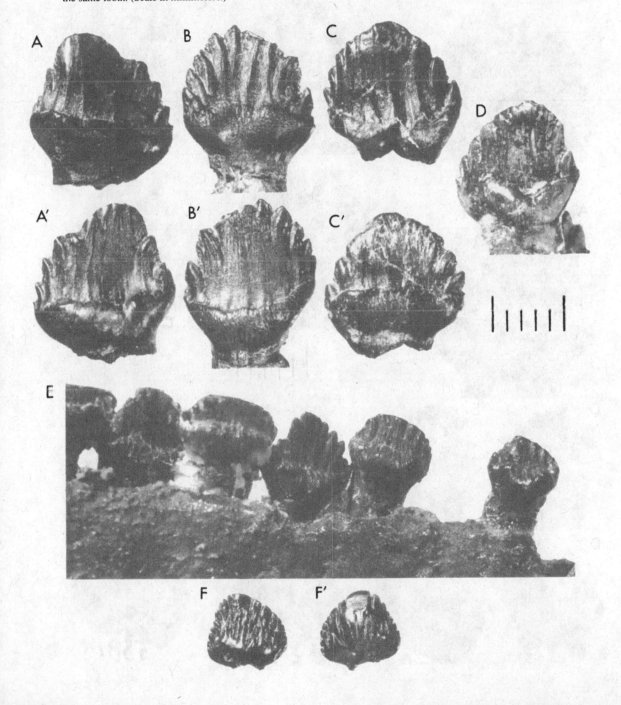

The latest Cretaceous deposits of North America (Lance, Hell Creek, and Scollard formations) have yielded a few nodosaurid specimens (Carpenter 1983; Carpenter and Breithaupt 1986), but YPM 4810 differs in some ways from any undoubted ankylosaur tooth (one derived from a skull), so its assignment to the suborder is at best doubtful. This tooth may belong to a pachycephalosaur, and *Palaeoscincus latus* is here regarded as a *nomen dubium* (following Coombs 1971).

Palaeoscincus asper Lambe (1902), based on an isolated tooth from the Judith River Formation below Berry Creek, Alberta, Canada, is another name that has

Figure 20.5. Teeth of AMNH 5381, referred to *Edmontonia rugosidens*. Additional teeth are present in the maxillae and dentaries of this specimen, but the jaws are attached to the skull by matrix and cannot at present be removed. Pairs of letters (A, A', etc.) are opposite crown views of the same tooth. (Scale in millimeters.)

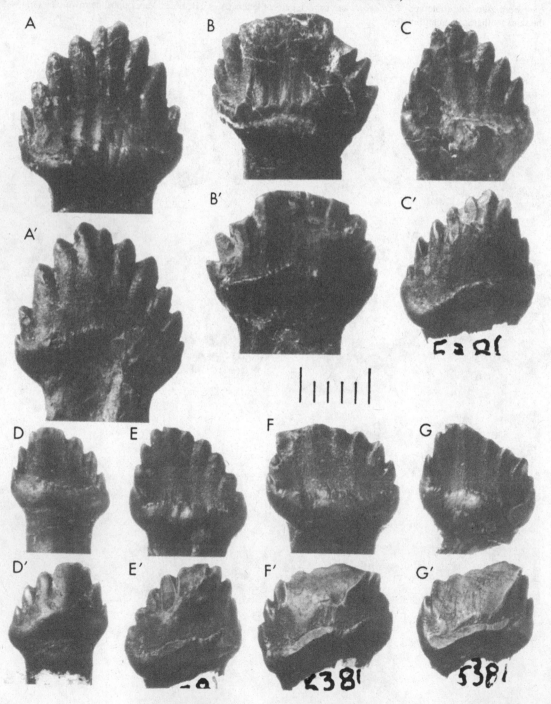

been applied only to the holotype (NMC 1349, Figs. 20.4F,F'). The tooth is indistinguishable from teeth removed from skulls of the ankylosaurid *Euoplocephalus tutus. Palaeoscinsus asper* is a junior synonym of *E. tutus* (following Coombs 1971).

The name *Palaeoscincus magoder* appears in a paper that reviews various dinosaur faunas, but which contains no formal descriptions of new taxa (Henning 1914). No type is designated, and the name is probably a misprint. Following Hay (1930) and Gilmore (1930), *P. magoder* is considered a *nomen nudum*.

Palaeoscincus rugosidens Gilmore (1930), the last named species of the genus, is discussed in the next section.

Panoplosaurus and Edmontonia

Pertinent to resolving the validity of *Palaeoscincus costatus* are specimens described under the names *Panoplosaurus mirus, Edmontonia longiceps*, and *E. rugosidens. Panoplosaurus mirus* Lambe (1919; Sternberg 1921) is based on a skull and jaws with substantial post-cranial material and armor from the Judith River Formation (Oldman Formation) of Alberta, Canada (holotype: NMC 2759). *Edmontonia longiceps* Sternberg (1928) is based on a skull, jaw, and partial skeleton from the Horseshoe Canyon Formation (members A and B of the Edmonton Formation) of Alberta, Canada (holotype: NMC 8531). Gilmore (1930) established *Palaeoscincus rugosidens* for a skull and partial skeleton from the Two Medicine Formation of Montana (holotype: NMNH 11868). Russell (1940) transferred the latter species to *Edmontonia* as *E. rugosidens* and referred a second specimen to the species (ROM 1215). In a review of all ankylosaurs, Coombs (1971) advocated synonomizing *Panoplosaurus* and *Edmontonia*, but retained three species (*Panoplosaurus mirus, P. longiceps*, and *P. rugosidens*), and removed several additional specimens to these taxa. Carpenter and Breithaupt (1986, also Carpenter this volume) again separated the two genera and retained all three species. Carpenter (Carpenter and Breithaupt 1986, Carpenter this volume) agrees with Coombs (1971) in the species assignments of referred specimens. *Panoplosaurus* and *Edmontonia* are distinguished by differences in armor, cranial scute pattern, skull proportions, and other details (Carpenter and Breithaupt 1986; Carpenter this volume). *E. longiceps* has an elongate skull with straight, converging sides as seen in dorsal view. *E. rugosidens* has a protruding orbital region and a pinched-in snout with near parallel sides. The latter two species also differ in the shape of the premaxillary scoop and in other details of the cranium (Coombs 1971; Carpenter and Breithaupt 1986; Carpenter this volume).

The holotype tooth of *Palaeoscincus costatus* is derived from one of these three species: *Panoplosaurus mirus, Edmontonia longiceps*, or *E. rugosidens*. Therefore, the validity of *P. costatus* is contingent upon the distinctness of teeth in these three species, and upon the ability to assign *P. costatus* unambiguously to one of them. Problems arise immediately. No teeth are available from the holotype of *Panoplosaurus mirus*, and there are no referred specimens. The fine rugose texturing on tooth crowns that inspired the specific name *Edmontonia rugosidens* may be a preservational phenomenon. Comparisons will be made with the following four specimens: NMC 8531, holotype of *E. longiceps* (ten measurable teeth derived from the skull and jaws, Figs. 20.4A–D); ROM 1215, referred to *E. longiceps* (seventeen measurable teeth still in the skull, Figs. 20.1, 20.4E); AMNH 5381 (seven measurable teeth from the skull and jaws, Fig. 20.5); and AMNH 5665 (four measurable teeth from the skull and jaw, Figs. 20.3A–D). The latter two specimens are referred to *E. rugosidens*. ROM 1215 provides the most complete dentition, with each maxilla having seven teeth in seventeen alveoli (41% of the maxillary teeth represented), although the first four teeth, presumably the smallest of the series, are missing from both maxillae (Fig. 20.1).

Comparisons

The holotype of *Palaeoscincus costatus* is a relatively small tooth (compare Figs. 20.3E,E' with the various *Edmontonia* teeth in Figs. 20.3, 20.4, and 20.5), but it is not necessarily from a juvenile. A plot of maximum length (measured anteroposteriorly) and maximum width (measured labio-lingually) of teeth from the four comparison skulls shows that ANSP 9263 falls within the range of values for two specimens: ROM 1215, referred to *Edmontonia longiceps*, and AMNH 5381, referred to *E. rugosidens* (Fig. 20.2). The smaller ranges for the other specimens may be a consequence of the small number of teeth available from each. The considerable positional variation in nodosaurid tooth size is dramatically apparent in teeth of AMNH 5381 (Figs. 20.5A,D). The dimensions of the *Palaeoscincus costatus* holotype, therefore, have little taxonomic significance.

Because ANSP 9263 is badly worn, an accurate cusp count is impossible. Using faint indications of grooves together with the cusps actually preserved yields an estimated count of eight or nine cusps. Counts for the comparison specimens follow: ROM 1215, seven to nine cusps; AMNH 5381 and NMC 8531, eight to eleven cusps; and AMNH 5665, nine or ten cusps. Cusp count is generally higher in larger teeth, and again ANSP 9263 could belong to either *Edmontonia rugosidens* or *E. longiceps*. Cusp count is not a reliable taxonomic character.

Cusps along one side of ANSP 9263 curve away from the central axis of the tooth (Fig. 20.3E, right half of tooth, 20.3E', left half of tooth). One tooth of AMNH 5665 has a slight outward angulation of some cusps (Fig. 20.3A, right side of tooth), but cusps on most teeth are nearly parallel to the central tooth axis (Figs. 20.3A–D). A distinct outward curve is present in cusps

Figure 20.6. **A** and **B**, teeth associated with the skull of AMNH 5895, holotype of *Ankylosaurus magniventris*. **C** and **D**, teeth of NMC 8880, referred to *Ankylosaurus magniventris*. **E–H**, teeth of AMNH 5214, referred to *Ankylosaurus magniventris*. Two badly worn and poorly preserved teeth are not figured. Pairs of letters (A, A', etc.) are opposite crown views of the same tooth. (Scale in millimeters.)

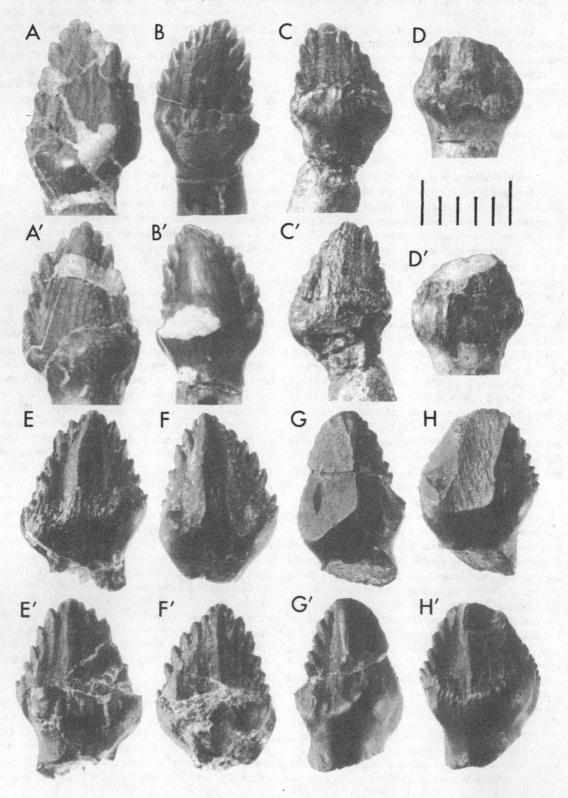

of one tooth of NMC 8531 (Fig. 20.4A', left side of tooth), and is less conspicuously developed on another (Fig. 20.4B', left side of tooth). One tooth of AMNH 5381 has the outward curve (Fig. 20.5G, right side of tooth). Therefore, this feature appears to be a positional variant that is sporadically present on some teeth of both *Edmontonia longiceps* and *E. rugosidens,* and it has little taxonomic value.

The cingulum of ANSP 9263 is smooth, probably because of stream abrasion. On one crown flank the cingulum is almost straight (Fig. 20.3E) and on the opposite flank, the cingulum is arched toward the crown apex (Fig. 20.3E'). An apically arched cingulum is present on several teeth of AMNH 5381 (Figs. 20.5A',B',C',D') and perhaps one tooth each of AMNH 5665 (Fig. 20.3C'), NMC 8531 (Fig. 20.4C) and ROM 1215 (badly worn tooth, not illustrated). The upward arch is not present on most teeth of ROM 1215 (Figs. 20.1, 20.2E). The arched cingulum thus appears to associate ANSP 9263 more closely with specimens assigned to *E. rugosidens,* yet the variation in cingulum shape among teeth from a single individual casts doubt on the reliability of this feature.

The status of *Palaeoscincus costatus*

Edmontonia longiceps and *E. rugosidens* cannot be consistently distinguished by dental characters. The single holotype tooth of *Palaeoscincus costatus* falls within the variation of both these species, and cannot be assigned to one or the other. Moreover, ANSP 9263 is similar to specimens from the Lower Cretaceous Cloverly Formation that are identified as *Sauropelta edwardsi* (compare Figs. 20.3E,E' with Figs. 20.3F,F',G,G'). Therefore, *Palaeoscincus costatus* is a *nomen dubium* (following Coombs 1971; Carpenter and Breithaupt 1986; *contra* Carpenter 1983).

Teeth of *Ankylosaurus*
Description

There are three specimens of the genus *Ankylosaurus:* AMNH 5895, holotype of *Ankylosaurus magniventris* Brown (1908), AMNH 5214, a specimen that includes a good skull (see plate 11 and text-figs. 1 and 2 in Coombs 1978), and NMC 8880, an undescribed skull. Teeth from each of these specimens are described below.

The two teeth associated with the holotype of *Ankylosaurus magniventris* were found in matrix filling the complex cranial sinuses (Figs. 20.6A,B). Neither the maxillae nor the mandibles are preserved, which casts some doubt on whether these teeth are in fact derived from the skull or are only fortuitously associated. One of the teeth appears to be unworn, although somewhat broken (Figs. 20.6A,A'), the other has a wear facet on the cingulum of one crown face (Fig. 20.6B'). The position of this wear facet, and the fact that it has anteroposteriorly directed striations, are totally anomalous within the Ankylosauria, and the facet may be a preparation scar. Assuming the slight curve bends the crown posteriorly,

then one tooth is a right and the other a left, but there is no way to determine if they are maxillary or dentary teeth. Both teeth are tall relative to their anteroposterior length, and have a cingulum that is weakly developed on one crown face, but is prominent and irregularly bulbous on the other. One crown face is shaped like a half-cone, making the tooth quite wide labio-lingually. On one tooth this crown face is smooth (Fig. 20.6B'), but on the second tooth it has faintly visible fluting (Fig. 20.6A'). The opposite crown face of each tooth is almost flat and has shallow, vertical fluting that roughly aligns with the gaps between the large upwardly inclined marginal cusps (Figs. 20.6A,B). This fluted crown face has an irregularly bulbous cingulum at the base (especially Fig. 20.6A). Crowns are vertically aligned with the axis of the root in both lateral and anterior views, and the roots are moderately inflated. These two teeth of AMNH 5895 are readily distinguished from those of virtually any other ankylosaur.

Five detached maxillary teeth are present in AMNH 5214, but their exact positions in the skull were not recorded during preparation. One dentary tooth (twenty-fourth from front in thirty-six alveoli, left dentary) is essentially identical to the maxillary teeth (Figs. 20.6F,F'). Tooth crowns have subequal lengths and heights and are relatively shorter than teeth of AMNH 5895. The posteriorly offset apical cusp aligns with a vertical ridge that is more prominent on one crown face (Figs. 20.6E,E',F,F',H'). Faint fluting of the crown flanks aligns with notches between marginal cusps on one crown face (Figs. 20.6E',F',H'), whereas the opposite crown face is essentially smooth (Figs. 20.6A,B). The base of the crown is swollen into a bulbous, irregular, lumpy cingulum on some crown faces (Figs. 20.6E', F',G'), and into a finely rugose, poorly defined cingulum on others (Figs. 20.6E,H'). The swollen base with its poorly defined cingulum may be quite broad labio-lingually, and on at least one tooth constitutes most of the mass of the crown (Figs. 20.6H,H'). On one tooth, a wear facet runs vertically down the prominent medium ridge (Fig. 20.6H). On a second tooth, the anterior(?) half of the crown is worn smooth along an oblique facet, but the posterior(?) half is essentially intact (Fig. 20.6G). Any tooth of AMNH 5214 can easily be distinguished from the two teeth of AMNH 5895. No tooth of AMNH 5214 has the distinctive recurved shape seen in teeth of AMNH 5895, and the best preserved specimens (Figs. 20.6E,E',F,F') are not as tall and slender, although one of the worn teeth (Figs. 20.6G,G') might have a crown more similarly proportioned to those of AMNH 5895. Also, the prominent central ridge on teeth of AMNH 5214 (Figs. 20.6E,F) differs from the smooth, rounded crown flank of AMNH 5895 (Fig. 20.6B').

NMC 8880 includes only two maxillary teeth, one of which is badly worn (Figs. 20.6C,C',D,D'). The more complete tooth has a crown that is slightly shorter relative to its width than teeth of AMNH 5995 and

AMNH 5214. The apical cusp is offset and the entire crown is angled posteriorly, although the magnitude of the slant may be exaggerated by breakage at the base of the crown (Figs. 20.6C,C′). One flank of the crown is rounded (Fig. 20.6C′) and the opposite flank is almost flat, as in teeth of AMNH 5895 (Fig. 20.6C). There is faint fluting along both flanks of the crown, and the enamel is covered by very fine texturing that may be a preservational feature. Both crown faces have an apically-arched, swollen cingulum. There is no median ridge on the crown flanks as in teeth of AMNH 5214. The second tooth of NMC 8880 has an irregularly lumpy cingulum as in the preceding specimens, but the oblique wear surface does not reach the cingulum as it does in AMNH 5214 (Fig. 20.6D′).

Teeth and species of *Ankylosaurus*

Each of the ten teeth described above can easily be assigned to a respective skull. There are some notable morphologic distinctions among the teeth. Teeth of AMNH 5895 are tall relative to anteroposterior length and have one flat and one rounded crown flank. Teeth of AMNH 5214 are relatively shorter and have a median ridge rather than rounded crown flank on one side. The single good tooth of NMC 8880 is somewhat intermediate, but is more like teeth of AMNH 5895 than of AMNH 5214. Based on cranial measurements, AMNH 5214 is about 5% smaller than AMNH 5895, and NMC 8880 is 20 to 25% larger (only a few homologous measurements are possible for the incomplete skull of AMNH 5895). Thus, the two specimens with the most distinctive teeth are closest in size, and ontogenetic variation is not a viable explanation for the differences.

There is considerable temptation to define a new species of *Ankylosaurus* based on the subtly distinct teeth of AMNH 5214. Against this temptation is the lingering uncertainty about the two teeth associated with the holotype, and incomplete knowledge of positional variation. There are thirty-five to thirty-six alveoli in each maxilla of AMNH 5214 with only five teeth preserved, and thirty-six to thirty-seven alveoli in each dentary with only one tooth preserved (no alveolar counts possible for the other specimens). Preliminary work on other material of these specimens lends no support to defining a second species. For example, there is as much variability in the pattern of cranial armor between the left and right sides of a single specimen as there is between any two specimens. The pyramidal horns at the posterolateral corners of the skull roof appear to be larger and more massive in AMNH 5214 and NMC 8880 than in the more fragmentary skull of AMNH 5895, but this feature groups the skulls in a manner different from tooth morphology. Considering the object lesson of *Palaeoscincus costatus* and the absence of supporting morphological distinctions in the skulls, the hypothesis of a second species of *Ankylosaurus* defined by tooth morphology is rejected. However, I have no illusions that this decision will remain unchallenged.

Summary

The five possible sources of variation in dinosaur teeth – positional, ontogenetic, intraspecific, taxonomic, and chimeric – are difficult to document and have not always been taken into account in establishing taxa based on isolated teeth. Teeth of *Edmontonia longiceps* and *E. rugosidens* overlap in morphology, and it is not possible to distinguish these species by dental characters. Teeth of *Panoplosaurus mirus* may also overlap morphologically with those of *Edmontonia* species. The holotype tooth of *Palaeoscincus costatus* falls within the variation range of both *E. longiceps* and *E. rugosidens,* and therefore both *Palaeoscincus* and *P. costatus* are *nomina dubia.* Teeth from two skulls of *Ankylosaurus* differ slightly from those of the *A. magniventris* holotype, but these dental characters are not an adequate basis to establish a second species.

Acknowledgments

My thanks to E. Gaffney (American Museum of Natural History), C. Schaff (Academy of Natural Sciences, Philadelphia), D. Russell and R. Day (National Museum of Canada), and A. Leitch and K. Seymour (Royal Ontario Museum) for providing access to and loans from collections at their respective institutions. K. Carpenter exchanged several ideas on ankylosaurs with the author. All illustrations are by the author.

References

Brown, B. 1908. The Ankylosauridae, a new family of armored dinosaurs from the Upper Cretaceous. *American Museum of Natural History Bulletin* 24:187–201.

Brown, B., and Schlaikjer, E. M. 1940. The structure and relationships of *Protoceratops. New York Academy of Science, Annals* 40:133–266.

Carpenter, K. 1983. Evidence suggesting gradual extinction of Latest Cretaceous dinosaurs. *Naturwissenschaften* 70:611–612.

Carpenter, K., and Breithaupt, B. 1986. Latest Cretaceous occurrence of nodosaurid ankylosaurs (Dinosauria, Ornithischia) in Western North America and the gradual extinction of the dinosaurs. *Journal of Vertebrate Paleontology* 6:251–257.

Coombs, W. P., Jr. 1971. *The Ankylosauria* (Ann Arbor, Michigan: Columbia University) Ph.D. thesis, University Microfilms no. 72-1291.

1978. The families of the ornithischian dinosaur order Ankylosauria. *Palaeontology* 21:143–170.

Crompton, A. W., and Charig, A. J. 1962. A new ornithischian from the Upper Triassic of South Africa. *Nature* 196:1074–1077.

Dodson, P. 1975. Taxonomic implications of relative growth in lambeosaurine hadrosaurs. *Systematic Zoology* 23:37–54.

1976. Quantitative aspects of relative growth and sexual dimorphism in *Protoceratops. Journal of Paleontology* 50:929–940.

Eaton, T. H., Jr. 1960. A new armored dinosaur from the Cretaceous of Kansas. *University of Kansas Paleontological Contributions, Vertebrata* 8:1–24.

Gilmore, C. W. 1930. On dinosaurian reptiles from the Two Medicine Formation of Montana. *United States National Museum, Proceedings* 77:1–39.

Hay, O. P. 1930. Second Bibliography and Catalogue of the Fossil Vertebrates of North America. *Carnegie Institution of Washington, Publication* 390, 1:1–916, 2:1–1074.

Henning, C. L. 1914. Ueber neuer Saurierfunde aus Kanada und deren geologische Position. *Naturwissenschaften* 2:769–776.

Horner, J. R. 1983. Cranial osteology of the type specimen of *Maisaura peeblesorum* (Ornithischia: Hadrosauridae), with discussion of its phylogenetic position. *Journal of Vertebrate Paleontology* 3:29–38.

Horner, J. R., and Makela, R. 1979. Nest of juveniles provides evidence of family structure among dinosaurs. *Nature* 282:296–298.

Lambe, L. M. 1902. On Vertebrata of the mid-Cretaceous of the Northwest Territory. 2. New genera and species from the Belly River Series (mid-Cretaceous). *Contributions to Canadian Paleontology* 3:25–81.

——— 1919. Description of a new genus and species (*Panoplosaurus mirus*) of armored dinosaur from the Belly River Beds of Alberta. *Royal Society of Canada, Transactions*, series 4, 3:313–363.

Leidy, J. 1856. Notice of the remains of extinct reptiles and fishes discovered by Dr. F. V. Hayden in the Badlands of the Judith River, Nebraska Territory. *Academy of Natural Sciences, Philadelphia, Proceedings* 8:72.

Lull, R. S. 1904. Fossil footprints of the Jura-Trias of North America. *Boston Society of Natural History, Memoirs* 5:461–557.

Marsh, O. C. 1892. Notes on Mesozoic vertebrate fossils. *American Journal of Science* 44:171–176.

Matthew, W. D. 1922. A super-dreadnaught of the animal world, the armored dinosaur *Palaeoscincus*. *Natural History* 22:333–342.

Nopcsa, F. B. 1929. Dinosaurierreste aus Siebenburgen. *Geologica Hungarica, Series Paleontologica* 1(4):1–76.

Ostrom, J. H. 1970. Stratigraphy and paleontology of the Cloverly Formation (Lower Cretaceous) of the Bighorn Basin area, Wyoming and Montana. *Bulletin of the Peabody Museum of Natural History* 35:1–234.

Russell, L. S. 1940. *Edmontonia rugosidens* (Gilmore), an armored dinosaur from the Belly River series of Alberta. *University of Toronto Studies, Geological Series* 43:3–28.

Sternberg, C. H. 1921. A supplementary study of *Panoplosaurus mirus*. *Royal Society of Canada, Transactions* 15:93–102.

——— 1928. A new armored dinosaur from the Edmonton Formation of Alberta. *Royal Society of Canada, Transactions* 22:93–106.

21 Ankylosaur systematics: example using *Panoplosaurus* and *Edmontonia* (Ankylosauria: Nodosauridae)

KENNETH CARPENTER

Abstract

Three species of nodosaurid ankylosaurs are present in the Upper Cretaceous of the Western Interior. These are *Panoplosaurus mirus* Lambe 1919, *Edmontonia longiceps* Sternberg 1928, and *Edmontonia rugosidens* (Gilmore 1930). Stratigraphically, *P. mirus* and *E. rugosidens* occur in the Middle Campanian Judith River and Two Medicine Formations of Alberta and Montana. *E. longiceps* occurs in the Campanian–Maastrichtian Horseshoe Canyon Formation. *Panoplosaurus* sp. is present in the Campanian Aguja Formation of Texas and the Campanian Fruitland or Kirtland Formations of New Mexico. *Edmontonia* sp. occurs in the Campanian Aguja Formation of Texas, and the Maastrichtian Laramie Formation of Colorado, the Lance Formation of Wyoming, and the Hell Creek Formation of South Dakota.

Panoplosaurus is characterized by: skull with tapered snout when viewed dorsally; reniform cranial armor; swollen, grooved vomer; tall neural pedicles; tall, slender neural spines; four co-ossified sacral vertebrae; co-ossified coracoid and scapula; absence of laterally projecting spines; keeled plates longer than wide.

Edmontonia is characterized by: skull with parallel sided snout when viewed dorsally; "smooth" cranial scutes; keeled vomer; neural pedicles and neural spines shorter than in *Panoplosaurus*; three co-ossified sacral vertebrae; coracoid not fused to scapula. Of the two species of *Edmontonia*, the stratigraphically older *E. rugosidens* is distinguished from the stratigraphically younger *E. longiceps* by the presence of postorbital prominences, divergent tooth rows and wide palate, a synsacrum that is longer than wide (hence less robust), and larger lateral body spines.

Introduction

Panoplosaurus mirus was named by Lawrence Lambe (1919) on the basis of a partial skeleton collected by Charles M. Sternberg from the Judith River (= Oldman) Formation of Alberta (Figs. 21.1, 21.2). In his descrip-

tion of the specimen, Lambe noted that the teeth resembled that of *Palaeoscincus costatus* Leidy 1856, a dinosaur named for a single tooth found by F. V. Hayden in the type area of the Judith River Formation in northcentral Montana. However, Lambe stated that the teeth of *Panoplosaurus* were smaller, higher crowned, and had finer denticulations along the margins (compare Fig. 21.3A with 21.3B,C). Lambe's preliminary description of *Panoplosaurus* was later supplemented with a description of the postcranium by C. M. Sternberg (1921).

In 1922, William Matthew wrote a popular article on another nodosaurid collected from the Judith River Formation of Alberta. This specimen, found by Levi Sternberg, consists of the anterior half of an articulated skeleton with armor preserved *in situ* (Figs. 21.4 and 21.5). Matthew referred the specimen to *Palaeoscincus* on the basis of teeth found with the skull (Figs. 21.3D–L). An unpublished partial manuscript in the archives of the American Museum of Natural History, possibly by W. Matthew and Barnum Brown (MS no date), indicates that the teeth were thought to differ sufficiently from the type tooth of *Palaeoscincus costatus* to warrant a new species. Unfortunately, the manuscript was never com-

Figure 21.1. Oblique view of the cervical and anterior armor of the holotype *Panoplosaurus mirus* preserved *in situ* (NMC 2759).

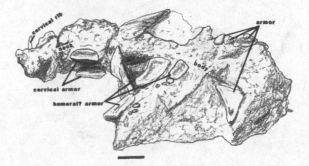

In Dinosaur Systematics: Perspectives and Approaches, *Kenneth Carpenter and Philip J. Currie, eds. Copyright © Cambridge University Press, 1990.*

pleted, and the teeth have never been illustrated. This caused a great deal of confusion in subsequent papers on nodosaurids by Sternberg (1928), Gilmore (1930), and Russell (1940). By their own admittance, none of them saw the teeth of Matthew's specimen and had to rely on his published observations.

C. M. Sternberg (1928) briefly described a partial skeleton from the Horseshoe Canyon (= Lower Edmonton)

Figure 21.2. Holotype skull of *Panoplosaurus mirus* (NMC 2759) in **A**, dorsal; **B**, ventral; and **C**, lateral views. Modified from Lambe 1919. (Scale bar = 10 cm. Abbreviations are defined at the end of the chapter.)

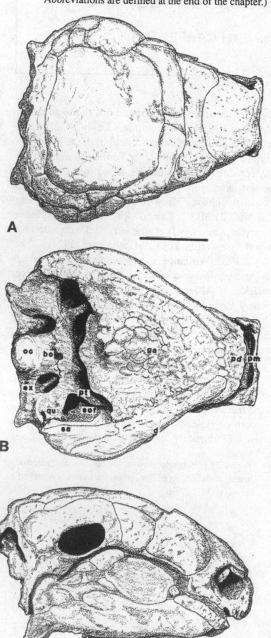

Formation of Alberta, which he named *Edmontonia longiceps*. Sternberg noted similarities between the skull of *E. longiceps* (Fig. 21.6) and that of *Panoplosaurus mirus*, but considered them distinct genera because of differences in the skull proportions, the teeth (compare Figs. 21.3B,C with 21.3M,N), and their different stratigraphic occurrence. According to the field notes of C. M. Sternberg, who found the specimen, much of the dorsal armor was preserved in life position. Unfortunately, this association was lost during preparation.

Charles Gilmore (1930) described a disarticulated specimen from the Two Medicine Formation of Montana. Comparison of his material with photographs in Matthew (1922) led Gilmore to conclude that his specimen was also *Palaeoscincus*, but he mistakenly thought that Matthew had identified his specimen from Alberta as *P. costatus*. Gilmore, therefore, proposed a new species, *Palaeoscincus rugosidens*, based on differences between the teeth of the Two Medicine specimen and the type of *P. costatus* (compare Figs. 21.3B,C with 21.3O). In proposing his new species, Gilmore noted that the skull of *E. rugosidens* (Fig. 21.7) resembled that of *Edmontonia longiceps* as figured by Sternberg, and questioned the validity of the genus *Edmontonia*.

Loris Russell (1940) reviewed the known nodosaurids when describing another partial skeleton from the Judith River Formation of Alberta (Figs. 21.3P,Q, 21.8). Russell agreed that the skulls of *Edmontonia longiceps* and *Palaeoscincus rugosidens* were similar enough to be considered the same genus. However, he felt that *Palaeoscincus costatus* was based on inadequate material and that the name should be restricted to the holotype. The preferred generic name, he argued, should be *Edmontonia* and not *Palaeoscincus*. As for *Panoplosaurus*, he felt that it was different enough to be retained as a distinct genus. Thus, Russell recognized two genera and three species: *Edmontonia longiceps*, *E. rugosidens*, and *Panoplosaurus mirus*. He also thought the specimen described by Matthew (1922) might represent yet another species of *Edmontonia* or a distinct genus, but was uncertain. He referred the specimen he described to *Edmontonia rugosidens*.

In a more recent review of the nodosaurids, Walter Coombs (1978) concluded that *Edmontonia* was a junior synonym of *Panoplosaurus*. He gave no justification for this, and as has been observed "In decided synonymy or otherwise, the onus of proof lies with him who wishes to establish it" (Charig and Reig 1970, p. 155). It therefore seems best not to accept the synonymies proposed by Coombs at this time.

Bakker (1988) has attempted a revision of Upper Cretaceous nodosaurids from North America, and has proposed the name *Denversaurus schlessmani* for a skull previously described by Carpenter and Breithaupt (1986) as *Edmontonia* sp. In addition, he has proposed a subgeneric name *Chassternbergia* for the skull described by Gilmore (1930).

Figure 21.3. Lingual and buccal views of Upper Cretaceous nodosaurid teeth from North America. **A**, holotype *Palaeoscincus costatus* (ANSP 9263); **B, C**, maxillary teeth of the holotype *Panoplosaurus mirus* (NMC 2759); **D, E, F**, dentary teeth of *Edmontonia rugosidens* (AMNH 5665); **G, H**, maxillary(?) teeth of *Edmontonia rugosidens* (AMNH 5665); **I, J, K, L**, maxillary teeth of *Edmontonia rugosidens* (AMNH 5665); **M**, maxillary tooth of the holotype *Edmontonia longiceps* (NMC 8531); **N**, dentary(?) tooth of the holotype *Edmontonia longiceps* (NMC 8531); **O**, maxillary tooth of the holotype *Edmontonia rugosidens* (USNM 11868); **P**, buccal view of the left maxillary teeth of *Panoplosaurus mirus* (ROM 1215); **Q**, lingual view of the left maxillary teeth of *Panoplosaurus mirus* (ROM 1215). (Scale bar = 1 cm.)

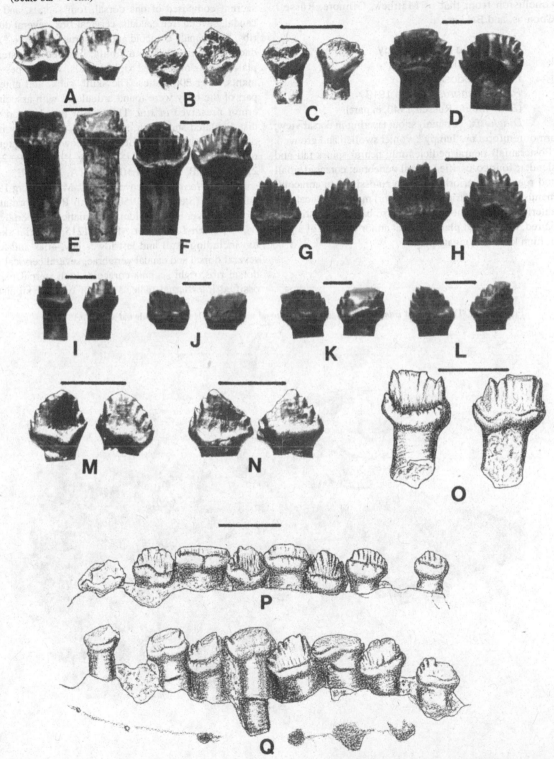

Because Gilmore, Russell, Coombs, and Bakker have all raised the issue of the taxonomic validity of *Palaeoscincus costatus, Palaeoscincus* sp. (of Matthew), *Palaeoscincus rugosidens, Panoplosaurus mirus, Edmontonia longiceps,* and *Edmontonia rugosidens,* all the material has been reexamined in light of new specimens which have been collected. This has led to a different conclusion from that of Matthew, Gilmore, Russell, Coombs, and Bakker.

Systematic paleontology

Order Ankylosauria
Family Nodosauridae
Panoplosaurus Lambe 1919
(*Edmontonia* Russell 1940, in part)
Diagnosis. (Cranial) snout tapering in dorsal view; armor reniform or "lumpy"; vomer swollen and grooved. (Postcranial) neural pedicles tall; neural spines tall and slender; four co-ossified sacral vertebrae; coracoid small and rounded anteriorly; scapula co-ossified to coracoid; manus apparently tridactyl. (Armor) medial cervical and anterior dorsal armor of transverse bands consisting of paired, low keeled plates; lateral armor consists of a pair of high keeled elongate plates.

Panoplosaurus mirus Lambe 1919
Panoplosaurus mirus Lambe 1919
[*Edmontonia rugosidens* (Gilmore 1930) in part]
Holotype. NMC 2759, skull with lower jaw, atlas and axis, a block 128 cm long, 60 cm wide with articulated cervicals and anterior dorsals, and armor preserved *in situ,* several mid- and posterior dorsals, a partial synsacrum composed of one dorsal, four sacrals, and one caudal, several free caudals, cervical ribs, several dorsal ribs, left scapula-coracoid and humerus, complete(?) tridactyl manus, tibia, and fibula, ossified intersternal plate, a pair of ossified xiphisternals, various pes elements, over 200 scutes. The skull, neck, and anterior part of the body were found articulated with associated armor preserved *in situ.* The rest of the skeleton was disarticulated and much of it eroded. Field notes indicate the skeleton was not upside down as frequently reported for ankylosaurs (Sternberg 1970). Figs. 21.1, 21.2, 21.3A,B, 21.9A, 21.10A, 21.11A, 21.12A,E.

Type locality. Discovered by C. M. Sternberg 1917, Quarry 69 (Sternberg 1950), Judith River Formation (= Belly River Beds = Oldman Formation), Alberta.

Referred specimens. ROM 1215, a partial skeleton including skull and left lower jaw, atlas and axis, several dorsal and caudal vertebrae, several cervical and dorsal ribs, right scapula-coracoid, both sternal plates, ossified intersternal plate, a pair of ossified xiphister-

Figure 21.4. The armor of *Edmontonia rugosidens* in dorsal view (AMNH 5665). (Scale bar = 20 cm.)

nals, both humeri, left ulna, three phalanges, over 200 scutes, 38 gastroliths. The skeleton was completely disarticulated when found. Figs. 21.3P,Q, 21.8, 21.12B,G. TMP 83.25.2, skull without lower jaws, Fig. 21.13. (See Fig. 21.14A for a reconstruction.)

Localities. ROM 1215 discovered by L. Sternberg 1935, Quarry 73 (Sternberg 1950), Judith River Formation, Alberta. TMP 83.25.2 found by J. Parsons in

1974, Judith River Formation, Dinosaur Provincial Park, Alberta.

Diagnosis. As for the genus.

Panoplosaurus sp.

Referred specimen. OMNH uncatalogued co-ossified scapulocoracoid.

Locality. Collected by J. Stovall and W. Langston, Aguja Formation, Big Bend region, Texas.

Edmontonia

Edmontonia Sternberg 1928

[*Panoplosaurus* (Lambe 1919) in part]

Diagnosis. (Cranial) snout with parallel or near parallel sides in dorsal view; "smooth" armor; keeled vomer. (Postcranial) neural pedicles short; neural spines shorter and more robust than in *Panoplosaurus;* three sacral vertebrae; coracoid large and almost square;

Figure 21.5. Skull of *Edmontonia rugosidens* (AMNH 5665) in **A**, dorsal; **B**, ventral; and **C**, lateral (left side reversed) views. The premaxilla and predentary are missing. (Scale bar = 10 cm.)

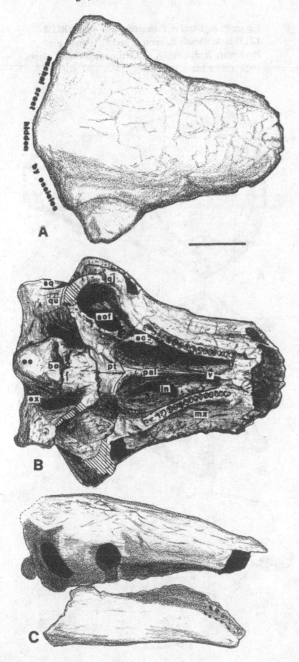

Figure 21.6. Skull of the holotype *Edmontonia longiceps* (NMC 8531) in **A**, dorsal; **B**, ventral; and **C**, lateral views. Predentary missing. Modified from Sternberg 1928. (Scale bar = 10 cm.)

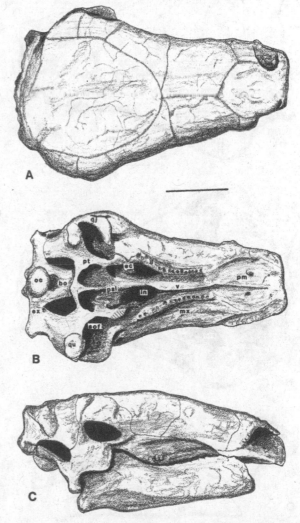

scapula not co-ossified with coracoid; manus tetradactyl. (Armor) medial cervical and anterior dorsal armor of transverse bands of subrectangular and oval low keeled plates; lateral armor of spines.

Edmontonia longiceps Sternberg 1928

Edmontonia longiceps Sternberg 1928
Panoplosaurus longiceps (Sternberg 1928)
Holotype. NMC 8531, skull with right mandible, 11 dorsal vertebrae, complete synsacrum composed of nine vertebrae (four dorsals, three sacrals, and two caudals), nine free caudals, four cervical ribs, 26 dorsal ribs, left humerus, radius, and ulna, right radius, right and left ilia, right and left ischia, both pubes (?), right

femur, right tibia, right and left fibula, numerous scutes. Found upside down with vertebrae and armor preserved *in situ,* but the rest of the skeleton disarticulated. Figs. 21.3M,N, 21.6, 21.9B, 21.12I, 21.15.

Type locality. Discovered by G. Paterson in 1924, Horseshoe Canyon Formation (= Lower Edmonton Formation), Alberta.

Diagnosis. No postorbital prominence; palate narrow; tooth rows moderately divergent. Synsacrum robust compared to *E. rugosidens.* Lateral body spines smaller than in *E. rugosidens.*

Figure 21.7. Skull of the holotype *Edmontonia rugosidens* (USNM 11868) in **A**, dorsal; **B**, ventral; and **C**, lateral views. Predentary is missing. Modified from Gilmore 1930. (Scale bar = 10 cm.)

Figure 21.8. Skull of *Panoplosaurus mirus* (ROM 1215) in **A**, dorsal; **B**, ventral; and **C**, lateral views. Predentary is missing. A and B modified from Coombs 1978. (Scale bar = 10 cm.)

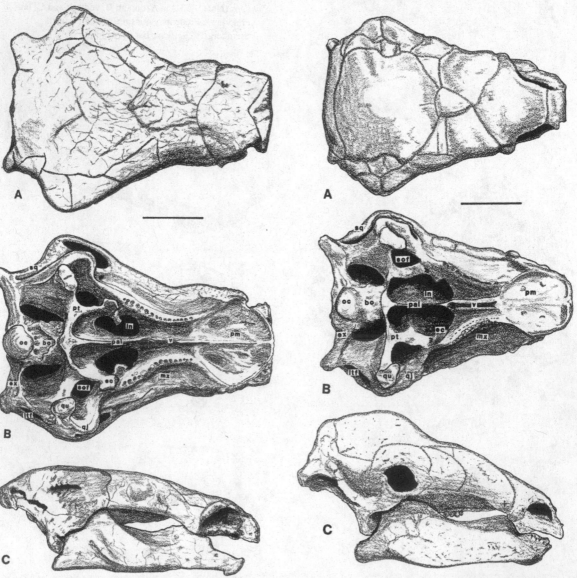

Edmontonia rugosidens (Gilmore 1930)

Palaeoscincus sp. Matthew 1922
Palaeoscincus rugosidens Gilmore 1930
Edmontonia rugosidens (Gilmore 1930)
Panoplosaurus rugosidens (Gilmore 1930)
E. (Chassternbergia) rugosidens Bakker 1988

Holotype. USNM 11868, skull with right mandible, five cervicals, ten dorsals, synsacrum (composed of four dorsals, three sacrals, and two caudals), 11 caudals, 17 ribs, partial right ilium, left and right ischia,

right pubis, 50 scutes. The skeleton was completely disarticulated and scattered when found. Figs. 21.3O, 21.7, 21.9C, 21.12G.

Type locality. Discovered by G. Sternberg 1928, from the Two Medicine Formation, Blackfeet Indian Reservation, Montana.

Referred specimens. AMNH 5381, partial skeleton including skull, six dorsals, one proximal, one distal caudal, numerous ribs, left humerus, ulna, partial radius, partial manus, fragments of the pelvis, tibia, partial

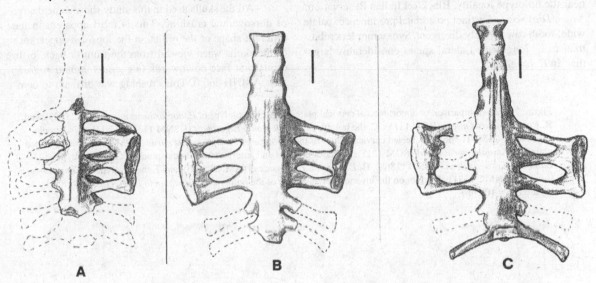

Figure 21.9. Synsacrum of: **A**, the holotype of *Panoplosaurus mirus* (NMC 2759); **B**, the holotype of *Edmontonia longiceps* (NMC 8531); and **C**, the holotype of *Edmontonia rugosidens* (USNM 11868). (Scale bars = 10 cm.)

Figure 21.11. Dorsal views of the skull and first two medial cervical plates of **A**, the holotype of *Panoplosaurus mirus* (NMC 2759); and **B**, *Edmontonia rugosidens* (AMNH 5665). (Scale bars = 10 cm.)

Figure 21.10. **A**, coracoid of *Edmontonia rugosidens* (AMNH 5665); **B**, co-ossified scapula and coracoid of the holotype of *Panoplosaurus mirus* (NMC 2759). (Scale bars = 10 cm.)

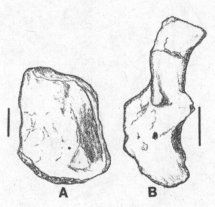

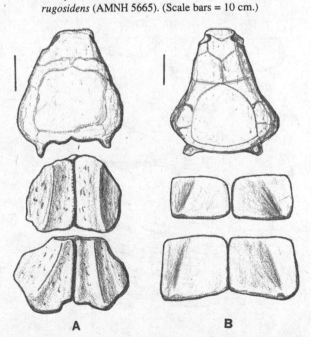

fibula (?), approximately 25 armor plates (Fig. 21.16). AMNH 5665, anterior half of an articulated skeleton with armor preserved *in situ;* postcranium includes seven cervicals, five dorsals, four partial ribs, right coracoid, scapula and foreleg (Figs. 21.3C–L, 21.4, 21.5, 21.10A, 21.11B, 21.12H, 21.14B, 21.17). ROM 5340, paired first medial scutes. MOR 433 bifurcate spine.

Localities. AMNH 5381 collected by B. Brown 1915, Quarry 45 (Sternberg 1950), Judith River Formation, Alberta. AMNH 5665 discovered by L. Sternberg in 1917, Quarry 76 (Sternberg 1950), Judith River Formation, Alberta. ROM 5340 from Judith River Formation, Alberta. MOR 433 collected by J. Horner 1986, near the holotype locality, Blackfeet Indian Reservation.

Diagnosis. Distinct postorbital prominence; palate wide; tooth rows widely divergent. Synsacrum less robust than in *E. longiceps.* Lateral spines considerably larger than in *E. longiceps.*

Edmontonia sp.
Denversaurus schlessmani Bakker 1988
Referred specimens. AMNH 3076, skull lacking lower jaws (Fig. 21.18). DMNH 468, skull lacking lower jaws and scutes (Fig. 21.19).

Localities. AMNH 3076 collected by B. Brown and R. Bird, 1940, Aguja Formation, Brewster County, Texas. DMNH 468 collected by P. Reinheimer, Lower Hell Creek Formation, Corson County, South Dakota (Carpenter and Breithaupt 1986).

Discussion
Cranial features

All the skulls used in this study show some degree of dorsoventral crushing. This is most apparent in the different shape of the orbits, in the lopsided appearance of the skulls when viewed from the front or back, or in orbits that face downwards (e.g., *Edmontonia rugosidens* AMNH 5665). This crushing was difficult to com-

Figure 21.12. Comparison of armor: medial cervical plates of **A**, the holotype of *Panoplosaurus mirus* (NMC 2759), and **B**, *Panoplosaurus mirus* (ROM 1215); **C**, the holotype of *Edmontonia rugosidens* (USNM 11868), and **D**, *Edmontonia rugosidens* (AMNH 5665); **E**, keeled cervical plates of the holotype of *Panoplosaurus mirus* (ROM 1215), and **F**, keeled plates of *Panoplosaurus mirus* (ROM 1215); posterior cervical and anterior lateral body spines of **G**, the holotype of *Edmontonia rugosidens* (USNM 11868), **H**, *Edmontonia rugosidens* (AMNH 5665), and **I**, the holotype of *Edmontonia longiceps* (NMC 8531); **J**, armor on the inverted pelvic region of the holotype of *Edmontonia longiceps* (NMC 8531). (Scale bar = 10 cm.)

pensate for in the restoration of the heads (Fig. 21.14). For example, it is possible that the head of *Edmontonia* in Fig. 21.14 is not deep enough posterior to the eyes. Other skulls are very badly crushed, which exaggerates skull width (e.g., *Panoplosaurus mirus* TMP 83.25.2, Fig. 21.13). Measurements for skulls used in this study are presented in Table 21.1.

Among *Panoplosaurus,* there is a wide range of skull shapes that cannot be accounted for by crushing.

NMC 2759 is a short, deep skull that is almost egg-shaped, whereas ROM 1215 and TMP 85.25.2 have a more "normal" (i.e., longer) nodosaurid skull shape. This difference may be accounted for by the shorter muzzle in NMC 2759. Because I am unable to find any differences among the postcrania, I suspect the shorter face is either a juvenile or a sexually dimorphic character.

As with all nodosaurids, *Panoplosaurus* and *Edmontonia* have dermal armor co-ossified with the skull.

Table 21.1 *Cranial measurements for* Panoplosaurus *and* Edmontonia

| | Panoplosaurus mirus | | | Edmontonia longiceps | | Edmontonia sp. | Edmontonia rugosidens | | |
	NMC 2759	ROM 1215	TMP 85.25.2	NMC 8531	AMNH 3076	DMNH 468	USNM 11868	AMNH 5381	AMNH 5665
skull length (cm)	35.5	40	44.5	49	39.5	49.6	47	45.6	48.5+
skull width (cm)	29.4	29.8	33[a]	29	22	34.6	34	40	43

[a]Estimate.

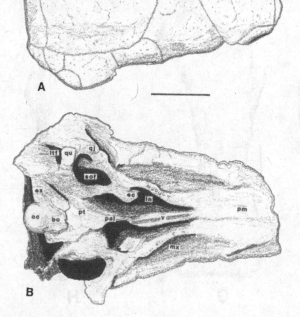

Figure 21.13. Skull of *Panoplosaurus mirus* (TMP 83.25.2). (Scale bar = 10 cm.)

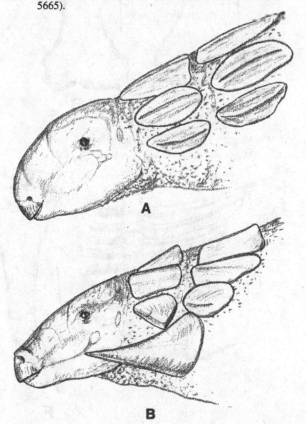

Figure 21.14. Comparative life restorations of the head and neck of **A**, the holotype of *Panoplosaurus mirus* (NMC 2759) and **B**, *Edmontonia rugosidens* (AMNH 5665).

Individual plates are delineated by a shallow groove in all specimens of *Panoplosaurus*. In *Edmontonia*, on the other hand, large skulls generally do not show individual plates distinctly (e.g., AMNH 5665, Fig. 21.5A; DMNH 468, Fig. 21.19), apparently as a result of fusion between adjacent plates in old individuals. The pattern of the scutes is similar between *Panoplosaurus* and *Edmontonia*, although the scutes of *Panoplosaurus* are distinctly more lumpy (reniform), whereas those of *Edmontonia* are smooth (compare Fig. 21.3 with Fig. 21.6). Furthermore, *Panoplosaurus* always has a distinct narrow scute along the rear dorsal edge of the skull. This scute is apparently absent or indistinct in *Edmontonia*.

Coombs (1978) has noted that the skulls of *Panoplosaurus* and *Edmontonia* are longer than wide. Further-more, the skulls of both are widest just behind the orbits, across the postorbital prominences. In *Panoplosaurus*, there is a gradation in the development of this prominence. It is best developed in ROM 1215, is moderately developed in TMP 85.25.2, and is least developed in NMC 2759 (Figs. 21.2A, 21.8, 21.13). A well developed postorbital prominence is apparently characteristic of *Edmontonia rugosidens*, but not of *Edmontonia longiceps* (compare Figs. 21.5, 21.7, 21.16 with Fig. 21.6), although additional specimens of the latter are needed to verify this.

A secondary palate is developed by the maxillaries in both *Panoplosaurus* and *Edmontonia*. The width of this secondary palate differs between the two species of *Edmontonia*. It is narrow in *E. longiceps* (Fig. 21.6B) and wide in *E. rugosidens* (Figs. 21.5B, 21.7B). The

Figure 21.15. Badly crushed postcrania of the holotype of *Edmontonia longiceps*, left humerus in **A**, anterior and **B**, posterior views; **C**, right ulna in lateral view; **D**, right radius in anterior view; **E**, pelvis in ventral view; **F**, right ischium in lateral view; right femur in **G**, posterior and **H**, anterior views; **I**, (?) right fibula. (Scale bar = 10 cm.)

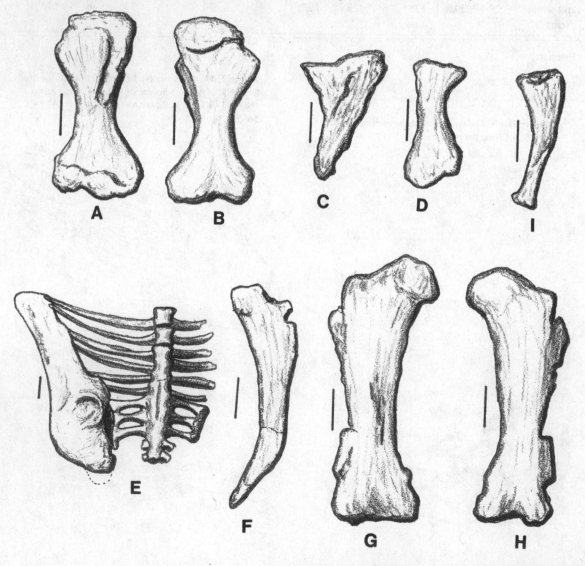

vomer (prevomer of Sternberg 1928) is swollen and grooved down the middle in the referred *Panoplosaurus* specimens (Figs. 21.8B, 21.13B). In both species of *Edmontonia,* however, the vomer is a simple keel (Figs. 21.5B, 21.6B, 21.7B, 21.18B, 21.19B).

Panoplosaurus and *Edmontonia* have a cingulum on each side of the teeth, although this is usually swollen and better developed on one side (Fig. 21.3). It is difficult to differentiate some of the teeth of *Panoplosaurus* from those of *Edmontonia* (see Coombs, this volume). Generally, the teeth of *Edmontonia* are larger than those of *Panoplosaurus,* have proportionally larger marginal

Figure 21.16. Skull of *Edmontonia rugosidens* (AMNH 5381). **A**, dorsal; **B**, ventral; and **C**, lateral (left lateral reversed) views. Predentary missing. (Scale bar = 10 cm.)

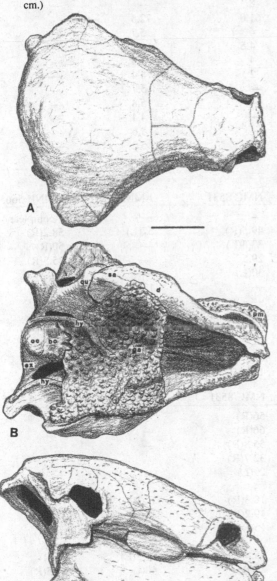

denticulations and a less prominent (swollen) cingulum (compare Figs. 21.3B,C,P with 21.3D–N). However, not all teeth of *Edmontonia* have a cingulum less prominent than *Panoplosaurus* (e.g., Fig. 21.3O).

Many teeth of *Panoplosaurus* and *Edmontonia* (Fig. 21.3, and numerous vials of isolated teeth from the Judith River of Montana, USNM 5940, 5944, 7733) exhibit thegosis wear (terminology of Thulborn 1974). This results in an oblique wear surface across the crown (Thulborn's type C). Occasionally, however, the wear facet may be along the margins of the crown (Thulborn's type D). It is not uncommon also to find wear on the lateral face of the cingulum (Thulborn's type A). It would appear that *Panoplosaurus* and *Edmontonia* had complicated jaw movement in feeding, although Coombs (pers. comm.) suggests the results are due to an imprecise alignment of teeth.

The holotype tooth of *Palaeoscincus costatus* seems to be more similar to some of the teeth of *Edmontonia* (AMNH 5665) than to any of the teeth of *Panoplosaurus* (compare Figs. 21.3A and 21.3D), although Coombs (pers. comm.) disagrees. This similarity is misleading because the cingulum of *P. costatus* is not prominent. However, as pointed out elsewhere (Carpenter and Breithaupt 1986) the cingulum in nodosaurids appears to be developed by enamel and only partially by the underlying dentine. The tooth of *Palaeoscincus costatus* lacks enamel, so, although it seems to resemble a tooth of *Edmontonia*, I cannot rule out that it once had a well developed enamel cingulum making it more like *Panoplosaurus*. The name *Palaeoscincus costatus* is a *nomen dubium* as advocated by Coombs (1971, this volume). My use of *Palaeoscincus* elsewhere (Carpenter 1983) is incorrect and refers to material listed above as *Edmontonia* sp. (DMNH 468).

Postcranial features

Postcrania of North American Late Cretaceous nodosaurids have been described for *Panoplosaurus mirus* by Sternberg (1921) and for *Edmontonia rugosidens* by Gilmore 1930. The postcranium of *E. longiceps* has not been described. Much of this material is incomplete or badly crushed making comparisons difficult. Measurements for specimens used in this study are presented in Table 21.2. A complete vertebral column is unknown for either *Panoplosaurus* or *Edmontonia.* What is known indicates that *Panoplosaurus* differs from *Edmontonia* in having taller neural pedicles and taller, more slender neural spines (compare Russell 1940, Plate 5, Figs. 1–3 with Gilmore 1930, Figs. 7–13).

A complete cervical series is known for the holotype of *Panoplosaurus mirus* (NMC 2759) and for *Edmontonia rugosidens* (AMNH 5665). Unfortunately, armor still covers most of the cervicals of the *Panoplosaurus* specimen making it impossible to get a vertebral count. Cervicals of the *Edmontonia* specimen suffer from crushing, reconstruction, and the absence of most

Table 21.2 *Postcranial measurements of* Panoplosaurus *and* Edmontonia *(maximum lengths in cm)*

	Panoplosaurus mirus		Edmontonia longiceps	Edmontonia rugosidens	
	NMC 2759	ROM 1215	NMC 8531	USNM 11868	AMNH 5665
atlas–axis (cm)	13.8	13.2	—	16.5	16.5
C-3	—	—	—	8.3	8.5
C-4	—	—	—	—	8
C-5	—	—	—	5.5	7
C-6	—	—	—	—	7?
C-7	—	—	—	6.6	9
D-a	—	6	7.3	6.5	—
D-b	—	—	7.8	8.3	—
D-c	—	—	7.5	6.8	—
D-d	—	—	6.8	—	—
D-e	—	—	8	—	—
synsacrum	36	—	61.8	72.5	—
CA-a	—	8.1	6	5.7	—
CA-b	—	7.2	4.5	5.2	—
CA-c	—	6.7	4	4.9	—
CA-d	—	6.6	2	4.4	—
CA-e	—	4.7	—	5.7	—
CA-f	—	—	—	5.8	—
CA-g	—	—	—	4.4	—
CA-h	—	—	—	5.6	—
CA-i	—	—	—	5.2	—
CA-j	—	—	—	5.6	—
CA-k	—	—	—	6.6	—
	NMC 2759	ROM 1215	NMC 8531	AMNH 5381	AMNH 5665
scapula	41(L)	40(R)	—	—	incomplete
humerus	43(L)	42(R)	49.5(R)	57(L)	58.2(R)
ulna	—	34.8(L)	33.5(L)	—	50(R)
radius	—	—	29.5	—	35.5(R)
	—	—	30(L)	—	—
mtc. I	9.6(L)	—	—	12.1(L)	13.5(R)
mtc. II	10.5(L)	—	—	—	9(R)
mtc. III	11(L)	—	—	12.2(L)	13(R)
mtc. IV	—	—	—	—	10.5(R)
ph. I–1	5.2(L)	—	—	10(L)	9.5(R)
ph. II–1	2.8(L)	—	—	—	15.8
ilium	—	—	114.5(L)	—	—
	NMC 2759	ROM 1215	NMC 8531	USNM 11868	AMNH 5665
ischium	—	—	56(R)	57(L)	—
femur	—	—	66(R)	—	—
tibia	38.5(R)	—	55.8(R)[a]	—	—
fibula	—	—	32.7(R)	—	—
	—	—	29(L)	—	—
medial cerv.					
plate – 1	19.8(R)	20(R)	18.3(R)	14(R)	12.1(R)
plate – 2	20.4(R)	—	19.2(R)	19(R)	18.3(R)
plate – 3	—	—	15(R)	—	22(R)
bifurcated spine, anterior edge	—	—	25.8(R)	24.2(L)	54(R)
large body spine	—	—	28(R)	40(L)	44.5(R)

[a]Measurement from Sternberg's field notes. R = right; L = left.

dorsal ribs (making it difficult to delineate the cervical–dorsal transition). *Edmontonia* appears to have seven or possibly eight cervicals, the centra of which become progressively more robust posteriorly. The neck of this specimen (AMNH 5665) gives the impression of being short (Fig. 21.4; see also figures in Matthew 1922). In actuality, it is no shorter than in *Sauropelta edwardsi* (Carpenter 1984, Fig. 21.3). As mounted, the scapula and coracoid of AMNH 5665 are too far forward as indicated by the fact that the anterior edge of the coracoid is between cervicals four and five. It should be nearer the cervical–dorsal transition, level with the seventh or eighth cervical position.

No complete set of dorsal vertebrae is known, although Gilmore (1930) reports at least 10 free dorsals in *Edmontonia* (USNM 11868). It is doubtful that this represents the entire series, because *Sauropelta* has 12 free dorsals (Carpenter 1984). The dorsal vertebrae of *Panoplosaurus* are poorly known, and it is not known if posterior dorsals have co-ossified ribs as in *Edmontonia* (Gilmore 1930, Fig. 11). Ribs are known for both *Panoplosaurus* (ROM 1215) and *Edmontonia* (AMNH 5665, AMNH 5381, NMC 8531, USNM 11868), but I cannot find any characters to separate them.

The synsacrum of *Panoplosaurus* is incompletely known, but Sternberg (1921) reports that it has four sacrals (Fig. 21.9A). The first sacral could be either a true sacral vertebra or the last dorsal, but its rib is co-ossified to the sacriocostal yoke (terminology of Gilmore 1930). The sacral ribs in *Panoplosaurus* are shorter relative to the width of the synsacrum than in *Edmontonia*. The synsacrum of *Edmontonia* is made up of four dorsals, three sacrals, and two caudals (Figs. 21.9B,C). The sacriocostal yoke in *Edmontonia* abuts the ilium medially with a slight underlap to strengthen the contact. Furthermore, the second sacral rib is scooped-out ventrally and may contribute to the medial wall of the acetabulum, or may brace the medial side of the ischium. The two species of *Edmontonia* can be differentiated on the basis of their synsacra, that of *E. longiceps* being wider relative to its length (hence more robust) than that of *E. rugosidens* (Fig. 21.9B, 21.9C). This difference is probably not correlated with size, as the more robust synsacrum belongs to the smaller individual (Table 21.2).

Only a few caudals are known for *Panoplosaurus* and *Edmontonia*, thus their number is unknown. Considering how conservative nodosaurids are, it is not

Figure 21.17. Manus elements of *Edmontonia rugosidens* (AMNH 5381): **A,B**, right metacarpal I; **C,D**, right metacarpal II; **E,F**, right phalanges I-1; **G,H**, (?)II-2 and 3; **I,J**, (?)III-2; **K,L**, unguals I; **M,N**, (?)II; and **O,P**,(?)III. A, C, E, G, I, K, M, O in dorsal views; B, D, F, H, J, L, N, P in palmar views.

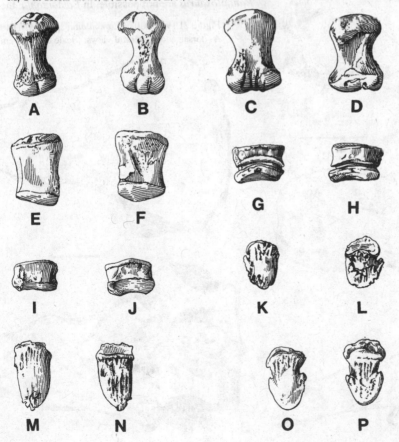

unreasonable to assume that both had long tails with 50 or more caudals as in *Sauropelta* (Carpenter 1984).

The scapula and coracoid are known for *Edmontonia* only in AMNH 5665. The coracoid is not co-ossified to the scapula, in marked contrast to *Panoplosaurus* (Figs. 21.10A,B). The posterior third of the scapula in AMNH 5665 is reconstructed, thus comparison with *Panoplosaurus* is difficult. It would appear, however, that the blade is much straighter in lateral profile. Compared to the size of the glenoid, the coracoid of *Edmontonia* is proportionally larger than its counterpart in *Panoplosaurus*. It is almost rectangular in shape, with a straight anterior edge, rather than rounded as in *Panoplosaurus* (compare Figs. 21.10A and B). Sternal plates are known for *Edmontonia* (AMNH 5665) and *Panoplosaurus* [ROM 1215, where they were mistakenly identified by Russell (1940) as clavicles]. The sternals of both animals are hatchet-shaped and paired, and resemble those of hadrosaurs rather than the single diamond-shape plate of ankylosaurids (Coombs 1978;

Carpenter 1982). An ossified internal plate and ossified xiphisternals are known for both *Panoplosaurus* (ROM 1215, Russell 1940, Pl. IV, Figs. 2–4, figured as the sternum) and *Edmontonia* (NMC 8531, AMNH 5665).

The forelimb of *Edmontonia* is indistinguishable from that of *Panoplosaurus*. In both, the humerus, ulna, and radius are robust (Figs. 21.15A–D). The radius of AMNH 5665 is uncrushed and shows a slight twist of the midshaft. The manus may be tridactyl in *Panoplosaurus* based upon NMC 2759 (see Lambe 1919, Plate 8) with a phalangeal formula 2:3:3:0:0. I am uncertain if this represents all of the digits, but if correct, it is the only quadrupedal dinosaur that has reduced the manus to three digits. Reduction of the pes to three digits is more common, occurring in *Stegosaurus* (Gilmore 1914), *Euoplocephalus* (Coombs 1986), and *Dyoplosaurus* (Carpenter 1982 as *Euoplocephalus*). Based on AMNH 5665, the manus is tetradactyl in *Edmontonia* and carpals are absent. Most of the phalanges in this specimen are missing and it is not possible to give an accurate phalangeal formula. Some individual elements of the *Edmontonia* manus are illustrated for the first time in Fig. 21.17. The metacarpals seem to be more robust than those illustrated by Lambe (1919) for *Panoplosaurus*.

A complete pelvis is known only for *Edmontonia* (NMC 8431, Fig. 21.15E), thus direct comparisons with *Panoplosaurus* are not possible. In *Edmontonia*, the first

Figure 21.18. Skull of *Edmontonia longiceps* (AMNH 3076) in **A**, dorsal; **B**, ventral; and **C**, lateral views. (Scale bar = 10 cm.)

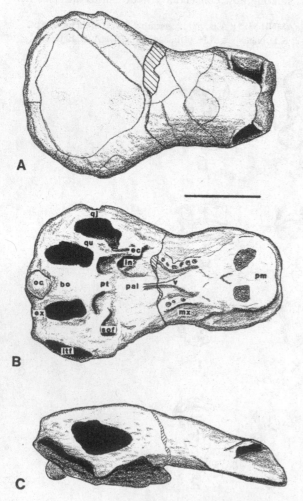

Figure 21.19. Skull of *Edmontonia* sp. (DMNH 468) in **A**, dorsal; and **B**, ventral views. (Scale bar = 10 cm.)

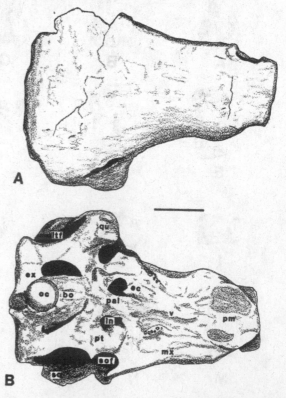

three synsacral dorsal ribs underlap the ilia but do not extend beyond the lateral margin. On the other hand, the fourth synsacral dorsal rib is short and just contacts the medial edge of the ilium. Compared to *Sauropelta edwardsi* (Coombs 1978, Fig. 14), the preacetabular portion of the ilium is longer and broader, thus resembling those of ankylosaurids. The posteriormost portion of the ilium in *Edmontonia* also differs from that of *Sauropelta* in that it projects backwards relatively farther. The pubis is known only for *Edmontonia* (Gilmore 1930, Fig. 16). The ischium is also only known for *Edmontonia;* it is long and anteriorly curved at the distal third of the shaft (Fig. 21.15F).

The hind limb and foot are poorly known in both *Panoplosaurus* and *Edmontonia*. The femur is known only for *Edmontonia* (NMC 8531, Figs. 21.15G,H), and in overall appearance resembles that of *Sauropelta* (Coombs 1978, Fig. 16), except that the greater trochanter is much lower than the femoral head. Only the proximal end of the tibia is known for *Edmontonia* (NMC 8531), although field notes of C. M. Sternberg report that the tibia was complete when found. The tibia of *Panoplosaurus* is flared at both ends and narrow in the middle resulting in an hourglass shape (Sternberg 1921, Plate 2, Fig. 3). The fibula of *Panoplosaurus* differs from that of *Edmontonia* in being straight, rather than curved (compare Sternberg 1921, Plate 2, Fig. 3 with Fig. 21.15I). This difference, however, may be due to crushing of both specimens (NMC 2759 and NMC 8531). Only a few metatarsals and phalanges are known for the pes of *Panoplosaurus* and *Edmontonia,* and these are substantially the same. Tarsals are apparently absent.

Armor

There are significant differences in the armor, especially in the cervical region, between *Panoplosaurus* and *Edmontonia*. As in extant crocodilians, the dermal armor of the cervical and shoulder regions differs considerably in shape and arrangement (Brazaitis 1973; Ross and Mayer 1983). Dermal armor, or osteoderms, have long been used in fossil lizard taxonomy (Gilmore 1928; Meszoely 1970; Sullivan 1979; Estes 1983), and by modern herpetologists (Strahm and Schwartz 1977). Osteoderms have also proven to be taxonomically useful for aetosaurs (Long and Ballew 1985), and for certain mammals (e.g., glyptodonts, Gillete and Ray 1981).

As with other ankylosaurs (*Dyoplosaurus* and *Sauropelta*, Carpenter 1982, 1984), the cervical armor of *Panoplosaurus* and *Edmontonia* consists of transverse bands separated by small irregular ossicles (Figs. 21.4, 21.14). In *Edmontonia,* and possibly *Panoplosaurus,* these small ossicles overlap the back edge of the skull (Figs. 21.4, 21.5C) where there is a small shelf or step (Fig. 21.2C). This zone of ossicles separates the first band of plates from the skull. In life, these ossicles permitted neck and head movement.

Three cervical bands of armor are present in both *Panoplosaurus* and *Edmontonia*. The first band in *Panoplosaurus* consists of three pairs of plates, the medial plates of which are suboval in shape, longer than wide, and with a low oblique keel subdividing the plates (Figs. 21.11A, 21.12A,B). Lateral to this, on each side of the neck, are a pair of elongated narrow plates, each with a high sharp keel (Figs. 21.1, 21.12E,F, 21.14A). Adjacent plates are separated from each other by a zone of small ossicles. *Edmontonia* also has three pairs of plates in the first transverse band. Medially, there is a pair of subrectangular to subtriangular plates, wider than long, and with a keel rising to a peak (Figs. 21.4, 21.11B, 21.12C,D). These plates are considerably smaller than their counterparts in *Panoplosaurus*. Lateral to these medial plates is a rectangular plate, longer than wide, with a low keel rising posteriorly to a peak (Figs. 21.4, 21.14B). And lateral to this plate is a triangular plate with a high keel projecting laterally from the neck. None of the adjacent plates are separated by ossicles, and the plates commonly fuse together (Gilmore 1930, Plate 5, Fig. 3; also seen in NMC 8531).

The second transverse band of plates in *Panoplosaurus* also has three pairs of plates. The medial pair consists of subtriangular, low keeled plates that curve slightly posterolaterally (Fig. 21.11A). As with the first cervical band, there is a pair of elongate, keeled plates separated by small ossicles on the side of the neck (Fig. 21.14A). In *Edmontonia* there are three pairs of plates and a pair of spikes making up the transverse armor band. The medial plates are similar to those of the first cervical band, only larger (Figs. 21.4, 21.11B). The plate immediately lateral to the medial plates is oval to suboval in shape with a low keel that in some specimens may rise to a small peak. Lateral to this plate is a subtriangular plate with a low keel dividing the plate in half. Lateral to this plate is a spiked plate projecting anterolaterally from the neck (Fig. 21.4, 21.14B).

Lambe reports a third cervical band is present in *Panoplosaurus* with a pair of medial plates "like those in front except that it was considerably shorter" (Lambe 1919, p. 47). Only a single partial plate is represented and its exact shape is uncertain. It was apparently as wide as the other pairs of medial plates and was probably flanked by two or three pairs of keeled plates. *Edmontonia* also has a third cervical armor band, but it is separated from the second band by a broad zone of ossicles (Fig. 21.4). This third band of plates occurs at the base of the neck and consists of two pairs of plates and a pair of spines (Fig. 21.4). The medial plates are the largest plates on the body. They are rectangular in shape, and each is subdivided by a low keel that extends obliquely across the plate. Lateral to the medial plates is a large oval, low keeled plate set at an oblique angle relative to the body midline. Lateral to this plate is a large spine projecting forward with a small fork on the dorsal edge. This bifurcation is seen in several specimens (Fig.

21.12G,H) and therefore cannot be pathological. The two species of *Edmontonia* can be separated on the basis of the bifurcated spines. Spines of *E. rugosidens* are long and project anteriorly, while those of *E. longiceps* are proportionally smaller and project more laterally (Figs. 21.12G,H,I). Sexual dimorphism may be present in the spines of *E. rugosidens,* as indicated by the proportional differences between USNM 11868 and AMNH 5665 (Figs. 21.12G,H).

Elsewhere, I have discussed the behavioral use of the bifurcated spine in *Edmontonia* (Carpenter in preparation). If this hypothesis is correct, then I suspect that the spine of AMNH 5665 with its larger secondary spine at the bifurcation is that of a male, and USNM 11868 and MOR 433 are those of females. However, because of geographical differences in the collecting sites, it is possible that the differences are subspecific. It is also possible the differences are ontogenetic because USNM 11868 and MOR 433 are smaller individuals.

The arrangement of armor on the back, hips, and tail of *Panoplosaurus* is not well known. A large block of matrix of the holotype (NMC 2759) still retains some of the body armor preserved *in situ* (Fig. 21.1). This indicates that much of the lateral body armor on the anterior portion of the body was a continuation of the lateral cervical armor (i.e., elongated, narrow, high keeled plates). Although this block of armor is difficult to interpret because it was damaged in a fire (R. Day, pers. comm. 1984), there appear to be two elongated keeled plates perpendicular to long axis of the body (Fig. 21.1). These seem to be located in the vicinity of the left forelimb and indicate that the forelegs were protected by a single row of keeled plates in a manner similar to *Dyoplosaurus* (Carpenter 1982). There are smaller keeled plates associated with both NMC 2759 and ROM 1215, but their arrangement on the body is unknown. I suspect that they were arranged in transverse rows and separated by small ossicles as in *Edmontonia*. These small keeled plates are longer than they are wide, which is the reverse of the condition in *Edmontonia*.

The dorsal armor of *Edmontonia* is known from AMNH 5665, which retains the armor *in situ* (Fig. 21.4). Sacral armor is preserved *in situ* in NMC 8531 (holotype of *E. longiceps*). Dorsally, the body armor consists of numerous transverse rows of small circular plates across the back and separated by small ossicles. The circular plates become transversely oval across the hips and may extend onto the tail. These transversely oval plates have low keels and are wider than they are long, thus distinguishing them from similar scutes in *Panoplosaurus*. Across the hips, the oval plates are arranged in staggered transverse rows, and none of them are tightly interlocked as in *Sauropelta* (Carpenter 1984). Oval plates similar to those across the pelvis of NMC 8531 are also known for USNM 11868 (Gilmore 1930, Plate 7, Fig. 1), as are the small circular plates. Thus, this interpretation of armor arrangement for the back

and hips is believed to be correct although based upon two different species of *Edmontonia*. The first dorsal band of plates is separated from the last cervical band by a zone of irregular ossicles. This zone contains a few circular keeled plates along the sides of the body.

Spines or tall keeled plates continue along the sides of the body from the bifurcated spine in *Edmontonia*. All of them project laterally (Fig. 21.4). The first lateral body spine is posterodorsal to the bifurcated spine. It is a large conical spine and is keeled dorsally (Figs. 21.12G,H,I). Posterior to this spine is a small conical spine, keeled along the anterodorsal edge. Occasionally the bases of these two spiked plates may fuse (Fig. 21.12G). The rest of the lateral body armor is poorly known, but apparently consists of high keeled plates that, based on the pelvis of NMC 8531, become progressively lower across the thighs.

Gilmore (1930) questioned whether the forward projecting spines of AMNH 5665 are in their correct position, because he felt that they would have encumbered the animal by snagging on vegetation. He suggested that the spines should project backwards. Matthew (1922), however, makes it quite clear that the armor is mounted in the position it was found, and because the specimen was dorsoventrally crushed the spines could not have been switched or rotated prior to burial. Nor do I believe that preparators would have caused the spines to switch. However, to test Gilmore's hypothesis I laid out the entire skeleton and armor of USNM 11868 with the assistance of Greg Paul. We found that due to the curvature and peculiar bevelling of the base, it is not possible to orient the large bifurcated shoulder spine laterally or posteriorly. Therefore, the forward projecting spines of AMNH 5665 are in their correct life position.

Conclusions

Nodosaurid ankylosaurs in North America can be divided into two genera and three species: *Panoplosaurus mirus*, *Edmontonia longiceps*, and *Edmontonia rugosidens*. *Denversaurus* is based on a badly crushed skull, and Bakker's (1988) diagnosis relies in part on his reconstruction of the skull in an uncrushed state. Thus, the "much greater posteriorward displacement of the orbits, postorbital boss [and] internal nares ..." (Bakker 1988, p. 19) are dependent upon the accuracy of his reconstruction. As may be seen in Fig. 21.19, the skull is dorsoventrally crushed, and skewed towards the left (in ventral view). This crushing pushed the palatine, quadrate, and pterygoid across the anterior portion of the internal nares, obscuring this region. This crushing has also collapsed the lateral temporal fenestra, and it is doubtful that Bakker has reconstructed this region correctly. If the size of the lateral temporal fenestra were increased in his reconstruction and the quadrate was more erect, as in less crushed *Edmontonia* skulls, the orbits and the postorbital prominences (= postorbital boss of Bakker) would not be displaced backwards.

Finally, "the greater spread of the basituberal rugosities over the basisphenoid and basioccipital" (Bakker 1988, p. 19) also occurs in the largest known *Edmontonia rugosidens* skull (AMNH 5665, Fig. 21.5) and seems to be size/age related. DMNH 468 most closely resembles *E. rugosidens* in the wide palate and divergent tooth rows. However, because of its crushed condition, it seems best to refer to it as *Edmontonia* sp.

Bakker's reasons for separating the subgenus *E. (Chassternbergia) rugosidens* from *Edmontonia* are "the much greater development of the derived characters of snout shortening and expansion of postorbital boss" (Bakker 1988, p. 18). While it is true that the snout is shorter in *E. rugosidens* than in *E. longiceps*, this is more apparent than real, and is due to the wide palate and divergent tooth rows in *E. rugosidens* (compare Figs. 21.5B and 21.6B). *E. rugosidens* is also characterized by postorbital prominences, but the degree of their development is variable (Figs. 21.5, 21.7, 21.16). Bakker also suggests that AMNH 5665 and AMNH 5381 may be new species of *E. (Chassternbergia)*, noting "smooth teeth would seem to be a good specific character" (Bakker 1988). In this, Bakker is wrong (Figs. 21.3I–L; Coombs this volume). Thus, there is no reason for accepting *E. (Chassternbergia)* as a valid subgenus.

The differences separating *Panoplosaurus* from *Edmontonia* may be found in the cranial armor, palate, vertebrae, manus(?), and body armor. The two species of *Edmontonia* may be separated by differences in the palate, synsacrum, and lateral spines. Furthermore, there is some indication of sexual difference in the armor, at least in *Edmontonia rugosidens*. Although the absence of statistical samples makes it difficult to prove, it would appear that females have smaller bifurcated spines than the males. The results of this study, in conjunction with those of ankylosaurids (e.g., Maryańska 1977), indicate that much of the skeleton can be used in the systematics of the Ankylosauria.

Acknowledgments

Thanks are due to the following individuals for permission to study material under their care and for providing casts of teeth: P. Currie and J. Danis (Tyrrell Museum of Palaeontology), E. Gaffney (American Museum of Natural History), N. Hotton and R. Purdy (U.S. National Museum), C. McGowan and K. Seymour (Royal Ontario Museum), D. Russell and R. Day (National Museum of Canada), and C. Smart (Academy of Natural Sciences, Philadelphia). Thanks also to N. Hotton for permitting me to remove from the skull of *Edmontonia rugosidens* the tooth originally figured from one side by Gilmore (1930). Access into the display case of AMNH 5665 was made possible through B. Werscheck, E. Gaffney, and the maintenance crew (AMNH), and to them all I owe a very special thanks. Figure 21.4, photographed by E. M. Fulda (Negative No. 310269), is courtesy of the Department of Library Services, American Museum of Natural History. Figure 21.17 is by E. Christman. This paper is dedicated to Walter Coombs for the numerous discussions and disagreements we have had about ankylosaurs in the past. These, plus his editorial comments, have greatly improved this manuscript.

Abbreviations for figures

bo – basioccipital
d – dentary
ec – ectopterygoid
ex – exoccipital
ga – gular armor
hy – hyoid
in – internal nares
ltf – lateral temporal fenestra
mx – maxilla
oc – occipital condyle
pal – palatine
pd – predentary
pm – premaxilla
pt – pterygoid
qj – quadratojugal
qu – quadrate
sa – surangular
sof – suborbital fenestra
sq – squamosal
v – vomer.

References

Bakker, R. 1988. Review of the Late Cretaceous nodosauroid Dinosauria. *Hunteria* 1(3):1–23.

Brazaitis, P. 1973. The identification of living crocodilians. *Zoologica* 58:59–101.

Carpenter, K. 1982. Skeletal and dermal armor reconstructions of *Euoplocephalus tutus* (Ornithischia: Ankylosauridae) from the Late Cretaceous Oldman Formation of Alberta. *Canadian Journal of Earth Sciences* 19:689–697.

1983. Evidence suggesting gradual extinction of Latest Cretaceous dinosaurs. *Naturwissenschaften* 70:611–612.

1984. Skeletal reconstruction and life restoration of *Sauropelta* (Ankylosauria: Nodosauridae) from the Cretaceous of North America. *Canadian Journal of Earth Sciences* 21:1491–1498.

In preparation. Functional and behavioral significance of armor in ankylosaurs. *In* Horner, J. (ed.), *Interpretation of Vertebrate Behavior from the Fossil Record.*

Carpenter, K., and Breithaupt, B. 1986. Latest Cretaceous occurrences of nodosaurid ankylosaurs (Dinosauria: Ornithischia) in western North America and the gradual extinction of the dinosaurs. *Journal of Vertebrate Paleontology* 6:251–257.

Charig, A., and Reig, R.. 1970. The classification of the Proterosuchia. *Biological Journal of the Linnaean Society* 2:125–171.

Coombs, W. 1971. *The Ankylosauria.* Ph.D. thesis (New York: Columbia University).

1978. The families of the ornithischian dinosaur order Ankylosauria. *Palaeontology* 21:143–170.

1986. A juvenile ankylosaur referable to the genus *Euoplocephalus* (Reptilia, Ornithischia). *Journal of Vertebrate Paleontology* 6:162–173.

Estes, R. 1983. Sauria terrestria, Amphisbaenia. *Handbuch der Paläoherpetologie* (Stuttgart: Fischer Verlag), part 10A:1–248.

Gillete, D., and Ray, C. 1981. Glyptodons of North America. *Smithsonian Contributions to Paleobiology* 40:1–255.

Gilmore, C. 1914. Osteology of the armored Dinosauria in the United States National Museum, with special reference to the genus *Stegosaurus*. *U.S. National Museum Bulletin* 89:1–143.

1928. Fossil lizards of North America. *National Academy of Sciences Memoir* 22:1–201.

1930. On dinosaurian reptiles from the Two Medicine Formation of Alberta. *United States National Museum Proceedings* 77, article 16:1–39.

Lambe, L. 1919. Description of a new genus and species (*Panoplosaurus mirus*) of an armoured dinosaur from the Belly River Beds of Alberta. *Royal Society of Canada Transactions* 13:39–50.

Leidy, J. 1856. Notice of the remains of extinct reptiles and fishes discovered by Dr. F. V. Hayden in the badlands of the Judith River, Nebraska Territory. *Academy of Natural Sciences of Philadelphia Proceedings* 8:72–73.

Long, R., and Ballew, K. 1985. Aetosaur dermal armor from the Late Triassic of southwestern North America, with special reference to material from the Chinle Formation of Petrified Forest National Park. *Museum of Northern Arizona Bulletin* 54:45–68.

Maryańska, T. 1977. Ankylosauridae (Dinosauria) from Mongolia. *Palaeontologia Polonica* 37:85–151.

Matthew, W. 1922. A super-dreadnaught of the animal world. The armoured dinosaur *Palaeoscincus*. *Natural History* 22:333–342.

Matthew, W., and Brown, B. (MS no date). Armor of *Palaeoscincus*. Unpublished manuscript. Archives of the Department of Vertebrate Paleontology, American Museum of Natural History.

Meszoely, C. 1970. North American fossil anquid lizards. *Museum of Comparative Zoology Bulletin* 139:87–150.

Ross, F., and Mayer, G. 1983. On the dorsal armor of the Crocodilia. *In* Rhodin, A., and Miyata, K. (eds.), *Advances in Herpetology and Evolutionary Biology* (Cambridge, MA: Museum of Comparative Zoology).

Russell, L. 1940. *Edmontonia rugosidens* (Gilmore), an armoured dinosaur from the Belly River Series of Alberta. *University of Toronto Studies, Geological Series* 43:3–27.

Sternberg, C. 1921. A supplementary study of *Panoplosaurus mirus*. *Royal Society of Canada Transactions* 15:93–102.

1928. A new armored dinosaur from the Edmonton Formation of Alberta. *Royal Society of Canada Transactions* 22:93–106.

1950. Notes and annotated list of quarries. *Geological Survey of Canada Map* 969A.

1970. Comments on dinosaurian preservation in the Cretaceous of Alberta and Wyoming. *National Museums of Canada, National Museum of Natural Sciences, Publications in Paleontology* 4:1–9.

Strahm, M., and Schwartz, A. 1977. Osteoderms in the anguid lizard subfamily Diploglossinae and their taxonomic importance. *Biotropica* 9:58–72.

Sullivan, R. 1979. Revision of the Paleogene genus *Glyptosaurus* (Reptilia, Anguidae). *American Museum of Natural History Bulletin* 163:3–72.

Thulborn, R. 1974. Thegosis in herbivorous dinosaurs. *Nature* 250:729–731.

IX Footprints

22 A name for the trace of an act: approaches to the nomenclature and classification of fossil vertebrate footprints

WILLIAM A. S. SARJEANT

Abstract

To apply a Linnaean-style classification to the footprints of dinosaurs and other fossil vertebrates is innately absurd, because they represent behavior and do not necessarily indicate affinity. They are the effects of the interaction of the animal with the sediment over which it was passing – sedimentary structures, not actual remains of living creatures. They can be identified only to the variable extent that the nature of the activity and the physical character of the sediment permit. Because changes of behavior and of position within the substrate can produce different structures, it is difficult to decide on the limits for taxa.

The history of past approaches to the naming of fossil vertebrate footprints is detailed. It is urged that names should henceforward be based essentially on morphology, not (or not entirely) on systematic affinity, and that changing behavior should be recognized by nomenclatural differentiation. It is further urged that no single classificatory hierarchy should be imposed but that alternative approaches, each appropriate to the nature of the research undertaken, should be considered equally acceptable.

Historical background

From the time of the first scientific discovery of fossil vertebrate footprints in Scotland during the late 1820's there has been an attempt, not only to identify the trackmakers, but also to give names to the tracks. On the basis of experiments in trackmaking by William Buckland (see Sarjeant 1974, pp. 268–269), those earliest-discovered tracks were considered to have been made by tortoises and the name *Testudo duncani* came to be applied to them. This name is conventionally quoted as having been published by Richard Owen in 1842, but neither Kuhn (1963, p. 14) nor I have been able to track it down. Probably it was mentioned in Owen's 1841 address to the British Association for the Advancement

of Science and taken as dating from that year, even though the name itself was not included in the published version of his address. In the atmosphere of taxonomic informality then current, this would have been acceptable. Certainly, when in 1850 Sir William Jardine made these same tracks the type for a new genus based wholly on footprints, he quoted the trivial name as *Chelichnus duncani* (Owen).

The history of vertebrate footprint nomenclature thus began obscurely and continued with confusion. The discovery by F. K. L. Sickler (1834) of some remarkable, hand-sized and clawed tracks in the German Triassic caused considerable excitement and much speculation concerning the nature of the trackmaker. J. F. Kaup (1835), unable to decide whether they were made by a mammal or a reptile, proposed two alternative generic names, *Chirotherium,* "hand-beast," or *Chirosaurus,* "hand-reptile" – in that order in his text, unfortunately, so that the first and less appropriate name had priority and has come to be used.

Victorian scholars found another problem with those names; they did not represent a completely accurate transliteration from the original Greek, which would have been better expressed as "*Cheirotherium*" or "*Cheirosaurus*" (see Jeffs 1894). Moreover, no illustrations of either type of track – *Chelichnus* or *Chirotherium* – were initially forthcoming. The type specimens were not identified and the place of their lodgment was not stated. This was before the time of formulation of the first *International Code of Zoological Nomenclature* and requirements for valid publication had not been agreed upon. Nor did matters improve much even when that *Code* was published. Until the third edition of the *Code* appeared in 1985, trace-fossils were to remain taxonomic outlaws and their classification a systematic "no-man's-land" in which each individual opinion had equal status.

Yet, long before 1985, paleoichnologists were

In Dinosaur Systematics: Perspectives and Approaches, *Kenneth Carpenter and Philip J. Currie, eds. Copyright © Cambridge University Press, 1990.*

striving to accord with the *Code*'s principles. For example, the first-published name for the German tracks, *Chirotherium*, came to be accepted in its original spelling, because these were decisions that accorded with those principles.

The earliest classification for vertebrate footprints was formulated by Edward Hitchcock (1836), on the basis of tracks in the red sandstones in the Connecticut Valley, U.S.A. Believing those tracks to have been made by birds, he termed them Ornithichnites and divided them into two groups, Pachydactyli and Leptodactyli, on the form of the digits. Within a year, however, he was recognizing that reptile tracks were also present (1837a,b). Terming the impressions overall as "Ichnites," he distinguished the reptile tracks as Sauroidichnites and formulated the category Tetrapodichnites for mammal tracks – a rather odd choice of name, because reptiles are also tetrapods!

The years 1841 to 1844 saw further modifications to Hitchcock's concepts. He decided that footprints should be treated separately from body fossils and erected for them a class Ichnolithes. This incorporated invertebrate, as well as vertebrate, tracks. On the basis of the number or absence of foot impressions, it was divided into four orders: Polypodichnites, Tetrapodichnites, Dipodichnites, and Apodichnites. By then, Hitchcock was aware that there had been bipedal dinosaurs – and, one feels, not altogether sure that he could always distinguish their footprints from those of birds! Nevertheless the Dipodichnites were divided into Sauroidichnites and Ornithoidichnites.

Hitchcock was soon busily engaged in formulating generic names for his ichnites. In 1845, there came *Anticheiropus, Palamopus, Eubrontes, Sillimanius,* and *Typopus;* in 1847, *Otozoum;* in 1848, *Helcura* and *Xiphopeza;* in 1855, *Gigandipus;* and in 1858, a flurry of names – *Antipus, Corvipes, Cheirotheroides, Cunichnoides, Exocampe, Grallator, Plectopterna,* and *Tarsodactylus.* Though those genera often contained only a single species, a few embraced several. In some instances, Hitchcock's taxa were unambiguously defined and adequately illustrated. More often they were not. Some differed in morphology because the trackmaker was behaving differently – walking swiftly or running instead of walking, so that only the front part of the palm or sole, or only the digits, were impressed. In other instances, the differences were a consequence of a differing substrate or a differing position within that substrate. Whether such differences warranted taxonomic recognition was to prove a matter for longstanding dispute.

The list exemplifies several of the possible approaches to the naming of fossil footprints. Some names are purely descriptive of the footprint (*Anticheiropus, Xiphopeza, Gigandipus, Tarsodactylus*); some are poetic (*Eubrontes,* "primitive thunder"); some suggest a possible affinity (*Corvipes,* "raven-foot") or behavior (*Grallator,* "one who walks on stilts"); and one pays tribute

to a distinguished scientist (*Sillimanius,* named for Benjamin Silliman, Sr.). Also to be noted are the earliest uses of three terminations that would come to be widely utilized for footprint names, "-opus," "-dactylus," and "-ipes." Hitchcock's earlier formulations, though not conceived of as generic names, were to provide another, "-ichnites."

Hitchcock's classification of footprints continued to evolve as his ideas changed and he was never averse to abandoning names when he ceased to think them appropriate. His final classification, published in 1865, retains little from the earlier ones. Footprints overall were placed into a new class Lithichnozoa. The Connecticut Valley prints, instead of being assembled into orders or families, were organized into seven groups, each allotted a Roman numeral (I, marsupialoid animals; II, pachydactylous birds; III, leptodactylous birds; IV, ornithoid lizards or batrachians; V, lizards; VI, batrachians; and VII, chelonians). Though category IV seems strange to modern eyes, this attempt to set up a sort of ichnofossil paraclassification – a classification parallel to, but separate from the standard Linnaean approach – is a valid approach to a difficult problem. Yet, though Hitchcock's work was revered, he has left behind a considerable legacy of taxonomic problems. None of his classifications was destined to survive – perhaps because his approaches had changed so fast.

Even at the generic level, Hitchcock's concepts did not attract universal support. Two German paleontologists reacted against his plethora of genera. Geinitz (1861) proposed a single collective genus, *Saurichnites,* to embrace all reptilian footprints. Pabst (1900) formulated an essentially non-Linnaean system, in which all the Permian footprints he was studying were given the umbrella name *Ichnium.* This was divided into units with binomens that were equivalent to families and these, in turn, into units with trinomens that might be equivalent either to genera (*Ichnium gampsodactylum*) or to species (e.g., *Ichnium gampsodactylum friedrichrodanum*). His higher categories are set forth below:

I. Brachydactylichnia (short-toed tracks)
1. Pachydactylichnia (heavy-toed tracks) *Ichnium pachydactylum*
2. Brachydactylichnia (short-toed tracks) *I. brachydactylum*
3. Anakolodactylichnia (shortest-toed tracks) *I. anakolodactylum*
4. Sphaerodactylichnia (lump-toed tracks) *I. sphaerodactylum*
5. Rhopalodactylichnia (club-toed tracks) *I. rhopalodactylum*

II. Dolichodactylichnia (long-toed tracks)
6. Akrodactylichnia (pointed-toed tracks) *I. acrodactylum*
7. Tanydactylichnia (elongate-toed tracks) *I. tanydactylum*
8. Dolichodactylichnia (long-toed tracks) *I. dolichodactylum*
9. Gampsodactylichnia (curved-toed tracks) *I. gampsodactylum*

This is a morphological classification, not one

based on affinity, and as such has its attractions. Unfortunately, Pabst's system was at once too cumbrous for a classification of footprints at one stratigraphical level, yet too specific to serve as basis for a wider classification. Though adopted for a while by a few workers in Germany and England, it soon ceased to be employed.

A contrasting approach, possible only in cases where there seemed a reasonable presumption that the animal making the tracks had been correctly identified, was to directly utilize the name of the trackmaker for the footprints. Owen's presumed allocation of the Dumfries-shire tracks to *Testudo* had been the earliest instance of this and certainly Owen included footprints, as well as bones, when he erected (1842) his rhynchocephalian species *Rhynchosaurus articeps*. The many European Late Jurassic and Early Cretaceous footprints considered, between about 1862 and the present, to have been made by the dinosaurs *Megalosaurus* or *Iguanodon*, and accordingly given those names, present more familiar examples.

This procedure may appear a sensible and convenient one. However, when the array of footprints in museum collections that have been allotted these names are examined, it becomes depressingly apparent that many are not even footprints of iguanodonts or megalosaurs, let alone of those particular genera! Of course, misidentification of fossils is a problem recurrent throughout paleontology. But the lack of any available standard for comparison, in the form of a defined footprint taxon with a clearly characterized morphology, meant that confidence in the naming of such impressions is simply not possible.

An intermediate procedure has been to formulate a name for the ichnogenus which, while distinctive in itself, nevertheless identifies with varying degrees of precision the animal presumed to have been the trackmaker. Examples are *Anchisauripus* Lull 1904, *Rhynchosauroides* Beasley 1911, *Emydichnium* and *Krokodilipus* Nopcsa 1923, *Aetosauripus* Weiss 1934, *Chelonipus* Rühle von Lilienstern 1939, *Rhynchocephalichnus, Coelurosaurichnus,* and *Saurischichnus* von Huene 1941, and *Nothosauripus* Kuhn 1958. This is at first sight a desirable and systematically informative procedure. However, it can cause its embarrassments, as when *Pteraichnus* Stokes 1957 was shown by Padian and Olsen (1984) to be the trail, not of a pterodactyl, but probably of a crocodile!

Richard Lull (1915) carried this approach further. He placed the Connecticut Valley footprints into families based on ichnogenera, e.g., Selenichnidae and Grallatoridae, and placed these in turn into osteologically-based orders and classes. For example, the footprint-based family Grallatoridae is placed in the infraorder Coelurosauria of the saurischian suborder Theropoda alongside the osteologically-based family Coelurosauridae, which includes the probable makers of the tracks. Such an approach, while systematically informative, has an ele-

ment of absurdity because it gives equal systematic status to the animal itself and to a random effect of its actions.

A compromise was adopted by Haubold (1971), whereby the ichnofossil families are distinguished as "morpho-families." But these are again assembled into osteologically-based orders and classes. Moreover, his usage of the morpho-families has been inconsistent. Within the superorder Ornithopoda, for example, one finds some dinosaur footprints assigned to the morpho-family Tetrapodosauridae, whereas others are placed into the osteologically-based family Iguanodontidae. Essentially, Haubold places footprints into families when confident he has identified the trackmaker sufficiently precisely, and uses morpho-families when he is confident only at a higher systematic level.

Hitchcock's ultimate classificatory approach was echoed in a review of fossil amphibian and reptile tracks by Nopcsa (1923). Six categories were recognized: (1) salamandroid and stegocephaloid; (2) lacertoid; (3) theromorphoid (dicynodontoid, theriodontoid); (4) rhynchosauroid; (5) crocodiloid; and (6) dinosauroid. Conceptually, this seems admirable; in practice, however, each of Nopcsa's groupings proved to comprise footprints of morphologically convergent but unrelated types (see criticisms by Abel 1935). Nevertheless, a modified version of this approach has been adopted by Lessertisseur (1955) and other French paleoichnologists. Indeed, it does afford the asset of allowing affinities to be indicated, without either conceptual absurdity or taxonomic rigidity. Its prime liability is its imprecision in defining either particular morphologies or particular behaviors.

The approach advocated by O. S. Vialov (1966) echoes the one utilized by Hitchcock between 1841 and 1844, but subsequently abandoned. Vialov's hierarchy of names embraces all sorts of animal tracks and trails under a single "umbrella" category Zooichnia (Vivichnia). The vertebrate traces are grouped as follows:

Amphibipedia
  Order Labyrinthopedida
  Order Caudipedida
    Suborder Salamandripedoidei
Reptilipedia
  Superorder Theromorphipedii
    Order Therapsidipedida
  Superorder Cotylosauripedii
    Order Procolophonipedida
  Superorder Chelonomorphipedii
    Order Testudipedida
  Superorder Lepidosauripedii
    Order Rhynchocephalipedida
    Order Lacertipedida
    Order Sauropterygipedida
    Order Pterosauripedida
    Order Saurischipedida
      Suborder Coelurosauripedoidei
    Order Ornithischipedida
      Suborder Ornithopedoidei

Order Thecodontipedida
  Suborder Pseudosuchipedoidei
  Suborder Parasuchipedoidei
Mammalipedia
  Order Carnivoripedida
  Order Perissodactipedida
  Order Artiodactipedida
    Suborder Pecoripedoidei
Avipedia

Within these categories are placed both new and existing ichnogenera and new ichnosubgenera, the new names being consistently based on presumed systematic affinity (*Avipeda, Hippipeda, Gazellipeda*) and not upon morphology.

Subsequently, Vialov (1972) added a second major category – the Vivisignia – to his scheme. Whilst his Vivichnia were considered always to afford direct indications of the nature of the animal's body, these latter did not. Instead, they were the products of physiological functions – coprolites, gastroliths, eggs and egg-capsules, and even effects of injuries! This expansion, though logical, has generated such conceptual problems that it has tended to militate against the acceptance of his overall classification. This is a pity, for Vialov's earlier classification, at least, has much to recommend it.

Even as late as the 1960's, trace fossils in general were not dealt with in any logical fashion under either of the available nomenclatural *Codes,* zoological or botanical. In theory at least, the earlier name – whether given to a trace or to a body fossil – had priority over the later. For example, the footprint genus *Rhynchosauroides* Beasley 1911 might have been treated as a junior synonym of the essentially osteological genus *Rhynchosaurus* Owen 1842 and *Chirotherium* Kaup 1835 regarded as a senior synonym of *Ticinosuchus* Krebs 1965, the foot structure of which accords exactly in proportions and dimensions with the type species of Kaup's genus, *C. barthi*. In practice, however, the separate names of footprint taxa continued to be retained even when their affinity was reasonably presumed, because it could rarely be demonstrated beyond doubt. However, there was a disinclination to formulate separate names when, as with the supposed *Iguanodon* footprints, a likely affinity had been perceived almost from the outset. No one has yet formally proposed an alternative name such as *Iguanodontipus*, desirable though this might be if clearly defined and based upon a well-chosen type series of footprints.

A separate code for ichnofossils?

Is there, indeed, any need to treat ichnofossils in the same fashion as body fossils? Sarjeant and Kennedy (1973) argued that they were not properly within the scope of any of the three nomenclatural codes – animal, plant, or bacteriological. Trace fossils are not body fossils. They are biogenic structures in sediments – and structures, moreover, that do not necessarily indicate the nature of

their maker with any precision. Those authors drafted a possible nomenclatural code, specifically for trace fossils and intended to be administered by an independent body comprised, not of zoologists or botanists, but of geologists. Unfortunately for the success of this proposal, it was referred by the International Palaeontological Union to T. P. Crimes for consideration – and Crimes was self-avowedly a taxonomic conservative, determined that biogenic and bioerosional structures produced by animals should continue to be treated under the zoological *Code*. The consequence was predictable. Instead of striving for taxonomic independence, paleontologists at large followed Crime's recommendations and have meekly accepted instead the series of minor modifications to that *Code* that were proposed by W. Häntzschel (see discussion in Basan 1979) and included into the revised *Code* published in 1979.

These provisions do have certain assets. They give ichnotaxa, for the first time, a defined status within zoological taxonomy and exempt them from priority problems in competition with "Linnaean" genera, making it explicit (Art. 23g, iii) that "a name established for an ichnotaxon does not compete in priority with one established for an animal, even for one that may have formed the ichnotaxon." Furthermore, they furnish a blanket exemption of ichnogenera from the need for a type species – a highly dubious benefit! – and they permit ichnotaxa to stand as types for families. (The *Code* does not, of course, attempt to govern the nomenclature of taxonomic groups of animals above the family-group level.)

The present position

The changes made recently to the *International Code of Zoological Nomenclature* mean that biogenic and bioerosional structures made by animals are no longer taxonomic outlaws. (The *International Code of Botanical Nomenclature*, in contrast, continues to make no provision whatsoever for plant trace fossils.) However, much else is left unresolved. Though ichnotaxa are avowedly different from the other "natural" animal taxa, they do not require any differentiation of category. They are still to be called families, genera, and species, instead of being made recognizable by being called ichnofamilies, ichnogenera, or ichnospecies.

The limits of acceptance are vague, because no philosophy is enunciated. If the activities of the same animal make different sorts of traces, should they or should they not be synonymized? If the traces of those different activities are to be awarded separate generic status, how are the ichnologically-based families to be defined – on biological affinity of the tracemakers, on similarity of activity, or on similarity between the sedimentary structures resulting from that activity, even though they might be produced differently? These, and other, problems arise because trace fossils are being forced into a *Code* not designed to accommodate them instead of being governed by one designed specifically for them.

Trace fossils are structures in sediments, certainly produced by biological activity, but having a form that is determined only in part by that activity and reflects only to a limited extent the morphology of the tracemaker. Their nature is governed by position within the sediment: was the imprint that one sees formed in the bedding plane one is examining or was it formed in a surface above or below the bedding plane? (In other words, is it a true trace, a subtrace, or a supertrace?) Moreover, its morphology is governed also by the physical and chemical conditions of the substrate and by its diagenetic history. These are factors of which geologists and paleontologists are aware but to which, even if aware of them, zoologists are indifferent. Though I realize I may be flogging a fossilized dinosaur, I continue to believe that a separate code of nomenclature for trace fossils is the ideal for which we should be striving.

Even if the status quo is accepted insofar as the rules governing the taxonomy of trace fossils are concerned, we who work on fossil footprints still need to agree on a number of matters, if we are not to be wholly bogged down in a morass of procedural uncertainty and bemused by perpetual nomenclatural disagreement.

First of all, we must decide on our aims. Is it our priority to identify the trackmaker or to define its activity? If the former is our sole purpose, then all tracks made by the same animal should be given the same name, whether they be walking tracks, running or galloping tracks, or leaping tracks, and whether they be clearly or poorly formed. We may feel, indeed – as did Richard Owen (1842) and Louis Dollo (1883) – that there is no need to give the footprint a name. If it is a track of *Rhynchosaurus* or *Iguanodon*, why not just name it so? If it is certainly the track of a hadrosaur but not identifiable to generic level, why not just style it a hadrosaur track, thus identifying it at least to familial level? If there be need to differentiate different behaviors – however minor the interest of such matters – it may be simple enough just to speak of a "hadrosaur running track" or a "hadrosaur leaping track."

Yet the results of these different activities, in terms of the sedimentary structures – the casts and molds of tracks and the individual footprints – may be markedly different. An animal walking slowly may, in favourable conditions, leave a perfect impression of the whole undersurface, and of parts of the lateral surfaces, of its foot – including all of its digits, with their claws, hooves or nails. As it moves faster, not only will less and less of the foot be impressed, but also the stresses will fall differently. The track pattern will change and the individual imprints will become less distinctive. The central digits sustaining the animal's weight – and, in particular, their claws, hooves, or nails – will be pressed down deeply into the sediment, while the lateral digits become impressed partially and incompletely, or not at all. The whole impression might permit identification to generic level; the fast walking track, at least to familial level;

but the running track might well only permit an identification to subordinal or even ordinal level.

How, then, to handle these changing situations in taxonomic terms? Do we name only walking tracks, where the information is complete? Yet as the substrate changes from fine and moist to coarser and harder, the information will inevitably become less complete even if the reptile is still walking. Can we then remain so confident in our identifications?

If our interest is largely or wholly in behavior, then it becomes more important to know that these are walking, running, or leaping tracks of particular types of vertebrates than to identify the trackmaker with precision. Such names as *Coelurosaurichnus* are satisfactory when recognition of affinity is the prime concern, but should we apply them to traces of *all* sorts of behavior by a coelurosaur? Upon what philosophy shall we base the naming of the constituent ichnospecies – on more precise recognition of affinity, or morphology, or behavior? A name based only on affinity might well come to embrace a variety of morphotypes produced by different behaviors, some of which might well prove indistinguishable from variants within other similarly based ichnotaxa.

If we are most interested in behavior as reflected by footprint morphology, names like Peabody's (1948) *Rotodactylus cursorius* become admirable, because they suggest that morphology and define the style of motion. Nor does the changing substrate present such a problem, because we are concerned more with what the reptile was doing than with what it was. However, such an ichnogenus might well embrace the tracks made by animals of different genera, families, or even suborders, that happen to be behaving in a similar fashion. If interest is in behavior, this is not necessarily a problem; if interest is in affinity, this may be a real liability.

The bigger and heavier the animal, the more likely we are to discover its tracks and to have displayed to us, from cast or from mould, the details of their morphology. Even on a soft, moist substrate, a small, light animal is likely to leave only a shallow impression; on a harder substrate, it might leave no impression at all. In a given environment, the lacertilians might outnumber the dinosaurs a hundred to one, yet only the dinosaurs might leave footmark evidence of their presence. Moreover, even if lacertilian tracks were to be found alongside the dinosaur tracks, they might be not nearly so informative, because only their claws might have left an impression, or only the claws and a few of the phalanges. If one is interested only in identifying the trackmaker, there is no point in naming such tracks. However, if one is interested in behavior, it becomes quite worthwhile. Jardine's *Actibates triassae* (1853) was named only on impressions of the tips of digits. Do we wish to retain such names, or should we regard them as valueless? In terms of recognition of affinity, they are indeed virtually valueless, but for characterizing a pattern of impressions produced by a particular behavior, they are extremely useful.

To add to our difficulties it is likely that, in many instances, we are not dealing with the footprints themselves, but with reflections of them, produced at horizons lower or higher in the sedimentary sequence than that upon which the animal actually walked. That horizon might well have been removed, as a consequence of resuspension of the sediment or of erosion during the next inrush of waters. The tracks may persist only as subtraces, dangerously capable of misinterpretation as actual footprints (see Heyler and Lessertisseur 1963; Sarjeant 1975; Leonardi (ed.) 1987) or as mere disturbances of the stratification, definable but difficult indeed to interpret (see Van der Lingen and Andrews 1969). At the 1986 International Symposium on Dinosaur Tracks and Traces in Albuquerque, New Mexico, Wann Langston (University of Texas, Austin) discussed "stacked" dinosaur footprints. The subtracks on successively lower surfaces beneath the actual stratum of impression showed a progressive enlargement and loss of detail, while the supertracks formed in layers of sediment accumulating within the mold became ever smaller and more shapeless. At the same meeting, Seilacher discussed the means by which subtracks and supertracks might be distinguished. He suggested that, in the fossil record, we are much more often dealing with subtracks than with actual footprints.

How should we deal with this situation? Should we name only those undoubted tracks and their natural casts, in which the fine detail of scales, etcetera, allows no doubt that these were true footprints? Certainly such impressions may tell us much more about foot anatomy than will ever be determinable from bones. The problem, however, is that footprints of such quality are extremely rare. If these are all that we feel able to name, then we will be left unable to give names to the vast majority of fossil tracks – not only the subtracks and supertracks, but also the many true footprints impressed into sediments too dry and hard to reflect such details or too liquid to retain them without distortion.

In our naming of fossil footprints, we must continually bear in mind that we are dealing with structures made in sediment, subject to a great deal of variation, and *not* with bones and teeth which, though certainly variable in detail, are essentially consistent in morphology even if subject to the vagaries of preservation. Just as a river or an ocean current will produce different effects on the sediment according to its speed, so also will a vertebrate according to its pace of movement and its purposes. The different structures it will produce are not fixed and definite, like bones. Instead they are highly mutable, determined not only by the animal's actions but also by the physical condition of the different layers of the substrate at the time of impression of the feet, the nature of the overlying, infilling sediment, and the digenetic history of the stratum.

Moreover, in studying footprints, our interest is quite as much in behavior as in affinity. Yes, it is nice to have completely and beautifully impressed tracks, which tell us exactly what animal made them. But is it not interesting also to know when that animal was running or leaping, and is it wrong to give different names to the different patterns of impressions produced by that changing behavior? When we are naming footprints, we are not naming animals; we are naming the traces of their actions.

The specialists on invertebrate ichnofossils recognize this quite clearly. The walking trace of a trilobite is called *Diplichnites*, its sideways-grazing trace *Dimorphichnus*, its swimming grazing grace *Monomorphichnus*. These structures do not tell us what sort of trilobite made the trace; nevertheless they can tell us quite a lot, not only about the trilobite's behavior, but also about its environment and the original condition of the sediment surface. When the trilobite furrows down into that surface, the mark is called *Cruziana* – and sometimes the tracemaker can be identified, but only sometimes. Its resting excavation, *Rusophycus*, may well give us enough details to enable us to name the tracemaker, at least to generic level; but again, its prime interest is in terms of behavior. Though the connection between these different structures has long since been perceived, the different ichnogeneric names are retained because of their utility as a means for succinctly characterizing both a morphology and a behavior.

Proposals concerning the naming of footprints

I suggest that we should treat – or rather, continue to treat – dinosaur and other vertebrate footprints in the same fashion as invertebrate tracks are treated. When we have complete enough information to identify the trackmaker, well and good. But let us nevertheless give the track a separate name, so as to emphasize that it is an ichnofossil, not a body fossil, that we are dealing with.

That name should be based upon morphology, not upon presumed affinity. When the track indicates a different behavior by the same animal, let us give it a different name. If we can identify the trackmaker only vaguely or not at all, let us nevertheless give a name to the track, provided that it be distinctive in morphology. Such a procedure has the asset that, when the behavior – e.g., running at speed – leaves impressions too incomplete for exact generic (or even familial or subordinal) identification, a name will be available that indicates unambiguously both the morphology and the behavior.

We must accept also that the variable nature of substrates will affect, not only the detail, but also the character of the tracks. A small animal, travelling easily and with fair speed over a hard substrate, may well produce a quite different footprint pattern from the same animal toiling arduously to traverse a water-saturated substrate into which its feet are sinking. It is neither necessary nor conceptually desirable to base footprint names entirely upon substrate character. However, it may be both necessary and desirable to utilize separate names when the changing substrate character has enforced major changes in the fashion of movement, and thus in the pattern of the tracks.

Moreover, we must recognize that there are circumstances between which the morphology of the trace fossil may not allow us to distinguish. If we are considering a broad and poorly formed dinosaur track, this may result either from the wetness of the surface over which the animal was travelling or from the fact that we are examining a subtrack and not an actual trail. So long as our name implies only the morphology of the impression, it will retain a clear and unambiguous meaning. Only if the name reflects an interpretation rather than an observation will we encounter problems and generate ambiguities. The diagnosis of an ichnotaxon should be based only upon its observable characteristics. The deductions that we make from those characteristics are, of course, highly significant. However, they are subjective, whereas the diagnosis should be objective.

It is important also for us to accept that the limits between trace fossil genera and species are, and will remain, arbitrary. Just as, in trilobite tracks, a walking trace may grade into a furrow, then into a resting excavation, or vice versa, so also may we expect to see the track of a single bipedal dinosaur changing from a walk to a trot and then to a run or, in the case of quadrupedal dinosaurs, from a walk to a gallop. Such changes may not yet have been reported from dinosaur tracks, but they are to be expected. Already, in the Permian, single trackways of smaller reptiles have been described in which the footprints show an acceleration from walking slowly, with four or five digits impressed, to running swiftly with only two digits impressed (e.g., Sarjeant 1971). When we encounter such trackways, we should not feel constrained to give a single name to the changing pattern, just because it was made by one animal. The names of ichnotaxa are applied to the molds made in the sediment, not to the trackmaker. If the pattern changes significantly, then the name should change.

What, then, is a "significant" change? Well, that is inevitably a matter for subjective judgment – but the judgment is no more subjective and arbitrary than the delimitation of so-called "natural" species in a continuously evolving morphological plexus!

Where behavioral patterns intergrade, one may prefer to give primacy to the name given to the most perfectly expressed footprints – the walking track, from which the fullest details of the foot are seen – and to note that this grades into running tracks of a type to which another ichnotaxon name might be given. However, though preferring that first name, one should not feel constrained to abandon the second name, which will remain useful when only the running tracks are seen and there can be no certainty concerning their correlation with that – or any other – type of walking track. Even when such certainty exists, the running track name will remain useful, since it succinctly characterizes the particular pattern and needs no explanatory expansion.

In contrast, I cannot consider it desirable that different names should be given simply on the basis of good or poor preservation. In a single track made on the edge of the water, some footprints may have been washed out by a tide returning briefly and marginally, while others may remain intact (see Wills and Sarjeant 1970, pl. 31, for an example). Such changes within a track certainly do not merit nomenclatorial differentiation. Nor do differences arising solely from the nature of the casting or molding sediment medium merit nomenclatorial recognition, if the character and pattern of the tracks is otherwise unaffected.

In instances where an ample material so permits (for example in the abundant ichnofauna of dinosaur tracks from the Middle Cretaceous of the Peace River Canyon, British Columbia; see Currie and Sarjeant 1979), changes in form of tracks may be seen to occur that result from modifications in foot morphology with growth. Such changes should be encompassed within the diagnosis of a single ichnogenus and species; they do not merit nomenclatorial differentiation above varietal level.

When a taxon based largely or exclusively on a manus is shown to accord with a taxon based largely or wholly on a pes or *vice versa*, the senior name should be utilized, in accordance with the principle of priority. Retention of a separate name for the manus or pes alone is never to be recommended.

In summary, I believe that, in the naming of footprints, the morphology of the mold, cast, subtrace, or supertrace is the matter of primary concern, the identification of the trackmaker being only of secondary importance. Significant changes in pattern deserve nomenclatural recognition, even if the same animal has made the tracks. An inability to identify accurately the trackmaker should not prevent us from giving a name to a distinctive type of trackway. We are not naming animals, we are naming biogenic sedimentary structures. Our concern must be quite as much with environmental and behavioral interpretation as with the systematic identity of the dinosaur, or other animal, whose feet produced those structures.

Proposals concerning footprint classification

With regard to the classification of trace fossils above the level of genus, I have already expressed my belief (Sarjeant and Kennedy 1973; Sarjeant 1975) that there is no need to impose a single scheme, because several different approaches may be desirable according to the particular purpose of the research. Indeed, as in the case of invertebrate traces, the nature of a track or the structure of a burrow may give only the vaguest indication, if any at all, of the identity of its maker. It is often much more crucial to understand the style of movement and feeding or the nature of the environment.

While it is, in general, easier to determine what sorts of animals made vertebrate footprints – at least at ordinal or familial, if not at generic or specific, level – the same argument still applies. Whilst knowledge of the trackmaker's affinity is almost always useful, the paleoecologist may have good reason to prefer to class his tracks in terms of the environment they indicate rather than in a Linnaean hierarchy of affinity. Indeed, a purely toponomic classification, such as that formulated by Martinsson (1970), may be found to be more useful.

If interest is only in the trackmaker's systematic affinity, then a modified Linnaean hierarchy, such as that advocated by Haubold (1971), may continue to be employed. However, there is no reason to view with disfavor, and still less to outlaw, non-Linnaean classifications of footprints that are designed, and highly suitable, for environmental and behavioral analysis. A classification based strictly on behavior, comparable to that formulated by Seilacher (1953) for marine trace fossils and developed for wider application by Vialov (1972), may well prove most suitable for many purposes.

In the case of dinosaurs in particular, good footprints usually enable the identification of the trackmaker, at least to familial level. Consequently, a classification based on systematic affinity can be utilized with ease.

This is emphatically *not* the case with other vertebrates. The tracks of small reptiles may not even suffice to enable the identification of major groups. How does one distinguish, for example, between the tracks of the different groups of lepidosaurs, and can one separate these from the footprints of eosuchians or even (in some instances) small amphibians? The footprints of many groups of birds – passerines and wading birds, for example – defy precise systematic identification; while the hoofprints of many artiodactyls and perissodactyls have changed little from the Oligocene to the present, though there have been tremendous changes in their overall morphology. Any philosophy for footprint classification must extend beyond dinosaur tracks. It must take into consideration such instances where systematic affinity can only be imprecisely determined and where the behavioral and environmental implications of the tracks assume an infinitely greater interest.

Certainly, we should remember at all times that we are naming, not the animal itself, but the trace of its act as reflected in the sediment and preserved in the fossil record.

References

Abel, O. 1935. *Vorzeitliche Lebensspuren* (Jena, Germany: Gustav Fischer).

Basan, P. 1979. Trace fossil nomenclature: the developing picture. *Palaeogeography, Palaeoclimatology, Palaeoecology* 28:143–146.

Beasley, H. C. 1911. Description of a group of footprints in the Storeton find of 1910. *Liverpool Geological Society Proceedings* 111:108–115.

Currie, P. J. and Sarjeant, W. A. S. 1979. Lower Cretaceous dinosaur footprints from the Peace River Canyon, British Columbia, Canada. *Palaeogeography, Palaeoclimatology, Palaeoecology* 28:103–115.

Dollo, L. 1883. Troisième note sur les dinosaures de Bernissart. *Bulletin du Musée Royal d'Histoire Naturelle de Belgique* 2:85–120.

Geinitz, H. B. 1861. *Dyas I* (Germany: Leipzig).

Haubold, H. 1971. *Ichnia amphibiorum et reptiliorum fossilium. Handbuch der Paläoherpetologie* (Stuttgart: Fischer Verlag).

Heyler, D., and Lessertisseur, J. 1963. Pistes de tétrapodes permiens de la région de Lodéve (Herault). *Mémoires du Muséum d'Histoire Naturelle*, new series 11:125–221.

Hitchcock, E. 1836. Ornithichnology. Description of the footmarks of birds (Ornithoidichnites) on New Red Sandstone in Massachusetts. *American Journal of Science* 29:307–340.

——— 1837a. Ornithichnites in Connecticut. *American Journal of Science* 31:174–175.

——— 1837b. Fossil footsteps in a sandstone and greywacke. *American Journal of Science* 32:174–176.

——— 1841. *Final Report on the Geology of Massachusetts.* Amherst and Northampton, Massachusetts 2:301–831.

——— 1843. Description of five new species of fossil footmarks, from the red sandstone of the valley of the Connecticut River. *Association of American Geologists and Naturalists Transactions* 254–264.

——— 1844. Report on ichnology, or fossil footmarks, with a description of several new species. *American Journal of Science* 47:292–322.

——— 1845. An attempt to name, classify and describe the animals that made the fossil footmarks of New England. *Association of American Geologists and Naturalists, Proceedings* 6:23–25.

——— 1847. Description of two new species of fossil footmarks found in Massachusetts and Connecticut, or of the animals that made them. *American Journal of Science*, series 2, 4:46–57.

——— 1848. An attempt to discriminate and describe the animals that made the fossil footmarks of the United States, and especially of New England. *American Academy of Arts and Sciences Memoirs*, series 2, 3:129–256.

——— 1855. Shark remains from the coal formation of Illinois and bones and tracks from the Connecticut River sandstone. *American Journal of Science*, series 2, 20:416–417.

——— 1858. *Ichnology of New England. A Report on the Sandstone of the Connecticut Valley, Especially Its Fossil Footmarks* (Boston: W. White).

——— 1865. *Supplement to the Ichnology of New England* (Boston: Wright & Potter).

Huene, F. H. von. 1941. Die Tetrapoden-Fährten im toskanischen Verrucano und ihre Bedeutung. *Neues Jahrbuch für Mineralogie, Geologie und Paläontologie* Abt. B 86:1–34.

International Trust for Zoological Nomenclature. 1985. *International Code of Zoological Nomenclature* [Berkeley and Los Angeles, California: University of California Press for International Trust for Zoological Nomenclature, in association with the British Museum (Natural History), London].

Jardine, Sir W. 1850. Note to Mr. Harkness' paper on "The position of the impressions of footsteps in the Bunter Sandstones of Dumfries-shire". *Annals and Magazine of Natural History; including Zoology, Botany and Geology*, series 2, 6:208–209.

——— 1853. *The Ichnology of Annandale; or, Illustrations of Foot-*

marks Impressed on the New Red Sandstone of Corn-
cockle Muir (Edinburgh: the author).

Jeffs, O. W. 1894. On a series of saurian footprints from the Che-
shire Trias (with a note on the Cheirotherium). *Geological Magazine*, new series 31:451–454.

Kaup, J. F. 1835. [Mitteilungen über Tierfahrten von Hildburg-
hausen] *Neues Jahrbuch für Mineralogie, Geognosie,
Geologie und Petrefaktenkunde* 327–328.

Krebs, B. 1965. *Ticinosuchus ferox* n.g. n.sp., ein neuer Pseudo-
suchier aus der Trias des Monte San Giorgio. *Schweizerische Paläontologische Abhandlungen* 81:1–411.

Kuhn, O. 1958. *Die Fährten der vorzeitlichen Amphibien und
Reptilien* (Bamberg, Germany: Meissenbach), pp. 1–64.

1963. *Ichnia tetrapodorum*. Pt. 101. *In* Westphal, F. (ed.),
Fossilium catalogus, I: Animalia (s'Gravenhage, Nether-
lands: Junk).

Leonardi, G. (ed.) 1987. *Glossary and Manual of Tetrapod Footprint Palaeoichnology* (Brasília, Brazil: Departamento
Nacional da Producão Mineral).

Lessertisseur, J. 1955. Traces d'activité animale et leur signifi-
cation paléobiologique. *Mémoires de la Société Géologique de France*, new series 74:1–150.

Lull, R. S. 1904. Fossil footprints of the Jura-Trias of North
America. *Boston Society of Natural History Memoirs*
5:461–557.

1915. Triassic life of the Connecticut Valley. *Connecticut
Geological and Natural History Survey Bulletin*
81:1–285.

Martinsson, A. 1970. Toponomy of trace fossils. *In* Crimes,
T. P. and Harper, J. C. (eds.), *Trace Fossils* (Liverpool:
Seel Howe Press), pp. 323–330.

Nopcsa, F. 1923. Die fossilen Reptilien. *Fortschritte der Geologie und Paläontologie* 2:1–210.

Owen, R. 1842. Description of an extinct lacertilian reptile,
Rhynchosaurus articeps Owen, of which the bones and
footprints characterize the upper New Red Sandstones
etc. *Cambridge Philosophical Society Transactions*
7:355–369.

Pabst, W. 1900. Beiträge zur Kenntnis der Tierfährten in den
Rothliegenden "Deutschlands." *Zeitschrift der Deutschen Geologischen Gesellschaft* 52:48–63.

Padian, K., and Olsen, P. E. 1984. The fossil trackway *Pteraichnus:* not pterosaurian, but crocodilian. *Journal of Paleontology* 58:178–184.

Peabody, F. E. 1948. Reptile and amphibian trackways from the
Lower Triassic Moenkopi Formation of Arizona and
Utah. *University of California Publications in Zoology*
27:295–468.

Rühle von Lilienstern, K. 1939. Fährten und Spuren im Chiro-
theriumsandstein von Südthüringen. *Fortschritte der
Geologie und Paläontologie* 12:290–387.

Sarjeant, W. A. S. 1971. Vertebrate tracks from the Permian of
Castle Peak, Texas. *Texas Journal of Science* 22:343–366.

1974. A history and bibliography of the study of fossil vertebrate footprints in the British Isles. *Palaeogeography,
Palaeoclimatology, Palaeoecology* 16:265–378.

1975. Fossil tracks and impressions of vertebrates. *In* Frey,
R. W. (ed.), *The Study of Trace-fossils* (New York:
Springer-Verlag), pp. 283–324.

Sarjeant, W. A. S., and Kennedy, W. J. 1973. Proposal of a code
for the nomenclature of trace-fossils. *Canadian Journal
of Earth Sciences* 10:460–475. (Code republished in
Palaeogeography, Palaeoclimatology, Palaeoecology
1979, 28:147–167).

Seilacher, A. 1953. Über die Methoden der Paläoichnologie. I.
Studien zur Paläoichnologie. *Neues Jahrbuch für Geologie und Paläontologie Abhandlungen* 96:421–452.

Sickler, F. K. L. 1834. *Sendschreiben an Dr. J. F. Blumenbach
über die hochst merckwürdigen, vor einigen Monaten
erst entdecken Reliefs der Fährten urweiltlicher grosser
und unbekannter Thiere in den Hessberger Sandstein-
brüchen bei der Stadt Hildburghausen* (Germany: Schul-
programm Gymnasiums Hildburghausen).

Stokes, W. L. 1957. Pterodactyl tracks from the Morrison Formation. *Journal of Paleontology* 31:952–954.

Van der Lingen, G. J., and Andrews, P. B. 1969. Hoofprint structures in beach sand. *Journal of Sedimentary Petrology*
39:350–357.

Vialov, O. S. 1966. *Sledy zhiznedeyatelnosti organizmow i ikh
paleontologicheskoe znachenie* (The traces of the vital
activity of organisms and their paleontological significance) (Ukraine, S.S.R.: Akademiya Nauk).

1972. The classification of the fossil traces of life. *Proceedings of the 24th International Geological Congress,
Montreal* 7:639–644.

Weiss, W. 1934. Ein Fährtenschicht im mittelfränkischen Blasen-
sandstein. *Jahresbericht und Miteilungen des Oberrheinischen Geologischen Vereins*, new series 23:5–11.

Wills, L. J., and Sarjeant, W. A. S. 1970. Fossil vertebrate and
invertebrate tracks from boreholes through the Bunter
Series (Triassic) of Worcestershire. *Mercian Geologist*
3:399–414.

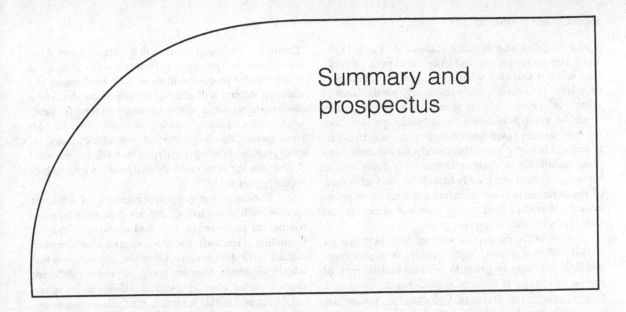

Summary and prospectus

It became apparent over the course of the Dinosaur Systematics Symposium that significant advances have been made in recent years in the number of species known, in understanding the anatomy and interrelationships of those species, in narrowing the gap between morphological and biological species, and in recognizing the significance of associated geological and paleontological data. In a paper delivered at the Symposium, Peter Dodson (1987) noted that there are 265 genera of dinosaurs recognized, forty percent of which have been described since 1969. For many attending the conference, this was the first time that it became evident that dinosaur research had finally reached a level of activity comparable to the surge at the turn of the century.

Nevertheless, the data base is still small, dinosaurian studies are still in their infancy, and the field is wide open for further development. In comparison with the number of species of vertebrates known to be alive today (in 1969, Mayr estimated that there are 8,600 extant species of birds, 3,700 species of mammals, 6,300 species of reptiles, and 2,500 species of amphibians), the number of dinosaur species known for their 140 million year history is insignificant. Although most dinosaurs were relatively large animals that were probably long-lived, there could have been thousands of species alive at any one time throughout the world, and the discovery of new forms (Chapters 8, 9, 14, 15, 19) is inevitably going to continue.

Paleontological species

It is critical that we gain a better understanding of what a paleontological species is (Chapter 5; Fox 1986). Species are the bricks of the foundation on which we build all of our theories of relationships. Refinement of the species concept will lead to refinement of our clas-

sifications. And it does not end there. The usefulness of dinosaurs in biostratigraphy is improved by correct identifications of species. Much of the evidence used to interpret dinosaur physiology is based on our understanding of species. Correct identification of species is critical as well in behavioral studies. Multidisciplinary research on paleoecology gives us the best understanding of how dinosaurs interacted with their world. Yet it makes a tremendous difference to our understanding of any ecosystem if taxa are oversplit or overlumped. A biological species (Chapter 5) is a population of interbreeding organisms that produce viable offspring, and this is the only level within our classification schemes that has a nonarbitrary definition. Even so, it is often difficult for neontologists to define extant species, because there are many gray areas (such as circular speciation). Paleontologists generally cannot apply the definition of biological species to fossils because evidence of breeding behavior and the viability of offspring cannot be observed in extinct animals. In some ways this simplifies the task of the paleontologist (Mayr 1969), but ultimately this is less satisfying because alternative, arbitrary definitions must be used at the species level. It is not surprising that the emphasis of many of the papers in this book is towards developing a better understanding of dinosaurs as biological species (but see Chapter 22 for an opposing viewpoint). With the imperfections of the fossil record, it is impossible to equate paleontological and biological species in most cases. But recent work on bone-beds and nesting sites suggest that the equation can be made for at least one species in each major group of dinosaurs. These species then can serve to show the range of variation expected in at least closely related forms.

With a large number of well preserved specimens, it is possible to define a fossil species. However, paleontologists rarely have complete specimens, and suites of specimens of the same species are even more uncommon. The concept of form species was an essential

In Dinosaur Systematics: Perspectives and Approaches, *Kenneth Carpenter and Philip J. Currie, eds. Copyright © Cambridge University Press, 1990.*

aspect of the earliest dinosaur studies (i.e., Leidy 1856), and binomial names are still assigned to parts of skeletons without knowing what the rest of the animal may look like. If species were named only on the basis of complete specimens, less than 10% of the dinosaurs would be valid. Sometimes, form species are more useful than species based on more complete material. This is particularly true since the introduction of bulk sampling techniques at microvertebrate sites. Teeth tend to be more common and readily identifiable than other parts of the skeleton, and are therefore more useful in paleoecological studies than nearly complete skeletons that lack skulls or other diagnostic parts.

One of the dangers of working with form species is that different names (at the species or higher taxonomic levels) may be given to different skeletal parts of the same species (Chapters 8, 20), thereby creating a false impression of diversity. Unfortunately, the options are usually less desirable. Unnamed skeletons tend to be ignored, even if they are significantly different from other specimens. In faunal studies, apparent diversity will be low. Partial specimens are sometimes assigned on speculation to established species, but this can also be misleading [Berman and McIntosh (1978) give an excellent account of the problems created when Marsh added a *Camarasaurus*-like skull to a restoration of *Apatosaurus*].

There will always have to be some give and take in the application of names to form species. And if the biological significance of a form species is not overstated, and form species are eliminated as more complete specimens are found, then this concept is quite sound.

The concept of distinguishing species by name on the basis of geographic and/or temporal separation is common. Often, this cannot be backed up by anatomical evidence, either because of preservational differences, or because there are no significant differences. But species can have long periods of existence, and can span stratigraphic boundaries. Furthermore, stratigraphic differences cannot be equated strictly with temporal differences, when in fact they can represent a combination of environmental differences (resulting in different lithologies) and time. The more widespread a formation is, the less likely it is that the formation boundaries are synchronous throughout a given geographic range. Time lines that cross stratigraphic boundaries are recognized rarely.

One would intuitively expect that relating dinosaur species to time lines would have more validity than to stratigraphic boundaries within any given area. However, this is not always the case. For example, Dinosaur Provincial Park and Devil's Coulee in southern Alberta are separated by only 300 kilometers, but represent different ecosystems. Some species (large and small theropods) may be shared by the two sites, suggesting those species were capable of existing in a range of ecosystems. However, it would appear that the lambeosaurine hadrosaurs and centrosaurine ceratopsians are specifically distinct at the two sites, even though radiometric dating

(Eberth et al. in press) of associated volcanic ashes show that the sites are synchronous. Further analysis in progress of ash beds at productive dinosaur sites in Montana and southern Alberta will almost certainly show that some dinosaur species had narrow environmental preferences, while others inhabited a variety of ecosystems. As with form species, the designation of new species based on stratigraphic, lithologic (environmental), or temporal differences has to be carefully analyzed for it to be biologically meaningful.

Similarly, the geographic ranges of different species are highly variable, and are dependent on environmental preferences, size and mobility, behavior (including migration), and other species characteristics that are difficult to analyze. Often the assumption is that widely separated sites are going to produce different species, even if no morphological differences are apparent. In support of this approach, *Pachyrhinosaurus canadensis* from Scabby Butte in southern Alberta is distinct from a new species of *Pachyrhinosaurus* (Langston et al. in preparation) from Grande Prairie, 500 kilometers to the north. Yet the latter does not appear to be distinct from a new specimen of *Pachyrhinosaurus* found 1,500 kilometers farther north in Alaska (Clemens pers. comm. 1988). It is possible that there are temporal and/or environmental differences between Scabby Butte and Grande Prairie to account for the presence of distinct species of *Pachyrhinosaurus*, but it is also evident that the new species had a wide geographic range. Questions like these can only be resolved by additional fieldwork and analysis. And because of the imperfections of the fossil record, we must always recognize that some evidence will never be found.

The recognition of fossil species is based primarily on morphological differences. With few exceptions (Chapter 22), only the hard parts of dinosaurs are preserved, and we do not know what kind of variation there may have been in the soft anatomy or coloration. Skin impressions are now known from a variety of dinosaurs (Osborn 1912; Lessem 1989), but they are still not common enough to be useful for detailed comparisons between specimens.

Footprints, including those with skin impressions, provide some evidence on the anatomy of the foot. Unfortunately, the same basic foot structure can be shared by many species, and therefore more than one species can make footprints that are morphologically identical (Chapter 22). Ironically, a single species (and even a single individual) can also produce a wide range of footprint morphotypes, depending on the type of activity the animal is engaged in and the type of substrate in which footprints are left. For these reasons, it is unlikely that ichnospecies can ever be related to biological species concepts. That does not mean that biological information cannot be extracted from footprints, which have proven very useful in biomechanical, physiological, behavioral, and other studies.

There are modern species that are morphologically identical, but are genetically distinct because of behavioral or other differences. Such differences cannot be seen directly in the fossil record of dinosaurs. Nevertheless, dinosaur behavior can sometimes be deduced from the fossil record (Horner and Gorman 1988), which may eventually give us some information on speciation.

Classification

Classification of organisms is an artificial, often arbitrary concept to help us sort and file the tremendous variety of living things, and to help us make some sense of the world. Understanding the interrelationships of – and developing classifications of – dinosaurs takes up a significant proportion of overall research effort (Chapters 9, 11, 13, 16, 17, 19, 21). Like filing systems, some classification systems are better than others. The best classification systems will reflect closely our understanding of the relationships of organisms. Within recent years, cladistics has had a major influence on the classification of dinosaurs (Chapters 1, 9, 11). The emphasis is on derived characters shared by animals. Because of the imperfect nature of the fossil record, the difficulties in recognizing character states produced by convergent and parallel evolution, and problems in quantifying genetic change, cladistic classifications are still subject to different interpretations by researchers. As a consequence of this, classifications proposed by different workers for the same groups of dinosaurs do not necessarily come to the same conclusions. As in all fields of science, differences in opinion indicate where additional research is needed.

Sereno (Chapter 1) compares and contrasts the clade and grade concepts. These might be considered the cladistic and traditional methods, as the former only recognizes monophyletic taxa, while the latter permits paraphyletic taxa. Unlike other fields of vertebrate paleontology, dinosaur systematists have been slow to use cladistic analysis (Chapters 1, 9). Ornithischian dinosaurs have received the most benefit from this approach (Milner and Norman 1984; Norman 1984a,b; Sereno 1984; Cooper 1985; Maryánska and Osmólska 1985; Sereno 1986), although some papers have appeared on saurischian cladistics (Paul 1984; Gauthier 1986). Brinkman and Sues (1987) used cladistic analysis to conclude that *Herrerasaurus* and *Staurikosaurus* do not fit into the Saurischia–Ornithischia dichotomy.

Problems in assessing variation

Morphological differences remain the most important way of assessing variation. In most cases, there are limitations to what can be done because few specimens may exist for any species. In other cases, large numbers of specimens, often exhibiting considerable variation, are difficult to work with because of their size. There is an inverse correlation between the size of a dinosaur and the amount of detailed work done on it. Papers on small theropods (Chapters 6, 7, 8) are more numerous than those

on sauropods (Chapter 4), even though the material is less common.

Biases introduced by preservational differences, collection and preparation techniques, and researchers are frequently encountered problems in the study of morphological variation. Even when there is an overlap of skeletal parts, differential crushing can make specimens of the same species look very different. Preservational biases can favor the fossilization of some parts of the body (for example, the massive femur of a hadrosaur is more likely to be preserved than its fragile skull bones), or individuals of certain sizes. Collecting biases are more common than most people realize, and can range from differences in training and search image, to selective collection of dinosaurs by type, size, or even skeletal part. As pointed out by Sternberg (1970), it is essential for dinosaur paleontologists to get into the field to assess the potential for preservational and collection biases. Another type of bias is discussed by Brett-Surman and Chapman (Chapter 12), who show how an individual's philosophy about taxonomy can influence the outcome.

Dinosaur fossils have always been considered as prizes for museums to display. Many hours are invested in the preparation and mounting of dinosaurs, which are frequently damaged by poorly trained temporary staff or hasty preparation schedules. Because many are collected for display value rather than research potential, they can be restored and mounted in such a way as to obscure details of the real bone. Published drawings and photographs sometimes include reconstructed sections as if they were real. When only casts are available for study, it may be impossible to distinguish the parts based on real bone from the reconstructed regions.

Genetic variability is expressed to different degrees in different parts of the skeleton. Some bones, such as the vertebrae, can be indistinguishable over a wide range of species. At the other end of the scale, variation in the shape of footprints can be enormous even within a single trackway of a single species (Chapter 22). Consequently, paleontologists have to assess the variation of each bone differently. Bones associated with feeding and locomotion are usually more variable than bones of the axial skeleton, and tend to be more useful in understanding speciation. However, they also tend to be more variable within any given species.

Analysis of variation

The emphasis of the majority of papers presented at the Dinosaur Systematics Conference was the variation (individual, ontogenetic, sexual) manifest in dinosaur species. As discussed above, there are many factors that must be considered when trying to determine if specimens belong to the same species or not. In reaching a conclusion, one has in fact set the parameters for individual variation within the species. The more specimens there are for any species, then the more sound are the conclusions related to analysis of variation.

Bivariate and multivariate analyses in morphological studies have been greatly sped up by computers, which have proven invaluable in sorting large fields of morphological data so that patterns can be seen (Chapter 17; Dodson 1975a,b,c,d). Similarly, shape analysis using computers (Chapters 2, 3, 12, 14) takes away some of the subjectivity in morphological comparisons. Morphometrics in general are useful in quantifying individual variation, sexual dimorphism, and ontogenetic change.

Variation has been recognized for a long time in protoceratopsian dinosaurs from central Asia (Brown and Schlaikjer 1940; Dodson 1975b; Maryánska and Osmólska 1975). The specimens were recovered from an environment that preserved eggs, hatchlings, juveniles, and all sizes of adults. Because there were so many specimens recovered from a single, relatively small locality (the Flaming Cliffs at Bayn Dzak), it was relatively easy to see the range of variation possible in a single population.

Most collecting sites for dinosaurs are not good for preserving juveniles (Jepsen 1964; Carpenter 1982). One of the richest collecting sites is Dinosaur Provincial Park in southern Alberta, where only a small percentage of thousands of specimens are juveniles. There was a bias in the depositional regime that strongly selected against the preservation of small animals, including juvenile dinosaurs (Currie 1987). Nevertheless, the skeletons of immature animals do exist, but are difficult to identify because of the amount of change that can take place during growth. In his monumental study on the lambeosaurine dinosaurs of the Park, Dodson (1975a) reviewed 12 species of crested hadrosaurs and concluded that they represented only three species assignable to two genera. All of the variation could be attributed to different forms of quantifiable variation.

Until recently, papers assessing ontogenetic change and sexual dimorphism in other dinosaurs (i.e., Casier 1960; Rozhdestvensky 1965; Russell 1970) have appeared infrequently. With the discovery of nesting sites and the analysis of bone-beds where whole herds of animals perished, the situation has improved dramatically. At this time, studies in progress are giving us a good understanding of all forms of variability for one or more species of each of the major groups of dinosaurs, including theropods (Chapters 6, 7, 10), hypsilophodonts (Horner and Weishampel 1988), hadrosaurs (Horner and Makela 1979; Currie and Horner 1988; Dilkes 1988; Horner and Weishampel 1988), iguanodonts (Norman 1987), ceratopsians (Chapters 16, 18; Currie and Dodson 1984; Tanke 1988; Rogers and Sampson 1989), and ankylosaurs (Maryánska 1971; Currie 1989).

As time goes on and more fieldwork and analysis is done on isolated bones and skeletons, bone-beds, footprint sites, and nesting sites, we will get a better handle on ontogenetic, sexual, and individual variation. This will help us to evaluate our understanding of dinosaurs as biological species, and will certainly help us refine our physiological, ecological, and other inferred levels of biological studies.

Prospectus

By the end of the conference, an overwhelming amount of data had been presented, of which this volume represents only a fraction of the information. A feeling emerged that dinosaur systematics had taken several steps forward, and that problems that appeared impossible to resolve a decade or two ago now seemed to be within striking distance. That is not to say that these are resolved, but simply that for the first time there is some hope that they might be. At the current levels of funding, it is unlikely that this will be an overnight process.

The discovery and recognition of embryos and hatchlings associated with eggs and nests from India (Mohabey 1987), Mongolia (Sochava 1972), Montana (Dilkes 1988; Horner and Weishampel 1988), and Alberta (Currie and Horner 1988), has provided us with good growth ranges for sauropods, hypsilophodonts, hadrosaurs, and possibly protoceratopsians. The recognition of juveniles has similarly allowed us to look at longer ontogenetic series for psittacosaurs (Chapter 15; Coombs 1982), several species of protoceratopsians (Maryánska and Osmólska 1975), ankylosaurs (Maryánska 1971), sauropods (Dong 1987), and ceratopsians (Dodson and Currie 1988).

The study of eggs and nestlings can also provide information on individual variation. One of the three hadrosaur embryos collected by the Tyrrell Museum of Palaeontology in 1987 was 10% smaller than the other two, even though it was taken from the same nest. However, the best hope for looking at individual and sexual variation comes from bone-beds, particularly those that appear to have been mass death events of single species.

Sites continue to be discovered where large numbers of articulated skeletons are found associated in single quarries. Continued preparation and study of this material will provide our greatest source of data on variability within a single species.

Articulated skeletons found in great numbers in a single area, especially if it represents rapid burial over a short interval, can provide almost as much information as the bone-beds. Good examples include *Protoceratops andrewsi* from the Djadokhta Formation of Mongolia and China, and the lambeosaurines of Dinosaur Provincial Park. As these are almost invariably attritional death assemblages, all ages can be represented, provided the animals were living in the area at all times of the year.

Bone-beds come in many forms. Monospecific bone-beds have already been discussed in detail. Multifaunal bone-beds, usually resulting from long term accumulations of animal remains from wide geographic areas, can provide large samples for the study of variation. However, the identity of elements at a species level is usually prohibitively difficult.

Microvertebrate sites are a special type of bone-bed where there are large concentrations of teeth and bones, each of which tend to be less than one centimeter in length. These can be good for providing information on juveniles of many of the dinosaurs (Chapter 8), which in turn can provide information on ontogenetic growth. Microvertebrate sites are usually formed by hydraulic sorting, but occasionally these are concentrates within coprolites. It is conceivable that scat analysis could provide information on particular prey preferences of a carnivore, and that because any single coprolite is produced over a very short period of time, this would give data on the instantaneous variation of a prey species.

Footprints are common elements of the fossil record in some areas (Gillette and Lockley 1989). Although much of the variation is caused by the interaction of the animal with the substrate, nevertheless real variation can also be seen in the footprints themselves. For example, Currie and Sarjeant (1979) recognized juveniles of the ornithopod *Amblydactylus* in the Lower Cretaceous rocks of the Peace River Canyon, as did Bird (1944) in the sauropod tracks at Glen Rose, Texas. If the footprints are associated with herding behavior, then one can be reasonably confident that some of the variation may be biological.

And museum collections cannot be neglected as a source of new and valuable information, as reanalysis of existing material continues to produce new species, further examples of intraspecies variation, new evidence of relationships, and new information on the biology of dinosaurs.

The new sources of information are enhanced by the maturation of dinosaur studies themselves. The way we look at the resources is more sophisticated than ever before. Cladistics, for example, has greatly refined our understanding of the relationships of dinosaur taxa. The computer continues to speed up and simplify virtually all aspects of dinosaurian research, while new techniques and equipment (CT scans, SEM, geochemistry) are producing data and results that were undreamed of even a decade ago. And more people are involved in the study of dinosaurs than ever before. In a few more years, it may be desirable to run the Second International Dinosaur Systematics Symposium, which will document all of those exciting changes that are taking place now!

References

Berman, D. S. and McIntosh, J. S. 1978. Skull and relationships of the Upper Jurassic sauropod *Apatosaurus* (Reptilia, Saurischia). *Carnegie Museum of Natural History, Bulletin* 8:1–35.

Bird, R. T. 1944. Did *Brontosaurus* ever walk on land? *Natural History* 53:60–67.

Brinkman, D., and Sues, H.-D. 1987. A staurikosaurid dinosaur from the Upper Triassic Ischigualasto Formation of Argentina and the relationships of the Staurikosauridae. *Palaeontology* 30:494–503.

Brown, B. and Schlaikjer, E. M. 1940. The structure and relationships of *Protoceratops*. *New York Academy of Sciences, Annals* 40:133–266.

Carpenter, K. 1982. Baby dinosaurs from the Late Cretaceous Lance and Hell Creek formations and a description of a new species of theropod. *University of Wyoming, Contributions to Geology* 20:123–134.

Casier, E. 1960. *Les iguanodons de Bernissart* (Brussels: Institut Royal des Sciences naturelles de Belgique).

Coombs, W. P., Jr. 1982. Juvenile specimens of the ornithischian dinosaur *Psittacosaurus*. *Palaeontology* 25:89–107.

Cooper, M. R. 1985. A revision of the ornithischian dinosaur *Kangnasaurus coetzeei* Haughton, with a classification of the Ornithischia. *South African Museum, Annals* 95:281–317.

Currie, P. J. 1987. Bird-like characteristics of the jaws and teeth of troodontid theropods (Dinosauria, Saurischia). *Journal of Vertebrate Paleontology* 7:72-81.

1989. Long distance dinosaurs. *Natural History* 6/89:60–65.

Currie, P. J. and Dodson, P. 1984. Mass death of a herd of ceratopsian dinosaurs. *In* Reif, W. E. and Westphal, F. (eds.), *Third Symposium on Mesozoic Terrestrial Ecosystems* short papers (Tübingen: Attempto Verlag), pp. 61–66.

Currie, P. J. and Horner, J. R. 1988. Lambeosaurine hadrosaur embryos (Reptilia: Ornithischia). *Journal of Vertebrate Paleontology* 8 (Supplement to no. 3):13A (Abstract).

Currie, P. J. and Sarjeant, W. A. S. 1979. Lower Cretaceous footprints from the Peace River Canyon, B.C., Canada. *Palaeogeography, Palaeoclimatology, Palaeoecology* 28:103–115.

Dilkes, D. W. 1988. Relative growth of the hind limb in the hadrosaurian dinosaur *Maiasaura peeblesorum*. *Journal of Vertebrate Paleontology* 8 (Supplement to no. 3):13A (Abstract).

Dodson, P. 1975a. Taxonomic implications of relative growth in lambeosaurine hadrosaurs. *Systematic Zoology* 24:37–54.

1975b. Quantitative aspects of relative growth and sexual dimorphism in *Protoceratops*. *Journal of Paleontology* 50:929–940.

1975c. Functional and ecological significance of relative growth in *Alligator*. *Journal of Zoology, London* 175:315–355.

1975d. Relative growth in two sympatric species of *Sceloporus*. *American Midland Naturalist* 94:421–450.

1987. Dinosaur Systematics Symposium, Tyrrell Museum of Palaeontology, Drumheller, Alberta, June 2–5, 1986. *Journal of Vertebrate Paleontology* 7:106–108.

Dodson, P. and Currie, P. J. 1988. The smallest ceratopsid skull – Judith River Formation of Alberta. *Canadian Journal of Earth Sciences* 25:926–930.

Dong, Z. M. 1987. Untitled section on saurischian dinosaurs. *In* Zhao, X. J. et al. (eds.), *Stratigraphy and Vertebrate Fossils of Xinjiang.* (Beijing: Institute of Vertebrate Paleontology and Paleoanthropology). [in Chinese]

Eberth, D. A., Thomas, R. G., and Deino, A. In press. Preliminary K-Ar dates from bentonites in the Judith River and Bearpaw formations (Upper Cretaceous) of Dinosaur Provincial Park, southern Alberta, Canada. *In* Mateer, N. J. and Chen, P. J. (eds.), *Aspects of Nonmarine Cretaceous Geology, Proceedings of the Conference on Nonmarine Cretaceous Correlations, Urumqi, China 1987* (Beijing: Ocean Press).

Fox, R. 1986. Paleoscene no. 1. Species in paleontology. *Geoscience in Canada* 13:73-84.

Gauthier, J. 1986. Saurischian monophyly and the origin of birds. *California Academy of Sciences, Memoirs* 8:1–47.

Gillette, D. D. and Lockley, M. G. (eds.), 1989. *Dinosaur Tracks and Traces* (New York: Cambridge University Press).

Horner, J. R. and Gorman, J. 1988. *Digging Dinosaurs* (New York: Workman Publishing).

Horner, J. R. and Makela, R. 1979. Nest of juveniles provides evidence of family structure among dinosaurs. *Nature* 282:297–298.

Horner, J. R. and Weishampel, D. B. 1988. A comparative embryological study of two ornithischian dinosaurs. *Nature* 332:256–257.

Jepsen, G. L. 1964. Riddles of the terrible lizards. *American Scientist* 52:227–246.

Leidy, J. 1856. Notices of remains of extinct reptiles and fishes, discovered by Dr. F. V. Hayden in the bad lands of Judith River, Nebraska Territory. *Academy of Natural Sciences of Philadelphia, Proceedings* 8:72–73.

Lessem, D. 1989. Skinning the dinosaur. *Discover* 10(3):38–44.

Maryańska, T. 1971. New data on the skull of *Pinacosaurus grangeri* (Ankylosauria). *Palaeontologia Polonica* 25:45–53.

Maryańska, T. and Osmólska, H. 1975. Protoceratopsidae (Dinosauria) of Asia. *Palaeontologia Polonica* 33:133–181.

1985. On ornithischian phylogeny. *Acta Palaeontologica Polonica* 30:137–150.

Mayr, E. 1969. *Principles of Systematic Zoology* (New York: McGraw-Hill).

Milner, A. and Norman, D. 1984. The biogeography of advanced ornithopod dinosaurs (Archosauria: Ornithischia) – a cladistic-vicariance model. *In* Reif, W. E. and Westphal, F. (eds.), *Third Symposium on Mesozoic Terrestrial Ecosystems*, short papers (Tübingen: Attempto Verlag), pp. 145–150.

Mohabey, D. M. 1987. Juvenile sauropod dinosaur from Upper Cretaceous Lameta Formation of Panchmahals District, Gujarat, India. *Journal of the Geological Society of India* 30:210–216.

Norman, D. B. 1984a. A systematic reappraisal of the reptile order Ornithischia. *In* Reif, W. E. and Westphal, F. (eds.), *Third Symposium on Mesozoic Terrestrial Ecosystems*, short papers (Tübingen: Attempto Verlag), pp. 157–162.

1984b. On the cranial morphology and evolution of ornithopod dinosaurs. *Zoological Society of London Symposium* 52:521–547.

1987. A mass-accumulation of vertebrates from the Lower Cretaceous of Nehden (Sauerland), West Germany. *Royal Society of London, Proceedings*, series B 230:215–255.

Osborn, H. F. 1912. Integument of the iguanodont dinosaur *Trachodon*. *American Museum of Natural History, Memoirs* 1:33–54.

Paul, G. 1984. The archosaurs: a phylogenetic study. *In* Reif, W. E. and Westphal, F. (eds.), *Third Symposium on Mesozoic Terrestrial Ecosystems*, short papers (Tübingen: Attempto Verlag), pp. 175–180.

Rogers, R. R., and Sampson, S. D. 1989. A drought-related mass death of ceratopsian dinosaurs (Reptilia: Ornithischia) from the Two Medicine Formation (Campanian) of Montana: Behavioral Implications. *Journal of Vertebrate Paleontology* 9 (Supplement to no. 3):36A.

Rozhdestvensky, A. K. 1965. Growth variability and some aspects of dinosaur systematics in Asia. *Paleontologischesky Zhurnal* 1965:95–103.

Russell, D. A. 1970. Tyrannosaurs from the Late Cretaceous of western Canada. *National Museum of Canada, Publications in Palaeontology* 1:1–34.

Sereno, P. C. 1984. The phylogeny of the Ornithischia: a reappraisal. *In* Reif, W. E. and Westphal, F. (eds.), *Third Symposium on Mesozoic Terrestrial Ecosystems*, short papers (Tübingen: Attempto Verlag), pp. 219–226.

1986. Phylogeny of the bird-hipped dinosaurs (Order Ornithischia). *National Geographic Research* 2:234–256.

Sochava, A. V. 1972. The skeleton of an embryo in a dinosaur egg. *Paleontological Journal* 6:527–531.

Sternberg, C. M. 1970. Comments on dinosaurian preservation in the Cretaceous of Alberta and Wyoming. *National Museums of Canada, Publications in Palaeontology* 4:1–9.

Tanke, D. 1988. Ontogeny and dimorphism in *Pachyrhinosaurus* (Reptilia, Ceratopsidae), Pipestone Creek, N.W. Alberta, Canada. *Journal of Vertebrate Paleontology* (Supplement to no. 3)7:27A.

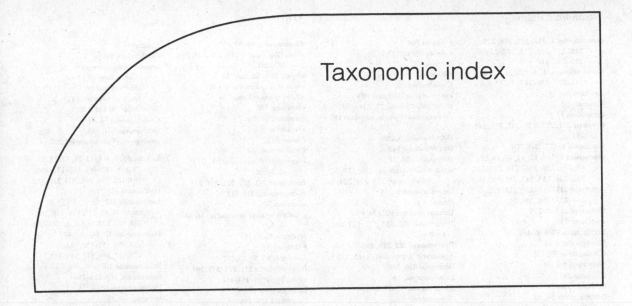

Taxonomic index

This index combines osteological, biological, and ichnological taxa. Species names are listed, followed by their generic assignment in parentheses. Some early ichnological species have no generic name; these stand alone. To make the index more useful, derivatives of higher taxonomic categories are also given.